WAVE MECHANICS FOR OCEAN ENGINEERING

Elsevier Oceanography Series
Series Editor: David Halpern (1993-)

FURTHER TITLES IN THIS SERIES
Volumes 1-7, 11, 15, 16, 18, 19, 21, 23, 29 and 32 are out of print.

8 E. LISITZIN
SEA-LEVEL CHANGES
9 R.H. PARKER
THE STUDY OF BENTHIC COMMUNITIES
10 J.C.J. NIHOUL (Editor)
MODELLING OF MARINE SYSTEMS
12 E.J. FERGUSON WOOD and R.E. JOHANNES
TROPICAL MARINE POLLUTION
13 E. STEEMANN NIELSEN
MARINE PHOTOSYNTHESIS
14 N.G.JERLOV
MARINE OPTICS
17 R.A. GEYER (Editor)
SUBMERSIBLES AND THEIR USE IN OCEANOGRAPHY AND OCEAN ENGINEERING
20 P.H. LEBLOND and L.A. MYSAK
WAVES IN THE OCEAN
22 P. DEHLINGER
MARINE GRAVITY
24 F.T. BANNER, M.B. COLLINS and K.S. MASSIE (Editors)
THE NORTH-WEST EUROPEAN SHELF SEAS: THE SEA BED AND THE SEA IN MOTION
25 J.C.J. NIHOUL (Editor)
MARINE FORECASTING
26 H.G. RAMMING and Z. KOWALIK
NUMERICAL MODELLING MARINE HYDRODYNAMICS
27 R.A. GEYER (Editor)
MARINE ENVIRONMENTAL POLLUTION
28 J.C.J. NIHOUL (Editor)
MARINE TURBULENCE
30 A. VOIPIO (Editor)
THE BALTIC SEA
31 E.K. DUURSMA and R. DAWSON (Editors)
MARINE ORGANIC CHEMISTRY
33 R.HEKINIAN
PETROLOGY OF THE OCEAN FLOOR
34 J.C.J. NIHOUL (Editor)
HYDRODYNAMICS OF SEMI-ENCLOSED SEAS
35 B. JOHNS (Editor)
PHYSICAL OCEANOGRAPHY OF COASTAL AND SHELF SEAS
36 J.C.J. NIHOUL (Editor)
HYDRODYNAMICS OF THE EQUATORIAL OCEAN
37 W. LANGERAAR
SURVEYING AND CHARTING OF THE SEAS
38 J.C.J. NIHOUL (Editor)
REMOTE SENSING OF SHELF-SEA HYDRODYNAMICS
39 T.ICHIYE (Editor)
OCEAN HYDRODYNAMICS OF THE JAPAN AND EAST CHINA SEAS
40 J.C.J. NIHOUL (Editor)
COUPLED OCEAN–ATMOSPHERE MODELS
41 H. KUNZENDORF (Editor)
MARINE MINERAL EXPLORATION
42 J.C.J NIHOUL (Editor)
MARINE INTERFACES ECOHYDRODYNAMICS
43 P. LASSERRE and J.M. MARTIN (Editors)
BIOGEOCHEMICAL PROCESSES AT THE LAND-SEA BOUNDARY
44 I.P. MARTINI (Editor)
CANADIAN INLAND SEAS
45 J.C.J. NIHOUL (Editor)
THREE-DIMINSIONAL MODELS OF MARINE AND ESTUARIN DYNAMICS
46 J.C.J. NIHOUL (Editor)
SMALL-SCALE TURBULENCE AND MIXING IN THE OCEAN
47 M.R. LANDRY and B.M. HICKEY (Editors)
COASTAL OCENOGRAPHY OF WASHINGTON AND OREGON
48 S.R. MASSEL
HYDRODYNAMICS OF COASTAL ZONES
49 V.C. LAKHAN and A.S. TRENHAILE (Editors)
APPLICATIONS IN COASTAL MODELING
50 J.C.J. NIHOUL and B.M. JAMART (Editors)
MESOSCALE SYNOPTIC COHERENT STRUCTURES IN GEOPHYSICAL TURBULENCE
51 G.P. GLASBY (Editor)
ANTARCTIC SECTOR OF THE PACIFIC
52 P.W. GLYNN (Editor)
GLOBAL ECOLOGICAL CONSEQUENCES OF THE 1982–83 EL NINO–SOUTHERN OSCILLATION
53 J. DERA (Editor)
MARINE PHYSICS
54 K. TAKANO (Editor)
OCEANOGRAPHY OF ASIAN MARGINAL SEAS
55 TAN WEIYAN
SHALLOW WATER HYDRODYNAMICS
56 R. CHARLIER and J. JUSTUS
OCEAN ENERGIES, ENVIRONMENTAL, ECONOMIC AND TECHNOLOGICAL ASPECTS OF ALTERNATIVE POWER SOURCES
57 P.C. CHU and J.C. GASCARD (Editors)
DEEP CONVECTION AND DEEP WATER FORMATION IN THE OCEANS
58 P.A. PIRAZZOLI
WORLD ATLAS OF HOLOCENE SEA-LEVEL CHANGES
59 T. TERAMOTO (Editor)
DEEP OCEAN CIRCULATION-PHYSICAL AND CHEMICAL ASPECTS
60 B. KJERFVE (Editor)
COASTAL LAGOON PROCESSES
61 P. MALANOTTE-RIZZOLI (Editor)
MODERN APPROACHES TO DATA ASSIMILATION IN OCEAN MODELING
62 H.W.A. BEHRENS, J.C. BORST, L.J. DROPPERT, and J.P. VAN DER MEULEN (Editors)
OPERATIONAL OCEANOGRAPHY
63 D. HALPERN (Editor)
SATELLITES, OCEANOGRAPHY AND SOCIETY

Elsevier Oceanography Series, 64

WAVE MECHANICS FOR OCEAN ENGINEERING

P. Boccotti

Faculty of Engineering
University of Reggio-Calabria
Feo di Vito
I-89060 Reggio-Calabria
Italy

2000
ELSEVIER
Amsterdam – Lausanne – New York – Oxford – Shannon – Singapore – Tokyo

ELSEVIER SCIENCE B.V.
Sara Burgerhartstraat 25
P.O. Box 211, 1000 AE Amsterdam, The Netherlands

First edition 2000

Library of Congress Cataloging in Publication Data
A catalog record from the Library of Congress has been applied for.

ISBN: 0 444 50380 3

∞ The paper used in this publication meets the requirements of ANSI/NISO Z39.48-1992 (Permanence of Paper).
Printed in The Netherlands.

To my parents

PREFACE

The book aims to cover in a unitary way both the deterministic and statistical topics of the mechanics of sea waves. Furthermore, it aims to highlight some recent progress on the dynamics of random wind-generated waves and on long-term wave statistics. Finally, it aims to give a fresh approach to traditional concepts. In this regard, some original proofs are given (see the conclusive notes of each chapter), new evidence from small scale field experiments is used to introduce crucial topics like *wave forces*, and some widely worked examples are given (e.g. an example of 17 pages for the calculation of the largest wave loads on a submerged tunnel during its lifetime).

The text is intended for researchers and graduate students, but the style is such that most of the book is suitable for undergraduate students. This is because the various formulae are proved from the fundamental equations, and the harder concepts are explained with both examples and sometimes also with short stories. Strictly speaking, it is assumed that the reader has knowledge of calculus and basic mechanics (e.g. volumes 1 and 2 of Calculus by T.M. Apostol, and Physics I by R. Resnick, D. Halliday, and K.S. Krane). The fundamentals of strength of materials are needed for a part of chapter 12.

Chapters 1 and 2 cover the dynamics of the periodic waves. Chapter 1 starts from the differential equations of motion to obtain the velocity potential. Chapter 2 applies the linear momentum equation and the energy equation to control volumes with surface waves. As a consequence, in chapter 1 you will find solutions for the velocity potential of progressive waves (with and without a current), and of waves interacting with vertical walls. While, in chapter 2 you will find the solutions to problems like those of shoaling, refraction, set-down, and group celerity.

Chapter 3 covers the beach processes. How large is the run-up? How is the planform evolution of a nourishment project? How is the deformation of a beach after the construction of a detached breakwater or a groin? These problems are dealt with analytically. The starting point is the formal solution for the beach planform evolution under the assumption of small curvature of the contour lines.

Chapters 4 and 5 cover the short-term wave statistics. Chapter 4 introduces the basic concepts (sea state, autocovariance of the surface displacement, frequency spectrum, and so on). It also explains how to obtain the autocovariance and the frequency spectrum from the time series data, and how to infer the nature of a sea state from these functions. Chapter 5 gives formal solutions and experimental evidence for the statistics of wave height and wave period in a sea state.

Chapters 6 and 7 cover the long-term wave statistics. Chapter 6 explains how to obtain the probability of exceedance of the significant wave height at a given location. Then, it introduces the concept of triangular equivalent storm: for each actual sea storm, a triangular storm exists with the same probability that the largest wave height exceeds any fixed threshold. Chapter 7 applies this concept to give the formal solutions for return periods of sea storms with some given characteristics. This chapter also discusses in depth the relationship between the return period, the lifetime, and the probability of occurrence.

Chapters 8, 9, and 10 cover the dynamics of the random wind-generated waves. Chapter 8 gives the velocity potential of a sea state to the first approximation in a Stokes' expansion. Also, the cases of wind waves interacting with vertical walls are dealt with in detail. This chapter points out some very big differences between wind-generated waves and periodic waves, in what concerns reflection and diffraction. Then, chapter 9 develops the recent theory of quasi-determinism of the highest waves in a sea state: if a wave with a given height H occurs at a fixed point $\boldsymbol{x}_o$, and H is very large with respect to the mean wave height at this point, we can expect the water surface and velocity potential near $\boldsymbol{x}_o$ to be very close to some well precise deterministic forms.

Chapter 10 deals with the more exciting topic. The highest waves in a random sea state belong to quasi-deterministic (three-dimensional) wave groups. The wave group is like a family and each individual wave is a member of this family, with a life cycle. The characteristics of these wave groups are deeply investigated in view of the applications (wave loads on structures). And, through this investigation, we arrive at a new insight into random wind-generated waves: these waves are higher on the time domain than on the space domain and they possess a sort of genetic code.

Chapters 11 and 12 cover the wave loads on offshore structures. Chapter 11 firstly deals with the so-called large bodies on which waves exert only the inertia force. Then it considers the small bodies on which waves also exert a significant drag force. The problem of a long structure (what is the maximum expected wave height on a long structure like a floating tunnel?) completes this chapter. Then chapter 12 is entirely devoted to worked examples of wave force estimates.

Chapter 13 covers the wave loads on coastal structures: first, wave forces on vertical breakwaters; then, stability of the rubble mound breakwaters under the action of sea states of given characteristics.

Chapter 14 deals with topics which call for an overall overview of offshore and coastal engineering. The first topic is comparison between the effects of tsunamis and of wind waves, from the open sea to the coast. The second topic is small scale models, and the third topic is wave measurements for the various needs.

These subjects, which I approached in an organic way for the first time in my Italian book *Idraulica Marittima* (UTET, 1997), serve as three courses to graduate students in civil engineering or ocean engineering, courses that my assistants and I give in the University of Reggio Calabria.

First course (wave theory and coastal structures)

a.1) dynamics of the periodic waves: chapters 1 and 2, except sects. 2.5.4 and 2.6.3 (set-down);

a.2) short-term wave statistics: chapters 4 and 5;

a.3) long-term wave statistics: chapter 6 except sect. 6.3 (directional wave prediction), chapter 7 till sect. 7.5 (design sea state for coastal structures), and Appendix A;

a.4) stability of coastal structures: chapter 13;

a.5) small scale models and wave measurements: sects. 14.2.1, 14.2.2 and 14.3.

Second course (advanced wave theory and offshore structures)

b.1) design wave for offshore structures: from sect. 7.6 to sect. 7.8;

b.2) dynamics of the random wind-generated waves: chapters 8, 9, and 10, except the formal proof from sect. 9.6 to sect. 9.10;

b.3) wave loads on offshore structures: chapters 11 and 12;

b.4) discussion on the small scale models: from sect. 14.2.3 to sect. 14.2.5.

Third course (beach processes)

c.1) variation of the mean water level from offshore to coast: sects. 2.5.4 and 2.6.3;

c.2) wave actions on coasts: chapter 3;

c.3) prediction of the wave height for given wave direction: sects. 6.3 and 7.9;

c.4) comparison between effects of tsunamis and effects of wind waves: sect. 14.1.

The first course is introductory to the others, so that a student can attend only course a), or a) and b), or a) and c), or a), b), and c).

The sections from 9.6 to 9.10 and Appendix B go beyond the three courses and are intended for researchers only. It is also advisable to limit a few of the more analytical parts. Specifically, for sect. 2.10 (shoaling and set-down of waves on current) one could give the introduction (sect. 2.10.1) and conclusion (sect. 2.10.9). Of course there is also the possibility to cut off a few parts, so as to shorten the courses, if necessary. In particular, course a) could omit the wave-current interaction (sects. 1.9 and 2.10), and/or the way to obtain the continuous spectrum (from sect. 4.4.3 to sect. 4.4.6). Finally, there are a few introductory sections which have been given for the sake of completeness, and which can be omitted if the relevant subjects are treated in other courses. These are: sects. 1.1-2 and 2.1-2 which give the concepts of fluid mechanics needed to develop the wave theory, and sects. 5.1-2 which introduce the random Gaussian processes.

Paolo Boccotti

Reggio Calabria, Italy
May 1999

ACKNOWLEDGEMENTS

I am very grateful to professors Enrico Marchi and Giulio Scarsi who, nearly twenty-five years ago, introduced me to research on sea waves.

Then I wish to express my gratitude to assistant professors Felice Arena, Giuseppe Barbaro, Pasquale Filianoti, and to engineer Francis Cirianni for their reading of the manuscript and some useful comments. F. Arena also deserves an acknowledgement for his help with the analysis of NDBC buoy data and satellite data. The graphics are due to engineer Vincenzo Fiamma who drew the figures from my file data and my sketches. They all were my students, and I am proud of each of them.

Finally, I very gratefully acknowledge the great courtesy and fruitful co-operation of UTET of Turin publisher of my Italian book *Idraulica Marittima* (1997) which was the basis of this new book.

Part of this work was done with the financial support of the Italian National Group for Prevention from Hydrogeological Disasters (under contract CNR-GNDCI No 98.00586).

P. B.

NOTATION

Some symbols used in only one section are not included in the following list.

a	wave amplitude
a	triangle height
$\mathbf{a} = (a_x, a_y, a_z)$	particle acceleration
A	area
b	width
b	threshold of the surface displacement
b	duration of the equivalent triangular storm
b	threshold or special value of the wave crest elevation
B	berm height
c	propagation speed
C_d	diffraction coefficient
C_{FR}	Fresnel integral with integrand cosine
C_{dg}, C_{in}	drag coefficient, inertia coefficient
C_1, C_2, C_3, C_4	safety factors
$\mathscr{C}$	wave crest elevation
d	still water depth
D	diameter
D	directional spreading function
$\overline{D}$	mean persistence
$\mathscr{D}$	duration of a storm or of the design sea state
e	potential + kinetic energy per unit mass
e	eccentricity
E	frequency spectrum
$\mathscr{E}$	mean wave energy per unit surface
$\mathscr{E}$	nondimensional frequency spectrum
EX	expected number per unit time
f	general function
$\mathbf{f} = (f_x, f_y, f_z)$	force per unit length
F	general function
F	horizontal force per unit extension of upright breakwater
F_o, F_v	horizontal force, vertical force
$\mathbf{F} = (F_x, F_y, F_z)$	force
$\mathscr{F}$	distribution function of the frequency spectrum
$\mathscr{F}_{\mathscr{R}}$	speed drop factor
g	acceleration of gravity
G	longshore diffusivity
G	general function

h	threshold or special value of the significant wave height
H	wave height
H_s	significant wave height
$\mathbf{i}_x, \mathbf{i}_y, \mathbf{i}_z$	unit vectors
$\mathscr{I}$	Irribarren number
k	wave number
$\mathscr{k}$	nondimensional wave number
K_E	Keulegan-Carpenter number
l	length of a structure
L	wavelength
L	lifetime of a structure
m	mass
m_j	jth order moment of the frequency spectrum
m_0	variance of the surface displacement of a sea state
M	moment of a force or of a force per unit length
M	determinant of a covariance matrix
n	parameter of the spreading direction function
n, N, $\mathscr{N}$	number being specified every time
$\mathbf{n} = (n_x, n_y, n_z)$	unit normal vector to a surface
p	pressure
p	probability density function
P	probability of exceedance
P	weight
P^*	weight in still water
$\mathscr{P}$	probability
Q	flow rate per unit length
r	polar coordinate (the r-curves are straight lines through the origin)
R	radius
R	return period
R_E	Reynolds number
R_o, R_v	reactions of the foundation
R_U	run-up
$\mathbf{R} = (\mathbf{R}_x, \mathbf{R}_y)$	radiation stress tensor
s	local propagation axis
S	directional wave spectrum
S	velocity head
S	total uplift pressure per unit length of breakwater
S_D, S_U	set-down, set-up
S_{FR}	Fresnel integral with integrand sine
$\mathscr{S}$	nondimensional directional spectrum

t	time
t_o	special time instant
T	wave period
T	time lag
T^*	abscissa of the absolute minimum of the autocovariance function
T_i	interarrival time of a random point process
$\mathscr{T}$	very large time interval
$\mathscr{T}$	greatest lag used for obtaining the spectrum from the autocovariance
u	dummy variable
u	current velocity, wind speed
u	parameter of the probability of H_s
$\mathbf{U} = (U_x, U_y, U_z)$	particle displacement
$\mathbf{v} = (v_x, v_y, v_z)$	particle velocity
V	random variable
w	dummy variable
w	nondimensional frequency
w	parameter of the probability of H_s
w_α, w_β	parameters of the probability of H_s for a given wave direction
W	volume
x	dummy variable
x	horizontal coordinate axis
$\mathbf{x}_o = (x_o, y_o)$	fixed point of the horizontal plane
X	ancillary variable related to H_s
$\mathbf{X} = (X, Y)$	vector whose initial point is $\mathbf{x}_o$
y	horizontal coordinate axis
Y	component of vector $\mathbf{X}$
Y	ancillary variable related to the probability of exceedance of H_s
$\mathscr{Y}$	fetch
z	vertical coordinate axis with the origin at the mean water level
α	angle between x-axis and direction of wave advance
α	quotient between wave height and r.m.s. surface displacement of a sea state
$\tilde{\alpha}$	quotient between wave height and H_s
A	Phillips's parameter
β	polar coordinate (the β-curves are circles centred at the origin)
β	quotient between wave crest elevation and r.m.s. surface displacement of a sea state
γ	specific weight
δ	delta function
Δ	variation of the mean water level due to the wave motion
Δp	actual pressure $+\rho g z$

ε	phase angle
ζ	vertical coordinate axis with the origin at the seabed
η η_{ph}	surface displacement fluctuating pressure head at points which remain always beneath the water surface
θ θ $\bar{\theta}$	angle between the y-axis and the direction of wave advance azimuth of the direction of wave advance angle between the y-axis and the dominant direction of the spectrum
λ λ	slope angle scale factor
μ	coefficient of friction
ν	kinematic viscosity (10^{-6} m^2 s^{-1} in calculations)
ξ ξ	distance offshore ratio between wave crest elevation and wave height
ρ	water density (1030 kg/m^3 in calculations)
σ	r.m.s. surface displacement of a sea state
τ τ	time lag between crest and trough threshold of interarrival time
ϕ Φ Φ $\mathbf{\Phi} = (\Phi_x, \Phi_y)$	velocity potential covariance of the surface displacement and the velocity potential norm of $\mathbf{\Phi}$ mean energy flux
χ_1, χ_2	shape parameters of the JONSWAP spectrum
ψ ψ^* $\mathbf{\Psi}$	autocovariance function narrow bandedness parameter covariance of surface displacements
ω	angular frequency

SUBSCRIPTS

0	deep water[1]	*in*	inertia
a	water (aqua)	*m*	abscissa maximum / minimum
A	triangle height	*mo*	model
b	breaking	*n*	nourished beach, natural, nominal
B	triangle base	*o*	horizontal[2]
c	current	*ou*	outgoing
C	centre	*p*	peak[3]
cl	closure	*ph*	pressure head
cr	crest	*pr*	prototype
d	diffraction	*R*	reaction
e	entering	*s*	solid[4]
f	seabed (fundus)	*so*	soil
F	Fourier	*st*	still water
FR	Fresnel	*s.w.*	same wave
G	group	*v*	vertical
h	high waves	*w*	wave

[1] Exceptions: m_0 stands for zeroth order moment of the frequency spectrum; in sects. 9.6-10 and in Appendix B, η_0, ψ_0 stand for $\eta(0)$, $\psi(0)$.

[2] Exception: x_o, y_o, t_o where the subscript o denotes a special value of x, y or t.

[3] Exception: h_p in sect. 7.10 where p denotes a given probability.

[4] Exceptions: H_s where s stands for "significant"; v_s in chap. 2 where s denotes the local propagation axis.

SYMBOLS

$< f(t) >$	=	temporal mean
$\overline{V}$	=	mean value of the random variable V
$\dot{f}$	=	derivative
f'	=	first order in the Stokes expansion, or other meanings being specified every time; the apex is also used to distinguish between the dummy variable and the independent variable of an indefinite integral
f''	=	second order in the Stokes expansion
$\underline{G}$	=	quantity relevant to the equivalent water volume

ABBREVIATIONS

e.t.s.	equivalent triangular storm
l.h.s.	left-hand side
MWL	mean water level
p.d.f.	probability density function
RC 1990, RC 1991, RC 1992, RC 1993, RC 1994	experiments in the natural laboratory of Reggio Calabria
r.m.s.	root mean square
r.h.s.	right-hand side

LIST OF CONTENTS

Chapter 1
Periodic wave pattern: the approach of differential calculus

1 1.1 The irrotational flow, the continuity equation, the Bernoulli equation
5 1.2 The differential equations of an irrotational flow with a free surface
7 1.3 Introduction to wave mechanics
9 1.4 Stokes' theory to the first order
13 1.5 Analysis of the linear dispersion rule
16 1.6 The flow field
19 1.7 Stokes' theory to the second order
21 1.8 Non-linearity effects
23 1.9 Wave-current interaction. Part I: velocity potential and wavelength
28 1.10 Preliminary remarks on three dimensional waves
29 1.11 Wave reflection
34 1.12 Wave diffraction
38 *Conclusive note*
38 *References*

Chapter 2
Periodic wave pattern: the control volume approach

39 2.1 The linear momentum equation for a control volume
42 2.2 The energy equation for a control volume
46 2.3 Radiation stress, mean energy flux, mean wave energy per unit surface
48 2.4 Formulae for radiation stress and mean energy flux of progressive waves
54 2.5 The problem of the control volume extending from deep to shallow water
58 2.6 Practical consequences of the control volume problem
61 2.7 A current associated with the wave motion
65 2.8 Wave refraction for an arbitrary configuration of the seabed
70 2.9 The group celerity
74 2.10 Wave-current interaction. Part II: shoaling and set-down
85 *Conclusive note*
86 *References*

Chapter 3
Wave effects on coasts

87 3.1 The control volume from the breaker line to the beach
88 3.2 The run-up
92 3.3 The longshore transport
94 3.4 The analytical approach to the problem of beach planform evolution
95 3.5 Problem of beach planform evolution: the case of contour lines parallel up to deep water
100 3.6 Problem of beach planform evolution: the case of contour lines parallel only within a certain distance from the shoreline
103 3.7 Planform evolution of a natural shoreline
104 3.8 Stability of a nourished beach
107 3.9 Planform evolution of beach nourishment projects
111 3.10 A useful simplification
113 3.11 Beach planform evolution caused by structures
116 *Conclusive note*
118 *References*

Chapter 4
Wind generated waves: basic concepts

119 4.1 The sea state
121 4.2 The theory of the sea states
123 4.3 Some basic relations in the theory of the sea states
126 4.4 How to obtain the input data of the theory
136 4.5 A mathematical form of the wind wave spectrum
140 4.6 Possibility of testing small scale models in sea or lakes
145 4.7 Inferring the nature of waves from the bandwidth
150 *Conclusive note*
151 *References*

Chapter 5
Analysis of the sea states: the time domain

153 5.1 Why the surface displacement represents a stationary Gaussian process
157 5.2 Joint probability of surface displacements
159 5.3 Rice's problem
161 5.4 Rice's logic
162 5.5 Corollaries of Rice's problem
166 5.6 Solved and still unsolved problems
167 5.7 The period of a very high wave and the wave height probability under general bandwidth assumptions
172 5.8 Experimental verification

175 5.9 Characteristic wave heights
177 5.10 The maximum expected wave height in a sea state of given duration
180 *Conclusive note*
181 *References*

Chapter 6
The wave climate

183 6.1 The function $H_s(t)$
186 6.2 The probability of the significant wave height
191 6.3 The probability of the significant wave height for a given direction of wave advance
196 6.4 Probabilities of the significant wave height for a few areas of the globe
199 6.5 The maximum expected wave height in a storm with a given history
200 6.6 The concept of "equivalent triangular storm"
204 6.7 Storm durations
206 *Conclusive note*
206 *References*

Chapter 7
Design waves and risk analysis

207 7.1 The return period of a sea storm where the significant wave height exceeds a fixed threshold
211 7.2 The significant wave height and its persistence vs the return period
213 7.3 The encounter probability
221 7.4 The chain: lifetime, encounter probability ⇒ return period ⇒ significant wave height
223 7.5 Coastal structures: the design sea state
226 7.6 The return period of a wave with a height exceeding a fixed threshold
228 7.7 The return period of a sea storm containing at least one wave higher than a fixed threshold
235 7.8 Offshore structures: the design wave
239 7.9 Calculations for different wave directions
242 7.10 Corollary of risk analysis: a general relation between the confidence interval and the sampling rate
246 *Conclusive note*
247 *References*

Chapter 8
Analysis of the sea states in the space-time

249 8.1 The concept of homogeneous wave field
251 8.2 The wave field in the open sea

254 8.3 The directional spectrum
257 8.4 Shoaling and refraction of the wind-generated waves
263 8.5 Reflection of the wind-generated waves
267 8.6 Diffraction of the wind-generated waves
270 8.7 Long-crested random waves: the link between periodic waves and wind-generated waves
274 8.8 Direct proportion between the maximum expected wave height and the diffraction coefficient
276 8.9 Space-time covariances
279 *Conclusive note*
279 *References*

Chapter 9
The theory of quasi-determinism

281 9.1 A sufficient condition for occurrence of a wave of given height very large
286 9.2 A necessary condition for occurrence of a wave of given height very large
288 9.3 The water surface on space-time, if a wave of given height very large occurs at a fixed point
291 9.4 The velocity potential if a wave of given height very large occurs at a fixed point
293 9.5 Theory's generality and consistency with Stokes' theory
294 9.6 Formal proof of the necessary condition. Part I: symbols and assumptions
296 9.7 Formal proof of the necessary condition. Part II: core of the proof
300 9.8 Formal proof of the necessary condition. Part III: the central inequality
303 9.9 Formal proof of the necessary condition. Part IV: conclusion
306 9.10 Corollary: the closed solution for the wave height distribution

Chapter 10
Uses and consequences of the quasi-determinism theory

311 10.1 The first way to employ the theory
317 10.2 A three dimensional wave group
318 10.3 The waves are higher on the time domain than on the space domain!
320 10.4 Effects of water depth and of spectrum shape on the wave group
324 10.5 Shoaling and refraction of the wave group
326 10.6 Explanation of the first big difference between sea waves and periodic waves
335 10.7 Explanation of the second big difference between sea waves and periodic waves
339 10.8 The second way to employ the theory
346 10.9 The "genetic code" of the sea waves
350 10.10 The determinism arises from within the random waves
359 *Conclusive note*
359 *References*

Chapter 11
Analysis of the wave forces on offshore structures

361 11.1 Wave forces on gravity offshore platforms
364 11.2 Local perturbation of the flow field at an offshore structure
367 11.3 Wave forces on submerged tunnels
373 11.4 The diffraction coefficients of the forces
383 11.5 Wave forces on space frame structures
387 11.6 The long-structure problem
392 *Conclusive note*
392 *References*

Chapter 12
Calculation of the wave forces on offshore structures

393 12.1 Calculation of the wave forces on a gravity offshore platform
397 12.2 Calculation of the wave forces on a space frame structure
402 12.3 Design of a submerged tunnel. I: calculation of the wave forces
412 12.4 Design of a submerged tunnel. II: the effect of currents
413 12.5 Design of a submerged tunnel. III: the risk of resonance

Chapter 13
Stability analysis of coastal structures

419 13.1 Wave pressure on a wall
427 13.2 Forces on a vertical breakwater
430 13.3 Design of vertical breakwaters
436 13.4 Further verifications of the vertical breakwaters
438 13.5 The Japanese practice
441 13.6 The problem of the rubble mound breakwaters
445 *Conclusive note*
445 *References*

Chapter 14
Topics calling for an overall overview of offshore and coastal engineering

447 14.1 A comparison between tsunami and wind waves from the open sea to the coast
450 14.2 Small scale models
455 14.3 Wave measurements
461 *Conclusive note*
461 *References*

Appendix A

Appendix to chapters 6 and 7: use of wave hindcast and wave measurements from satellites

463 A.1 Long term wave statistics from satellite data
464 A.2 Wave hindcast
470 A.3 Trend in the wave climate and its effects on engineering
473 *References*

Appendix B

Appendix to chapters 9 and 10: the wave group of the maximum expected crest elevation, and the wave group of the maximum expected crest-to-trough height

475 B.1 The first version of the quasi-determinism theory
480 B.2 Corollaries of the first version
482 B.3 The relationship between the two versions of the theory
485 *Conclusive note*
485 *References*

Chapter 1
PERIODIC WAVE PATTERN: THE APPROACH OF DIFFERENTIAL CALCULUS

1.1 The irrotational flow, the continuity equation, the Bernoulli equation

The condition

$$\frac{\partial v_z}{\partial y}=0, \quad \frac{\partial v_y}{\partial z}=0, \tag{1.1}$$

where v_y, v_z are the velocity components, is necessary and sufficient for a *two-dimensional* y-z motion of a rigid body to be irrotational. Such a condition is only sufficient for a two-dimensional y-z water flow. Indeed a given water mass does not have a form, and hence it can undergo such a deformation that condition (1.1) is not satisfied and the flow is still irrotational [see fig. 1.1]. The necessary and sufficient condition for a two-dimensional y-z motion of a small volume of water $\mathrm{d}x\,\mathrm{d}y\,\mathrm{d}z$ not to be rotational is

$$\frac{\partial v_z}{\partial y}=\frac{\partial v_y}{\partial z}. \tag{1.2}$$

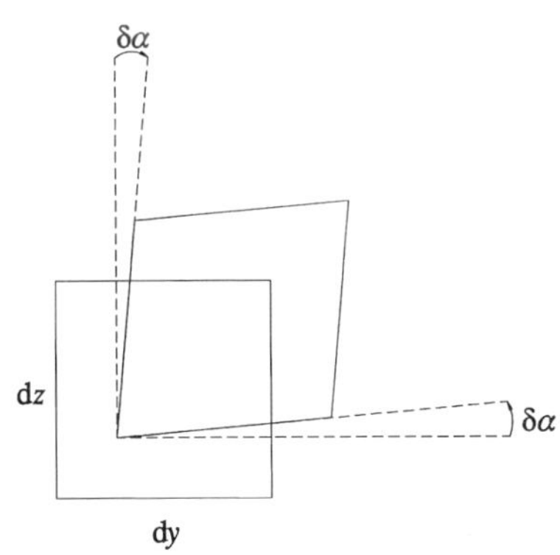

Fig. 1.1 Possible deformation of a small water mass in an irrotational flow.

Such a condition guarantees that a clockwise rotation $\delta\alpha$ of face $\mathrm{d}x\,\mathrm{d}z$ is balanced by an equal anticlockwise rotation $\delta\alpha$ of face $\mathrm{d}x\,\mathrm{d}y$ (or vice versa an anticlockwise rotation of $\mathrm{d}x\,\mathrm{d}z$ is balanced by an equal clockwise rotation of $\mathrm{d}x\,\mathrm{d}y$).

We shall see that wave motions are generally assumed to be irrotational, and that in a very common case the orbits of the particles are circular. Obviously the two facts (irrotational motion and circular orbit) do not contradict each other. To understand this point we can think of the Ferris wheel in an amusement park. The head of the man in fig. 1.2 moves along a circular trajectory of radius R, and his feet also move along a circular trajectory of radius R. Nevertheless the man does not rotate, and indeed his head always remains above his feet. The orbits of the various particles forming the man are circular, but the man does not rotate simply because those orbits have different centres. We shall see that the motion of water particles in a wave is similar to the motion of the man's particles on the wheel.

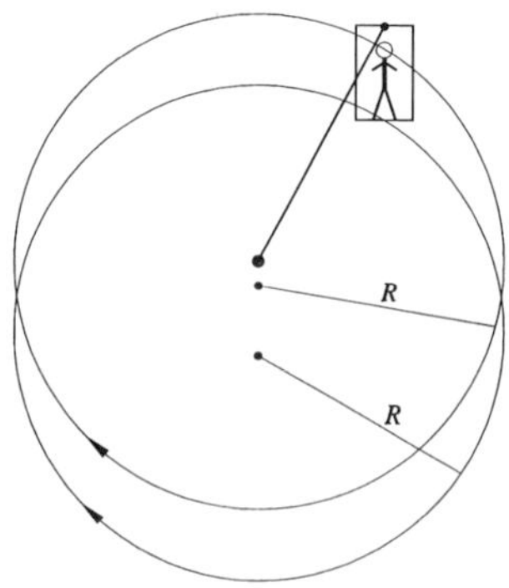

Fig. 1.2 The head and feet of the man on the Ferris wheel move along circular trajectories with the same radius and different centres.

If condition (1.2) is satisfied, that is to say, if the motion is irrotational, a function $\phi(y, z, t)$ does exist, for which:

$$v_y \equiv \frac{\partial \phi}{\partial y}, \quad v_z \equiv \frac{\partial \phi}{\partial z}. \tag{1.3}$$

With this function, namely "velocity potential", the continuity equation takes the form

$$\frac{\partial^2 \phi}{\partial y^2} + \frac{\partial^2 \phi}{\partial z^2} = 0, \tag{1.4}$$

and the Bernoulli equation, the form

$$p + \rho g z + \rho \frac{\partial \phi}{\partial t} + \frac{1}{2}\rho\left[\left(\frac{\partial \phi}{\partial y}\right)^2 + \left(\frac{\partial \phi}{\partial z}\right)^2\right] = f(t), \tag{1.5}$$

where $f(t)$ is an arbitrary function of time. The Bernoulli equation says: the l.h.s. of (1.5) can vary with time, but at any fixed time it does not vary from one point to another.

Let us briefly recall the reasoning leading to the continuity equation and Bernoulli equation. Having to deal with a y-z flow the abstraction of two-dimensional space is convenient, and hence we shall resort to a small "volumc" $\mathrm{d}y\,\mathrm{d}z$ in order to prove equations (1.4) and (1.5). Naturally, if one prefers the more concrete three-dimensional approach, one can resort to a small volume $\mathrm{d}x\,\mathrm{d}y\,\mathrm{d}z$. In that case, all terms in the following equations must be multiplied by $\mathrm{d}x$, which does not modify the final result.

To prove the continuity equation we shall write that the water mass entering the small volume $\mathrm{d}y\,\mathrm{d}z$ in the small time interval $\mathrm{d}t$ is equal to the water mass going out of the same volume in the same time interval. The result is

$$\rho\left(v_y - \frac{\partial v_y}{\partial y}\frac{\mathrm{d}y}{2}\right)\mathrm{d}z\,\mathrm{d}t + \rho\left(v_z - \frac{\partial v_z}{\partial z}\frac{\mathrm{d}z}{2}\right)\mathrm{d}y\,\mathrm{d}t =$$

$$= \rho\left(v_y + \frac{\partial v_y}{\partial y}\frac{\mathrm{d}y}{2}\right)\mathrm{d}z\,\mathrm{d}t + \rho\left(v_z + \frac{\partial v_z}{\partial z}\frac{\mathrm{d}z}{2}\right)\mathrm{d}y\,\mathrm{d}t, \tag{1.6}$$

where the l.h.s. gives the entering mass and the r.h.s. gives the outgoing mass [see fig. 1.3*a*]. Equation (1.4) proceeds straightforwardly from (1.6) on cancelling a few terms and using definition (1.3).

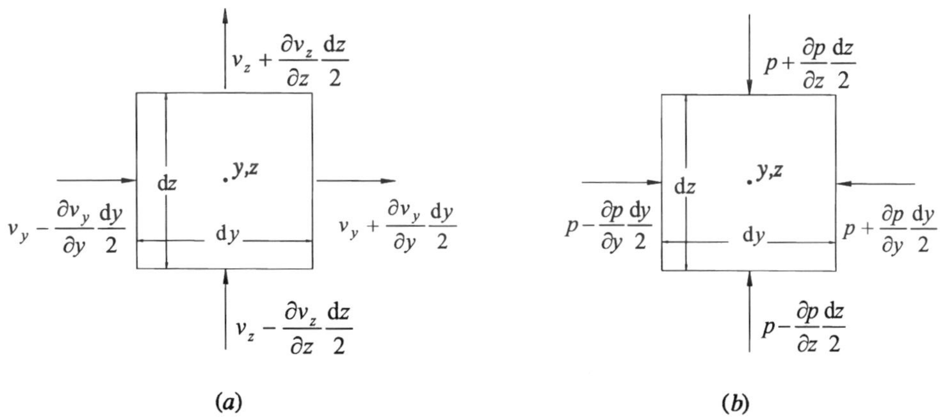

Fig. 1.3 (*a*) Reference scheme for the continuity equation. (*b*) Reference scheme for the Bernoulli equation.

As to the Bernoulli equation, it proceeds on applying Newton's second law to any small volume $\mathrm{d}y\,\mathrm{d}z$ under ideal flow assumptions, that is under the assumption that the shear stress is negligible [see fig. 1.3*b*]. The result is

$$\left(p-\frac{\partial p}{\partial y}\frac{\mathrm{d}y}{2}\right)\mathrm{d}z-\left(p+\frac{\partial p}{\partial y}\frac{\mathrm{d}y}{2}\right)\mathrm{d}z=\rho\,\mathrm{d}y\,\mathrm{d}z\,a_y, \tag{1.7a}$$

$$\left(p-\frac{\partial p}{\partial z}\frac{\mathrm{d}z}{2}\right)\mathrm{d}y-\left(p+\frac{\partial p}{\partial z}\frac{\mathrm{d}z}{2}\right)\mathrm{d}y-\rho g\,\mathrm{d}y\,\mathrm{d}z=\rho\,\mathrm{d}y\,\mathrm{d}z\,a_z. \tag{1.7b}$$

Acceleration **a** of the particle being at point y, z (the centre of the small volume) at time t is given by

$$\mathbf{a}=\frac{\mathbf{v}_P(t+\mathrm{d}t)-\mathbf{v}_P(t)}{\mathrm{d}t},$$

where $\mathbf{v}_P$ denotes the velocity of the specific particle (called P) which at time t is at point y, z. Therefore $\mathbf{v}_P(t)$ coincides with the velocity $\mathbf{v}$ at point y, z at time t; while $\mathbf{v}_P(t+\mathrm{d}t)$ is the velocity at time $t+\mathrm{d}t$ at the point being occupied by particle P. Since at time $t+\mathrm{d}t$ particle P is at point

$$y+v_y\,\mathrm{d}t,\quad z+v_z\,\mathrm{d}t,$$

we have

$$\mathbf{v}_P(t+\mathrm{d}t)=\mathbf{v}(y+v_y\,\mathrm{d}t,\ z+v_z\,\mathrm{d}t,\ t+\mathrm{d}t)=\mathbf{v}+\frac{\partial\mathbf{v}}{\partial y}v_y\,\mathrm{d}t+\frac{\partial\mathbf{v}}{\partial z}v_z\,\mathrm{d}t+\frac{\partial\mathbf{v}}{\partial t}\mathrm{d}t,$$

and hence

$$\mathbf{a}=\frac{\partial\mathbf{v}}{\partial y}v_y+\frac{\partial\mathbf{v}}{\partial z}v_z+\frac{\partial\mathbf{v}}{\partial t}. \tag{1.8}$$

With this formula and definition (1.3), the equations (1.7*a-b*) are reduced to

$$\frac{\partial}{\partial y}\left[p+\frac{1}{2}\rho\left(\frac{\partial\phi}{\partial y}\right)^2+\frac{1}{2}\rho\left(\frac{\partial\phi}{\partial z}\right)^2+\rho\frac{\partial\phi}{\partial t}\right]=0,$$

$$\frac{\partial}{\partial z}\left[p+\rho gz+\frac{1}{2}\rho\left(\frac{\partial\phi}{\partial y}\right)^2+\frac{1}{2}\rho\left(\frac{\partial\phi}{\partial z}\right)^2+\rho\frac{\partial\phi}{\partial t}\right]=0,$$

from which follows that the sum

$$p+\rho gz+\frac{1}{2}\rho\left(\frac{\partial\phi}{\partial y}\right)^2+\frac{1}{2}\rho\left(\frac{\partial\phi}{\partial z}\right)^2+\rho\frac{\partial\phi}{\partial t}$$

does not vary if we move along y for a fixed z and t, nor if we move along z for a fixed y and t; which can be expressed in the form (1.5).

Before concluding this introductory section, we need to explain the reason for the unusual y-z reference frame in place of the more common x-z. The reason is that, throughout the book, y will be the direction of wave propagation (at least whenever this direction is constant), and clearly if the wave travels along the y-axis, the flow is two-dimensional y-z.

1.2 The differential equations of an irrotational flow with a free surface

1.2.1 *The free surface equation*

Let us assume that the two-dimensional irrotational flow has a free surface, and let us call $\eta(y, t)$ the vertical displacement of this free surface. Our first goal is to put in mathematical form what we have just said, namely "$\eta(y, t)$ is the surface displacement".

In order to express in mathematical form that $\eta(y, t)$ is the vertical elevation of the air-water boundary, we shall resort to the small control volume of fig. 1.4. Specifically we shall say that the water mass entering this volume in a small time interval $\mathrm{d}t$ is equal to the sum of the water mass leaving the volume and of the water mass piling up in the volume in the same interval $\mathrm{d}t$. Doing so, we shall implicitly specify the physical meaning of $\eta(y, t)$.

The water mass entering the small volume in a small time interval $\mathrm{d}t$ is

$$m_e = \int_{-d}^{\eta} \rho \frac{\partial \phi}{\partial y} \, \mathrm{d}z \, \mathrm{d}t .$$

The water mass going out of the small volume is

$$m_{ou} = \int_{-d}^{\eta+\mathrm{d}\eta} \rho \left(\frac{\partial \phi}{\partial y} + \frac{\partial^2 \phi}{\partial y^2} \mathrm{d}y \right) \mathrm{d}z \, \mathrm{d}t \quad \text{with} \quad \mathrm{d}\eta = \frac{\partial \eta}{\partial y} \mathrm{d}y . \tag{1.9}$$

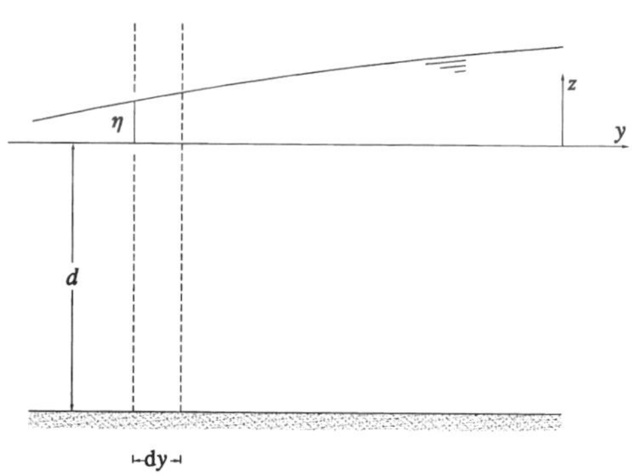

Fig. 1.4 The small volume used for obtaining equation (1.13).

The water mass piling up is

$$\Delta m = \rho \left(\frac{\partial \eta}{\partial t} \mathrm{d}t \right) \mathrm{d}y .$$

In these equations, η and $\dfrac{\partial \phi}{\partial y}$ are, respectively, the surface displacement and the horizontal particle velocity at the left side of the control volume.

Equation (1.9) can be rewritten in the form

$$m_{ou} = \int_{-d}^{\eta} \rho \left(\frac{\partial \phi}{\partial y} + \frac{\partial^2 \phi}{\partial y^2} \mathrm{d}y \right) \mathrm{d}z \, \mathrm{d}t + \rho \left[\left(\frac{\partial \phi}{\partial y} \right)_{z=\eta} + \left(\frac{\partial^2 \phi}{\partial y^2} \right)_{z=\eta} \mathrm{d}y \right] \frac{\partial \eta}{\partial y} \mathrm{d}y \, \mathrm{d}t , \tag{1.10}$$

so that the equation

$$m_{ou} - m_e + \Delta m = 0$$

yields

$$\int_{-d}^{\eta} \frac{\partial^2 \phi}{\partial y^2} \mathrm{d}z + \left(\frac{\partial \phi}{\partial y} \right)_{z=\eta} \frac{\partial \eta}{\partial y} + \frac{\partial \eta}{\partial t} = 0 , \tag{1.11}$$

apart from some negligible terms. Let us consider the integral on the l.h.s. of this equation. Thanks to the continuity equation (1.4), we have

$$\int_{-d}^{\eta} \frac{\partial^2 \phi}{\partial y^2} \mathrm{d}z = - \int_{-d}^{\eta} \frac{\partial^2 \phi}{\partial z^2} \mathrm{d}z = - \left(\frac{\partial \phi}{\partial z} \right)_{z=\eta} + \left(\frac{\partial \phi}{\partial z} \right)_{z=-d} , \tag{1.12}$$

where the term $(\partial \phi / \partial z)_{z=-d}$ is zero because it represents the vertical particle velocity at the bottom, which is zero if the bottom is horizontal. Therefore equation (1.11) is reduced to

$$\left(\frac{\partial \phi}{\partial z} \right)_{z=\eta} = \left(\frac{\partial \phi}{\partial y} \right)_{z=\eta} \frac{\partial \eta}{\partial y} + \frac{\partial \eta}{\partial t} \tag{1.13}$$

which is the general equation of the free surface.

In the proof we have assumed the bottom to be horizontal. But the same result holds even if the bottom is sloping. In this more general case d denotes the bottom depth at the left side of the small control volume, so that the bottom depth at the right side is $d + (\mathrm{d}d / \mathrm{d}y) \, \mathrm{d}y$. Therefore, the lower limit in the integral on the r.h.s. of equation (1.10) changes, and (1.11) takes a new term:

$$\int_{-d}^{\eta} \frac{\partial^2 \phi}{\partial y^2} \mathrm{d}z + \left(\frac{\partial \phi}{\partial y} \right)_{z=\eta} \frac{\partial \eta}{\partial y} + \left(\frac{\partial \phi}{\partial y} \right)_{z=-d} \frac{\mathrm{d}d}{\mathrm{d}y} + \frac{\partial \eta}{\partial t} = 0 . \tag{1.14}$$

Moreover the boundary condition becomes

$$\left(\frac{\partial \phi}{\partial z}\right)_{z=-d} = -\left(\frac{\partial \phi}{\partial y}\right)_{z=-d} \frac{\mathrm{d}d}{\mathrm{d}y}$$

which taken together with (1.12) and (1.14) leads us again to the equation (1.13) of the free surface.

1.2.2 *The system of equations*

On the whole, in an irrotational two-dimensional flow with a free surface, functions ϕ and η must satisfy the following system of equations:

$$g\eta + \left(\frac{\partial \phi}{\partial t}\right)_{z=\eta} + \frac{1}{2}\left[\left(\frac{\partial \phi}{\partial y}\right)^2 + \left(\frac{\partial \phi}{\partial z}\right)^2\right]_{z=\eta} = \frac{1}{\rho} f(t), \tag{1.15a}$$

$$\left(\frac{\partial \phi}{\partial z}\right)_{z=\eta} = \left(\frac{\partial \phi}{\partial y}\right)_{z=\eta} \frac{\partial \eta}{\partial y} + \frac{\partial \eta}{\partial t}, \tag{1.15b}$$

$$\frac{\partial^2 \phi}{\partial y^2} + \frac{\partial^2 \phi}{\partial z^2} = 0, \tag{1.15c}$$

$$\left(\frac{\partial \phi}{\partial z}\right)_{z=-d} = 0. \tag{1.15d}$$

The first one says that the pressure is zero on the free surface (atmospheric pressure being taken as reference pressure); it proceeds straightforwardly from the Bernoulli equation. The second one is the general equation of the free surface, which we have just obtained. The third and fourth ones are, respectively, the continuity equation and the condition of the solid boundary which have been rewritten in order to get a general overview of the whole system of equations. As to the solid boundary condition, it has been specified for a horizontal bottom, which is the case being examined in the rest of this chapter.

1.3 Introduction to wave mechanics

A vertical plate swinging in a periodic way at one end of a channel generates waves on the free surface. If we take a photo of the water surface we get a picture of the surface displacement η as a function of abscissa y along the propagation axis (the channel's axis). Function $\eta(y)$ at a fixed instant represents the waves on the space domain [see fig. 1.5*a*]. If we record the surface displacement at a fixed point as a function of time t, we get the waves on the time domain [see fig. 1.5*b*].

From figs. 1.5*a*-*b* of the waves on the space domain and on the time domain, we get the definitions of the basic parameters: wave height H which represents the difference in height between crest and trough; wavelength L which represents the

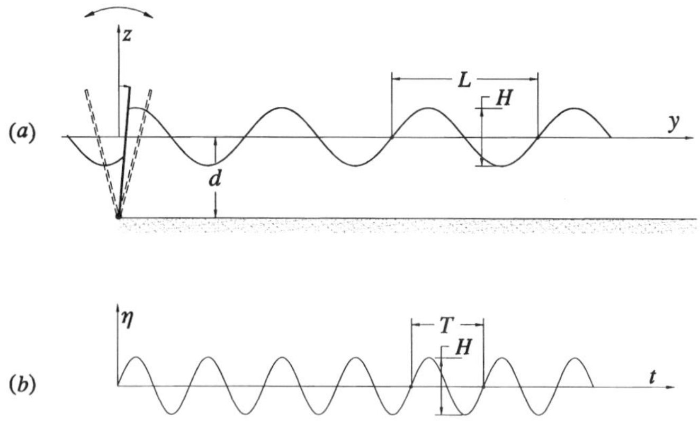

Fig. 1.5 (*a*) Waves on the space domain. (*b*) Waves on the time domain.

distance between the two extreme zeros of the wave; wave period T which represents the time lag between the two extreme zeros of the wave. Besides these three parameters, it is convenient to define

(i) the wave amplitude

$$a \equiv H/2,$$

(ii) the angular frequency

$$\omega \equiv 2\pi/T,$$

(iii) the wave number

$$k \equiv 2\pi/L.$$

The scheme of fig. 1.6 should be useful to understand the wave motion. Each point in the figure moves along a circular orbit, with constant speed. The time taken to cover the orbit (circumference) is T, and the figure shows two instant pictures taken at a time interval of $T/4$ from each other. The line connecting the points represents a wave. We see the wave advance of $L/4$ in a time interval of $T/4$, and this means that the propagation speed (celerity) of the wave is

$$c = L/T.$$

The speed v of each point is generally different from c, indeed

$$v = 2\pi R/T$$

(R is the radius of the circular orbits). The particle speed v is not only generally different from the propagation speed c of the wave, but we can even vary v arbitrarily without modifying c: it suffices to fix period T and length L (which depends only upon the distance between the circumferences) and let radius R vary.

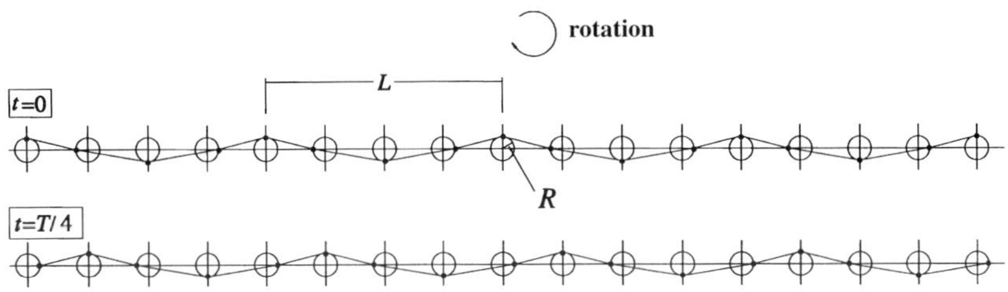

Fig. 1.6 Each point moves along a circular orbit of radius R in a time T; the line connecting the points is a wave whose propagation speed is L/T.

Finally, a last item before passing to the mathematical treatment of the wave motions. Let us imagine that we make a blob of two colours (e.g. red and black) within the water mass in a wave. If we put the red on top of the black, generally we should see that the red remains over the black, just as we see that the man's head remains above his feet on the wheel in fig. 1.2. In other words, we should see that the blob does not rotate, which denotes irrotational flow.

1.4 Stokes' theory to the first order

1.4.1 *A physical approach*

Let us fix the period and the swinging amplitude, and let us set the wavemaker in motion. Waves with a height H_1 will form.

Let us stop the wavemaker, and let us set the engine in a different manner: same period, smaller swinging amplitude. Then let us start again. Waves with a height H_2 smaller than H_1 will form. The wave period will be the same as before. Indeed the wave period proves to be the same as the swinging period of the wavemaker.

Let us repeat the process many times: each time with the same period T and with wave heights smaller and smaller. Doing so, in the first wave generations, which are the ones with the greater heights, we shall note asymmetry between the wave crest and trough: the crest will be steeper than the trough. Then, we shall find that a gradual lowering of the wave height, under the same period, leads to waves with smaller asymmetry: the wave approaches a sinusoid with a wavelength which depends on d and T.

Summarizing:

$$\text{as } H \to 0 \ (d \text{ and } T \text{ fixed}): \ \eta(y,t) = \frac{H}{2}\cos\left(\frac{2\pi}{L}y - \frac{2\pi}{T}t\right), \tag{1.16}$$

where, for the moment, wavelength L is unknown. Function (1.16) represents a periodic wave of length L on the space domain and it represents a periodic wave of period T on the time domain. The negative sign in the cosine implies that the wave travels along the y-axis (with a positive sign the wave would travel in the opposite direction).

Generating waves with smaller and smaller heights (and a fixed period) we shall also note that the velocity components, at any fixed depth, will tend to fluctuate in space-time like $\eta(y, t)$: v_y in phase with η, and v_z with some phase angle. Moreover, the particle speed will prove to be proportional to the wave height. In this, the phenomenon is similar to the one depicted in fig. 1.6. In that figure the wave height is equal to $2R$, and the particle speed is equal to $2\pi R/T$. Therefore, if we reduce R with the same T, the wave height and the particle speed are reduced in the same way.

From these observations on particle velocity, we can draw the following identikit of the velocity potential:

$$\text{as } H \to 0 \ (d \text{ and } T \text{ fixed}): \ \phi(y, z, t) =$$

$$= H f_1(z;\, d, T, L) \cos\left(\frac{2\pi}{L} y - \frac{2\pi}{T} t + \varepsilon\right) + f_2(t), \tag{1.17}$$

where $f_1(z;\, d, T, L)$ denotes a function of z, in which parameters d, T and L may be present. For the moment, functions f_1 and f_2 and phase angle ε are unknown.

1.4.2 *The system of linear equations*

The equations (1.16) and (1.17) show that both η and ϕ are infinitesimal of order H. In particular, the fact that η is infinitesimal enables us to rewrite equations (1.15*a*-*b*) in the form

$$g\eta + \left(\frac{\partial \phi}{\partial t}\right)_{z=0} + \left(\frac{\partial^2 \phi}{\partial z \partial t}\right)_{z=0} \eta + \frac{1}{2}\left[\left(\frac{\partial \phi}{\partial y}\right)^2 + \left(\frac{\partial \phi}{\partial z}\right)^2\right]_{z=0} +$$

$$+ \frac{1}{2}\left\{\frac{\partial}{\partial z}\left[\left(\frac{\partial \phi}{\partial y}\right)^2 + \left(\frac{\partial \phi}{\partial z}\right)^2\right]\right\}_{z=0} \eta = \frac{1}{\rho} f(t), \tag{1.18a}$$

$$\left(\frac{\partial \phi}{\partial z}\right)_{z=0} + \left(\frac{\partial^2 \phi}{\partial z^2}\right)_{z=0} \eta = \left[\left(\frac{\partial \phi}{\partial y}\right)_{z=0} + \left(\frac{\partial^2 \phi}{\partial z \partial y}\right)_{z=0} \eta\right] \frac{\partial \eta}{\partial y} + \frac{\partial \eta}{\partial t}, \tag{1.18b}$$

where the value of a function at $z = \eta$ has been expressed as: [value of the function at $z = 0$] + [value of the derivative with respect to z, at $z = 0$] $\times\, \eta$. Neglecting the terms of orders smaller than or equal to H^2, equations (1.18*a*-*b*) are reduced to

$$\left(\frac{\partial \phi}{\partial t}\right)_{z=0} = -g\eta + \frac{1}{\rho} f(t), \tag{1.19a}$$

$$\left(\frac{\partial \phi}{\partial z}\right)_{z=0} = \frac{\partial \eta}{\partial t}. \tag{1.19b}$$

At this point, the problem is how to obtain functions f_1, f_2, length L and phase ε (provided they exist) in such a way that the two forms (1.16) and (1.17) of η and ϕ satisfy equations (1.19*a-b*) as well as (1.15*c-d*). The problem has only one solution, as will be evident from the following analysis.

1.4.3 *Solution for* ϕ

Substituting the forms of η and ϕ in (1.19*a*), we have

$$H\, f_1(0;\; d, T, L)\; \omega\; \sin(ky - \omega t + \varepsilon) + \frac{\mathrm{d}}{\mathrm{d}t} f_2(t) = -g\frac{H}{2}\cos(ky - \omega t) + \frac{1}{\rho} f(t),$$

and the only way for the function of y and t on the l.h.s. of this equation to be equal to the function of y and t on the r.h.s., for every y and t, is that

$$f_1(0;\; d, T, L) = -\frac{1}{2} g\omega^{-1}, \tag{1.20a}$$

$$\varepsilon = \frac{\pi}{2}, \tag{1.20b}$$

$$f_2(t) = \frac{1}{\rho}\int_0^t f(t')\,\mathrm{d}t'. \tag{1.20c}$$

(Clearly, an arbitrary constant can be added to $f_2(t)$, but such a constant proves to be wholly insignificant for what follows.) Thanks to (1.20*b-c*) the identikit (1.17) of ϕ becomes more precise:

$$\phi(y, z, t) = -H\, f_1(z;\; d, T, L)\; \sin(ky - \omega t) + \frac{1}{\rho}\int_0^t f(t')\,\mathrm{d}t'. \tag{1.21}$$

The more general form of f_1 so that this function of ϕ satisfies (1.15*c*) is

$$f_1(z; d, T, L) = A\exp(kz) + B\exp(-kz),$$

with A and B *arbitrary constants*. The values of these constants can be obtained thanks to condition (1.20*a*) and equation (1.15*d*). The result is

$$A = -\frac{1}{2} g\omega^{-1}\frac{\exp(kd)}{\exp(kd) + \exp(-kd)}, \quad B = -\frac{1}{2} g\omega^{-1}\frac{\exp(-kd)}{\exp(kd) + \exp(-kd)},$$

so that (1.21) becomes

$$\phi(y, z, t) = g\frac{H}{2}\omega^{-1}\frac{\cosh[k(d+z)]}{\cosh(kd)}\sin(ky - \omega t) + \frac{1}{\rho}\int_0^t f(t')\mathrm{d}t'. \tag{1.22}$$

1.4.4 *The linear dispersion rule*

At this stage there is only equation (1.19*b*) to be satisfied, and only one parameter is left unknown: L or $k \equiv 2\pi / L$. The use of functions (1.16) and (1.22) in equation (1.19*b*) yields

$$k \tanh(kd) = \frac{\omega^2}{g}, \tag{1.23}$$

that is, in terms of L and T rather than of k and ω,

$$L = \frac{gT^2}{2\pi} \tanh\left(\frac{2\pi d}{L}\right) \tag{1.24}$$

which relates L to d and T. Such a relation is most commonly used in ocean engineering, both in the form (1.23) and in the form (1.24), and it is called the *linear dispersion rule.*

1.4.5 *The effect of function* $f(t)$

Since $f(t)$ is an arbitrary function of time, the velocity potential (1.22) is indeterminate. But the functions which are of interest, that is to say $\mathbf{v}(y, z, t)$, and $p(y, z, t)$, prove to be independent of $f(t)$ and thus they are definite. In particular, the components of vector $\mathbf{v}$ proceed through (1.3) and prove to be

$$v_y(y, z, t) = g\frac{H}{2}\omega^{-1}k\frac{\cosh[k(d+z)]}{\cosh(kd)}\cos(ky - \omega t), \tag{1.25a}$$

$$v_z(y, z, t) = g\frac{H}{2}\omega^{-1}k\frac{\sinh[k(d+z)]}{\cosh(kd)}\sin(ky - \omega t). \tag{1.25b}$$

As to the pressure, it is obtained by means of the Bernoulli equation (1.5). The result is

$$p(y, z, t) = -\rho g z + \rho g\frac{H}{2}\frac{\cosh[k(d+z)]}{\cosh(kd)}\cos(ky - \omega t)$$

(where the terms of orders smaller than or equal to H^2 have been neglected), and hence the fluctuating pressure head proves to be

$$\eta_{ph}(y, z, t) = \frac{H}{2}\frac{\cosh[k(d+z)]}{\cosh(kd)}\cos(ky - \omega t). \tag{1.26}$$

Clearly the formulae for p and η_{ph} are valid for $z \le \eta$, and the formula for η_{ph} requires in addition z to be smaller than or equal to zero. Indeed (1.26) presupposes that the static pressure is $-\rho g z$. To this purpose, note that, in the whole text, η_{ph} denotes the fluctuating pressure head only at points which are always beneath the water surface, where (1.26) is of course valid.

1.5 Analysis of the linear dispersion rule

1.5.1 *Calculation of L from d and T*

It is convenient to define

$$L_0 = \frac{gT^2}{2\pi}, \tag{1.27}$$

and rewrite (1.24) in the form

$$L = L_0 \tanh\left(\frac{2\pi d}{L}\right). \tag{1.28}$$

Since the range of tanh$x$ is (0,1), it follows that

$$L < L_0,$$

with L approaching L_0 on deep water.

The calculation of L can be done through an iterative approach, by means of

$$L_i = L_0 \tanh\left(\frac{2\pi d}{L_{i-1}}\right), \tag{1.29}$$

for $i = 1, 2, 3$ and so on. In this way the following inequalities are obtained:

$$\begin{cases} L_i < L & \text{if } i \text{ is an odd number,} \\ L_i > L & \text{if } i \text{ is an even number.} \end{cases}$$

To verify these inequalities, one has simply to analyse the quotient L_i/L that proceeds from (1.28), (1.29):

$$\frac{L_i}{L} = \tanh\left(\frac{2\pi d}{L_{i-1}}\right) \Big/ \tanh\left(\frac{2\pi d}{L}\right).$$

Indeed, being $L_0 > L$, it follows: $L_1 < L$ and consequently $L_2 > L$ which in turn implies $L_3 < L$, and so on. Sequence (1.29) converges, that is the difference $|L_i - L_{i-1}|$ approaches zero as i grows.

1.5.2 *Calculation of d/L from d/L_0*

From (1.28) it follows that

$$\frac{d}{L}\tanh\left(2\pi\frac{d}{L}\right) = \frac{d}{L_0}. \tag{1.30}$$

The value of d/L which satisfies this equation is equal to the special value $\tilde{x}$ for which the function

$$f(x) \equiv x \tanh(2\pi x) \tag{1.31}$$

is equal to the given value d/L_0. Function (1.31) is strictly increasing, it being the product of two functions, x and $\tanh(2\pi x)$, which are strictly increasing. Therefore, only one abscissa corresponds to a given ordinate [cf. fig.1.7]. To obtain $\tilde{x}$, one calculates $f(x)$ at regular increments Δx until $f(x) > d/L_0$ is found. At this point, one goes back a step Δx, then fixes a smaller Δx, and continues until the abscissa $\tilde{x}$ is reached with the desired degree of precision.

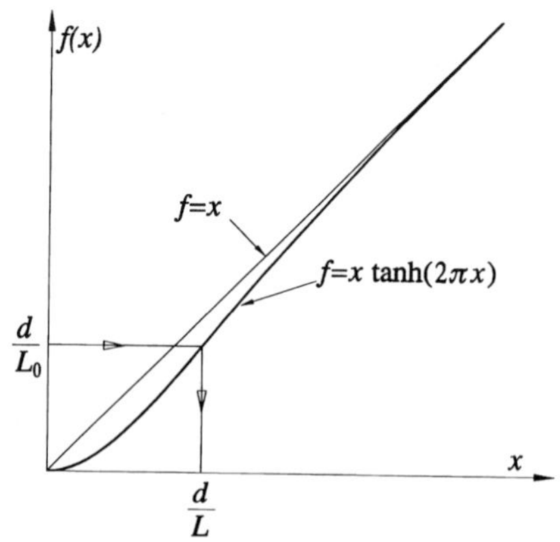

Fig. 1.7 Input: d/L_0. Output: d/L.

In the following there are a number of parameters which depend on kd, and we shall represent these parameters as functions of d/L_0. This is because for every $kd \equiv 2\pi d/L$ there is exactly one d/L_0 and vice versa.

1.5.3 *The concept of deep water*

From (1.29) we have

$$L_1 = 0.996 L_0 \quad \text{if} \quad \frac{d}{L_0} = \frac{1}{2},$$

which yields

$$0.996 L_0 < L < L_0 \quad \text{if} \quad \frac{d}{L_0} = \frac{1}{2}.$$

This means that, for $d/L_0 = \frac{1}{2}$, the difference between L and L_0 is smaller than 0.4%. The difference between L and L_0 gets even smaller for $d/L_0 > \frac{1}{2}$. For this reason, the waves on water depth $d > \frac{1}{2} L_0$ are called waves on deep water.

1.5.4 *A simplification valid for waves on deep water*

The modes of attenuation from the free surface to the seabed are

$$A1(z) \equiv \frac{\cosh[k(d+z)]}{\cosh(kd)},$$

$$A2(z) \equiv \frac{\sinh[k(d+z)]}{\cosh(kd)}.$$

The first one is the mode of the horizontal velocity and the pressure fluctuation (cf. equations 1.25*a* and 1.26); the second one is the mode of the vertical velocity (cf. equation 1.25*b*).

The two modes can be rewritten in the form

$$A1(z) = \frac{\exp(kd)\exp(kz) + \exp(-kd)\exp(-kz)}{\exp(kd) + \exp(-kd)},$$

$$A2(z) = \frac{\exp(kd)\exp(kz) - \exp(-kd)\exp(-kz)}{\exp(kd) + \exp(-kd)},$$

from which, bearing in mind that $|kz| \leq kd$, we get

$$\lim_{kd\to\infty} A1(z) - \exp(kz) = 0, \quad \lim_{kd\to\infty} A2(z) - \exp(kz) = 0.$$

Therefore, for high values of kd, both $A1$ and $A2$ can be assumed to be equal to $\exp(kz)$. Such an assumption leads to errors on $A1$ and $A2$ within 0.04 if $d/L_0 = 0.5$, within 0.02 if $d/L_0 = 0.6$, and within 0.01 if $d/L_0 = 0.7$. In appraising these errors, bear in mind that $A1$ and $A2$ range between 0 and 1.

1.5.5 *Analytical exercise: get an explicit form for the wavelength*

From the definition of wave celerity ($c = L/T$) and the definition of L_0, we have

$$\frac{c^2}{gd} = \frac{L/L_0}{2\pi d/L},$$

from which, using the linear dispersion rule, we get

$$\frac{c}{\sqrt{gd}} = \sqrt{\frac{\tanh(kd)}{kd}}. \tag{1.32}$$

Since

$$\tanh x = x - \frac{x^3}{3} + O(x^5) \qquad \text{as } x \to 0, \tag{1.33}$$

where O means *of the order*, and since

$$\sqrt{1+x} = 1 + \frac{1}{2}x + \mathrm{O}(x^2) \qquad \text{as } x \to 0,$$

it follows that

$$\sqrt{\frac{\tanh x}{x}} = 1 - \frac{x^2}{6} + \mathrm{O}(x^4) \qquad \text{as } x \to 0,$$

which enables us to rewrite (1.32) in the form

$$\frac{c}{\sqrt{gd}} = 1 - \frac{(kd)^2}{6} + \mathrm{O}(kd)^4 \qquad \text{as } kd \to 0. \tag{1.34}$$

At this stage, let us express kd in terms of d/L_0. To this end let us rewrite (1.30) in the form

$$kd \tanh(kd) = 2\pi\, d/L_0,$$

from which, using (1.33), we get

$$(kd)^2 = 2\pi \frac{d}{L_0} + \mathrm{O}\left(\frac{d}{L_0}\right)^2 \qquad \text{as } \frac{d}{L_0} \to 0$$

that enables us to rewrite (1.34) in the form

$$\frac{c}{\sqrt{gd}} = 1 - \frac{2\pi}{6}\frac{d}{L_0} + \mathrm{O}\left(\frac{d}{L_0}\right)^2 \qquad \text{as } \frac{d}{L_0} \to 0.$$

Hence, the formula

$$c \cong \left(1 - \frac{2\pi}{6}\frac{d}{L_0}\right)\sqrt{gd},$$

and consequently

$$L \cong \left(1 - \frac{2\pi}{6}\frac{d}{L_0}\right)\sqrt{gd}T. \tag{1.35}$$

For $d/L_0 \leq 0.20$, (1.35) leads to an error on L within 0.25%, that is negligible. Equation (1.35) is useful since it covers the range of small d/L_0 where the sequence (1.29) converges more slowly.

1.6 The flow field

1.6.1 *Velocity and acceleration*

From equations (1.25*a*-*b*) of v_y and v_z, one can draw the summary of fig.1.8*a*. Such a figure shows the particle velocity in a wave, on the space domain, at a fixed time instant. The horizontal particle velocity is positive under a wave crest and negative under a wave trough. The vertical particle velocity is positive between a crest and the following trough, while it is negative between a trough and the following crest. This is correct given that the water surface is rising between a crest and the following trough: the trough has just passed and the crest is arriving.

The summary is re-proposed in fig.1.8*b*, on the time domain, at a fixed point. We see that the horizontal particle velocity is positive when a wave crest passes over the fixed point, and is negative when a wave trough passes. We also see that the vertical velocity is negative on the time interval between a crest and the following trough, and in fact during this interval the water surface at the fixed point is going down. On the contrary, the vertical velocity is positive during the interval between a trough and the following crest, and in fact during this time interval the water surface at the fixed point must rise.

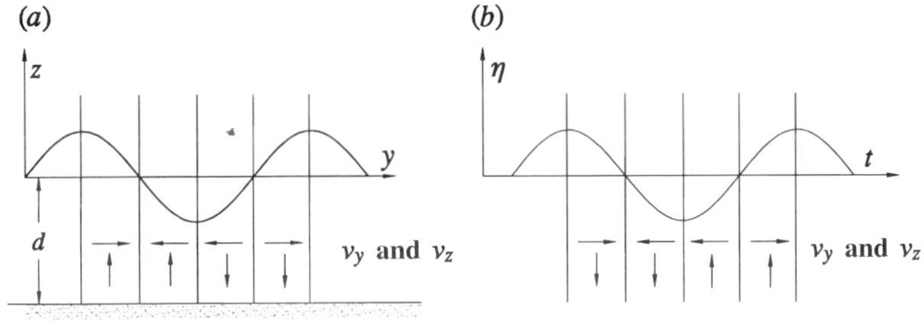

Fig. 1.8 Direction of v_y and v_z: (*a*) at a fixed time instant; (*b*) at a fixed point.

From (1.25*a-b*) of the particle velocity we draw the particle acceleration, by means of equation (1.8). Neglecting the terms of order H^2, the result is

$$a_y(y,z,t) = g\frac{H}{2}k\frac{\cosh[k(d+z)]}{\cosh(kd)}\sin(ky-\omega t),$$

$$a_z(y,z,t) = -g\frac{H}{2}k\frac{\sinh[k(d+z)]}{\cosh(kd)}\cos(ky-\omega t).$$

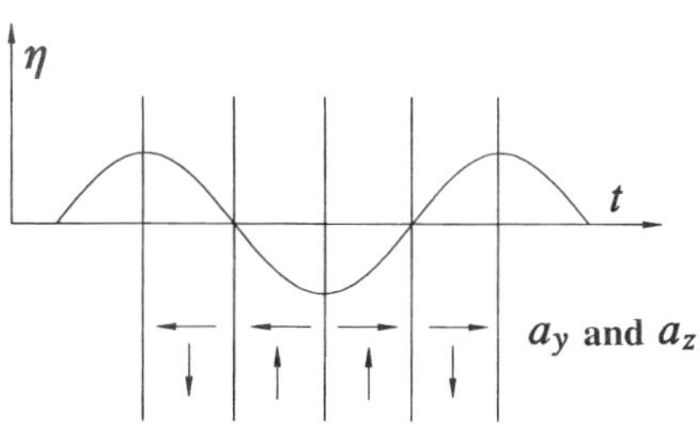

Fig. 1.9 Direction of a_y and a_z at a fixed point.

Fig. 1.9 shows the particle acceleration on the time domain, at a fixed point. We see the horizontal acceleration is negative on the time interval between a crest and the following trough, which is consistent with the fact that the horizontal velocity on this time interval at the fixed point passes from maximum to minimum. As to the vertical acceleration, it is positive in the interval between a zero down-crossing and a zero up-crossing of the surface displacement; and this is consistent with the fact that under a zero down-crossing the vertical velocity reaches a minimum and under a zero up-crossing it reaches a maximum.

1.6.2 *Particle displacement*

Let us fix a water particle P; as an example let us think of the particle that at $t = 0$ is at point $y = 0, z$ (z being arbitrarily fixed). Let us call $\mathbf{v}_P(t)$ and $\mathbf{U}(t)$ respectively the velocity and the position of the particle relative to its initial position $(y = 0, z)$. The velocity $\mathbf{v}_P$ can be expressed in the form

$$\mathbf{v}_P(t) = \mathbf{v}(0, z, t) + \left[\frac{\partial \mathbf{v}}{\partial y}(0, z, t)\right] U_y(t) + \left[\frac{\partial \mathbf{v}}{\partial z}(0, z, t)\right] U_z(t),$$

where the first term on the r.h.s. is of order H, while the second one and the third one are of order H^2. Therefore the velocity $\mathbf{v}_P(t)$ is equal to $\mathbf{v}(0, z, t)$ apart from some difference of order H^2, and hence the particle displacement is given by

$$U_y(t) = \int_0^t v_y(0, z, t')\,\mathrm{d}t',$$

$$U_z(t) = \int_0^t v_z(0, z, t')\,\mathrm{d}t',$$

that is, with the formulae (1.25*a*-*b*) for v_y and v_z,

$$U_y(t) = g\frac{H}{2}\omega^{-2}k\frac{\cosh[k(d+z)]}{\cosh(kd)}\sin(\omega t),$$

$$U_z(t) = g\frac{H}{2}\omega^{-2}k\frac{\sinh[k(d+z)]}{\cosh(kd)}[\cos(\omega t) - 1].$$

Vector $\mathbf{U}(t)$ represents an elliptic orbit whose horizontal and vertical diameters are respectively

$$D_o = \frac{H}{\tanh(kd)}\frac{\cosh[k(d+z)]}{\cosh(kd)}, \tag{1.36}$$

$$D_v = H\frac{\sinh[k(d+z)]}{\sinh(kd)}.$$

Three comments. *First*: if we change the initial abscissa y, the phases of U_y and U_z generally change, but the orbit does not change, that is D_o and D_v do not change. *Second*: as $kd \to \infty$, for any fixed kz, D_o and D_v tend to coincide with each other, and consequently the orbit becomes circular. *Third*: the horizontal particle displacement (D_o) reduces itself, from the water surface to the seabed, according to an hyperbolic law [see (1.36)], while the horizontal displacement at the wavemaker reduces itself linearly. The consequence is that the first two or three waves after the wavemaker show some differences from what is expected by (1.16).

1.7 Stokes' theory to the second order

1.7.1 *The system of equations exact to the second order*

Surface displacement and velocity potential are expressed in the form

$$\eta \equiv \eta' + \eta'' + \mathrm{o}(H^2), \quad \phi \equiv \phi' + \phi'' + \mathrm{o}(H^2), \tag{1.37}$$

where η' and ϕ' are the terms of order H, the formulae for which are respectively (1.16) and (1.22), η'' and ϕ'' are the terms of order H^2 which we shall obtain in what follows, and $\mathrm{o}(H^2)$ is for *terms of orders smaller than* H^2, that is terms of order H^3, H^4, and so on.

From definition (1.37) we have

$$\left(\frac{\partial \phi}{\partial t}\right)_{z=\eta} = \left(\frac{\partial \phi'}{\partial t}\right)_{z=0} + \left(\frac{\partial \phi''}{\partial t}\right)_{z=0} + \left(\frac{\partial^2 \phi'}{\partial z \partial t}\right)_{z=0} \eta' + \mathrm{o}(H^2).$$

This is the form exact to the order H^2 of one of the terms of the system of equations (1.15). Similarly, we can also write the form exact to the order H^2 of the other terms of the aforesaid system of equations. The result is

$$\boxed{g\eta'} + g\eta'' + \boxed{\left(\frac{\partial \phi'}{\partial t}\right)_{z=0}} + \left(\frac{\partial \phi''}{\partial t}\right)_{z=0} + \left(\frac{\partial^2 \phi'}{\partial z \partial t}\right)_{z=0} \eta' + \frac{1}{2}\left(\frac{\partial \phi'}{\partial y}\right)^2_{z=0} + \frac{1}{2}\left(\frac{\partial \phi'}{\partial z}\right)^2_{z=0} =$$
$$= \boxed{\frac{1}{\rho} f(t)}, \tag{1.38a}$$

$$\boxed{\left(\frac{\partial \phi'}{\partial z}\right)_{z=0}} + \left(\frac{\partial \phi''}{\partial z}\right)_{z=0} + \left(\frac{\partial^2 \phi'}{\partial z^2}\right)_{z=0} \eta' = \left(\frac{\partial \phi'}{\partial y}\right)_{z=0} \frac{\partial \eta'}{\partial y} + \boxed{\frac{\partial \eta'}{\partial t}} + \frac{\partial \eta''}{\partial t}, \tag{1.38b}$$

$$\boxed{\frac{\partial^2 \phi'}{\partial y^2}} + \frac{\partial^2 \phi''}{\partial y^2} + \boxed{\frac{\partial^2 \phi'}{\partial z^2}} + \frac{\partial^2 \phi''}{\partial z^2} = 0, \tag{1.38c}$$

$$\boxed{\left(\frac{\partial \phi'}{\partial z}\right)_{z=-d}} + \left(\frac{\partial \phi''}{\partial z}\right)_{z=-d} = 0, \tag{1.38d}$$

where the framed terms form the linear equations which are satisfied if η' and ϕ' are given by (1.16) and (1.22). Therefore the framed terms can be cancelled.

1.7.2 *Solution for* η'' *and* ϕ''

In order to solve the system (1.38*a-d*) of the two unknown functions η'' and ϕ'' we differentiate with respect to time all terms of (1.38*a*), multiply by g all terms of (1.38*b*), and finally add (1.38*b*) to (1.38*a*); in doing so we eliminate η'', that is we obtain an equation with only the unknown function ϕ'', and the known functions η' and ϕ'. Then, substituting η' and ϕ' by their expressions (1.16) and (1.22), we get

$$\left(\frac{\partial^2 \phi''}{\partial t^2}\right)_{z=0} + g\left(\frac{\partial \phi''}{\partial z}\right)_{z=0} = \frac{3}{8} H^2 \omega^3 \left[1 - \frac{1}{\tanh^2(kd)}\right] \sin[2(ky - \omega t)]. \quad (1.39)$$

Therefore function ϕ'' must satisfy this equation proceeding from (1.38*a-b*) as well as the equations (1.38*c-d*). The solution is a function of the kind

$$\phi''(y, z, t) = A\cosh[2k(d + z)]\sin[2(ky - \omega t)] + Bt + Cy,$$

where A, B and C are *unknown constants*. (One more arbitrary constant can be added to this function, but this new constant is insignificant for what follows.) Substituting this expression of ϕ'' in (1.39) we get

$$A = \frac{3}{32} H^2 \omega \frac{1}{\sinh^4(kd)}.$$

At this stage, having obtained the expression of ϕ'' (apart from constants B and C), we can obtain the expression of η'' by means of (1.38*a*). The result is

$$\eta''(y, t) = \frac{H^2}{16}\frac{\omega^2}{g}\left\{-\frac{1}{\sinh^2(kd)} + F_1(kd)\cos[2(ky - \omega t)]\right\} - \frac{B}{g},$$

where

$$F_1(kd) \equiv 3 - \frac{1}{\tanh^2(kd)} + 3\frac{\cosh(2kd)}{\sinh^4(kd)}. \quad (1.40)$$

Constant B can then be obtained given that the mean surface displacement is zero, so that the water mass in the tank is the same under the wave motion as when it is calm.

The conclusion is

$$\eta''(y, t) = \frac{H^2}{16}\frac{\omega^2}{g} F_1(kd)\cos[2(ky - \omega t)], \quad (1.41a)$$

$$\phi''(y,z,t) = \frac{3}{32} H^2 \omega \frac{1}{\sinh^4(kd)} \cosh[2k(d+z)] \sin[2(ky-\omega t)] + \\ -\frac{H^2}{16}\omega^2 \frac{1}{\sinh^2(kd)} t + Cy, \tag{1.41b}$$

where constant C will be obtained in sect. 2.7.

1.8 Non-linearity effects

$F_1(kd)$ [function (1.40)] is positive all over its domain [see fig. 1.10], and, as a consequence, the sum of η'' and η' leads to a wave profile like that of fig. 1.11: the crest sharpens and the trough flattens. Thus the non-linear theory succeeds in predicting the characteristic asymmetry between crest and trough (see sect. 1.4.1).

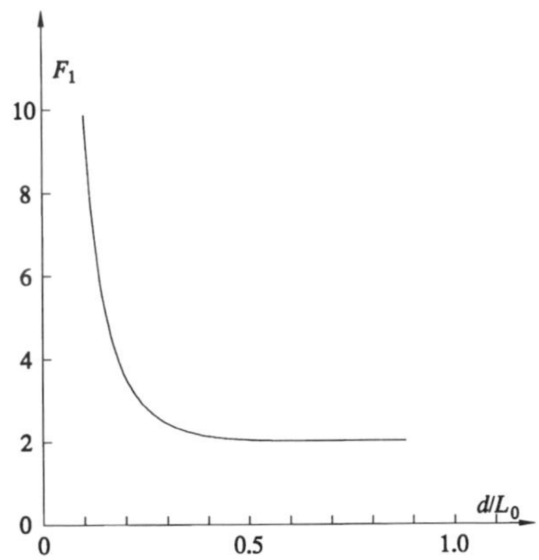

Fig. 1.10 The function F_1 defined by (1.40).

The quotient between the amplitude of the nonlinear term η'' and the amplitude of the linear term η' is

$$\frac{2\pi}{8}\frac{H}{L_0} F_1(kd),$$

and hence it becomes greater as H/L_0 becomes greater and/or as d/L_0 becomes smaller (fig. 1.10 shows that $F_1(kd)$ grows as d/L_0 gets smaller).

The non-linearity effects also attack the pressure head waves. Let us see how.

The second order component of the fluctuating pressure head can be obtained by means of (1.5) and is related to ϕ' and ϕ'' by the equation

$$\eta''_{ph} = -\frac{1}{g}\frac{\partial \phi''}{\partial t} - \frac{1}{2g}\left[\left(\frac{\partial \phi'}{\partial y}\right)^2 + \left(\frac{\partial \phi'}{\partial z}\right)^2\right] \tag{1.42}$$

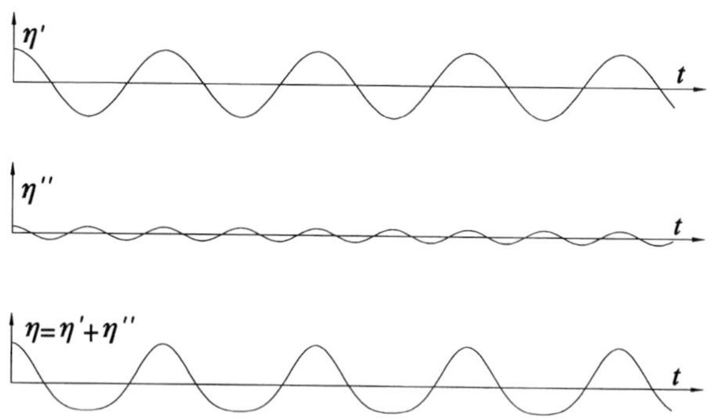

Fig. 1.11 The second order term η'' makes the wave crest steeper, and flattens the wave trough.

which we rewrite in the form

$$\eta''_{ph} \equiv \eta''_{ph1} + \eta''_{ph2},$$

where η''_{ph1} is the term depending on ϕ'', and η''_{ph2} is the term depending on $(\partial\phi'/\partial y)^2$ and $(\partial\phi'/\partial z)^2$:

$$\eta''_{ph1}(y,z,t) = \frac{1}{16}H^2\frac{\omega^2}{g}\left\{\frac{3}{\sinh^4(kd)}\cosh[2k(d+z)]\cos[2(ky-\omega t)] + \frac{1}{\sinh^2(kd)}\right\}, \tag{1.43a}$$

$$\eta''_{ph2}(y,z,t) = -\frac{1}{4}H^2\frac{\omega^2}{g}\frac{1}{\tanh(kd)\sinh(2kd)}\{\cosh^2[k(d+z)]\cos^2(ky-\omega t) + \\ + \sinh^2[k(d+z)]\sin^2(ky-\omega t)\}. \tag{1.43b}$$

Let us consider now what happens if we add the non-linear term $\eta''_{ph} \equiv \eta''_{ph1} + \eta''_{ph2}$ to the linear term η'_{ph} (given by 1.26). The effect of η''_{ph1} is similar to the effect of η'': it enhances the crest elevation and reduces the trough depth. On the contrary, η''_{ph2} reduces the crest elevation and enhances the trough depth. As $kd \to \infty$, $\eta''_{ph1}/\eta''_{ph2}$ approaches zero. As $kd \to 0$, $\eta''_{ph1}/\eta''_{ph2}$ tends to infinity. Therefore, as $kd \to \infty$, η''_{ph2} prevails on η''_{ph1}; vice versa as $kd \to 0$. Specifically, for $d/L_0 > 0.20$, η''_{ph2} prevails and hence the pressure head waves undergo a deformation like that of fig. 1.12; while η''_{ph1} prevails if $d/L_0 < 0.20$ and hence on this domain the pressure head waves undergo a deformation of the same kind as that of the surface waves, even if in a more attenuated form because of the effect of η''_{ph2}.

Finally, a comparison of (1.43*a*-*b*) with (1.26) reveals that the non-linear term of the fluctuating pressure head decreases, from the water surface to the seabed, more

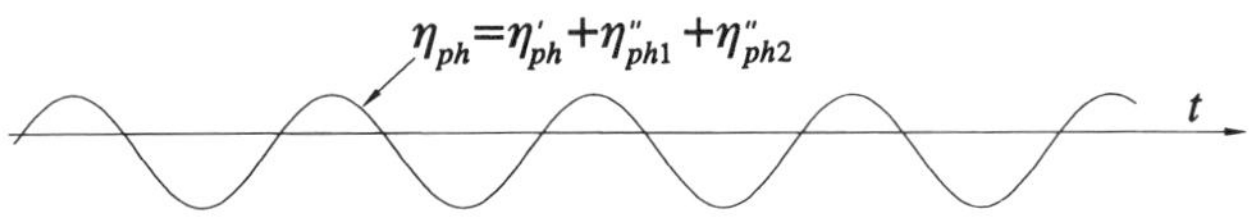

Fig. 1.12 Pressure head waves on deep water where the second nonlinear term (η''_{ph2}) prevails over the first nonlinear term (η''_{ph1}).

rapidly than the linear term. Therefore the incidence of the non-linear term is greater near the water surface than near the bottom.

1.9 Wave-current interaction. Part I: velocity potential and wavelength

1.9.1 *Solution for the velocity potential*

Let us consider the case of a periodic wave travelling on a current of velocity u, in the limit $H \to 0$ for fixed d, T and u. Let us assume the current to be parallel to the wave direction. The current direction can be the same or opposite to the direction of wave advance, that is u can be positive or negative.

The surface displacement will have the same form (1.16) valid in absence of a current, except for a different relation between wave number and water depth and wave period; that is

$$\eta(y,t) = \frac{H}{2}\cos(k_c y - \omega t),$$

where k_c is the wave number generally different from k, which must be determined. This new wave number will depend not only on d and T, but also on u. As to the velocity potential, it will be the sum of two terms: one of the uniform current and one like (1.22) [the velocity potential of a wave without current]. Accordingly we write

$$\phi(y,z,t) = uy + A\,\frac{\cosh[k_c(d+z)]}{\cosh(k_c d)}\sin(k_c y - \omega t) + F(t),$$

where A is a dimensional constant and $F(t)$ a function of time, both of which are to be determined. Without the current ($u=0$), A is equal to $g\dfrac{H}{2}\omega^{-1}$ [cf.(1.22)]. Naturally, with the current, A will depend on u.

Let us seek k_c, A and $F(t)$ such that η and ϕ satisfy the differential equations (1.15*a*-*b*-*c*-*d*) of general validity. As to (1.15*c*-*d*), it is easy to verify that they are satisfied whatever the k_c, A and $F(t)$. Let us now pass to the two most exacting equations, that is to say (1.15*a*-*b*).

This time $\partial\phi/\partial y$ is the sum of the finite term u and of a term of order H being due to the wave. Therefore the terms

$$\left(\frac{\partial\phi}{\partial y}\right)^2_{z=\eta} \quad \text{and} \quad \left(\frac{\partial\phi}{\partial y}\right)_{z=\eta}\frac{\partial\eta}{\partial y}$$

respectively of (1.15*a*) and of (1.15*b*) are no longer negligible to Stokes' first order, and these equations yield

$$g\,\frac{H}{2}\cos(k_c y-\omega t)-A\omega\cos(k_c y-\omega t)+\dot{F}(t)+\frac{1}{2}\,u^2+uAk_c\cos(k_c y-\omega t)=\frac{1}{\rho}\,f(t), \tag{1.44a}$$

$$Ak_c\tanh(k_c d)\sin(k_c y-\omega t)=-u\,\frac{H}{2}\,k_c\,\sin(k_c y-\omega t)+\frac{H}{2}\,\omega\sin(k_c y-\omega t), \tag{1.44b}$$

where $\dot{F}(t)$ denotes the derivative of $F(t)$. Equation (1.44*a*) is satisfied if and only if

$$A=g\,\frac{H}{2}\,(\omega-uk_c)^{-1},$$

$$F(t)=-\frac{1}{2}\,u^2t+\frac{1}{\rho}\int_0^t f(t')\,\mathrm{d}t',$$

and therefore

$$\phi(y,z,t)=uy+g\,\frac{H}{2}\,(\omega-uk_c)^{-1}\,\frac{\cosh[k_c(d+z)]}{\cosh(k_c d)}\,\sin(k_c y-\omega t)-\frac{1}{2}\,u^2t+\frac{1}{\rho}\int_0^t f(t')\,\mathrm{d}t'. \tag{1.45}$$

(Clearly, an arbitrary constant can be added to $F(t)$, but it has no effect on the wave mechanics.) Finally, (1.44*b*) is satisfied if and only if

$$k_c\tanh(k_c d)=\frac{(\omega-uk_c)^2}{g}. \tag{1.46}$$

This is the equation that relates k_c to d, T and u, and for $u=0$ it coincides correctly with the linear dispersion rule.

1.9.2 *Solution for the wavelength*

If we multiply by d both sides of (1.46) and use definition (1.27) of L_0, we can rewrite (1.46) in the form

$$\frac{d}{L_c}\tanh\left(2\pi\,\frac{d}{L_c}\right)=a\left(\frac{d}{L_c}-b\right)^2, \tag{1.47}$$

where L_c denotes the wavelength on current:

$$L_c \equiv 2\pi/k_c,$$

and where

$$a \equiv \frac{1}{d/L_0}\left(\frac{u}{c_0}\right)^2,$$

$$b \equiv \frac{d}{L_0}\Big/\left(\frac{u}{c_0}\right),$$

with $c_0 \equiv L_0/T$ and $u \neq 0$. [Note, the r.h.s. of (1.47) assumes the indeterminate form $0 \cdot \infty$ as $u \to 0$, and it approaches d/L_0. Thus (1.47) reduces itself correctly to (1.30) as u approaches zero.]

The d/L_c satisfying (1.47) is equal to the positive value of x (provided that it exists) such that function

$$f_1(x) \equiv x \tanh(2\pi x)$$

is equal to function

$$f_2(x) \equiv a(x-b)^2.$$

Therefore, in order to get the wavelength L_c on the current, we must seek the d/L_c such that

$$f_1\left(\frac{d}{L_c}\right) = f_2\left(\frac{d}{L_c}\right). \tag{1.48}$$

The two functions $f_1(x)$ and $f_2(x)$ are represented in fig. 1.13 for $d/L_0 = 0.2$ and a few values of u/c_0. The graph of $f_2(x)$ is a parabola whose ordinate at $x = 0$ is equal to d/L_0 whatever the u/c_0. The vertex of the parabola falls at $x = b$, and thus it falls on the positive x-axis or on the negative x-axis, according to whether u is greater or smaller than zero.

If u is greater than zero, (1.48) has two solutions: $(d/L_c)_1$ smaller than b and $(d/L_c)_2$ greater than b [cf. fig. 1.13*a*]. We discard the second solution, that is $(d/L_c)_2$, if we assume d/L_c to be a continuous function of u, so that

$$\lim_{u\to 0} \frac{d}{L_c} = \frac{d}{L}. \tag{1.49}$$

Indeed $(d/L_c)_1 \to d/L$ as $u \to 0$; while $(d/L_c)_2$ being greater than b, tends to infinity.

If u is smaller than zero, we have three cases:

(i) weak negative current: (1.48) admits two solutions,

(ii) transition from weak to strong negative current: (1.48) admits only one solution;

(iii) strong negative current: (1.48) does not admit any solution, which means that the wave is not able to travel against the stream.

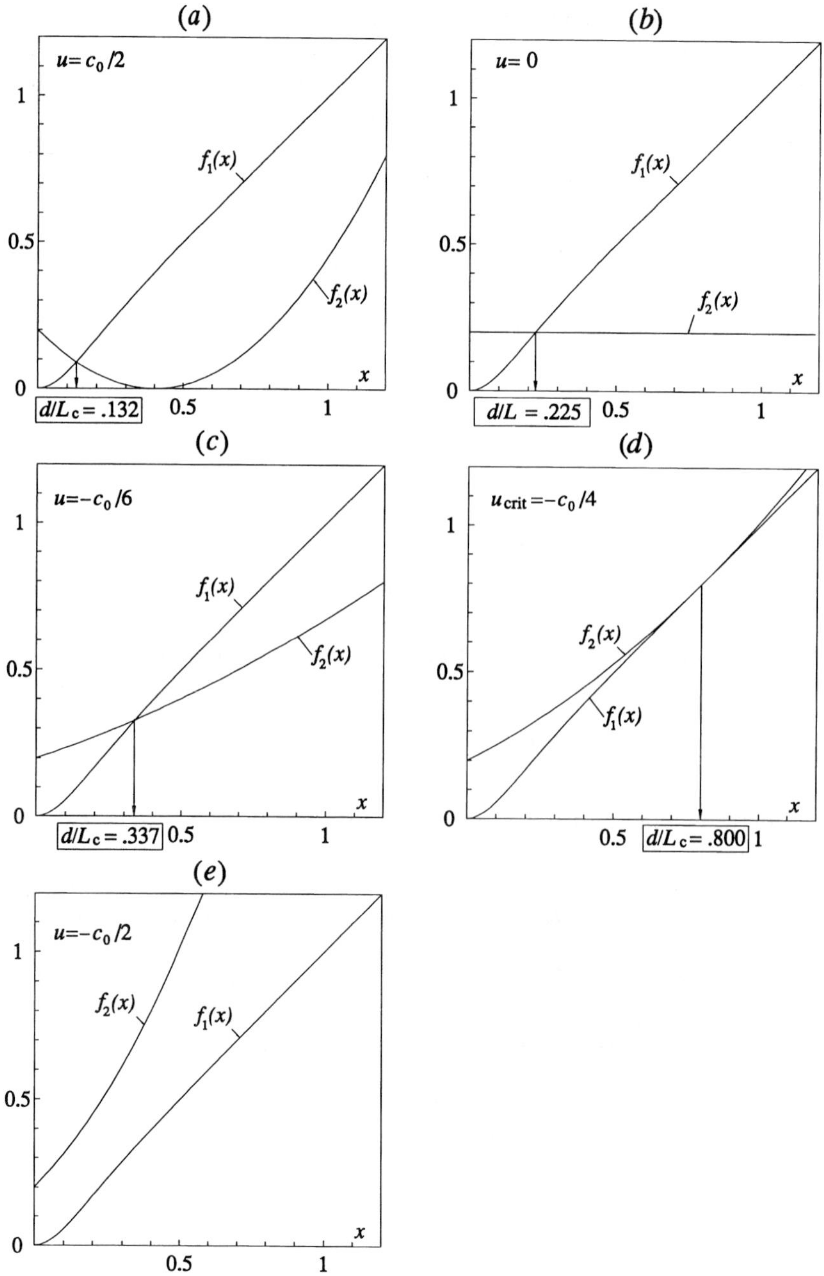

Fig. 1.13 The wavelength on given water depth $d/L_0 = 0.2$ for: (*a*) a positive current; (*b*) no current; (*c*) a low negative current; (*d*) the critical negative current; (*e*) a strong negative current, for which the equation (1.47) does not admit any solution (the wave is not able to travel against the stream).

The three cases are represented in figs. 1.13*c-d-e*. In the case of the weak negative current, the solution to be discarded is again the second one because the first solution satisfies condition (1.49) as u approaches zero, whereas the second solution tends to infinity. The transition from weak to strong negative current occurs for a critical value of $|u/c_0|$, which we shall call $|u/c_0|_{\text{crit}}$, that depends on d/L_0.

As discussed above, as $u \to 0$, the vertex of the parabola representing $f_2(x)$ falls at an infinitely large x. Specifically, as $u \to 0_+$, the vertex of the parabola is at $x \to +\infty$, while as $u \to 0_-$, the vertex of the parabola is at $x \to -\infty$. Moreover, as $x \to 0$, the curvature of the parabola approaches zero. Therefore, as $u \to 0$ the parabola approaches a horizontal line of ordinate d/L_0, on any finite interval including the origin $x = 0$, [see fig. 1.13*b*].

The whole set of pictures 1.13*a-e* shows that

$$\frac{d}{L_c} < \frac{d}{L} \Leftrightarrow L_c > L \quad \text{if} \quad u > 0,$$

$$\frac{d}{L_c} > \frac{d}{L} \Leftrightarrow L_c < L \quad \text{if} \quad u < 0.$$

Thus, according to intuition, the wavelength grows if waves and current have the same direction, while the wavelength shortens if waves and current have opposite directions.

1.9.3 *The critical negative current*

Let us discuss further the case of the negative current ($u < 0$). Firstly we consider what is the value of $|u/c_0|_{\text{crit}}$ for the basic case of $d/L_0 \to \infty$. Since $L_c < L$ if $u < 0$, it follows

$$\frac{d}{L_c} \to \infty \quad \text{as} \quad \frac{d}{L_0} \to \infty.$$

Therefore equation (1.47) reduces itself to

$$\frac{d}{L_c} = a\left(\frac{d}{L_c} - b\right)^2 \quad \text{as} \quad \frac{d}{L_0} \to \infty,$$

and is satisfied by only one value of d/L_c, if

$$ab = \frac{u}{c_0} = -\frac{1}{4}.$$

As a consequence:

$$\left|\frac{u}{c_0}\right|_{\text{crit}} = \frac{1}{4} \quad \text{as} \quad \frac{d}{L_0} \to \infty.$$

For the case of a finite d/L_0, we have to seek the u/c_0 such that equation (1.47) is satisfied by only one d/L_c. This operation can be executed numerically. In doing so one finds that $|u/c_0|_{\text{crit}}$ grows from 0 to the asymptotic value $\frac{1}{4}$, as d/L_0 grows. For $d/L_0 > 0.10$, $|u/c_0|_{\text{crit}}$ takes the asymptotic value.

1.10 Preliminary remarks on three dimensional waves

The surface displacement of a wave whose propagation direction makes an arbitrary angle θ with y-axis is given by

$$\eta(x, y, t) = \frac{H}{2} \cos(kx \sin\theta + ky \cos\theta - \omega t). \tag{1.50}$$

To prove this, let us imagine a point moving with a uniform speed L/T along a straight line making an angle θ with the y-axis. If the point starts from $x = 0$, $y = 0$ at time $t = 0$, its position is given by

$$x_P = \frac{L}{T} t \sin\theta, \quad y_P = \frac{L}{T} t \cos\theta,$$

so that the surface displacement at this point proves to be

$$\eta(x_P, y_P, t) = \frac{H}{2} \cos\left(\frac{2\pi}{T} t \sin^2\theta + \frac{2\pi}{T} t \cos^2\theta - \frac{2\pi}{T} t\right) = \frac{H}{2},$$

and hence it keeps constant in time, which confirms that the trajectory and speed of the wave is coincident with the trajectory and speed of the point.

The velocity potential associated with surface displacement (1.50) is

$$\phi(x, y, z, t) = g \frac{H}{2} \omega^{-1} \frac{\cosh[k(d+z)]}{\cosh(kd)} \sin(kx \sin\theta + ky \cos\theta - \omega t). \tag{1.51}$$

Here we can readily verify that the two functions (1.50) and (1.51) satisfy equations (1.19*a*-*b*) [provided that $f(t) = 0$ in (1.19*a*)] as well as the boundary condition at the bottom (equation 1.15*d*). Note: these equations, having been obtained for the two-dimensional flow y-z, retain their validity even for the three-dimensional flow x-y-z, as can be easily inferred from the fact that they depend only on z not on y. As to $f(t) = 0$, we have already seen $\mathbf{v}$ and p do not change whatever the $f(t)$ and therefore it is justified and advisable to put directly $f(t) = 0$ in the equation (1.19*a*).

Of the whole system of the linear flow equations, that is the system consisting of (1.15*c*-*d*) and (1.19*a*-*b*), the only equation which needs to be adjusted from the two-dimensional to the three-dimensional flow is (1.15*c*). For the three-dimensional flow it becomes

$$\frac{\partial^2\phi}{\partial x^2} + \frac{\partial^2\phi}{\partial y^2} + \frac{\partial^2\phi}{\partial z^2} = 0,$$

which is satisfied by function (1.51).

1.11 Wave reflection

1.11.1 *General solution for η and ϕ*

Let us consider the flow field if a wave train attacks an infinitely-long vertical breakwater. Let us assume that the breakwater is along line $y = 0$, and the direction of the incident waves makes an angle θ with the y-axis [see fig.1.14].

On the grounds of some intuitive considerations we could at once say that specular reflection will occur and that the height and period of the reflected waves will be equal to the height and period of the incident waves; but we believe it to be instructive to prove such intuitive knowledge. Therefore we assume that the direction of the reflected waves makes an unknown angle $\tilde{\theta}$ with the y-axis, and in addition we allow the possibility that the reflected waves have a height $\tilde{H}$ and a period $\tilde{T}$ different from height H and period T of the incident waves.

The η and ϕ of the incident waves are given by (1.50) and (1.51), and the η and ϕ of the reflected waves are given by the same equations (1.50-51) with $\tilde{H}$, $\tilde{\omega}$, $\tilde{k}$, and $\tilde{\theta}$ in place of H, ω, k and θ:

reflected waves

$$\eta(x,y,t) = \frac{\tilde{H}}{2}\cos(\tilde{k}x\sin\tilde{\theta} + \tilde{k}y\cos\tilde{\theta} - \tilde{\omega}t + \varepsilon), \tag{1.52a}$$

$$\phi(x,y,z,t) = g\frac{\tilde{H}}{2}\tilde{\omega}^{-1}\frac{\cosh[\tilde{k}(d+z)]}{\cosh(\tilde{k}d)}\sin(\tilde{k}x\sin\tilde{\theta} + \tilde{k}y\cos\tilde{\theta} - \tilde{\omega}t + \varepsilon). \tag{1.52b}$$

We cannot exclude some phase angle between the reflected and the incident waves, and this is why in the expressions of the reflected waves we have put a phase angle ε which must be determined.

The flow field before the wall is given by the sum of the incident waves (equations 1.50-51 of η and ϕ) and the reflected waves (equations 1.52*a*-*b*):

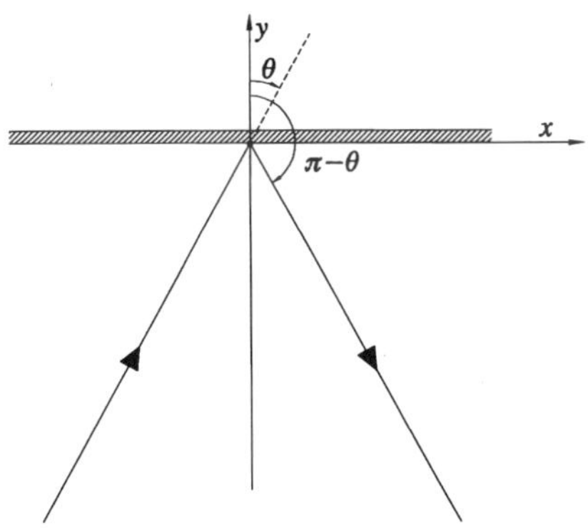

Fig. 1.14 Reflection: reference scheme.

$$\eta(x,y,t) = \frac{H}{2}\cos(kx\sin\theta + ky\cos\theta - \omega t) + \frac{\tilde{H}}{2}\cos(\tilde{k}x\sin\tilde{\theta} + \tilde{k}y\cos\tilde{\theta} - \tilde{\omega}t + \varepsilon), \tag{1.53a}$$

$$\phi(x,y,z,t) = g\frac{H}{2}\omega^{-1}\frac{\cosh[k(d+z)]}{\cosh(kd)}\sin(kx\sin\theta + ky\cos\theta - \omega t) +$$

$$+g\frac{\tilde{H}}{2}\tilde{\omega}^{-1}\frac{\cosh[\tilde{k}(d+z)]}{\cosh(\tilde{k}d)}\sin(\tilde{k}x\sin\tilde{\theta} + \tilde{k}y\cos\tilde{\theta} - \tilde{\omega}t + \varepsilon). \tag{1.53b}$$

The boundary condition is

$$\left(\frac{\partial\phi}{\partial y}\right)_{y=0} = 0, \tag{1.54}$$

that is

$$H\omega^{-1}k\cos\theta\,\frac{\cosh[k(d+z)]}{\cosh(kd)}\cos(kx\sin\theta - \omega t) =$$

$$= -\tilde{H}\tilde{\omega}^{-1}\tilde{k}\cos\tilde{\theta}\,\frac{\cosh[\tilde{k}(d+z)]}{\cosh(\tilde{k}d)}\cos(\tilde{k}x\sin\tilde{\theta} - \tilde{\omega}t + \varepsilon), \tag{1.55}$$

and it is satisfied for every x, z and t if and only if

$$\tilde{\omega} = \omega \Rightarrow \tilde{k} = k,\ \ \tilde{H} = H,\ \ \tilde{\theta} = \pi - \theta,\ \ \varepsilon = n2\pi \quad \text{with} \quad n = 0,\ 1,\ 2... \tag{1.56}$$

Because of the equalities (1.56), the two functions (1.53*a-b*) can be rewritten in the form

$$\eta(x,y,t) = H\cos(kx\sin\theta - \omega t)\cos(ky\cos\theta), \tag{1.57a}$$

$$\phi(x,y,z,t) = gH\omega^{-1}\frac{\cosh[k(d+z)]}{\cosh(kd)}\sin(kx\sin\theta - \omega t)\cos(ky\cos\theta). \tag{1.57b}$$

Verify that functions (1.53*a-b*) satisfy the linear flow equations, that is to say the system consisting of the equations (1.15*c-d*) and (1.19*a-b*), with $f(t) = 0$. Apply the following; if a pair of functions η_1, ϕ_1 satisfy a homogeneous linear system, and a second pair of functions η_2, ϕ_2 satisfy the same system, then the sum $\eta_1 + \eta_2$, $\phi_1 + \phi_2$ also satisfies this system of equations. Here η_1, ϕ_1 are functions (1.50-51), η_2, ϕ_2 are functions (1.52*a-b*) and $\eta_1 + \eta_2$, $\phi_1 + \phi_2$ are functions (1.53*a-b*).

To prove (1.56), bear in mind that two cosine functions are equal to each other all over their domains, if and only if they have the same frequency, same amplitude and phase angle of some multiple of 2π. Then, examining the l.h.s. and r.h.s. of (1.55) as functions of t for any fixed pair x, z, you will easily realize that $\tilde{\omega}$ must be equal to ω, which implies $\tilde{k} = k$. Similarly, examining the l.h.s. and r.h.s. of (1.55) as functions of x for any fixed pair z, t, you will easily realize that $\tilde{k}\sin\tilde{\theta}$ must be equal to $k\sin\theta$, which implies $\sin\tilde{\theta} = \sin\theta$. The rest of the proof follows straightforwardly. Note: if the breakwater were not at $y = 0$ but at some parallel line, the phase angle ε would be generally different from $n2\pi$.

1.11.2 *The orthogonal attack*

In the basic case of $\theta = 0$, in which the wave attacks the breakwater orthogonally, the flow becomes two-dimensional y-z and the formulae for η and ϕ reduce themselves to

$$\eta(y,t) = H\cos(\omega t)\cos(ky),$$

$$\phi(y,z,t) = -gH\omega^{-1}\frac{\cosh[k(d+z)]}{\cosh(kd)}\sin(\omega t)\cos(ky),$$

and hence the velocity components are

$$v_y(y,z,t) = gH\omega^{-1}k\frac{\cosh[k(d+z)]}{\cosh(kd)}\sin(\omega t)\sin(ky),$$

$$v_z(y,z,t) = -gH\omega^{-1}k\frac{\sinh[k(d+z)]}{\cosh(kd)}\sin(\omega t)\cos(ky).$$

Three instant pictures of this basic case are given in fig. 1.15.

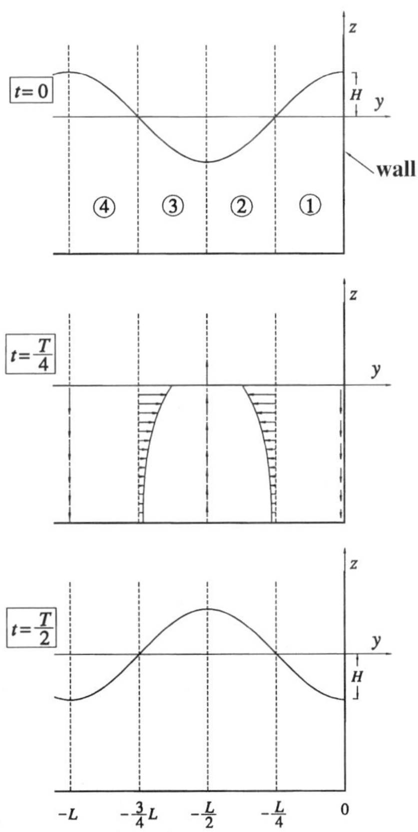

Fig. 1.15 Three *snapshots* of the wave field before an upright breakwater attacked orthogonally.

At time $t = 0$, both v_y and v_z are zero everywhere, given that both v_y and v_z are proportional to $\sin(\omega t)$. At that time ($t = 0$) the surface displacement gets up to its maximum (positive or negative) at each location. In particular, at the wall, the surface displacement at $t = 0$ is equal to the crest-to-trough height of the incident wave.

At time $t = T/4$ the surface displacement is zero everywhere given that η is proportional to $\cos(\omega t)$. The horizontal velocity has its negative maximum at $y = -L/4$, and it has its positive maximum at $y = -\dfrac{3}{4}L$. The vertical velocity has its negative maximum at the wall ($y = 0$) and at $y = -L$, and it has its positive maximum at $y = -L/2$.

At time $t = T/2$ the surface displacement is opposite with respect to the surface displacement at $t = 0$. Thus a consistent picture emerges, where:

(i) at time $t = 0$, the water surface is higher than the MWL in the compartments ① and ④, and is lower than the MWL in the compartments ② and ③ (fig. 1.15);

(ii) vice versa, at time $t = T/2$ the water surface is lower than the MWL in the compartments ① and ④, and is higher than the MWL in the compartments ② and ③;

(iii) consistently, at the intermediate time instant $t = T/4$, the water flows from the compartments ① and ④ towards the compartments ② and ③.

There are some points (nodes) where the surface displacement is always zero, and where the horizontal velocity attains its absolute maximum. These points are at $\frac{1}{4}, \frac{3}{4}, \frac{5}{4}, \ldots$ wavelength from the wall. Then there are the antinodes, at $0, \frac{1}{2}, 1, \frac{3}{2}, \ldots$ wavelength from the wall, where the wave height (on the time domain) and the vertical velocity attain their absolute maximum.

The wave height at the antinodes is $2H$ (that is, the height of the wave on the time domain), which is twice the wave height that would be there without the wall. The velocity maxima are also twice greater than the maxima in the absence of the wall.

1.11.3 *The pressure distribution on the breakwater*

Whatever the angle θ of the waves, the maximum pressure at any fixed section of the breakwater, according to Stokes' first order, is given by

$$p(z) = -\rho g z + \rho g H \frac{\cosh[k(d+z)]}{\cosh(kd)}$$

which proceeds from equation (1.57*b*) of ϕ, and can be rewritten in the equivalent form (apart from a term of order H^2):

$$p(z)\begin{cases} = -\rho g z + \rho g H \dfrac{\cosh[k(d+z)]}{\cosh(kd)} & \text{if} \quad z \le 0, \\ = \rho g (H - z) & \text{if} \quad 0 \le z \le H. \end{cases} \tag{1.58}$$

The wave pressure, that is the difference between (1.58) and the static pressure, is shown in fig. 1.16*a*.

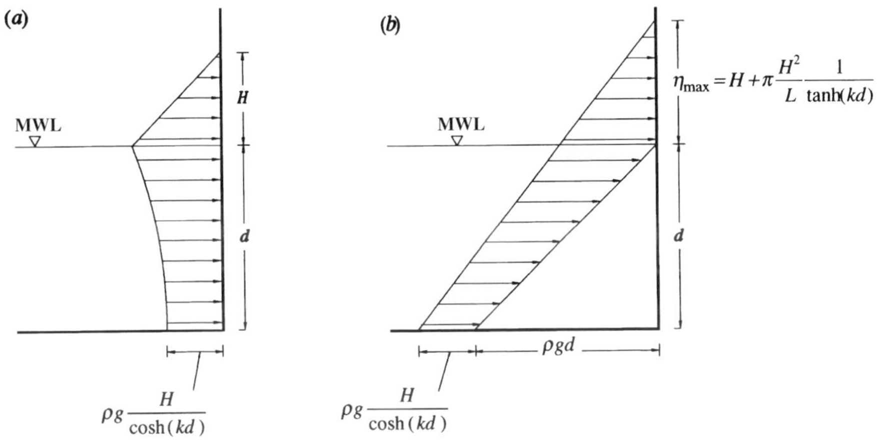

Fig. 1.16 The pressure exerted on an upright breakwater by a wave crest: (*a*) Stokes' linear theory; (*b*) Saintflou's model.

Fig. 1.16*b* shows Saintflou's model (1928). According to this model, we have to evaluate the two ends of the pressure distribution, that is

(i) the elevation where the pressure becomes zero, which is the highest elevation reached by the water surface;

(ii) the pressure at the lowest point of the wall.

We then get the pressure distribution by linear interpolation between these two ends. Since the incidence of the second order term diminishes from the free surface to the seabed [cf. sect. 1.8], the second order term is taken into account in the evaluation of η_{max} [the maximum surface displacement at the wall]:

$$\eta_{max} = H + \pi \frac{H^2}{L} \frac{1}{\tanh(kd)};$$

while the pressure at the lowest point of the wall is calculated with the linear theory that is with (1.58):

$$p(z = -d) = \rho g d + \rho g \frac{H}{\cosh(kd)}.$$

1.12 Wave diffraction

1.12.1 *The solution of Penny and Price (1952)*

Let us consider now the flow field if a wave train attacks a semi-infinite vertical breakwater. The breakwater is along the line $y = 0$, its origin is at $x = 0$ and its thickness is negligible with respect to the wavelength. The flow field that would be there without the breakwater is the one given by formulae (1.50-51) for η and ϕ.

The surface displacement and the velocity potential to Stokes' first order, in polar coordinates, are given by

$$\eta(r, \beta, t) = \frac{H}{2}\left[F(r, \beta; \omega, \theta)\cos(\omega t) + G(r, \beta; \omega, \theta)\sin(\omega t)\right], \qquad (1.59a)$$

$$\phi(r, \beta, z, t) = g\frac{H}{2}\omega^{-1}\frac{\cosh[k(d+z)]}{\cosh(kd)}\left[G(r, \beta; \omega, \theta)\cos(\omega t) - F(r, \beta; \omega, \theta)\sin(\omega t)\right], \qquad (1.59b)$$

where

$$F(r, \beta; \omega, \theta) \equiv A(u_1)\cos q_1 + A(u_2)\cos q_2 - B(u_1)\sin q_1 - B(u_2)\sin q_2, \qquad (1.60a)$$

$$G(r, \beta; \omega, \theta) \equiv A(u_1)\sin q_1 + A(u_2)\sin q_2 + B(u_1)\cos q_1 + B(u_2)\cos q_2, \qquad (1.60b)$$

$$A(u) \equiv \frac{1}{2}\left[1 + S_{FR}(u) + C_{FR}(u)\right], \quad B(u) \equiv \frac{1}{2}\left[S_{FR}(u) - C_{FR}(u)\right], \qquad (1.60c)$$

$$S_{FR}(u) \equiv \int_0^u \sin\left(\frac{\pi}{2}x^2\right)\mathrm{d}x, \quad C_{FR}(u) \equiv \int_0^u \cos\left(\frac{\pi}{2}x^2\right)\mathrm{d}x, \qquad (1.60d)$$

$$u_1 \equiv 2\sqrt{kr/\pi}\sin\left[\frac{1}{2}\left(\beta + \theta - \frac{\pi}{2}\right)\right], \quad u_2 \equiv -2\sqrt{kr/\pi}\sin\left[\frac{1}{2}\left(\beta - \theta + \frac{\pi}{2}\right)\right], \qquad (1.60e)$$

$$q_1 \equiv kr\sin(\beta + \theta), \quad q_2 \equiv -kr\sin(\beta - \theta), \qquad (1.60f)$$

cf. fig. 1.17 for the symbols.

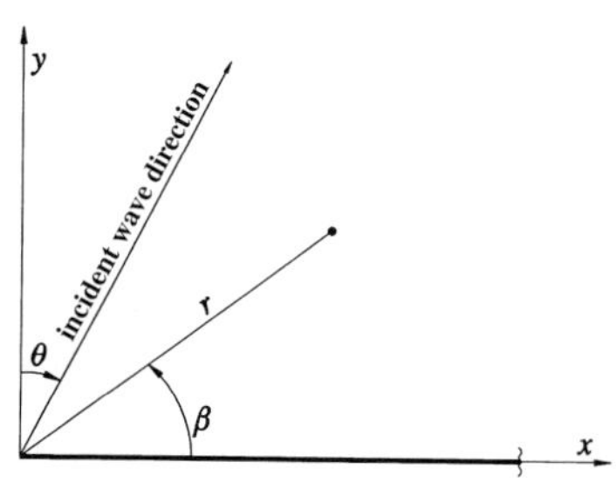

Fig. 1.17 Reference scheme for the interaction between waves and a semi-infinite breakwater.

Let us arbitrarily fix a point r, β and let us write F and G in place of $F(r, \beta; \omega, \theta)$ and $G(r, \beta; \omega, \theta)$. The surface displacement on the time domain, at the fixed point, has its maxima and minima at time instants t_m such that

$$\omega t_m = \arctan\left(\frac{G}{F}\right). \tag{1.61}$$

Equation (1.61) admits two solutions:

$$\sin(\omega t_{m1}) = \frac{G}{\sqrt{F^2+G^2}}, \quad \cos(\omega t_{m1}) = \frac{F}{\sqrt{F^2+G^2}}$$

and

$$\sin(\omega t_{m2}) = -\frac{G}{\sqrt{F^2+G^2}}, \quad \cos(\omega t_{m2}) = -\frac{F}{\sqrt{F^2+G^2}}$$

which, once substituted in (1.59*a*), give

$$\eta(t_{m1}) = \frac{H}{2}\sqrt{F^2+G^2}, \quad \eta(t_{m2}) = -\frac{H}{2}\sqrt{F^2+G^2}. \tag{1.62}$$

Therefore t_{m1} is the time instant of the crest and t_{m2} the time instant of the trough at the fixed point. Clearly, instants t_{m1} and t_{m2} generally change from a point to another since they depend on functions F and G. The wavefronts [see fig. 1.18] are the lines connecting points with the same value of t_{m1} (or t_{m2}).

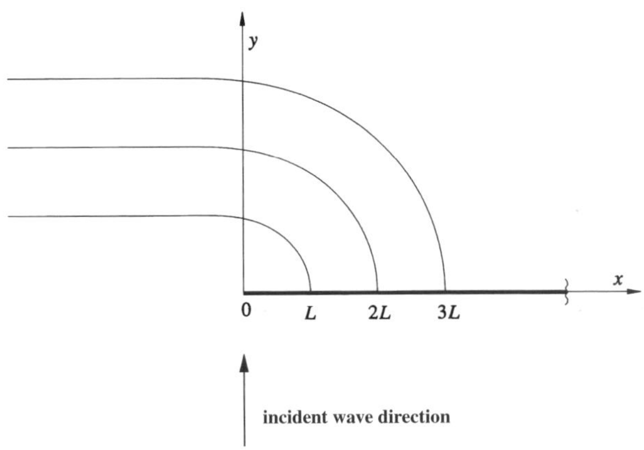

Fig. 1.18 Wavefronts behind a semi-infinite breakwater.

1.12.2 *The diffraction coefficient*

From (1.62) it follows that the wave height (that is, the height of the wave on the time domain) is

$$H(r, \beta) = H\sqrt{F^2+G^2}.$$

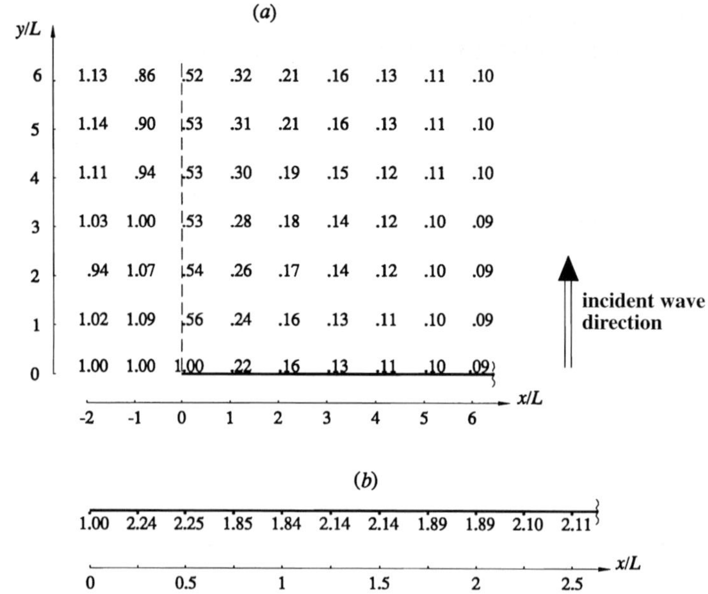

Fig. 1.19 Diffraction coefficient if the incident waves approach the wall orthogonally: (*a*) behind the breakwater; (*b*) along the wave beaten wall.

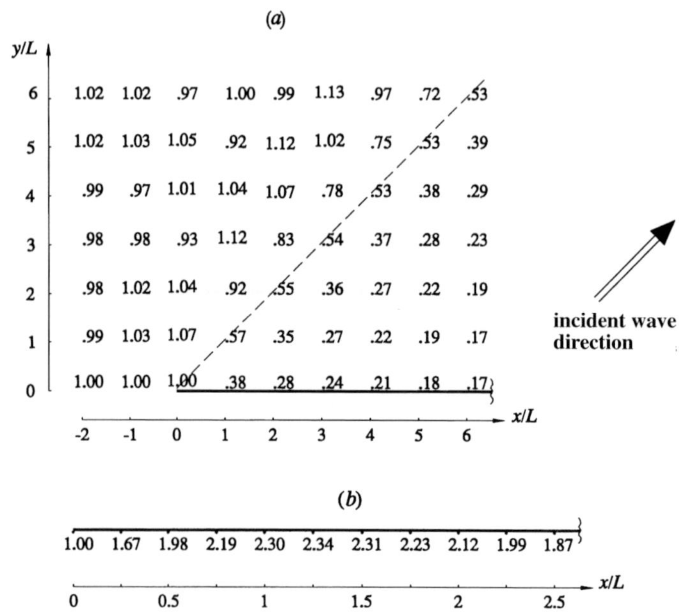

Fig. 1.20 Diffraction coefficient for a case of inclined attack: (*a*) behind the breakwater; (*b*) along the wave beaten wall.

As a consequence, the diffraction coefficient, which is defined as the quotient between the wave height at a given point and the height of the incident waves, is given by

$$C_d(r,\beta) = \sqrt{F^2 + G^2}.$$

The C_d for two different angles of the wave direction are given in figs. 1.19 and 1.20. Of course, the diffraction coefficient at the sheltered side of the breakwater gets smaller and smaller with the distance from the tip of the breakwater. At the wave-beaten side of the breakwater ($\beta = 2\pi$), C_d takes on a maximum at about $\frac{1}{3}$ wavelength, or at somewhat more than a wavelength according to whether the wave attack is wall-orthogonal ($\theta = 0$) or inclined at 45°. In both cases the maximum C_d is nearly 2.3, that is the maximum wave height at the wall is 2.3 times greater than the height of the incident wave.

The calculation of wave diffraction essentially consists in evaluating the Fresnel integrals [defined by (1.60*d*)]. In the numerical evaluation of these integrals one should bear in mind the nature of the function to be integrated. Such a function, that is $\sin(\pi x^2/2)$ or $\cos(\pi x^2/2)$, is a pseudo-sinusoid that progressively shrinks as x grows. If x_i and x_{i+1} are the abscissas of two consecutive maxima of function $\sin(\pi x^2/2)$ [see fig. 1.21], then they are given by

$$\frac{\pi}{2} x_i^2 = i2\pi + \frac{\pi}{2},$$

$$\frac{\pi}{2} x_{i+1}^2 = (i+1)2\pi + \frac{\pi}{2}$$

(with i positive integer or 0), so that

$$\frac{2}{x_{i+1}} < x_{i+1} - x_i < \frac{2}{x_i}. \tag{1.63}$$

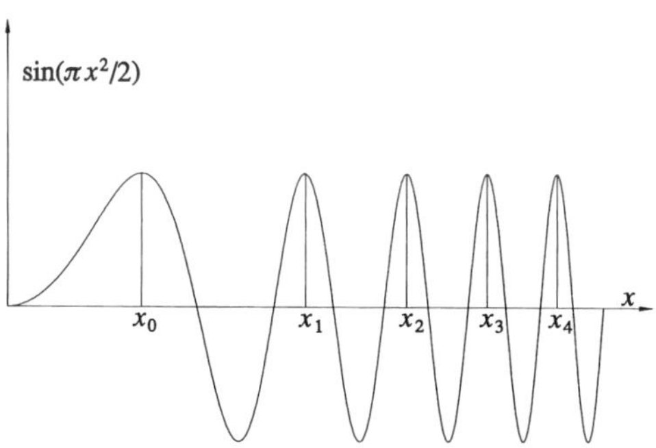

Fig. 1.21 The function $\sin(\pi x^2/2)$ the integral of which (Fresnel integral) must be evaluated to obtain the diffraction coefficients.

Such an inequality is useful in the choice of the integration step Δx. As an example if one wishes to evaluate the integral

$$\int_0^u \sin\left(\frac{\pi}{2}x^2\right)\mathrm{d}x$$

with at least 12 steps for each pseudo-sinusoid, thanks to (1.63) it suffices that one fixes

$$\Delta x = \frac{1}{12}\frac{2}{|u|}.$$

In the calculation one should bear in mind that u may be positive or negative. Since the integrand is an even function, it is convenient to take $|u|$ as the upper limit of integration, and then multiplying the result by $u/|u|$.

Conclusive note

The theory of the periodic waves was introduced by Airy (1845) and was fully developed by Stokes (1847). The first graphic method for the wavelength L_c on a current should be that of Jonsson et al. (1970). The wave-current interaction is usually treated with a reference frame moving with the current velocity. Our choice for the fixed reference frame is in view of the developments of sect. 2.10. The fixed reference frame appears more suitable for obtaining the formal solution for the shoaling and set-down of waves on a current.

Analytical solutions for the interaction waves-vertical wall are available also for the wall of finite length and the breakwater gap. The first one (Montefusco, 1968) gives the velocity potential as a series expansion of products of *odd* Mathieu functions. The second one (Sobey and Johnson, 1986) gives the velocity potential as a series expansion of products of *even* Mathieu functions.

References

Airy G. B., 1845 Tides and waves. *Encyc. Metrop.* Art. 192, 241-396.

Jonsson I. G., Skougaard C. and Wang J. D., 1970 Interaction between waves and currents. *Proc.* 12th *Conf. Coastal Eng. ASCE*, 489-507.

Montefusco L., 1968 The diffraction of a plane wave by an isolated breakwater. *Meccanica* 3, 156-166.

Penny W. G. and Price A. T., 1952 The diffraction theory of sea waves by breakwaters. *Phil. Trans. Roy. Soc. London* A 244, 236-253.

Saintflou G., 1928 Essai sur les digues maritimes, verticales. *Annales Ponts et Chaussées* 98, 4.

Sobey R.J. and Johnson R.J., 1986 Diffraction patterns near narrow breakwater gaps. *J. Waterways, Port, Coastal and Ocean Eng.* 112, 512-528.

Stokes G. G., 1847 On the theory of oscillatory waves. *Trans. Cambridge Phil. Soc.* 8, 441-455.

Chapter 2

PERIODIC WAVE PATTERN: THE CONTROL VOLUME APPROACH

2.1 The linear momentum equation for a control volume

2.1.1 *The case of control volume W within the water mass*

Let us fix a volume within the water mass, and call this volume W and its boundary A. Then let us imagine we can colour the water mass that at a fixed time t occupies W, and call the coloured water mass m.

Newton's second law for the mass m is

$$\sum_i \mathbf{F}_i(t) = \frac{1}{\mathrm{d}t}\left[\int_m \mathbf{v}(t+\mathrm{d}t)\,\mathrm{d}m - \int_m \mathbf{v}(t)\,\mathrm{d}m\right], \tag{2.1}$$

where $\mathbf{F}_i$ are all the forces acting on m, including body forces. Mass m at time t occupies volume W, while at time $t+\mathrm{d}t$ it occupies the volume

$$W + \int_A \mathbf{v}(t)\cdot\mathbf{n}\,\mathrm{d}A\,\mathrm{d}t,$$

where $\mathbf{n}$ is the unit vector which has a direction normal to surface A, positive outward. Referring to fig. 2.1, the volume occupied by the coloured water at time $t+\mathrm{d}t$ is equal to the volume occupied by the coloured water at time t, plus the darkened volume minus the dotted volume. Therefore

$$\int_m \mathbf{v}(t+\mathrm{d}t)\mathrm{d}m = \int_W \rho\mathbf{v}(t+\mathrm{d}t)\,\mathrm{d}W + \int_A \rho\mathbf{v}(t+\mathrm{d}t)\,\mathbf{v}(t)\cdot\mathbf{n}\,\mathrm{d}A\,\mathrm{d}t,$$

where the first integral on the r.h.s. of this equation gives the linear momentum of the water occupying volume W at time $t+\mathrm{d}t$, and the next integral adds the linear momentum of the coloured water which has flown out of W in the small time $\mathrm{d}t$, and subtracts the linear momentum of the non-coloured water which has entered W in the small time $\mathrm{d}t$.

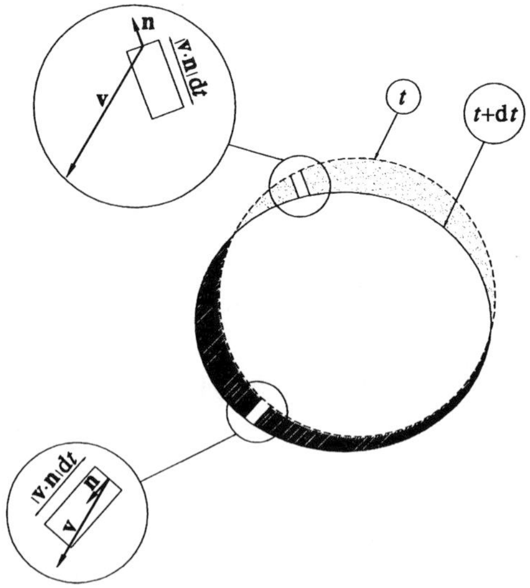

Fig. 2.1 Reference scheme for equation (2.2): W is the volume within the dashed surface; A is the dashed surface; m is the water mass that occupies W at time t.

Hence, we can rewrite (2.1) in the form

$$\sum_i \mathbf{F}_i(t) = \frac{1}{\mathrm{d}t}\left[\int_W \rho\mathbf{v}(t+\mathrm{d}t)\,\mathrm{d}W + \int_A \rho\mathbf{v}(t+\mathrm{d}t)\,\mathbf{v}(t)\cdot\mathbf{n}\,\mathrm{d}A\,\mathrm{d}t - \int_W \rho\mathbf{v}(t)\,\mathrm{d}W\right],$$

that is (apart from some negligible terms)

$$\sum_i \mathbf{F}_i = \int_W \rho\frac{\partial\mathbf{v}}{\partial t}\,\mathrm{d}W + \int_A \rho\mathbf{v}\mathbf{v}\cdot\mathbf{n}\,\mathrm{d}A. \tag{2.2}$$

As it is well known, W is called *control volume* and A is called *control surface*, and (2.2) is the linear momentum equation for a control volume.

2.1.2 *The case of control volume $\widetilde{W}$ piercing the free surface*

Let us consider a control volume bounded by the seabed, by the free surface at a fixed time t, and by four vertical planes [see fig. 2.2*a*]. In this case, equation (2.2) can be rewritten in the form

$$\sum_i \mathbf{F}_i = \int_W \rho\frac{\partial\mathbf{v}}{\partial t}\,\mathrm{d}W + \int_{F.S.} \rho\mathbf{v}\mathbf{v}\cdot\mathbf{n}\,\mathrm{d}A + \int_{\tilde{A}} \rho\mathbf{v}\mathbf{v}\cdot\mathbf{n}\,\mathrm{d}A, \tag{2.3}$$

where $F.S.$ denotes the free surface, and $\tilde{A}$ denotes the four vertical planes. For the step from (2.2) to (2.3), we have resolved the control surface A into three parts:

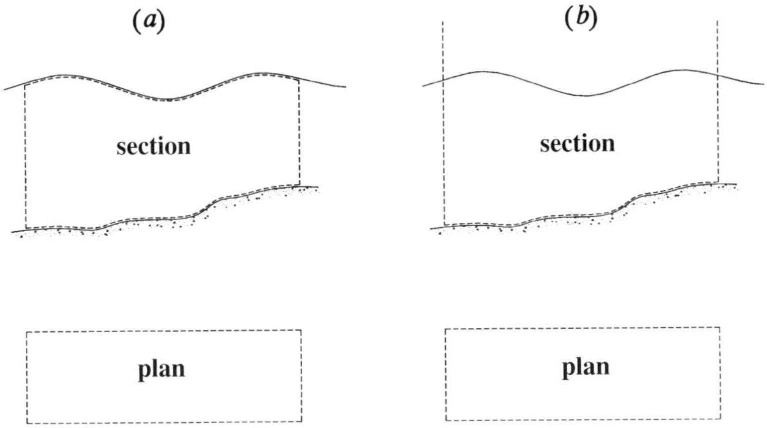

Fig. 2.2 (a) Usual control volume W. (b) Open control volume $\widetilde{W}$.

the free surface, the four vertical planes, and the seabed which does not give any contribution to the integral since on the seabed $\mathbf{v}\cdot\mathbf{n}=0$.

Here it is more convenient to reason in terms of $\widetilde{W}$ the volume bounded by the seabed and by four infinitely high planes. Since

$$\frac{\mathrm{d}}{\mathrm{d}t}\int_{\widetilde{W}}\rho\mathbf{v}\,\mathrm{d}W=\int_{W}\rho\frac{\partial\mathbf{v}}{\partial t}\,\mathrm{d}W+\int_{F.S.}\rho\mathbf{v}\mathbf{v}\cdot\mathbf{n}\,\mathrm{d}A, \tag{2.4}$$

equation (2.3) is rewritten in the form

$$\sum_i\mathbf{F}_i=\frac{\mathrm{d}}{\mathrm{d}t}\int_{\widetilde{W}}\rho\mathbf{v}\,\mathrm{d}W+\int_{\tilde{A}}\rho\mathbf{v}\mathbf{v}\cdot\mathbf{n}\,\mathrm{d}A \tag{2.5}$$

that is the linear momentum equation for a control volume piercing the free surface.

For a periodic wave motion, equation (2.5) implies that

$$<\sum_i\mathbf{F}_i>=<\int_{\tilde{A}}\rho\mathbf{v}\mathbf{v}\cdot\mathbf{n}\,\mathrm{d}A>, \tag{2.6}$$

where the symbol $<\cdot>$ denotes the average with respect to time:

$$<f(t)>\equiv\frac{1}{T}\int_{t_o}^{t_o+T}f(t)\,\mathrm{d}t\quad(\text{arbitrary } t_o). \tag{2.7}$$

Equation (2.6) shows that the mean value of the resultant force acting on the control volume is equal to the mean value of the net efflux of linear momentum from the control volume.

In the case of an *ideal flow*, the forces $\mathbf{F}_i$ in equation (2.6) reduce themselves to

(i) the weight $(-P\mathbf{i}_z)$ of the water mass in $\widetilde{W}$;

(ii) the force $\mathbf{F}_f$ that the seabed and/or some solid bodies exert on the water mass;

(iii) the pressure force acting on the lateral surface $\tilde{A}$:

$$-\int_{\tilde{A}} p\mathbf{n}\,\mathrm{d}A,$$

where the minus is due to the fact that the unit vector $\mathbf{n}$ is oriented toward the outside of the control volume. Therefore, under ideal flow assumptions, (2.6) is rewritten in the form

$$< -P\mathbf{i}_z + \mathbf{F}_f > = < \int_{\tilde{A}} p\mathbf{n} + \rho\mathbf{v}\mathbf{v}\cdot\mathbf{n}\mathrm{d}A >. \tag{2.8}$$

To check (2.4), note that

$$\int_{\widetilde{W}} \rho\mathbf{v}(t+\mathrm{d}t)\,\mathrm{d}W = \int_{W} \rho\mathbf{v}(t+\mathrm{d}t)\,\mathrm{d}W + \int_{F.S.} \rho\mathbf{v}(t+\mathrm{d}t)\mathbf{v}(t)\cdot\mathbf{n}\mathrm{d}t\,\mathrm{d}A,$$

$$\int_{\widetilde{W}} \rho\mathbf{v}(t)\,\mathrm{d}W = \int_{W} \rho\mathbf{v}(t)\,\mathrm{d}W.$$

2.2 The energy equation for a control volume

2.2.1 *The energy per unit mass*

Newton's second law, when applied to a small water mass $\mathrm{d}m$, can be written in the form

$$F_x = \mathrm{d}m\,\frac{\mathrm{d}v_x}{\mathrm{d}t},$$

$$F_y = \mathrm{d}m\,\frac{\mathrm{d}v_y}{\mathrm{d}t},$$

$$F_z = \mathrm{d}m\,\frac{\mathrm{d}v_z}{\mathrm{d}t} + \mathrm{d}m\,g,$$

where $\mathbf{dv}$ represents the velocity variation of the given small mass $\mathrm{d}m$ in the small time interval $\mathrm{d}t$, and where $\mathbf{F}$ here is the resultant of all the forces, except weight, acting on the small mass. Multiplying both sides of the first equation by v_x, both sides of the second equation by v_y, and both sides of the third equation by v_z, and adding up these equations we arrive at

$$F_x v_x + F_y v_y + F_z v_z = \mathrm{d}m\left[v_x\frac{\mathrm{d}v_x}{\mathrm{d}t} + v_y\frac{\mathrm{d}v_y}{\mathrm{d}t} + v_z\frac{\mathrm{d}v_z}{\mathrm{d}t} + g v_z\right]$$

which is rewritten in the form

$$\mathbf{F}\cdot\mathbf{v} = \mathrm{d}m\frac{\mathrm{d}e}{\mathrm{d}t}, \tag{2.9}$$

where e is the energy (potential and kinetic) per unit mass:

$$e \equiv \frac{1}{2} v^2 + gz. \tag{2.10}$$

Let us subdivide the coloured mass m, which at time t occupies the volume W, into many small masses $\mathrm{d}m$, and let us apply equation (2.9) to each of these masses ($\mathrm{d}m$ being the mass of a small volume $\mathrm{d}x\ \mathrm{d}y\ \mathrm{d}z$). Therefore we have N equalities like (2.9), one for each of the small masses $\mathrm{d}m$. Then adding up these N equalities, we obtain

$$\sum_{i=1}^{N} (\mathbf{F} \cdot \mathbf{v})_i = \sum_{i=1}^{N} \mathrm{d}m \left(\frac{\mathrm{d}e}{\mathrm{d}t} \right)_i. \tag{2.11}$$

Pay attention to the symbols! $\mathrm{d}\mathbf{v}/\mathrm{d}t$ represents the quotient between the velocity variation of the *given small mass* in a small time interval $\mathrm{d}t$, and $\mathrm{d}t$ itself. While $\partial \mathbf{v}/\partial t$ represents the quotient between the velocity variation at a *fixed point* in a small time interval $\mathrm{d}t$, and $\mathrm{d}t$ itself. Thus in the case of $\mathrm{d}\mathbf{v}/\mathrm{d}t$, the velocity variation depends on two causes: the fact that the time instant is slightly changed and the fact that also the position in space is slightly changed. While in the case of $\partial \mathbf{v}/\partial t$ the velocity variation depends only on the fact that the time instant is slightly changed. There is the same kind of difference also between $\mathrm{d}e/\mathrm{d}t$ and $\partial e/\partial t$.

2.2.2 *The case of the control volume W within the water mass*

The explicit form of the l.h.s. of (2.11)

Let us assume that the *shear stresses are negligible*. Under this hypothesis, each small volume $\mathrm{d}x\,\mathrm{d}y\,\mathrm{d}z$ is subjected only to two components of F_x: the first one acting on the left face $\mathrm{d}y\,\mathrm{d}z$ and the second one acting on the right face $\mathrm{d}y\,\mathrm{d}z$. Let us consider the contribution to the sum

$$\sum_{i=1}^{N} (F_x v_x)_i$$

from a row of small volumes along a line parallel to the x-axis [see fig. 2.3]. This contribution, which we call partial sum, is

$$\text{partial sum} = \mathrm{d}y\ \mathrm{d}z\,(p_1 - p_2)v_{x1} + \mathrm{d}y\ \mathrm{d}z\,(p_2 - p_3)v_{x2} + \ldots + \mathrm{d}y\ \mathrm{d}z\,(p_n - p_{n+1})v_{xn},$$

which can be rewritten in the form

$$\text{partial sum} = \mathrm{d}y\ \mathrm{d}z\,[p_1 v_{x1} + p_2 (v_{x2} - v_{x1}) + \ldots + p_n (v_{xn} - v_{xn-1}) - p_{n+1} v_{xn}],$$

which in its turn can be rewritten in the form

$$\text{partial sum} = [(p_1 v_{x1} - p_{n+1} v_{xn})\,\mathrm{d}y\,\mathrm{d}z] + \left[\sum_{i=2}^{n} p_i \left(\frac{\partial v_x}{\partial x} \right)_i \mathrm{d}x\,\mathrm{d}y\,\mathrm{d}z \right]. \tag{2.12}$$

Hence, adding up the contributions of all the rows of small volumes $\mathrm{d}x\,\mathrm{d}y\,\mathrm{d}z$, we obtain

$$\sum_{i=1}^{N}(F_x v_x)_i = -\int_A p v_x\,\mathbf{n}\cdot\mathbf{i}_x\,\mathrm{d}A + \int_W p\,\frac{\partial v_x}{\partial x}\,\mathrm{d}W,$$

where the two integrals on the r.h.s. correspond, respectively, to the terms within the first square parentheses, and to the summation within the second square parentheses on the r.h.s. of (2.12).

Through the same reasoning we can show that

$$\sum_{i=1}^{N}(F_y v_y)_i = -\int_A p v_y\,\mathbf{n}\cdot\mathbf{i}_y\,\mathrm{d}A + \int_W p\,\frac{\partial v_y}{\partial y}\,\mathrm{d}W,$$

$$\sum_{i=1}^{N}(F_z v_z)_i = -\int_A p v_z\,\mathbf{n}\cdot\mathbf{i}_z\,\mathrm{d}A + \int_W p\,\frac{\partial v_z}{\partial z}\,\mathrm{d}W,$$

and, consequently, we arrive at

$$\sum_{i=1}^{N}(\mathbf{F}\cdot\mathbf{v})_i = -\int_A p\mathbf{v}\cdot\mathbf{n}\,\mathrm{d}A + \int_W p\left(\frac{\partial v_x}{\partial x}+\frac{\partial v_y}{\partial y}+\frac{\partial v_z}{\partial z}\right)\mathrm{d}W,$$

which is reduced to

$$\sum_{i=1}^{N}(\mathbf{F}\cdot\mathbf{v})_i = -\int_A p\mathbf{v}\cdot\mathbf{n}\,\mathrm{d}A, \tag{2.13}$$

given that $\dfrac{\partial v_x}{\partial x}+\dfrac{\partial v_y}{\partial y}+\dfrac{\partial v_z}{\partial z}$ is equal to zero (continuity equation).

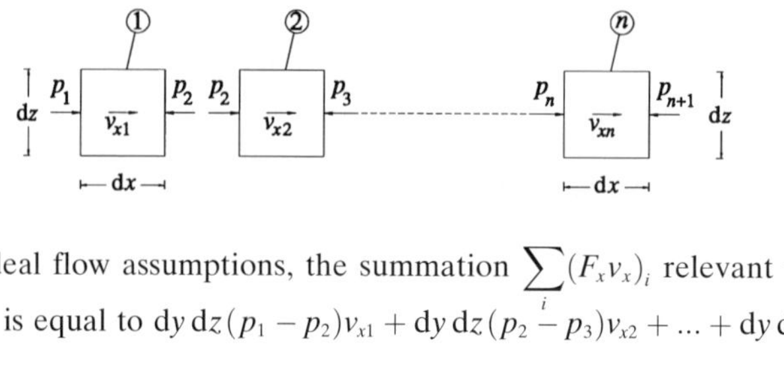

Fig. 2.3 Under ideal flow assumptions, the summation $\sum_i (F_x v_x)_i$ relevant to the row of small volumes 1, 2, ..., n is equal to $\mathrm{d}y\,\mathrm{d}z(p_1-p_2)v_{x1}+\mathrm{d}y\,\mathrm{d}z(p_2-p_3)v_{x2}+\ldots+\mathrm{d}y\,\mathrm{d}z(p_n-p_{n+1})v_{xn}$.

The explicit form of the r.h.s. of (2.11)

The r.h.s. of (2.11) can be rewritten in the form

$$\sum_{i=1}^{N} \mathrm{d}m \left(\frac{\mathrm{d}e}{\mathrm{d}t}\right)_i = \frac{1}{\mathrm{d}t}\left[\int_m e(t+\mathrm{d}t)\,\mathrm{d}m - \int_m e(t)\,\mathrm{d}m\right], \tag{2.14}$$

and hence in the form

$$\sum_{i=1}^{N} \mathrm{d}m \left(\frac{\mathrm{d}e}{\mathrm{d}t}\right)_i = \int_W \rho \frac{\partial e}{\partial t}\,\mathrm{d}W + \int_A \rho e\, \mathbf{v}\cdot\mathbf{n}\,\mathrm{d}A, \tag{2.15}$$

where the logical step from (2.14) to (2.15) is the same as that which we have already done from (2.1) to (2.2).

New formulation of (2.11)

Equations (2.13) and (2.15) enable us to rewrite (2.11) in the form

$$\int_A (p+\rho e)\,\mathbf{v}\cdot\mathbf{n}\,\mathrm{d}A = -\int_W \rho \frac{\partial e}{\partial t}\,\mathrm{d}W. \tag{2.16}$$

2.2.3 *The case of the control volume $\widetilde{W}$ piercing the free surface*

Let us pass to volume $\widetilde{W}$ whose definition has been given in sect. 2.1.2. Since

$$\frac{\mathrm{d}}{\mathrm{d}t}\int_{\widetilde{W}} \rho e\,\mathrm{d}W = \int_W \rho \frac{\partial e}{\partial t}\,\mathrm{d}W + \int_{F.S.} \rho e\,\mathbf{v}\cdot\mathbf{n}\,\mathrm{d}A,$$

(2.16) is rewritten in the form

$$\int_A (p+\rho e)\mathbf{v}\cdot\mathbf{n}\,\mathrm{d}A - \int_{F.S.} \rho e\,\mathbf{v}\cdot\mathbf{n}\,\mathrm{d}A = -\frac{\mathrm{d}}{\mathrm{d}t}\int_{\widetilde{W}} \rho e\,\mathrm{d}W. \tag{2.17}$$

Here,

(i) replacing ρe with $(p+\rho e)$ in the integral over *F.S.*, which is simply a formal step since pressure p is zero on the free surface,

(ii) subdividing the control surface into three components: $\tilde{A}$ which is formed by the four vertical planes, the free surface *F.S.*, and the seabed, we reduce (2.17) to

$$\int_{\tilde{A}} (p+\rho e)\mathbf{v}\cdot\mathbf{n}\,\mathrm{d}A = -\frac{\mathrm{d}}{\mathrm{d}t}\int_{\widetilde{W}} \rho e\,\mathrm{d}W. \tag{2.18}$$

Finally, averaging with respect to time, we arrive at

$$< \int_{\tilde{A}} (p + \rho e)\mathbf{v} \cdot \mathbf{n}\, \mathrm{d}A > = 0 \tag{2.19}$$

which is exact under the ideal flow assumption given that we have neglected the shear stresses.

2.3 Radiation stress, mean energy flux, mean wave energy per unit surface

2.3.1 *Radiation stress*

As an exercise, try to apply equation (2.8) to the control volume of fig. 2.4. From the r.h.s. of this equation, you will get four double integrals, one for each of the vertical planes forming the lateral surface $\tilde{A}$. In particular, the integral relevant to the left y-parallel plane will be

$$< \int_{y_1}^{y_2} \int_{-d(x_1,\, y)}^{\eta(x_1,\, y,\, t)} p(x_1, y, z, t)(-\mathbf{i}_x) + \rho\, \mathbf{v}(x_1, y, z, t)[-v_x(x_1, y, z, t)]\, \mathrm{d}z\, \mathrm{d}y > .$$

At this stage you will realize that the definition of the vectors

$$\mathbf{R}_x \equiv < \int_{-d}^{\eta} p\mathbf{i}_x + \rho\, \mathbf{v}\, v_x\, \mathrm{d}z > , \tag{2.20a}$$

$$\mathbf{R}_y \equiv < \int_{-d}^{\eta} p\mathbf{i}_y + \rho\, \mathbf{v}\, v_y\, \mathrm{d}z > \tag{2.20b}$$

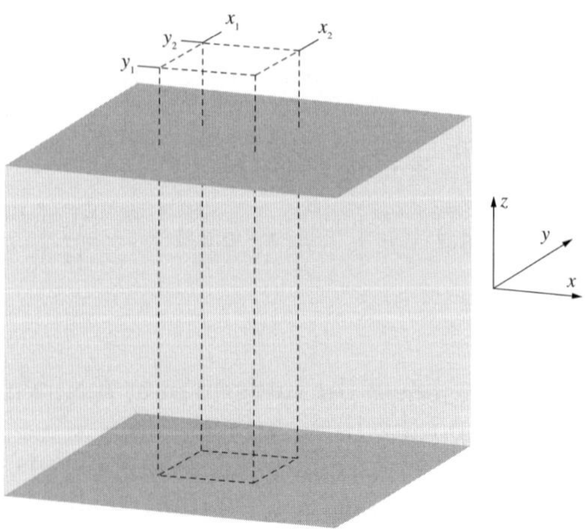

Fig. 2.4 For this control volume, the equations (2.8) and (2.19) are rewritten in the form (2.21) and (2.24), respectively.

leads to a much more compact expression. Indeed you will be able to write the result of the exercise in the form

$$<-P\mathbf{i}_z+\mathbf{F}_f>=\int_{x_1}^{x_2}-\mathbf{R}_y(x,y_1)+\mathbf{R}_y(x,y_2)\,\mathrm{d}x+\int_{y_1}^{y_2}-\mathbf{R}_x(x_1,y)+\mathbf{R}_x(x_2,y)\,\mathrm{d}y. \quad (2.21)$$

To get this compact form, you must also use the fact that

$$<\int_{y_1}^{y_2}\int_{-d}^{\eta}\ldots\mathrm{d}z\,\mathrm{d}y>=\int_{y_1}^{y_2}<\int_{-d}^{\eta}\ldots\mathrm{d}z>\mathrm{d}y,$$

since the limits of integration y_1 and y_2 are independent of time t.

The x, y components of equation (2.21) are

$$<F_{fx}>=\int_{x_1}^{x_2}-R_{yx}(x,y_1)+R_{yx}(x,y_2)\,\mathrm{d}x+\int_{y_1}^{y_2}-R_{xx}(x_1,y)+R_{xx}(x_2,y)\,\mathrm{d}y, \quad (2.22a)$$

$$<F_{fy}>=\int_{x_1}^{x_2}-R_{yy}(x,y_1)+R_{yy}(x,y_2)\,\mathrm{d}x+\int_{y_1}^{y_2}-R_{xy}(x_1,y)+R_{xy}(x_2,y)\,\mathrm{d}y. \quad (2.22b)$$

As for the z-component, it is reduced to

$$<F_{fz}>=<P>,$$

since it can be shown that R_{xz} and R_{yz} are zero in a wave motion.

The tensor

$$\mathbf{R}\equiv\begin{pmatrix}R_{xx} & R_{xy}\\ R_{yx} & R_{yy}\end{pmatrix}$$

is called *radiation stress tensor.*

Usually, the radiation stress tensor is defined as the difference between the tensor $\mathbf{R}$ just defined and the same tensor under static conditions (where R_{xy} and R_{yx} are zero and R_{xx} and R_{yy} are equal to $\frac{1}{2}\rho g d^2$). Thus the definition we adopt here is somewhat different from the usual definition of radiation stress. But it looks more convenient especially for giving the formal solution of sect. 2.10.

2.3.2 *Mean energy flux*

As a further exercise, try to apply equation (2.19) to the control volume of fig. 2.4. Also in this case you will obtain four double integrals, one for each of the vertical planes forming the lateral surface $\tilde{A}$. In particular, the integral relevant to the left y-parallel side will be

$$<\int_{y_1}^{y_2}\int_{-d(x_1,y)}^{\eta(x_1,y,t)}[p(x_1,y,z,t)+\rho e(x_1,y,z,t)][-v_x(x_1,y,z,t)]\,\mathrm{d}z\,\mathrm{d}y>.$$

So you will realize that, defining the new vector

$$\mathbf{\Phi} \equiv < \int_{-d}^{\eta} (p + \rho e)\mathbf{v} \mathrm{d}z > , \tag{2.23}$$

you will be able to greatly compact the resulting expression, which will be reduced to

$$\int_{x_1}^{x_2} - \Phi_y(x, y_1) + \Phi_y(x, y_2)\,\mathrm{d}x + \int_{y_1}^{y_2} - \Phi_x(x_1, y) + \Phi_x(x_2, y)\,\mathrm{d}y = 0. \tag{2.24}$$

The vector $\mathbf{\Phi}$ is called mean energy flux.

2.3.3 *Mean wave energy per unit surface*

Besides tensor $\mathbf{R}$ and vector $\mathbf{\Phi}$, it is convenient to define

$$\mathscr{E} \equiv \text{[mean energy per unit surface, in the wave motion]} + \\ - \text{[energy per unit surface in static condition]},$$

that is

$$\mathscr{E} = < \int_{-d}^{\eta} \rho e \,\mathrm{d}z > - \int_{-d}^{0} \rho g z \,\mathrm{d}z . \tag{2.25}$$

The usefulness of this definition will become apparent in sect. 2.9.

2.4 Formulae for radiation stress and mean energy flux of progressive waves

2.4.1 *Progressive waves on gentle slopes*

What has been shown in sects. 2.1-2-3 holds generally, that is for waves travelling over arbitrarily shaped bottoms, or even for waves interacting with some structures (provided that $\mathbf{F}_f$ includes also the forces exerted by these structures). In the rest of this chapter we shall deal only with the progressive waves, that is with the waves which do not interact with structures. As to the seabed, we shall allow it to be generally non-horizontal, but we shall assume the changes in water depth to take place very gradually (see fig. 2.5 and the relevant caption for the meaning of *very gradually*).

Referring to the aforesaid fig. 2.5, in the area $\Delta x\, \Delta y$ where the water depth is assumed to be constant, from Stokes' theory, we have

$$\eta(s,t) = \frac{H}{2}\cos(ks - \omega t) + \frac{H^2}{16}\frac{\omega^2}{g} F_1(kd)\cos[2(ks - \omega t)] + \mathrm{o}(H^2), \tag{2.26a}$$

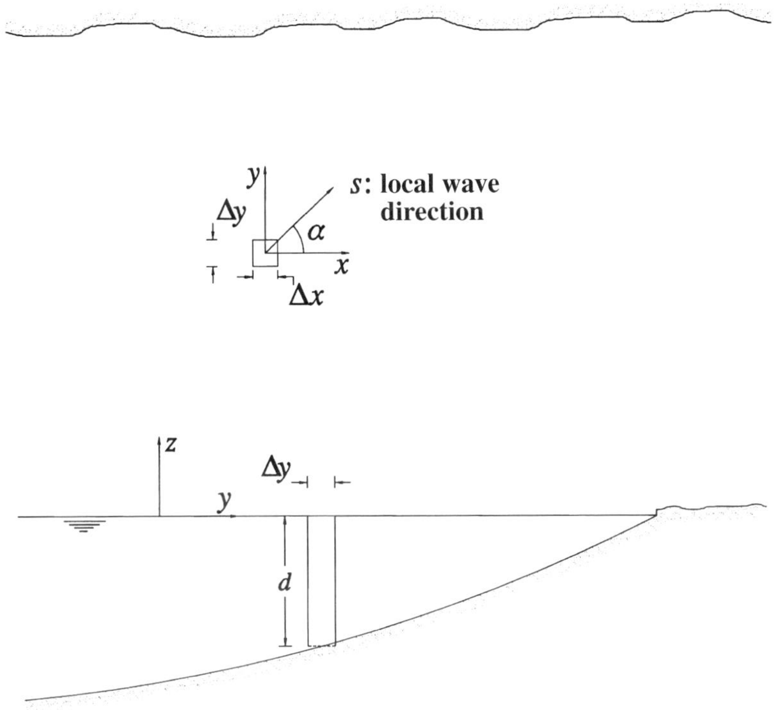

Fig. 2.5 The assumption is made that the bottom has a small slope, so that the water depth is regarded as almost constant in any area with sides Δx, Δy of a few wavelengths.

$$\phi(s,z,t) = g\,\frac{H}{2}\,\omega^{-1}\,\frac{\cosh[k(d+z)]}{\cosh(kd)}\,\sin(ks-\omega t) + \frac{3}{32}\,H^2\,\omega\,\frac{1}{\sinh^4(kd)}\,\cdot$$
$$\cdot\cosh[2k(d+z)]\,\sin[2(ks-\omega t)] - \frac{H^2}{16}\,\omega^2\,\frac{1}{\sinh^2(kd)}\,t + Cs + \mathrm{o}(H^2), \qquad (2.26b)$$

where s is the local wave direction, $F_1(kd)$ is given by (1.40) and C is a constant of order H^2 which is still unknown (it will be obtained only in sect. 2.7). Clearly, the velocity components v_x and v_y, being necessary to the radiation stress tensor $\mathbf{R}$, the mean energy flux $\boldsymbol{\Phi}$ and the mean wave energy per unit surface $\mathscr{E}$, proceed from the velocity potential (2.26*b*) through the equations

$$v_x = \frac{\partial\phi}{\partial s}\cos\alpha, \quad v_y = \frac{\partial\phi}{\partial s}\sin\alpha,$$

where α is the angle that the local wave direction makes with the x-axis [cf. again fig. 2.5].

2.4.2 *The formulae for* R_{xy} *and* R_{yx}

From definitions (2.20*a-b*) we have

$$R_{xy} = R_{yx} = \frac{1}{T}\int_0^T \int_{-d}^{\eta} \rho\, v_x\, v_y\, \mathrm{d}z\, \mathrm{d}t, \tag{2.27}$$

that is

$$R_{xy} = R_{yx} = \rho\frac{1}{T}\int_0^T \int_{-d}^{0} \left\{g\,\frac{H}{2}\,\omega^{-1}k\,\frac{\cosh[k(d+z)]}{\cosh(kd)}\cos(ks-\omega t)\right\}^2 \sin\alpha\,\cos\alpha\, \mathrm{d}z\, \mathrm{d}t. \tag{2.28}$$

The integrand having order H^2 can give at the most a contribution of order H^3 on the interval of integration $(0, \eta)$. This is why in the step from (2.27) to (2.28), the upper limit of integration has been changed from η to 0.

Bearing in mind that

$$\frac{1}{T}\int_0^T \cos^2(ks-\omega t)\,\mathrm{d}t = \frac{1}{2},$$

from (2.28) we obtain

$$R_{xy} = R_{yx} = \frac{1}{16}\rho g H^2\left[1+\frac{2kd}{\sinh(2kd)}\right]\sin\alpha\,\cos\alpha. \tag{2.29}$$

2.4.3 *The formulae for* R_{xx} *and* R_{yy}

From definitions (2.20*a-b*) we have

$$R_{xx} = \frac{1}{T}\int_0^T \int_{-d}^{\eta} p + \rho v_x^2\, \mathrm{d}z\, \mathrm{d}t,$$

$$R_{yy} = \frac{1}{T}\int_0^T \int_{-d}^{\eta} p + \rho v_y^2\, \mathrm{d}z\, \mathrm{d}t$$

which we rewrite in the form

$$R_{xx} = I_1 + I_2 + I_{v_x},$$

$$R_{yy} = I_1 + I_2 + I_{v_y},$$

on defining

$$I_{v_x} \equiv \frac{1}{T}\int_0^T \int_{-d}^{\eta} \rho v_x^2\, \mathrm{d}z\, \mathrm{d}t,$$

$$I_{v_y} \equiv \frac{1}{T}\int_0^T \int_{-d}^{\eta} \rho v_y^2\, \mathrm{d}z\, \mathrm{d}t,$$

$$I_1 \equiv \frac{1}{T}\int_0^T \int_0^{\eta} p\, \mathrm{d}z\, \mathrm{d}t,$$

$$I_2 \equiv \frac{1}{T}\int_0^T\int_{-d}^0 p\,\mathrm{d}z\,\mathrm{d}t. \tag{2.30}$$

The only difference between I_{v_x} and R_{xy} is that the integrand is v_x^2 instead of $v_x v_y$. Since $v_x v_y = v_s^2 \cos\alpha\sin\alpha$ whereas $v_x^2 = v_s^2\cos^2\alpha$, it follows that I_{v_x} has the same expression as R_{xy} with only $\cos^2\alpha$ in place of $\cos\alpha\sin\alpha$. Similarly, I_{v_y} has the same expression as R_{xy} with $\sin^2\alpha$ in place of $\cos\alpha\sin\alpha$:

$$I_{v_x} = \frac{1}{16}\rho g H^2\left[1+\frac{2kd}{\sinh(2kd)}\right]\cos^2\alpha,$$

$$I_{v_y} = \frac{1}{16}\rho g H^2\left[1+\frac{2kd}{\sinh(2kd)}\right]\sin^2\alpha.$$

Let us pass to I_1 and I_2. The pressure proceeds from the velocity potential (2.26*b*) through the Bernoulli equation, and to Stokes' second order proves to be

$$p = -\rho g z + \rho g\frac{H}{2}\frac{\cosh[k(d+z)]}{\cosh(kd)}\cos(ks-\omega t) + \frac{3}{16}\rho H^2\omega^2\frac{1}{\sinh^4(kd)}\cosh[2k(d+z)]\cdot$$

$$\cdot\cos[2(ks-\omega t)] + \rho\frac{H^2}{16}\omega^2\frac{1}{\sinh^2(kd)} - \frac{1}{2}\rho(v_s^2+v_z^2). \tag{2.31}$$

For getting I_1 exact to the order H^2, it is sufficient to take into account the terms of order H of the pressure:

$$I_1 = \frac{1}{T}\int_0^T\int_0^\eta -\rho g z + \rho g\frac{H}{2}\cos(ks-\omega t)\,\mathrm{d}z\,\mathrm{d}t,$$

where we have used the fact that

$$H\frac{\cosh[k(d+z)]}{\cosh(kd)} = H + \mathrm{o}(H) \quad \text{if} \quad |z|\,\mathrm{O}(H).$$

Evaluating the integral, we obtain

$$I_1 = \frac{1}{16}\rho g H^2.$$

At this stage only I_2 remains to be evaluated, for which we obtain

$$I_2 = \frac{1}{2}\rho g d^2 + \rho\frac{H^2}{16}\omega^2\frac{1}{\sinh^2(kd)}d - \frac{1}{16}\rho g H^2. \tag{2.32}$$

The three terms on the r.h.s. of this equation derive from, respectively, the first, the fourth and the fifth terms on the r.h.s. of (2.31). As for the second and the third terms on the r.h.s. of (2.31), they give no contribution to I_2.

The conclusion is that R_{xx}, being the sum of I_1, I_2 and I_{v_x}, proves to be

$$R_{xx} = \frac{1}{2}\rho g d^2 + \frac{1}{16}\rho g H^2 \left[\frac{2kd}{\sinh(2kd)}(1+\cos^2\alpha) + \cos^2\alpha\right],$$

and R_{yy}, being the sum of I_1, I_2 and I_{v_y}, proves to be

$$R_{yy} = \frac{1}{2}\rho g d^2 + \frac{1}{16}\rho g H^2 \left[\frac{2kd}{\sinh(2kd)}(1+\sin^2\alpha) + \sin^2\alpha\right]. \tag{2.33}$$

2.4.4 *The formula for* $\mathbf{\Phi}$

The x-component of vector $\mathbf{\Phi}$ (defined by 2.23) is given by

$$\Phi_x = \frac{1}{T}\int_0^T \int_{-d}^{\eta} \left[p + \rho g z + \frac{1}{2}\rho(v_s^2 + v_z^2)\right] v_s \cos\alpha \, dz \, dt, \tag{2.34}$$

where the integrand is of order H^2, and thus we can reduce the interval of integration from $(-d, \eta)$ to $(-d, 0)$ making at the most an error of an order smaller than H^2. Therefore we rewrite (2.34) in the form

$$\Phi_x = \int_{-d}^{0} \frac{1}{T}\int_0^T \left[\rho g \frac{H}{2} \frac{\cosh[k(d+z)]}{\cosh(kd)} \cos(ks - \omega t)\right] v_s \cos\alpha \, dt \, dz,$$

from which we easily arrive at

$$\Phi_x = \frac{1}{8}\rho g H^2 \frac{c}{2}\left[1 + \frac{2kd}{\sinh(2kd)}\right]\cos\alpha.$$

Similarly, we get

$$\Phi_y = \frac{1}{8}\rho g H^2 \frac{c}{2}\left[1 + \frac{2kd}{\sinh(2kd)}\right]\sin\alpha, \tag{2.35}$$

$$\Phi_z = 0,$$

from which we conclude that vector $\mathbf{\Phi}$ has the following norm:

$$\Phi = \frac{1}{8}\rho g H^2 \frac{c}{2}\left[1 + \frac{2kd}{\sinh(2kd)}\right], \tag{2.36}$$

and has the same direction as the local s-axis.

2.4.5 *The formula for $\mathscr{E}$*

With the definition (2.10) of e, equation (2.25) becomes

$$\mathscr{E} = \frac{1}{T}\int_0^T\int_0^\eta \rho g z \, \mathrm{d}z \, \mathrm{d}t + \frac{1}{T}\int_0^T\int_{-d}^\eta \frac{1}{2}\rho\left(v_s^2 + v_z^2\right) \mathrm{d}z \, \mathrm{d}t,$$

where the interval of integration of the second double integral can be reduced from to $(-d, \eta)$ to $(-d, 0)$, since the integrand is of order H^2. Then, it can be readily verified that each of the two double integrals (when divided by T) is equal to $\frac{1}{16}\rho g H^2$, so that

$$\mathscr{E} = \frac{1}{8}\rho g H^2. \tag{2.37}$$

2.4.6 *Comment: even the radiation stress tensor can be obtained with only the linear wave theory*

In the analysis leading to the formulae for tensor $\mathbf{R}$, vector $\mathbf{\Phi}$ and scalar $\mathscr{E}$, we have resorted once to the terms of order H^2 of η and ϕ; specifically when we have evaluated integral I_2 which was necessary to get the formulae for R_{xx} and R_{yy}. And even in this occasion, we have used only the nonlinear terms of ϕ, not those of η. Here below we show that, thanks to an alternative procedure, we can also obtain I_2 without resorting to the nonlinear terms of ϕ. Our goal is to prove that the nonlinear wave theory *is not necessary* to deal with phenomena like the transformation of waves on shallow water, the variation of the mean water level and the longshore drift, whose treatment is based on tensor $\mathbf{R}$ and vector $\mathbf{\Phi}$.

If we change the order of the two integrals in equation (2.30) of I_2, we can rewrite this equation in the form

$$I_2 = \int_{-d}^{0} <p> \mathrm{d}z, \tag{2.38}$$

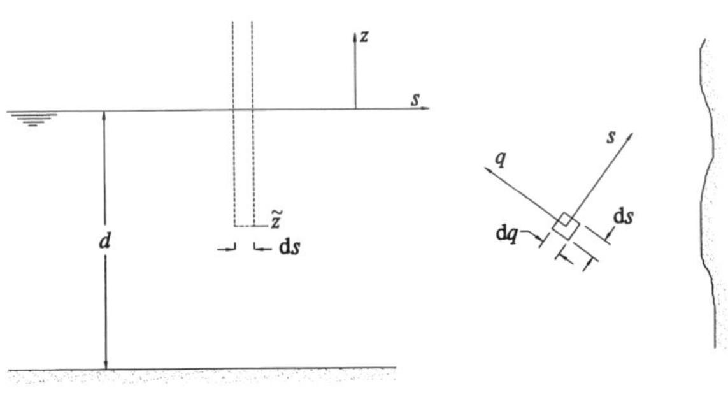

Fig. 2.6 The small control volume used to obtain $< p >$ at a given depth beneath the mean water level.

where $<p>$ represents the average pressure being dependent on z. To get $<p>$ we can use (2.31) of p (and this is the straight path of sect. 2.4.3), or we can also apply the linear momentum equation to the small control volume of fig. 2.6. The z-component of this equation when averaged over a wave period yields

$$\rho g \tilde{z}\, \mathrm{d}s\, \mathrm{d}q + < p >_{z=\tilde{z}} \mathrm{d}s\, \mathrm{d}q = -\rho < v_z^2 >_{z=\tilde{z}} \mathrm{d}s\, \mathrm{d}q, \tag{2.39}$$

where the two terms on the l.h.s. are respectively the mean weight and the mean vertical force on the base, and the term on the r.h.s. is the mean flux of linear momentum (z-component). Being $\tilde{z}$ arbitrary, equation (2.39) is reduced to

$$< p > = -\rho g z - \rho < v_z^2 > .$$

Here $< v_z^2 >$, exact to the order H^2, can be obtained with only the linear form of ϕ, and the result is

$$< p > = -\rho g z - \rho g^2 \frac{H^2}{8} \omega^{-2} k^2 \frac{\sinh^2[k(d+z)]}{\cosh^2(kd)} + \mathrm{o}(H^2).$$

Now, substitute the r.h.s. of this equation into (2.38). Evaluating the integral and using the linear dispersion rule, you will re-obtain formula (2.32) for I_2 without having resorted to the nonlinear terms of ϕ.

2.5 The problem of the control volume extending from deep to shallow water

2.5.1 *Application of the energy equation*

Let us consider the control volume of fig. 2.7, on a seabed with some x-parallel contour lines. Since the cross section of the bottom is constant along the x-axis, also the wave height, the wave direction and the mean characteristics of the wave

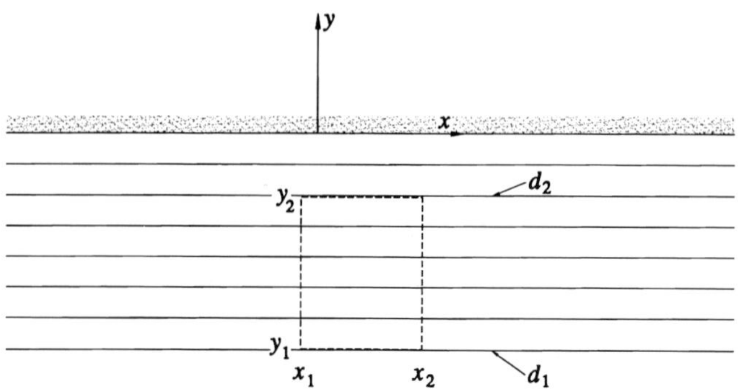

Fig. 2.7 The control volume extending from deep to shallow water, for the basic condition of straight contour lines.

motion are independent of x and depend only on y. The period is the same everywhere, indeed the period of the waves on the time domain at any point is equal to the period of the wavemaker (recalling that the periodic waves are generated by the periodic swing of a wavemaker). We shall call H_1 and α_1 respectively the wave height and the angle between the wave direction and x-axis on depth d_1; and we shall call H_2 and α_2 the wave height and the angle between the wave direction and x-axis on depth d_2.

Given that vector $\mathbf{\Phi}$ is constant along the x-axis, the compact notation $\Phi_y(y)$ is used in place of $\Phi_y(x, y)$, and (2.24) is reduced to

$$\Phi_y(y_2) = \Phi_y(y_1). \tag{2.40}$$

Hence, using formula (2.35) for Φ_y, we have

$$H_2^2 c_2 \left[1 + \frac{(2kd)_2}{\sinh(2kd)_2}\right] \sin\alpha_2 = H_1^2 c_1 \left[1 + \frac{(2kd)_1}{\sinh(2kd)_1}\right] \sin\alpha_1 .$$

Finally, if the offshore side of the control surface is in deep water and the nearshore side is in a water depth d, we have

$$H^2 \tanh(kd) \left[1 + \frac{2kd}{\sinh(2kd)}\right] \sin\alpha = H_0^2 \sin\alpha_0, \tag{2.41}$$

where, of course, H and k are wave height and wave number on water depth d.

2.5.2 *Application of the linear momentum equation* (x-*component*)

Given that

(i) R_{xx} and R_{yx} are constant with respect to x, for a fixed y;

(ii) F_{fx} is zero (because of the ideal flow assumption, and because the bottom slope is zero along the x-axis);

(2.22a) is reduced to

$$R_{yx}(y_2) = R_{yx}(y_1)$$

which, together with (2.29) of R_{yx}, yields

$$H_2^2 \left[1 + \frac{(2kd)_2}{\sinh(2kd)_2}\right] \sin\alpha_2 \cos\alpha_2 = H_1^2 \left[1 + \frac{(2kd)_1}{\sinh(2kd)_1}\right] \sin\alpha_1 \cos\alpha_1 .$$

If the plane $y = y_1$ of the control surface is in deep water and the plane $y = y_2$ is in a water depth d, we have

$$H^2 \left[1 + \frac{2kd}{\sinh(2kd)}\right] \sin\alpha \cos\alpha = H_0^2 \sin\alpha_0 \cos\alpha_0 . \tag{2.42}$$

2.5.3 *Solution for angle α and height H on given water depth d*

The two equations (2.41) and (2.42) enable us to obtain the two unknowns, that is to say H and α. From (2.41) we get the expression for H^2 and use it in (2.42). After these two operations, we obtain

$$\cos\alpha = \cos\alpha_0 \tanh(kd) = \cos\alpha_0\, c/c_0 \tag{2.43}$$

that is the Snell law of refraction which here has been re-obtained within fluid mechanics. Such an equation enables us to obtain angle α on water depth d, once angle α_0 on deep water is known (bear in mind that, if the wave travels landward on deep water, it travels landward also on shallow water, and vice versa; so that α and α_0 belong to the same quadrant).

Referring to the basic case in which the wave travels landward, angles α_0 and α range between 0 and π, and thus

$$\sin\alpha_0 = \sqrt{1 - \cos^2\alpha_0},$$

$$\sin\alpha = \sqrt{1 - \cos^2\alpha},$$

that is, for (2.43),

$$\sin\alpha = \sqrt{1 - \tanh^2(kd)\cos^2\alpha_0}. \tag{2.44}$$

At this stage, with $\cos\alpha$ and $\sin\alpha$ being known, we can operate on either (2.41) or (2.42) to obtain H. The result is

$$H = H_0 \sqrt{\frac{\sinh(2kd)}{\tanh(kd)[\sinh(2kd) + 2kd]}}\;\sqrt[4]{\frac{1-\cos^2\alpha_0}{1-\tanh^2(kd)\cos^2\alpha_0}}. \tag{2.45}$$

This equation enables us to get wave height H on water depth d, once height H_0 and angle α_0 on deep water are known.

2.5.4 *Application of the linear momentum equation (y-component)*

Given that both R_{yy}, R_{xy}, and $<f_{fy}>$ [the mean force exerted by the seabed per unit length of contour lines] are constant with respect to x, (2.22*b*) reduces itself to

$$<f_{fy}> = -R_{yy}(y_1) + R_{yy}(y_2). \tag{2.46}$$

Equation (2.46) can be satisfied only if the water depth undergoes some variations of order H^2, which means that, under the wave motion, the water depth becomes $d + \Delta$ (d being the depth of still water) with $\Delta O(H^2)$ depending on d. With these variations of the water depth, we have

$$<f_{fy}> = \int_{y_1}^{y_2} \rho g (d+\Delta)\frac{\mathrm{d}d}{\mathrm{d}y}\,\mathrm{d}y = \frac{1}{2}\rho g (d_2^2 - d_1^2) + \rho g \int_{d_1}^{d_2} \Delta\,\mathrm{d}d,$$

$$R_{yy}(y_1) = \frac{1}{2}\rho g\, d_1^2 + \rho g\, d_1\, \Delta_1 + \frac{1}{16}\rho g\, H_1^2\, [\cdot]_1,$$

$$R_{yy}(y_2) = \frac{1}{2}\rho g\, d_2^2 + \rho g\, d_2\, \Delta_2 + \frac{1}{16}\rho g\, H_2^2\, [\cdot]_2,$$

where

$$[\cdot] \equiv \left[\frac{2kd}{\sinh(2kd)}(1+\sin^2\alpha) + \sin^2\alpha\right].$$

Hence, (2.46) yields

$$\int_{d_1}^{d_2} \Delta\, \mathrm{d}d = d_2\Delta_2 - d_1\Delta_1 + \frac{1}{16}H_2^2[\cdot]_2 - \frac{1}{16}H_1^2[\cdot]_1$$

which can be rewritten in the form

$$\int_{d_2}^{d_1} d\,\frac{\mathrm{d}\Delta}{\mathrm{d}d}\,\mathrm{d}d = \frac{1}{16}H_2^2[\cdot]_2 - \frac{1}{16}H_1^2[\cdot]_1.$$

If the plane $y = y_1$ of the control surface is in deep water and the plane $y = y_2$ is in a water depth d, the last equation yields

$$\int_d^\infty d'\,\frac{\mathrm{d}\Delta'}{\mathrm{d}d'}\,\mathrm{d}d' = \frac{1}{16}H^2\left[\frac{2kd}{\sinh(2kd)}(1+\sin^2\alpha) + \sin^2\alpha\right] - \frac{1}{16}H_0^2\,\sin^2\alpha_0, \tag{2.47}$$

where Δ' denotes the variation of the mean water level on water depth d'.

It is convenient to express wave height H and angle α on water depth d in terms of H_0 and α_0 on deep water. Doing so, we rewrite (2.47) in the form

$$\int_d^\infty d'\,\frac{\mathrm{d}\Delta'}{\mathrm{d}d'}\,\mathrm{d}d' = \frac{1}{16}H_0^2\,F_2(kd), \tag{2.48}$$

defining

$$F_2(kd) \equiv \frac{\sinh(2kd)}{\tanh(kd)\,[\sinh(2kd) + 2kd]}\sqrt{\frac{1-\cos^2\alpha_0}{1-\tanh^2(kd)\cos^2\alpha_0}}\left\{\frac{2kd}{\sinh(2kd)}\cdot\right.$$

$$\left.\cdot\,[2-\tanh^2(kd)\cos^2\alpha_0] + 1 - \tanh^2(kd)\cos^2\alpha_0\right\} - \sin^2\alpha_0. \tag{2.49}$$

Differentiating both sides of (2.48) with respect to d, we obtain

$$\frac{\mathrm{d}\Delta}{\mathrm{d}d} = -\frac{1}{d}\frac{1}{16}H_0^2\frac{\mathrm{d}}{\mathrm{d}d}F_2(kd), \tag{2.50}$$

which can be given the nondimensional form

$$\frac{\mathrm{d}(\Delta/H_0)}{\mathrm{d}(d/L_0)} = -\frac{2\pi}{16}\frac{H_0}{L_0}\frac{1}{d/L_0\,\tanh(kd)}\left[1-\frac{kd}{\sinh(kd)\cosh(kd)+kd}\right]\dot{F}_2(kd), \tag{2.51}$$

where $\dot{F}_2(kd)$ is the derivative of $F_2(kd)$ with respect to kd, which is automatically deducible from (2.49).

Equation (2.51) gives the derivative of Δ/H_0 as a function of d/L_0. Hence, the function Δ/H_0 can be obtained by numerical integration. Clearly, we get Δ/H_0 less a constant, and this constant is then obtained from the principle of conservation of mass in the whole basin.

For the step from (2.50) to (2.51), use the chain rule:

$$\frac{\mathrm{d}}{\mathrm{d}d} F_2(kd) = \left(\frac{\mathrm{d}k}{\mathrm{d}d} d + k \right) \dot{F}_2(kd),$$

and obtain the formula for $\mathrm{d}k/\mathrm{d}d$ from the linear dispersion rule.

2.6 Practical consequences of the control volume problem

2.6.1 *Shoaling and refraction*

Fig. 2.8 shows the quotient H/H_0 as a function of d/L_0 [equation (2.45)] for $\alpha_0 = \pi/2$, that is for the wave direction on deep water being orthogonal to the contour lines. In this case the wave direction keeps orthogonal to the contour lines at any depth. Indeed we can readily verify by means of (2.43) that α is equal to $\pi/2$ whichever the d, if $\alpha_0 = \pi/2$. We shall call *shoaling curve* the graph of H/H_0 as a function of d/L_0, for $\alpha_0 = \pi/2$. This function has a minimum of 0.91 for d/L_0 of about 0.15; it gradually approaches 1 for increasing d/L_0; and it should tend to infinity as d/L_0 approaches zero. We use the conditional mood ("it should tend") since, as we shall see in the next section, the wave cannot survive on a water depth smaller than a critical threshold.

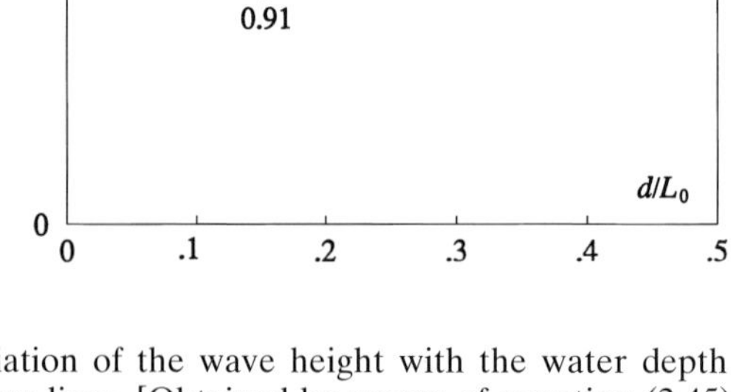

Fig. 2.8 Shoaling: variation of the wave height with the water depth if the wave direction is orthogonal to the contour lines. [Obtained by means of equation (2.45).]

If the wave direction on deep water makes an angle $\alpha_0 \neq \pi/2$ with the contour lines, then α approaches $\pi/2$ as the water depth decreases. Indeed from (2.43) we deduce that the smaller $|\cos\alpha|$ is than $|\cos(\alpha_0)|$, the smaller is d/L_0. In other words, whichever the wave direction on deep water, the wave direction on shallow water tends to become orthogonal to the shoreline. The variation of the wave direction on shallow water is called *wave refraction*, and it can be easily foreseen by means of (2.43) [see fig. 2.9 for a numerical example].

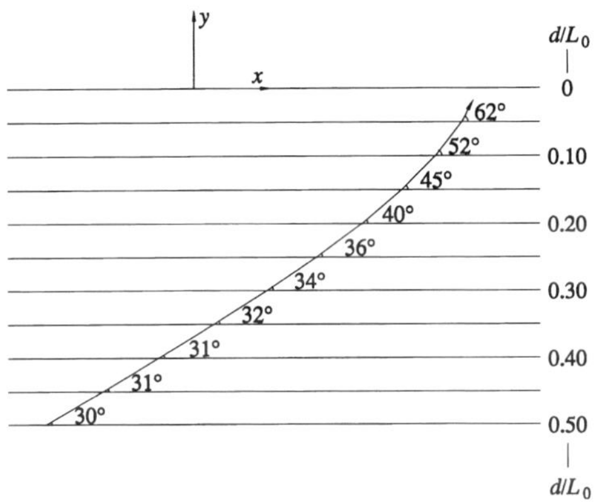

Fig. 2.9 Refraction: variation of the wave direction from deep water to shoreline. [Obtained by means of equation (2.43).]

We shall call *shoaling-refraction* curves the graphs of H/H_0 vs d/L_0 for the general cases of $\alpha_0 \neq \pi/2$. Two examples are given in fig. 2.10. We see that, for a fixed d/L_0, the larger $|\pi/2 - \alpha_0|$ is, the smaller is H/H_0. Indeed from (2.45) it follows that $H/H_0 \to 0$ as $\alpha_0 \to 0$ or $\alpha_0 \to \pi$ for any fixed d/L_0.

Sometimes we happen to see some high waves far off the coast and the sea nearly calm near the beach. Usually, this phenomenon is just due to the fact that H/H_0 gets very small on shallow water since α_0 is close to 0 or π (0 or π, according to whether the wave attack is from the right side or from the left side). Indeed, in these cases we see the wave direction on deep water is nearly parallel to the coastline.

2.6.2 *Where the wave breaks*

When the ratio between the wave height and the water depth exceeds a critical threshold, the wave breaks. Here we shall assume 0.8 as a typical value for this critical ratio.

A too large ratio between the wave height and the water depth is not the only cause of wave breaking. Another possible cause is the excess of wave steepness:

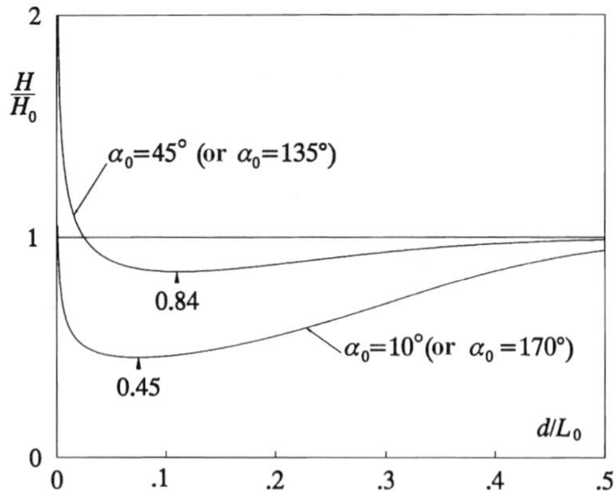

Fig. 2.10 Shoaling-refraction: variation of the wave height with the water depth for two distinct wave directions on deep water. [Obtained by means of equation (2.45).]

the wave breaks if the ratio H/L exceeds a critical threshold. Here, following Miche (1944) we shall assume this threshold to be equal to $0.14 \tanh(kd)$.

In summary, the two breaking conditions are

$$\begin{aligned} &1^{\text{st}} \text{ breaking condition:} \quad H = 0.8d, \\ &2^{\text{nd}} \text{ breaking condition:} \quad H = 0.14 \tanh(kd) L \end{aligned} \tag{2.52}$$

which can be rewritten in the form

$$1^{\text{st}} \text{ breaking condition:} \quad \frac{H}{H_0} = \frac{0.8}{H_0/L_0} \frac{d}{L_0}, \tag{2.53a}$$

$$2^{\text{nd}} \text{ breaking condition:} \quad \frac{H}{H_0} = \frac{0.14}{H_0/L_0} \tanh^2(kd). \tag{2.53b}$$

It follows that the wave is stable, that is, it does not break, if

$$\frac{H}{H_0} < \text{Min}\left(\frac{0.8}{H_0/L_0} \frac{d}{L_0}, \frac{0.14}{H_0/L_0} \tanh^2(kd)\right), \tag{2.54}$$

where the symbol $\text{Min}(a, b)$ stands for "the smaller of a or b". For a given H_0/L_0, the r.h.s. of (2.54) is a function of d/L_0 we shall call the *limit of stability*. The breaking depth d_b is that for which the shoaling-refraction curve intersects the limit of stability.

The limit of stability consists of two branches: branch A (the one on the smaller d/L_0) is given by (2.53*a*), and branch B is given by (2.53*b*). If the shoaling-refraction curve intersects branch A, the breaking occurs because the wave height is too large with respect to the water depth. While, if the shoaling-refraction curve intersects branch B, the wave breaking occurs because the wave steepness is too large. A case of wave breaking due to an excessively large H/d is shown in fig. 2.11, and a case of wave breaking due to an excessive wave steepness is shown in fig. 2.12.

Of course, the solutions we have obtained in the limit as $H \to 0$ for fixed d and T lead to the worst predictions near the breaking depth. However, the prediction for what concerns shoaling and refraction proves to be rather effective up to the breaker line.

2.6.3 *The set-down*

The two figures 2.11 and 2.12 show also the variation Δ of the mean water level. We see Δ is negative and its absolute value grows as the water depth decreases. In other words, the mean water level goes down from deep to shallow water. The lowering is rather small (please note that the scale factor of Δ/H_0 in figs 2.11 and 2.12 is ten times smaller than the scale factor of H/H_0). The graph of Δ/H_0 has been truncated at the breaking depth, since after the wave breaks the flow substantially changes and Δ starts growing, as we shall see in sect. 3.2. Thus the lowering at the breaking depth proves to be the largest one, and its absolute value is called set-down (S_D).

As to the calculation, Δ/H_0 is obtained, less a constant, by numerical integration of (2.51). Then the principle of conservation of mass enables us to obtain the constant as well. Indeed, for this principle, the lowering of the mean water level on shallow water must be counterbalanced by an increase on deep water. However, the extent of the sea on deep water $\left(d/L_0 > \dfrac{1}{2}\right)$ is so large that even a very low increase is sufficient to compensate for the lowering on shallow water. In concrete terms, this means the asymptotic value of Δ/H_0 as $d/L_0 \to \infty$ $\left(\text{in practice for } d/L_0 > \dfrac{1}{2}\right)$ can be assumed to be equal to zero.

2.7 A current associated with the wave motion

2.7.1 *The mean flow rate*

Let us go on assuming the contour lines to be parallel to the x-axis, and let us deal with the mass transport due to the wave motion. The mean rate of flow per unit length through a vertical plane orthogonal to the y-axis is

$$Q \equiv < \int_{-d}^{\eta} v_y \, \mathrm{d}z > ,$$

and proves to be the sum of two terms of order H^2:

$$Q_w \equiv < \int_0^{\eta} v_y' \mathrm{d}z > , \tag{2.55a}$$

$$Q_c \equiv < v_y'' > d, \tag{2.55b}$$

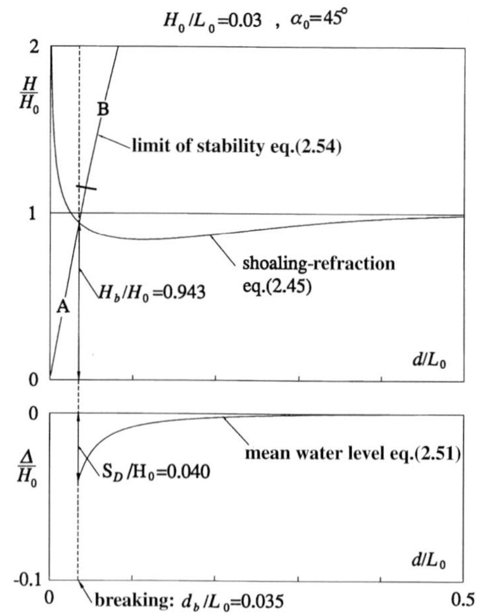

Fig. 2.11 Solution to the problem of the breaking depth. Here the wave breaks because the ratio between wave height and water depth is too large.

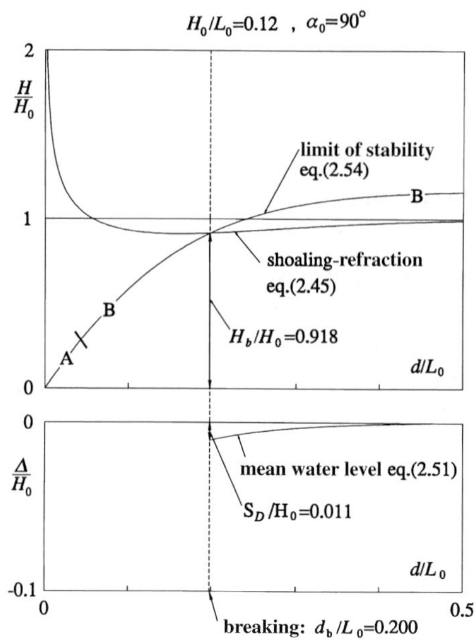

Fig. 2.12 Solution to the problem of the breaking depth. Here the wave breaks because the ratio between wave height and wavelength is too large.

where Q_w is known, while Q_c is unknown since

$$< v_y'' > = C \sin \alpha , \tag{2.56}$$

C being the still unknown constant in the formula (2.26*b*) for the velocity potential.

Q_w is due to the horizontal particle displacement of the wave motion (hence the subscript w which stands for *wave*). Really in sect. 1.6.2 we have seen that, in half a wave period, a water particle is subject to a positive horizontal displacement D_o of order H, and in the following half period it is subject to a compensating negative horizontal displacement $-D_o$. But if we expand the solution as far as the second order, we find that the sum of the positive horizontal displacement and the negative horizontal displacement is no longer 0, and indeed it is equal to a positive term of order H^2. In other words, the water particle moves forward along the wave direction, and the connected discharge per unit length is equal to Q_w. As to Q_c, it is the flow rate per unit length of a steady current (hence the subscript c which stands for *current*) which is necessary to ensure the conservation of mass.

2.7.2 *In front of a coast*

In the case of a coast, that is the case of fig. 2.7, the mean flow rate Q must be zero on any water depth (otherwise, we would have a progressive increase or decrease of the water mass nearshore). Accordingly, we have the situation depicted in fig. 2.13*a*: the wave motion produces a mass transport (average discharge Q_w) towards the surf zone where a steady seaward current originates, whose flow rate Q_c is the opposite of Q_w.

Analytically:

$$Q_c = -Q_w ,$$

and, from this equality and (2.55*a*-*b*), we get

$$< v_y'' > = -\frac{1}{d} g \frac{H^2}{8} \omega^{-1} k \sin \alpha .$$

Since we are able to calculate H and α on any fixed water depth d once H_0 and α_0 are given, the last equation enables us to get $< v_y'' >$ and (at last!) the constant C of the velocity potential (C being related to $< v_y'' >$ by 2.56).

2.7.3 *In a strait connecting two seas*

Let us now deal with a strait connecting two seas, we assume the strait to be a straight channel of constant width, passing from deep water $\left(d/L_0 > \frac{1}{2}\right)$ to shallow water, and once again to deep water. In this case the wave motion produces a mass transport from one sea to the other, and only a local current arises on shallow water [cf. fig. 2.13*b*]. This current is necessary to ensure the conservation of mass,

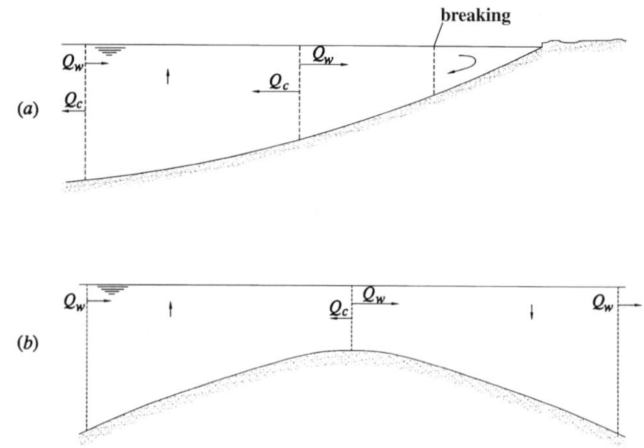

Fig. 2.13 (*a*) Before a coast: the average discharge Q_w of the wave motion is counterbalanced by a steady current seaward. (*b*) In a channel with a saddle: the variation of the average discharge Q_w, along the longitudinal axis, is counterbalanced by some steady currents near the saddle.

given that Q_w varies along the channel axis because of the changes in wave height and wavelength.

Specifically, Q_c on water depth d must counterbalance the difference between Q_w of deep water and Q_w of depth d:

$$Q_c = Q_{w0} - Q_w,$$

that is

$$< v_y'' > d = \left[< \int_0^\eta v_y' \mathrm{d}z > \right]_{\text{deep water}} - \left[< \int_0^\eta v_y' \mathrm{d}z > \right]_{\text{water depth } d}, \tag{2.57}$$

which yields

$$< v_y'' > = \frac{1}{d} \left(\frac{2\pi}{8} \frac{H_0^2}{T} - g \frac{H^2}{8} \omega^{-1} k \right).$$

The constant C of the velocity potential is then obtained thanks to (2.56).

On examining figures 2.13*a-b*, one should bear in mind that the major portion of Q_w is near the water surface, while Q_c is the flow rate of a uniform current (indeed we know $< v_y'' >$ is constant with respect to z). One should also bear in mind that the vertical flow connected with the variations of Q_w and Q_c along the y-axis gives rise to negligible vertical velocities, of order $H^2 \tan\lambda$, with $\tan\lambda$ being very small because of the assumption of a gentle slope.

2.8 Wave refraction for an arbitrary configuration of the seabed

2.8.1 *The differential equation of the wave orthogonal*

In sect. 2.5 we solved the problem of the control volume extending from deep to shallow water, for the basic case of straight contour lines. Here we deal with the same problem for the case of arbitrary contour lines. To this end, it is convenient to solve preliminarily the refraction problem.

Let us fix a point P in the horizontal plane, and let us define the natural coordinates: s with the local wave direction and q orthogonal to s. The inclination of a small stretch $\mathrm{d}q$ of wavefront varies in a small time interval $\mathrm{d}t$ of

$$\mathrm{d}\alpha = \frac{c\,\mathrm{d}t - \left(c + \dfrac{\partial c}{\partial q}\,\mathrm{d}q\right)\mathrm{d}t}{\mathrm{d}q} = -\frac{\partial c}{\partial q}\,\mathrm{d}t, \tag{2.58}$$

where α and c denote respectively the angle of the wavefront and the propagation speed at point P [see fig. 2.14]. Since

$$\mathrm{d}s = c\,\mathrm{d}t,$$

(2.58) is rewritten as

$$\frac{\mathrm{d}\alpha}{\mathrm{d}s} = -\frac{1}{c}\frac{\partial c}{\partial q}.$$

Here it is convenient to express $\partial c/\partial q$ in terms of the derivatives $\partial c/\partial x$ and $\partial c/\partial y$ (x and y being as usual the fixed axes). Since

$$\frac{\partial c}{\partial q}\,\mathrm{d}q = \frac{\partial c}{\partial x}(-\mathrm{d}q\sin\alpha) + \frac{\partial c}{\partial y}\,\mathrm{d}q\cos\alpha,$$

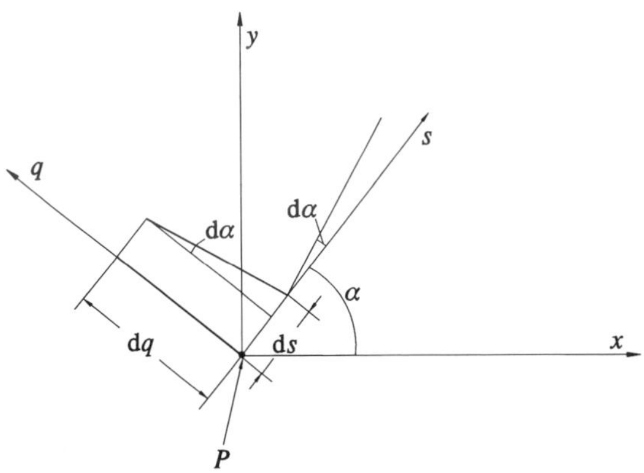

Fig. 2.14 Refraction: the short stretch $\mathrm{d}q$ of wave crest, covering the distance $\mathrm{d}s$, rotates through $\mathrm{d}\alpha$.

it follows that

$$\frac{d\alpha}{ds} = \frac{1}{c}\left(\frac{\partial c}{\partial x}\sin\alpha - \frac{\partial c}{\partial y}\cos\alpha\right). \tag{2.59}$$

As we shall see later, equation (2.59) enables us to get the wave orthogonal, that is the curve whose tangent vector gives the local wave direction. Specifically, given the wave direction on deep water and the wave period, we can obtain the wave orthogonal passing at any fixed point P.

2.8.2 *A more convenient form of the differential equation*

Applying the chain rule, we rewrite (2.59) in the form

$$\frac{d\alpha}{ds} = \frac{1}{c}\left(\frac{dc}{dd}\frac{\partial d}{\partial x}\sin\alpha - \frac{dc}{dd}\frac{\partial d}{\partial y}\cos\alpha\right).$$

Hence, obtaining the formula for dc/dd from (1.28) of L, we arrive at

$$\frac{d\alpha}{ds} = \frac{2k}{\sinh(2kd) + 2kd}\left(\frac{\partial d}{\partial x}\sin\alpha - \frac{\partial d}{\partial y}\cos\alpha\right) \tag{2.60}$$

which is more suitable for the finite difference technique.

To appreciate the effectiveness of (2.60) one should simply try to calculate $d\alpha/ds$ with the finite difference technique at any node i, j (coordinates: x_i, y_j) of a grid (Δx, Δy), for a flat bottom such that

$$\frac{\partial d}{\partial y} = \text{const}, \quad \frac{\partial d}{\partial x} = 0.$$

With equation (2.59) one finds

$$\left(\frac{d\alpha}{ds}\right)_{ij} = -\frac{1}{c_{ij}}\frac{c_{i,j+1} - c_{i,j-1}}{2\Delta y}\cos\alpha_{ij},$$

while with equation (2.60) one finds

$$\left(\frac{d\alpha}{ds}\right)_{ij} = -\frac{2k_{ij}}{\sinh(2kd)_{ij} + (2kd)_{ij}}\frac{d_{i,j+1} - d_{i,j-1}}{2\Delta y}\cos\alpha_{ij} =$$

$$= -\frac{2k_{ij}}{\sinh(2kd)_{ij} + (2kd)_{ij}}\,\text{const}\cos\alpha_{ij}.$$

Here it can be easily verified that

(i) the value obtained by means of (2.59) is approximate because c is not a linear function of d;

(ii) the value obtained by means of (2.60) is exact.

For an arbitrary bottom, the use of (2.60) with the finite difference technique leads to only one error which is due to the fact that generally d is not a linear function of x and/or y. While, the use of (2.59) leads to two errors: the first one because d is not a linear function of x and/or y, and the second one because c is not a linear function of d.

2.8.3 *Solution to the differential equation*

For a wave orthogonal, α, x and y are functions of s. We have

$$\alpha(s+\mathrm{d}s) = \alpha(s) + \frac{\mathrm{d}\alpha}{\mathrm{d}s}\,\mathrm{d}s + \mathrm{o}(\mathrm{d}s), \tag{2.61a}$$

$$x(s+\mathrm{d}s) = x(s) + \frac{\mathrm{d}x}{\mathrm{d}s}\,\mathrm{d}s + \frac{1}{2}\frac{\mathrm{d}^2x}{\mathrm{d}s^2}\,\mathrm{d}s^2 + \mathrm{o}(\mathrm{d}s^2), \tag{2.61b}$$

$$y(s+\mathrm{d}s) = y(s) + \frac{\mathrm{d}y}{\mathrm{d}s}\,\mathrm{d}s + \frac{1}{2}\frac{\mathrm{d}^2y}{\mathrm{d}s^2}\,\mathrm{d}s^2 + \mathrm{o}(\mathrm{d}s^2). \tag{2.61c}$$

Moreover we have

$$\frac{\mathrm{d}x}{\mathrm{d}s} = \cos\alpha, \qquad \frac{\mathrm{d}y}{\mathrm{d}s} = \sin\alpha,$$

and consequently

$$\frac{\mathrm{d}^2x}{\mathrm{d}s^2} = -\sin\alpha\,\frac{\mathrm{d}\alpha}{\mathrm{d}s}, \qquad \frac{\mathrm{d}^2y}{\mathrm{d}s^2} = \cos\alpha\,\frac{\mathrm{d}\alpha}{\mathrm{d}s}.$$

Substituting these expressions of $\mathrm{d}x/\mathrm{d}s$, $\mathrm{d}y/\mathrm{d}s$, $\mathrm{d}^2x/\mathrm{d}s^2$ and $\mathrm{d}^2y/\mathrm{d}s^2$ in equations (2.61*a*-*b*-*c*), and resorting to the finite difference technique we obtain

$$\alpha(s+\Delta s) = \alpha(s) + \frac{\mathrm{d}\alpha}{\mathrm{d}s}\,\Delta s, \tag{2.62a}$$

$$x(s+\Delta s) = x(s) + \cos\alpha\Delta s - \frac{1}{2}\sin\alpha\,\frac{\mathrm{d}\alpha}{\mathrm{d}s}\,\Delta s^2, \tag{2.62b}$$

$$y(s+\Delta s) = y(s) + \sin\alpha\,\Delta s + \frac{1}{2}\cos\alpha\,\frac{\mathrm{d}\alpha}{\mathrm{d}s}\,\Delta s^2. \tag{2.62c}$$

In conclusion, given the angle α_0 and an origin on deep water, the wave orthogonal is calculated with finite increments Δs through (2.62*a*-*b*-*c*). The derivative $\mathrm{d}\alpha/\mathrm{d}s$ in these equations is given by (2.60), and the partial derivatives $\partial d/\partial x$ and $\partial d/\partial y$ in (2.60) are evaluated by means of the finite difference technique.

2.8.4 *Calculation of the wave orthogonals: an example*

Let us consider the contour lines of fig. 2.15. The water depth is symmetrical with respect to the y-axis, that is:

$$d(-x,y) = d(x,y).$$

Therefore, it suffices to indicate the form of $d(x, y)$ only for $x \geq 0$. Referring to the symbols of the figure we have

$$\text{shoreline} \begin{cases} y = R\sin\beta - \sqrt{R^2 - x^2} & \text{if } \; x \leq R\cos\beta, \\ y = 0 & \text{if } \; x \geq R\cos\beta. \end{cases}$$

As to the water depth:

$$d(x, y) \begin{cases} = \tilde{d}(x, y) & \text{if } \; y \leq R\sin\beta - x\tan\beta, \\ = \tilde{d}\left(\dfrac{R\sin\beta - y}{\tan\beta}, y\right) & \text{if } \; y \geq R\sin\beta - x\tan\beta, \end{cases} \tag{2.63}$$

where

$$\tilde{d}(x, y) \equiv \left[\sqrt{x^2 + (R\sin\beta - y)^2} - R\right]\tan\lambda.$$

Preliminarily, obtain the formulae for $\partial d/\partial x$ and $\partial d/\partial y$ from (2.63). Then these formulae will be used to compute $\mathrm{d}\alpha/\mathrm{d}s$ by means of (2.60).

Take $R = 600\,\mathrm{m}$, $\beta = 30°$, $T = 10\,\mathrm{s}$, and $\alpha_0 = 90°$. Then assume a value of the increment $\Delta s\,(\Delta s = 10\,\mathrm{m} \approx L_0/15$ is small enough). Finally, to obtain a wave orthogonal, fix a point x_0, y_0 of this orthogonal on deep water.

Three sequences are dealt with:

$$\alpha_n,\, x_n, \text{ and } y_n, \quad \text{for } n = 0,\, 1,\, 2,\, \ldots$$

Each n is associated with a new point of the orthogonal: x_n, y_n are the coordinates of this point, and α_n is the angle that the orthogonal makes with the x-axis at this point. The three sequences proceed straightforwardly from equations (2.62*a*-*b*-*c*):

$$\alpha_{n+1} = \alpha_n + \left(\frac{\mathrm{d}\alpha}{\mathrm{d}s}\right)_n \Delta s,$$

$$x_{n+1} = x_n + \cos\alpha_n\,\Delta s - \frac{1}{2}\sin\alpha_n \left(\frac{\mathrm{d}\alpha}{\mathrm{d}s}\right)_n \Delta s^2,$$

$$y_{n+1} = y_n + \sin\alpha_n\,\Delta s + \frac{1}{2}\cos\alpha_n \left(\frac{\mathrm{d}\alpha}{\mathrm{d}s}\right)_n \Delta s^2.$$

Procedure. First, by means of (2.60), compute $\left(\dfrac{\mathrm{d}\alpha}{\mathrm{d}s}\right)_0$, that is $\dfrac{\mathrm{d}\alpha}{\mathrm{d}s}$ at point x_0, y_0. Then, compute α_1, x_1, and y_1. Then compute $\left(\dfrac{\mathrm{d}\alpha}{\mathrm{d}s}\right)_1$, that is $\dfrac{\mathrm{d}\alpha}{\mathrm{d}s}$ at point x_1, y_1, and so on.

The derivatives $\partial d/\partial x$ and $\partial d/\partial y$ prove to be discontinuous at the dashed lines (see fig. 2.15*a*), and consequently also the curvature of the wave orthogonal is discontinuous at these lines. A set of wave orthogonals for the given input data is shown in fig. 2.15*b*, and a new set for $\alpha_0 = 45°$ is shown in fig. 2.15*c*.

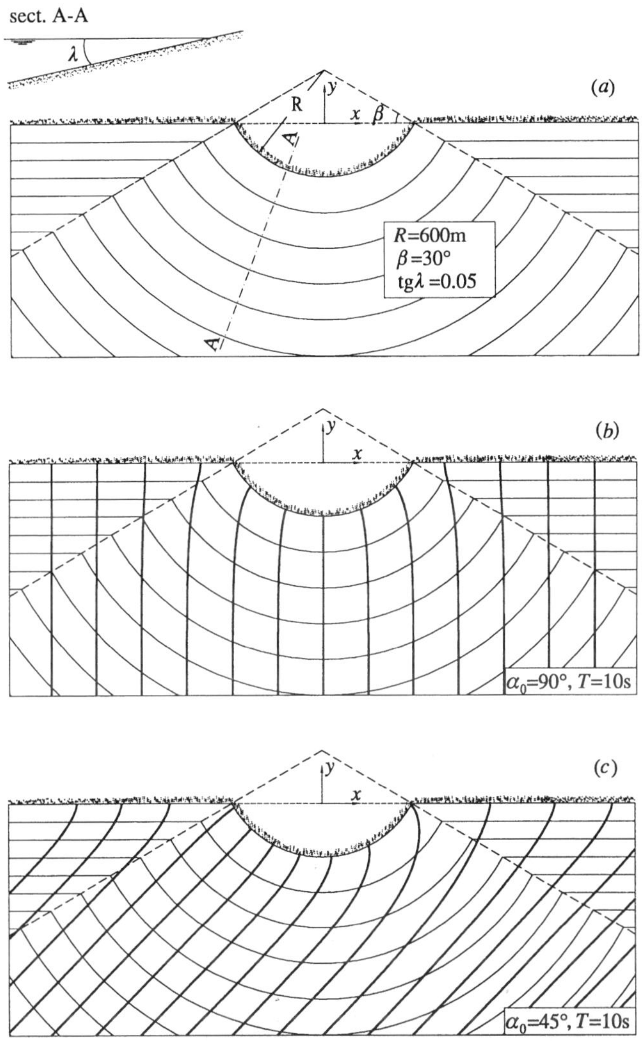

Fig. 2.15 (*a*) Contour lines before a promontory. (*b*) and (*c*) Wave refraction diagrams for two distinct wave directions on deep water. [Obtained by means of equations (2.60) and (2.62*a*-*b*-*c*).]

2.8.5 *The relation between the wave height and the distance between two orthogonals*

Let us consider a control volume whose horizontal section is bounded by two adjacent wave orthogonals (② and ③ in fig. 2.16) and two short stretches of wavefront (⓪ and ① in fig. 2.16).

Equation (2.19), for this control volume, yields

$$\Phi_1 b_1 - \Phi_0 b_0 = 0,$$

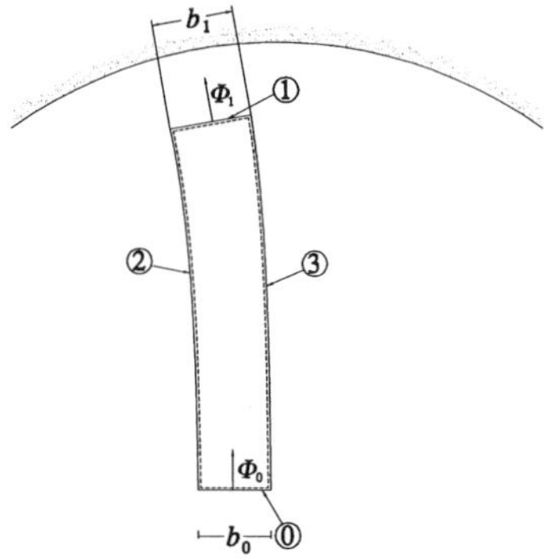

Fig. 2.16 If the contour lines are not straight, the control volume from deep to shallow water is bounded by two wave orthogonals and two short stretches of wavefront.

where b_0 and b_1 are the widths of the two stretches of wavefront and Φ_0 and Φ_1 are the norms of $\mathbf{\Phi}$. Hence, with the formula (2.36) for Φ, we obtain

$$H = H_0 \sqrt{\frac{\sinh(2kd)}{\tanh(kd)\,[\sinh(2kd) + 2kd]}}\sqrt{\frac{b_0}{b_1}},$$

where d, H and k are water depth, wave height and wave number at ①, and H_0 is the wave height at ⓪ on deep water. Clearly, this result requires b_1 to be small enough, for the wave height to be nearly constant on the stretch of wavefront.

2.9 The group celerity

2.9.1 *The time taken by the wave motion to cover an initially calm basin*

Fig. 2.17*a* shows a wave flume of constant depth. Let us assume the wavemaker is switched on at time $t = 0$. We wonder: how much time is needed for the waves to get to a fixed point at a distance y_o from the wavemaker. We call t_o this time.

To answer the question, we take a control volume extending from $y = 0$ (wavemaker) to $y = y_o$. In the time interval $0 < t < t_o$, the flow in the control volume is not periodic. Indeed, if we took some photographs at a regular time interval T (equal to the wave period) we would catch an evolving situation. Initially, we would see some waves close to the wavemaker with the rest of the tank being still calm. Then, we would see the wave zone widens gradually. As said above, this means the flow is not periodic and, consequently, equation (2.19) is not valid. On the contrary, equation (2.18) retains its validity, since it holds whether or not the flow is periodic.

Integrating (2.18) with respect to t on the interval $(0, t_o)$, we obtain

$$\int_0^{t_o}\int_{\tilde{A}} (p + \rho e)\mathbf{v}\cdot\mathbf{n}\,\mathrm{d}A\,\mathrm{d}t = \left[\int_{\widetilde{W}} \rho e\,\mathrm{d}W\right]_{t=0} - \left[\int_{\widetilde{W}} \rho e\,\mathrm{d}W\right]_{t=t_o}. \qquad (2.64)$$

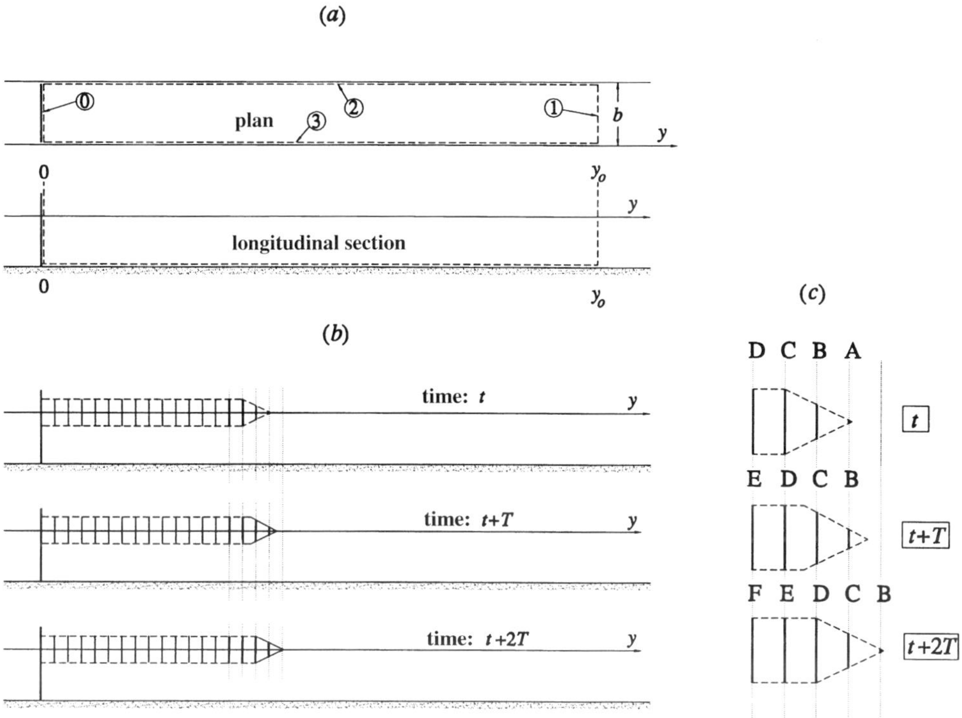

Fig. 2.17 (*a*) Plan view and longitudinal section of a long wave flume. (*b*) Three pictures taken a wave period from each other, while the wave motion advances on an initially still basin (the waves are sketched as vertical segments). (*c*) Details of the group's head.

Development of the l.h.s. of (2.64)

The lateral surface $\tilde{A}$ of the control volume consists of four parts: the cross-section ⓪ at the wavemaker; the cross-section ① at y_o; and the two sides ② and ③ of the wave flume. Planes ② and ③ do not provide any contribution to the integral on the l.h.s. of (2.64) since the scalar product $\mathbf{v} \cdot \mathbf{n}$ at these planes is zero. Neither plane ① gives any contribution to the integral, since $v = 0$ at this plane till instant t_o. Plane ⓪ remains. Here, at the origin of the wave flume, for the whole interval $(0,\ t_o)$ there is a wave of height H and period T, so that

$$\text{l.h.s. of (2.64)} = b \int_0^{t_o} \int_{-d}^{\eta} [(p + \rho e)(-v_y)]_{y=0} \, \mathrm{d}z \, \mathrm{d}t. \tag{2.65}$$

Assuming that t_o is very large with respect to T, we rewrite (2.65) as

$$\text{l.h.s. of (2.64)} = t_o \, b < \int_{-d}^{\eta} [(p + \rho e)(-v_y)]_{y=0} \mathrm{d}z >= -t_o \, b \, (\Phi_y)_{y=0}. \tag{2.66}$$

From which, using formula (2.35) for Φ_y, we arrive at the final form

$$\text{l.h.s. of (2.64)} = -t_o\, b\, \frac{1}{8}\, \rho g H^2\, \frac{c}{2} \left[1 + \frac{2kd}{\sinh(2kd)}\right].$$

Development of the r.h.s. of (2.64)

The r.h.s. of (2.64) represents the difference between the energy being present in the control volume at time $t = 0$ when the water is still calm, and the energy at time $t = t_o$ when the waves occupy the whole control volume. Therefore, we can write

$$\text{r.h.s. of (2.64)} = -b y_o\, \mathscr{E}, \tag{2.67}$$

and, using the formula (2.37) for $\mathscr{E}$,

$$\text{r.h.s. of (2.64)} = -b y_o\, \frac{1}{8}\, \rho g H^2.$$

Solution

Having got the expressions of the two sides of (2.64), we obtain

$$t_o = \frac{y_o}{\dfrac{c}{2}\left[1 + \dfrac{2kd}{\sinh(2kd)}\right]}. \tag{2.68}$$

2.9.2 *Comment on the mathematical solution*

As said, equation (2.66) calls for t_o to be very large with respect to the wave period; and similarly equation (2.67) calls for y_o to be very large with respect to the wavelength. As to (2.67), we also point out that $\mathscr{E}$ is not only the wave energy per unit surface, averaged over a wave period, but it is also the wave energy per unit surface, averaged over a wavelength at a fixed time instant. This second meaning has been applied when obtaining (2.67).

In simple words, the assumption leading to (2.66) and (2.67) is the same one that we use if we write

$$\int_0^{x_o} \sin^2 x\, dx = x_o\, \frac{1}{\pi} \int_0^{\pi} \sin^2 x\, dx,$$

which of course is exact if x_o is equal to $n\pi$ (with n an arbitrary integer), but it becomes exact as $x_o \to \infty$ even if x_o is not equal to $n\pi$.

The assumption of a very large t_o has been made implicitly also in another step of our reasoning. Specifically, it was made when we assumed the water at $y = y_o$ to be calm till t_o, and from t_o forth a wave of height H to occur at this location. In fact, at y_o, the water will be calm till an instant t_o'; then for two or three wave periods an unsteady flow will take place, with the wave height gradually increasing; then, from an instant t_o'' forth, the wave height will take on its final value H. But, if t_o is very large with respect to T, the duration of the transient gets very small with respect to t_o, and thus it can be neglected as we have done.

2.9.3 *The meaning of group celerity*

Equation (2.68) reveals that the propagation speed of a wave motion on a calm basin is

$$c_G = \frac{c}{2}\left[1 + \frac{2kd}{\sinh(2kd)}\right] \tag{2.69}$$

(this being the quotient y_o/t_o). Here it can be readily verified that c_G is generally smaller than c, and on deep water (the case illustrated by fig. 2.17) c_G is half the c. Let us see the reason for this.

Fig. 2.17*b* shows three instant pictures of the wave flume taken an interval T from each other. The waves are sketched as vertical segments: the height of the segment is equal to the wave height and the interval between two consecutive segments is equal to the wavelength. Each single wave advances by a wavelength L in a wave period T, so that its propagation speed is L/T [see the detail of the group's head in fig. 2.17*c*]. It is not so for the wave group which advances by a wavelength in two wave periods, so that its propagation speed is $c/2$. The propagation speed of the group is smaller than the propagation speed of each single wave, simply because each single wave goes to die at the group head. In particular, in the first picture, wave A is going to die; then in the third picture, two periods later, wave B is going to die; then it will be the turn of C, D and so on. [Of course the envelope front in fig. 2.17 has been somewhat simplified. For a more detailed representation see Miles (1962).]

It will be understood that subscript G is for *group*, and that c_G is the *group celerity* which plays a crucial role in the mechanics of the wind generated waves. Indeed, we shall see in chapter 10 that the wind-generated waves give rise to groups consisting of three or four waves, which travel with the celerity c_G. We shall see that in this case the waves are born at the group's tail, grow to a maximum height while reaching the envelope centre, and then go to die at the group's head. Such a phenomenon occurs because c_G, the propagation speed of the envelope, is smaller than the celerity c of each single wave.

The difference between spontaneous wave groups forming in the sea storms and the group of the periodic waves is great. It suffices to note that the group of the periodic waves, unlike the sea wave groups, grows longer and longer and its waves are already grown up at birth. Indeed, the periodic waves are to the wind generated waves as a simple animal species is to the human species. But, clearly the study of the elementary species is very helpful to understand the more sophisticated species. In particular we shall see that the quotient c_G/c for the sea wave groups is the same as for the elementary group of periodic waves.

2.10 Wave-current interaction. Part II: shoaling and set-down

2.10.1 *Preliminary analysis: only current without waves*

Let us consider a marine strait. To fix our ideas let us think of the Straits of Messina. Along the longitudinal axis, the water depth reduces itself to a minimum. In particular, in the Straits of Messina, the water depth is at a minimum (nearly 100 m) somewhat to the north of Messina. Often, some currents take place where the strait has its lowest depth. These currents in the Straits of Messina are due to the flow from the Ionian Sea to the Tyrrhenian Sea and vice versa.

Let us think of a strait as a straight channel of constant width, with a minimum water depth at $y = 0$ and with water depth tending to infinity as $y \to \pm\infty$. As in the problem of shoaling-refraction, let us assume the bottom slope to approach zero.

Let us analyse first the case of a current without waves, with a discharge Q per unit length. Referring to fig. 2.18 we call

S the difference between the still water level and the actual water level;

d the depth of the still water;

$\tilde{d} \equiv d - S$ the water depth;

$u = Q/\tilde{d}$ the velocity of the current.

Under ideal flow assumptions, the Bernoulli equation implies

$$S = \frac{u^2}{2g}. \tag{2.70}$$

As a consequence, u, d and Q are related to each other by

$$u = \frac{Q}{d - u^2/2g}$$

which can be rewritten in the form

$$\frac{ud}{Q} = 1 + \left(\frac{Q}{Q^*}\right)^2 \left(\frac{ud}{Q}\right)^3, \tag{2.71}$$

with

$$Q^* \equiv \sqrt{2g\,d}\;d.$$

Equality (2.71) admits two positive solutions for ud/Q, provided that

$$|Q| < Q_{\max} = \frac{2}{3\sqrt{3}}\,Q^*. \tag{2.72}$$

These are well known concepts from hydraulics. Clearly, here we are interested only in the first solution, that is in the smaller value of ud/Q satisfying (2.71). (In this connection, note that the second solution implies $\tilde{d} \to 0$ as $Q \to 0$.) Therefore, we must seek the lowest positive x for which the two functions

$$f_1(x) \equiv x,$$
$$f_2(x) \equiv 1 + \left(\frac{Q}{Q^*}\right)^2 x^3$$

are equal to each other. This particular x coincides with the sought value of ud/Q.

In the following section we shall deal with the problem of wave shoaling on a current. In preparation for this analysis it is useful to do a couple of preliminary exercises.

First exercise: verify the linear momentum equation for a control volume between two arbitrary ends y_1 and y_2 [see fig. 2.18]. What we have to do is to evaluate $< f_{fy} >$ and R_{yy}, and verify whether or not equality (2.46) is fulfilled.

Having assumed that the bottom slope approaches zero, the pressure distribution approaches the static form, so that we have

$$< f_{fy} > = \int_{y_1}^{y_2} \rho g \tilde{d} \frac{\mathrm{d}d}{\mathrm{d}y} \mathrm{d}y = \int_{d_1}^{d_2} \rho g \tilde{d} \, \mathrm{d}d, \tag{2.73}$$

$$R_{yy} = \int_{-\tilde{d}}^{0} -\rho g z + \rho u^2 \mathrm{d}z. \tag{2.74}$$

And, with these formulae for $< f_{fy} >$ and R_{yy}, equation (2.46) yields

$$\int_{d_1}^{d_2} \rho g \tilde{d} \mathrm{d}d = -\int_{-\tilde{d}_1}^{0} (-\rho g z + \rho u_1^2) \, \mathrm{d}z + \int_{-\tilde{d}_2}^{0} (-\rho g z + \rho u_2^2) \, \mathrm{d}z$$

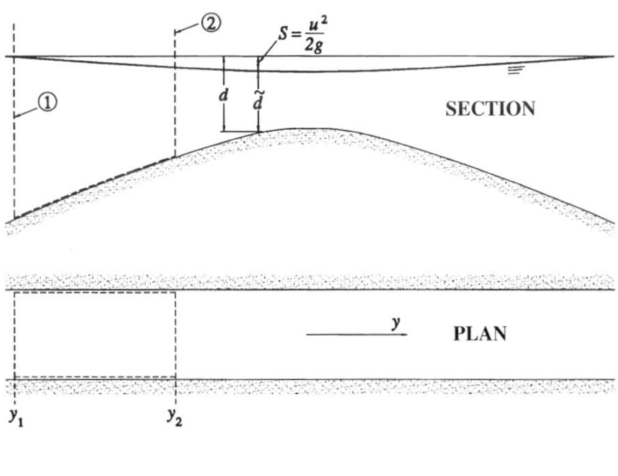

Fig. 2.18 Reference scheme for a steady current on a channel of varying depth. The bottom slope is assumed to approach zero.

which can be reduced to

$$-\int_{d_1}^{d_2} S\,\mathrm{d}d = d_2S_2 - d_1S_1 - 1.5S_2^2 + 1.5S_1^2 .$$

This equality can be proved numerically. To this end, you should fix two water depths d_1 and d_2 and a discharge Q satisfying inequality (2.72) for each d in (d_1, d_2), and you should evaluate S by means of (2.70) and (2.71).

Second exercise: verify the energy equation for a control volume between two arbitrary ends y_1 and y_2. What we have to do is to verify whether or not equality (2.40) is fulfilled.

In this case (current alone) (2.40) implies that

$$\int_{-\tilde{d}_1}^{0} \left\{-\rho g z + \rho\left[g(z - S_1) + \frac{1}{2}u_1^2\right]\right\} u_1\,\mathrm{d}z = \int_{-\tilde{d}_2}^{0} \left\{-\rho g z + \rho\left[g(z - S_2) + \frac{1}{2}u_2^2\right]\right\} u_2\,\mathrm{d}z , \tag{2.75}$$

where the l.h.s. is Φ_y on water depth d_1, and the r.h.s. is Φ_y on water depth d_2. The equality (2.75) is actually fulfilled since the two integrands are zero, which can be readily proved with the use of definition (2.70) of S.

It should be noted that the origin of the z-axis at each of the two ends of the control volume has been taken at the level of the free surface. In doing so, one is able to express the pressure simply as $-\rho g z$ both at y_1 and at y_2. But, pay attention: the potential energy must be expressed with one fixed reference level. That is why the potential energy per unit volume is $\rho g(z - S_1)$ at y_1, and $\rho g(z - S_2)$ at y_2; the fixed level being that of the free surface on infinite water depth.

2.10.2 *Effects of a variation of the mean water level on the linear momentum equation*

The presence of waves on the current causes a variation Δ of the mean water level. Such a variation will be of order H^2 will depend on d and will be zero on infinite depth. For these statements we refer to what we have shown in sect. 2.5.4 for waves without current.

The variation Δ of the mean water level yields some variations of order H^2 of $< f_{fy} >$ and R_{yy}. Specifically, the mean force exerted by the seabed changes from that given by (2.73) to

$$< f_{fy} > = \rho g \int_{y_1}^{y_2} (\tilde{d} + \Delta)\,\frac{\mathrm{d}d}{\mathrm{d}y}\,\mathrm{d}y = \rho g \int_{d_1}^{d_2} (\tilde{d} + \Delta)\,\mathrm{d}d ,$$

and hence the variation is

$$\delta < f_{fy} > = \rho g \int_{d_1}^{d_2} \Delta\,\mathrm{d}d ,$$

that is, in a form more suitable for the next developments,

$$\delta < f_{fy} > = \rho g \int_{d_2}^{d_1} d \frac{\mathrm{d}\Delta}{\mathrm{d}d} \mathrm{d}d + \rho g \left(d_2 \, \Delta_2 - d_1 \Delta_1 \right). \tag{2.76}$$

As to the radiation stress, it changes from (2.74) to

$$R_{yy} = \int_{-\tilde{d}}^{\Delta} -\rho g z + \rho g \Delta + \rho (u + \delta u)^2 \mathrm{d}z ,$$

where δu is the variation of the current velocity:

$$\delta u = - \frac{\Delta}{\tilde{d}} u , \tag{2.77}$$

which ensures the constancy of the discharge. And with this δu, the variation of the radiation stress proves to be

$$\delta R_{yy} = \rho g (d - 3S) \Delta . \tag{2.78}$$

2.10.3 *Effects of a variation of the mean water level on the energy equation*

Because of the variation of the mean water level, Φ_y becomes

$$\Phi_y = \int_{-\tilde{d}}^{\Delta} \left[-\rho g (z - \Delta) + \rho g (z - S) + \frac{1}{2} \rho (u + \delta u)^2 \right] (u + \delta u) \, \mathrm{d}z ,$$

so that the variation is

$$\delta \Phi_y = \int_{-\tilde{d}}^{\Delta} \left[-\rho g (z - \Delta) + \rho g (z - S) + \frac{1}{2} \rho (u + \delta u)^2 \right] (u + \delta u) \, \mathrm{d}z +$$
$$- \int_{-\tilde{d}}^{0} \left[-\rho g z + \rho g (z - S) + \frac{1}{2} \rho u^2 \right] u \, \mathrm{d}z .$$

Hence, using (2.77) of δu and neglecting the terms of orders smaller than H^2, we obtain

$$\delta \Phi_y = \rho g (Q - 2Su) \Delta . \tag{2.79}$$

2.10.4 *Particle velocities and pressure fluctuations in waves on currents*

We need to know wave pressure and particle velocity due to the wave motion, exact to the first order. We need also to know the mean values of such wave pressure and particle velocity (only the horizontal component), exact to the second order. We shall call v_{wy} the horizontal particle velocity due to the wave motion, to distinguish it from v_y which is the actual horizontal particle velocity (being due to waves and current).

As for the expressions of Δp, v_{wy}, and v_z exact to the first order, they straightforwardly proceed from the velocity potential (1.45) (bearing in mind that the water depth here is $\tilde{d}$):

$$\Delta p = \rho g \frac{H}{2} \frac{\cosh[k_c(\tilde{d}+z)]}{\cosh(k_c\tilde{d})} \cos(k_c y - \omega t) + o(H), \tag{2.80a}$$

$$v_{wy} = g\frac{H}{2}(\omega - uk_c)^{-1}k_c \frac{\cosh[k_c(\tilde{d}+z)]}{\cosh(k_c\tilde{d})} \cos(k_c y - \omega t) + o(H), \tag{2.80b}$$

$$v_z = g\frac{H}{2}(\omega - uk_c)^{-1}k_c \frac{\sinh[k_c(\tilde{d}+z)]}{\cosh(k_c\tilde{d})} \sin(k_c y - \omega t) + o(H). \tag{2.80c}$$

As for $<\Delta p>$ exact to the second order, we reason as we did in sect. 2.4.6 on the small control volume of fig. 2.6, the result being that

$$<\Delta p> = -\rho <v_z^2> = -\rho g^2 \frac{H^2}{8}(\omega - uk_c)^{-2}k_c^2 \frac{\sinh^2[k_c(\tilde{d}+z)]}{\cosh^2(k_c\tilde{d})} + o(H^2).$$

Finally, as for $<v_{wy}>$ exact to the second order, we reason as we did in sect. 2.7.3 for obtaining equation (2.57), the result being that

$$<v_{wy}> = \frac{1}{\tilde{d}}\left[\frac{2\pi}{8}\frac{H_0^2}{T} - g\frac{H^2}{8}(\omega - uk_c)^{-1}k_c\right] + o(H^2).$$

In concrete terms, a nonzero $<v_{wy}>$ is necessary to compensate the variation of the mass transport of the waves from deep to shallow water.

To verify the formula (2.80*a*) use the Bernoulli equation (1.5) with the formula (1.45) for the velocity potential. Δp will prove to be the sum of two terms: the first one obtained from $-\rho\partial\phi/\partial t$, the second one obtained from $-\frac{1}{2}\rho(\partial\phi/\partial y)^2$. Equation (2.80*a*) will proceed from addition of these two terms and simplification.

2.10.5 *Effects of the wave motion on the linear momentum equation*

Possibility of applying depth $\tilde{d}$ rather than $\tilde{d}+\Delta$

In sect. 2.10.2 we have evaluated how the terms of the linear momentum equation change due to a variation Δ of the mean water level. Now we are going to see how they change due to the wave motion on the current. To this end, we shall reason as if the water depth was $\tilde{d}$ rather than $\tilde{d}+\Delta$, and, in doing so, we shall make an error of an order smaller than H^2.

Effects on $< f_{fy} >$

The variation $\delta < f_{fy} >$ is zero:

$$\delta < f_{fy} > = 0\,, \tag{2.81}$$

since $< \Delta p >$ is zero at the bottom depth.

Effects on R_{yy}

R_{yy} for the current alone is given by (2.74). With the waves on the current, R_{yy} becomes

$$R_{yy} = \frac{1}{T}\int_0^T \int_{-\tilde{d}}^{\eta} -\rho g z + \Delta p + \rho (u + v_{wy})^2 \mathrm{d}z\, \mathrm{d}t,$$

so that the variation is

$$\delta R_{yy} = \frac{1}{T}\int_0^T \int_{-\tilde{d}}^{0} \Delta p + 2\rho u v_{wy} + \rho v_{wy}^2\, \mathrm{d}z\, \mathrm{d}t + \frac{1}{T}\int_0^T \int_0^{\eta} -\rho g z + \Delta p + \rho (u + v_{wy})^2 \mathrm{d}z\, \mathrm{d}t.$$

Hence, using the formulae of sect. 2.10.4 and neglecting the terms of orders smaller than H^2, we arrive at

$$\delta R_{yy} = \rho g H^2 \left[\frac{1}{16} + \frac{1}{8} g (\omega - u k_c)^{-2} k_c \frac{k_c \tilde{d}}{\cosh^2 (k_c \tilde{d})} \right] + \rho g \frac{\pi}{2} \frac{u}{gT} H_0^2. \tag{2.82}$$

2.10.6 *Effects of the wave motion on the energy equation*

In sect. 2.10.3 we have evaluated how the terms of the energy equation change due to a variation Δ of the mean water level. Now, let us consider how they change due to the wave motion.

The mean energy flux of a wave on a current on water depth $\tilde{d}$ is

$$\Phi_y = \frac{1}{T}\int_0^T \int_{-\tilde{d}}^{\eta} \left\{ -\rho g z + \Delta p + \rho g (z - S) + \frac{1}{2} \rho \left[(u + v_{wy})^2 + v_z^2 \right] \right\} (u + v_{wy})\, \mathrm{d}z\, \mathrm{d}t,$$

so that the variation with respect to the Φ_y of the current alone is

$$\delta \Phi_y = \frac{1}{T}\int_0^T \int_{-\tilde{d}}^{\eta} \left(\Delta p + \rho u\, v_{wy} + \frac{1}{2} \rho v_{wy}^2 + \frac{1}{2} \rho v_z^2 \right) (u + v_{wy})\, \mathrm{d}z\, \mathrm{d}t.$$

Hence, using the formulae of sect. 2.10.4 and neglecting the terms of orders smaller than H^2, we arrive at

$$\delta\Phi_y = \rho g H^2 \left\{ \frac{1}{16} g (\omega - u k_c)^{-2} u k_c \frac{k_c \tilde{d}}{\cosh^2 (k_c \tilde{d})} + \right.$$

$$\left. + \frac{1}{32} g \frac{2k_c\tilde{d} + \sinh(2k_c\tilde{d})}{\cosh^2 (k_c\tilde{d})} (\omega - uk_c)^{-1} \left[1 + uk_c(\omega - uk_c)^{-1}\right] + \frac{1}{8} u \right\} + \rho g \frac{\pi}{2} S \frac{H_0^2}{T}. \tag{2.83}$$

2.10.7 *The system of equations of the mean water level and wave height*

The variations $\delta < f_{fy} >$, δR_{yy} and $\delta\Phi_y$ must satisfy equations (2.40) and (2.46), that is:

$$\delta < f_{fy} > + \delta R_{yy}(y_1) = \delta R_{yy}(y_2), \tag{2.84a}$$

$$\delta\Phi_y(y_1) = \delta\Phi_y(y_2). \tag{2.84b}$$

Each of these variations consists of two parts, the first one being due to the change in the mean water level and the second one being due to the wave motion. These variations have been obtained in sects. 2.10.2-6, and their formulae have been gathered in table 2.1 for the reader's convenience.

If we take y_1 on infinite water depth, where u, S and Δ are zero, and y_2 on any given depth d, equations (2.84*a-b*) give rise to

$$\int_d^\infty d' \frac{\mathrm{d}\Delta'}{\mathrm{d}d'} \mathrm{d}d' = -\frac{H_0^2}{16} - 3S\Delta + H^2 \left[\frac{1}{16} + \frac{1}{8} g (\omega - uk_c)^{-2} k_c \frac{k_c\tilde{d}}{\cosh^2(k_c\tilde{d})} \right] + \frac{\pi}{2} \frac{u}{gT} H_0^2, \tag{2.85a}$$

$$(Q - 2Su)\Delta + H^2 \left\{ \frac{1}{16} g (\omega - uk_c)^{-2} uk_c \frac{k_c\tilde{d}}{\cosh^2(k_c\tilde{d})} + \frac{1}{32} g \frac{2k_c\tilde{d} + \sinh(2k_c\tilde{d})}{\cosh^2(k_c\tilde{d})} (\omega - uk_c)^{-1} \cdot \right.$$

$$\left. \cdot \left[1 + uk_c(\omega - uk_c)^{-1}\right] + \frac{1}{8} u \right\} + \frac{\pi}{2} S \frac{H_0^2}{T} = g \frac{H_0^2}{16} \omega^{-1}, \tag{2.85b}$$

where Δ is the variation of the mean water level on water depth d, and Δ' is the variation of the mean water level on water depth d'.

The two unknown functions in equations (2.85*a-b*) are $\Delta(d)$ and $H(d)$. The wave number $k_c \equiv 2\pi/L_c$ proceeds from (1.46) with $\tilde{d}$ as the water depth. As to $\tilde{d}$, S and u, they are functions of d, which can be specified once the discharge Q per unit length is given [cf. sect. 2.10.1].

The novelty with respect to the waves without current will not have passed unnoticed. Without the current, the energy equation in the two-dimensional flow contains the only unknown H, and the linear momentum equation contains the two unknowns H and Δ. Therefore, we obtain H from the energy equation and then we

Table 2.1 *Summary of the variations of* $<f_{fy}>$, R_{yy} *and* Φ_y

VARIATION	DUE TO	EQUATION	EXPRESSION
$\delta < f_{fy} >$	Δ	2.76	$\rho g \int_{d_2}^{d_1} d \frac{\mathrm{d}\Delta}{\mathrm{d}d} \mathrm{d}d + \rho g (d_2 \Delta_2 - d_1 \Delta_1)$
	$v_{wy}, v_z, \Delta p$	2.81	0
δR_{yy}	Δ	2.78	$\rho g (d - 3S) \Delta$
	$v_{wy}, v_z, \Delta p$	2.82	$\rho g H^2 \left[\frac{1}{16} + \frac{1}{8} g (\omega - u k_c)^{-2} k_c \frac{k_c \tilde{d}}{\cosh^2(k_c \tilde{d})} \right] + \rho g \frac{\pi}{2} \frac{u}{gT} H_0^2$
$\delta \Phi_y$	Δ	2.79	$\rho g (Q - 2Su) \Delta$
	$v_{wy}, v_z, \Delta p$	2.83	$\rho g H^2 \left\{ \frac{1}{16} g (\omega - u k_c)^{-2} u k_c \frac{k_c \tilde{d}}{\cosh^2(k_c \tilde{d})} + \frac{1}{32} g \frac{2 k_c \tilde{d} + \sinh(2 k_c \tilde{d})}{\cosh^2(k_c \tilde{d})} (\omega - u k_c)^{-1} \cdot \left[1 + u k_c (\omega - u k_c)^{-1} \right] + \frac{1}{8} u \right\} + \rho g \frac{\pi}{2} S \frac{H_0^2}{T}$

straightforwardly obtain Δ from the linear momentum equation. Whereas, *with the current*, we have got a system of two equations in the two unknowns H^2 and Δ.

The step from (2.84*a-b*) to (2.85*a-b*) should be done in three stages. First stage: $\delta R_{yy}(y_1)$ and $\delta \Phi_y(y_1)$ are substituted by the expressions of table 2.1 with $d = d_1$, $H = H_1$, $S = S_1$, $u = u_1$, $\Delta = \Delta_1$; $\delta R_{yy}(y_2)$ and $\delta \Phi_y(y_2)$ are substituted by the expressions of table 2.1 with $d = d_2$, $H = H_2$, $S = S_2$, $u = u_2$, $\Delta = \Delta_2$; $\delta < f_{fy} >$ is copied as it is from table 2.1. Second stage: a simplification can be done, in particular all terms with $d_1 \Delta_1$ and $d_2 \Delta_2$ can be cancelled. Third stage: y_1 is taken on infinite water depth so that $\Delta_1 \to 0$, $u_1 \to 0$, $S_1 \to 0$, $H_1 \to H_0$, and y_2 is taken on water depth d so that $\Delta_2 = \Delta$, $u_2 = u$, $S_2 = S$, $H_2 = H$.

As an exercise, verify that, for $u = 0$, equations (2.85*a*) and (2.85*b*) are reduced respectively to (2.47) and (2.45) with $\alpha = \alpha_0 = \pi/2$. To this end, you should bear in mind that $k_c \tilde{d} \to kd$ as $u \to 0$. Then, to re-obtain (2.45) and (2.47) exactly, you should make a few formal steps like multiplying and dividing by $\sinh(kd)$.

2.10.8 *Solution to the system of equations*

The second equation of system (2.85*a-b*) is of the type

$$H^2 = A + B\Delta, \tag{2.86}$$

where

$$A \equiv \left[g\frac{H_0^2}{16}\omega^{-1} - \frac{\pi}{2} S \frac{H_0^2}{T} \right] \Big/ \{\cdot\}, \quad B \equiv [2Su - Q]/\{\cdot\},$$

$$\{\cdot\} \equiv \left\{ \frac{1}{16} g (\omega - uk_c)^{-2} uk_c \frac{k_c \tilde{d}}{\cosh^2(k_c \tilde{d})} + \frac{1}{32} g \frac{2k_c\tilde{d} + \sinh(2k_c\tilde{d})}{\cosh^2(k_c\tilde{d})} (\omega - uk_c)^{-1} \cdot \right.$$

$$\left. \cdot \left[1 + uk_c(\omega - uk_c)^{-1}\right] + \frac{1}{8} u \right\}.$$

The first equation is of the type

$$\int_d^\infty d' \frac{\mathrm{d}\Delta'}{\mathrm{d}d'} \mathrm{d}d' = C + DH^2 + E\Delta,$$

with

$$C \equiv -\frac{H_0^2}{16} + \frac{\pi}{2}\frac{u}{g}\frac{H_0^2}{T}, \quad D \equiv \frac{1}{16} + \frac{1}{8} g(\omega - uk_c)^{-2} k_c \frac{k_c\tilde{d}}{\cosh^2(k_c\tilde{d})}, \quad E \equiv -3S.$$

If we multiply all terms of the first equation by D and add the result to the second equation, we obtain

$$\int_d^\infty d' \frac{\mathrm{d}\Delta'}{\mathrm{d}d'} \mathrm{d}d' = F + G\Delta, \tag{2.87}$$

where

$$F \equiv C + AD, \quad G \equiv BD + E.$$

Now our system consists of equation (2.86) of the two unknowns H^2 and Δ, and equation (2.87) of Δ alone. The terms A, B, F, and G in these equations are functions of d.

To solve (2.87), let us consider a sequence of growing depths $d_1, d_2, \ldots, d_N$, with $d_1 = d$. Then, from (2.87) we have

$$\int_{d_{i-1}}^{d_i} d' \frac{\mathrm{d}\Delta'}{\mathrm{d}d'} \mathrm{d}d' = (F_{i-1} + G_{i-1}\Delta_{i-1}) - (F_i + G_i\Delta_i),$$

where F_i, G_i are for F, G on water depth d_i. Provided that the difference $d_i - d_{i-1}$ is sufficiently small, the integral on the l.h.s. of this equation is equal to

$$\frac{1}{2}(d_{i-1} + d_i)(\Delta_i - \Delta_{i-1}),$$

so that the equation yields

$$\Delta_{i-1} = \frac{(d_{i-1} + d_i)\Delta_i - 2F_{i-1} + 2(F_i + G_i \Delta_i)}{2G_{i-1} + d_{i-1} + d_i}, \tag{2.88}$$

which is the solution to the problem. Indeed, provided that d_N is sufficiently large, we assume $\Delta_N = 0$ and, by means of (2.88), we can obtain Δ_{N-1}, Δ_{N-2}, and so on, up to the sought Δ on water depth d. Then, once Δ is known, we obtain also H by means of (2.86). The next section shows the details of this procedure.

2.10.9 *Summary and example of calculation*

The input data are H_0, T, and Q. The goal is to obtain the wave height H and the variation Δ of the mean water level on a given water depth d (d being the depth of the still water).

Preliminarily, check if condition (2.72) is fulfilled: the absolute value of the given flow rate Q per unit length must be smaller than

$$\frac{2}{3\sqrt{3}} \sqrt{2gd}\, d,$$

otherwise the problem does not admit a solution.

Then fix a sequence of growing depths d_1, d_2, ..., d_N with $d_1 = d$. We suggest taking $d_i - d_{i-1} = \text{costant} \le d/20$, and $d_N \ge 20d$. Then execute the following operations for each d_i.

(i) Compute the velocity u_i of the current, by means of (2.71). You have to seek the lowest positive x satisfying the equality

$$x = 1 + \left(\frac{Q}{Q_i^*}\right)^2 x^3, \quad \text{with} \quad Q_i^* = \sqrt{2gd_i}\, d_i.$$

This particular x, we call x_i, coincides with the product $u_i d_i / Q$, so that

$$u_i = \frac{Q}{d_i} x_i.$$

(ii) Evaluate S_i and $\tilde{d}_i$:

$$S_i = \frac{u_i^2}{2g}, \quad \tilde{d}_i = d_i - S_i.$$

(iii) By means of (1.47) evaluate the wavelength L_{ci}. You have to seek the lowest positive x satisfying the equality

$$x \tanh(2\pi x) = a_i (x - b_i)^2,$$

with

$$a_i \equiv \frac{1}{\tilde{d}_i / L_0} \left(\frac{u_i}{c_0}\right)^2, \quad b_i \equiv \frac{\tilde{d}_i}{L_0} \Big/ \left(\frac{u_i}{c_0}\right).$$

This particular x, we call x_i', is equal to $\tilde{d}_i/L_{ci}$, so that

$$L_{ci} = \tilde{d}_i \,/\, x_i'.$$

If the current is negative, x_i' might not exist. In that case the wave is not able to travel against the stream.

(iv) Evaluate the wave number

$$k_{ci} = 2\pi \,/\, L_{ci}.$$

(v) Compute the values of the parameters defined in sect. 2.10.8:

$$A_i \equiv \left[g\frac{H_0^2}{16}\,\omega^{-1} - \frac{\pi}{2}\,S_i\,\frac{H_0^2}{T} \right] \Big/ \{\cdot\}_i, \quad B_i \equiv [2S_i u_i - Q]/\{\cdot\}_i,$$

$$\{\cdot\}_i \equiv \left\{ \frac{1}{16}\,g\,(\omega - u_i k_{ci})^{-2} u_i k_{ci}\,\frac{k_{ci}\tilde{d}_i}{\cosh^2(k_{ci}\tilde{d}_i)} + \frac{1}{32}\,g\,\frac{2k_{ci}\tilde{d}_i + \sinh(2k_{ci}\tilde{d}_i)}{\cosh^2(k_{ci}\tilde{d}_i)}\,(\omega - u_i k_{ci})^{-1} \cdot \right.$$

$$\left. \cdot \left[1 + u_i k_{ci}(\omega - u_i k_{ci})^{-1}\right] + \frac{1}{8}\,u_i \right\},$$

$$C_i \equiv -\frac{H_0^2}{16} + \frac{\pi}{2}\,\frac{u_i}{g}\,\frac{H_0^2}{T}, \quad D_i \equiv \frac{1}{16} + \frac{1}{8}\,g\,(\omega - u_i k_{ci})^{-2} k_{ci}\,\frac{k_{ci}\tilde{d}_i}{\cosh^2(k_{ci}\tilde{d}_i)}, \quad E_i \equiv -3S_i,$$

$$F_i \equiv C_i + A_i D_i, \quad G_i \equiv B_i D_i + E_i.$$

At this stage, assume $\Delta_N = 0$ and compute Δ_{N-1} through (2.88). Hence, go on with equation (2.88) and compute Δ_{N-2}, Δ_{N-3}, and so on, up to Δ_1 which is the sought Δ on the given water depth d. Finally, use (2.86) to obtain wave height H on the given water depth d:

$$H = \sqrt{A_1 + B_1 \Delta_1}.$$

You should check the result, assuming a smaller value of the step $d_i - d_{i-1}$ and a greater value of d_N, and then repeat the procedure. The convergence is quick and, generally, the suggested values of $d_i - d_{i-1}$ and d_N prove to be suitable. The procedure entails no serious difficulty.

Fig. 2.19 shows the result of a calculation for

$$H_0 = 5\,\text{m}, \quad T = 10\,\text{s}.$$

The figure shows wave height H and variation Δ of the mean water level on $d = 15\,\text{m}$ versus the discharge of the current. As we can see, if the current is positive the wave height is smaller than if the current is null ($Q = 0$). Conversely, if the current is negative, the wave height grows. These are some intuitive phenomena, that now we can predict by means of the solution exact to the order H^2.

The two functions $H(Q)$ and $\Delta(Q)$ are cut off on the left side, on the discharge of the critical negative current (against which the wave cannot run). We see that near this critical discharge, the wave height grows quickly. In addition to this, the wavelength gets smaller (cf. sect. 1.9.2), so that we realize that the wave gets very steep and breaks. Indeed, the dashed piece of curve indicates the range where the wave exceeds the critical steepness.

On the right side the two functions are cut off at $Q = Q_{\max}$ (the maximum discharge for the fixed water depth) which is 99 $m^3\ s^{-1}/m$ for the water depth of 15 m.

As to the wave set-down, we see that it is larger than in absence of the current.

Conclusive note

The problem of shoaling was originally solved by Burnside (1915). The radiation stress tensor was introduced by Longuet-Higgins and Stewart (1960, 1961), and the existence and size of the set-down were predicted by these authors (1962). The current associated with the wave motion was predicted by Whitham (1962). The numerical calculation of the wave orthogonals by means of (2.59) probably started with Griswold (1963).

The solution which is commonly used for the shoaling and set-down of waves on current was given by Jonsson et al. (1970). This solution is of order H^2; the effects of the higher order terms were then considered by Jonsson and Arneborg (1995), and also the effects of the velocity distribution with depth were considered by Jonsson et al. (1978). Jonsson et al. (1970) concluded that the conservative

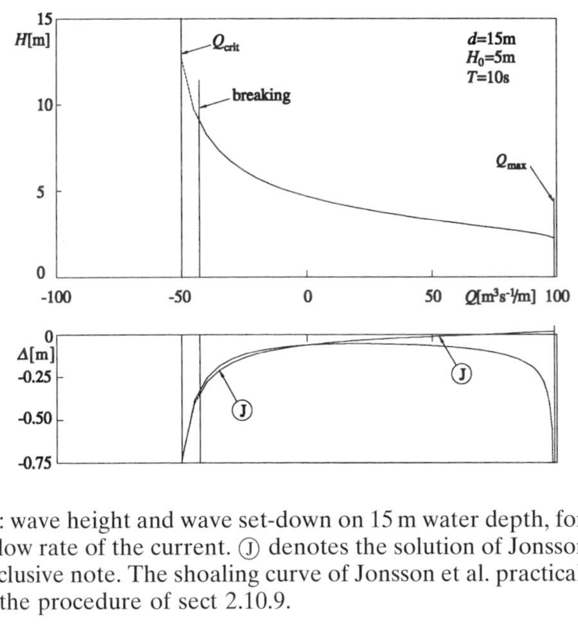

Fig. 2.19 Example: wave height and wave set-down on 15 m water depth, for given conditions on deep water, vs the flow rate of the current. Ⓙ denotes the solution of Jonsson et al. (1970), which is quoted in the conclusive note. The shoaling curve of Jonsson et al. practically coincides with the one obtained from the procedure of sect 2.10.9.

equations are not applicable for direct calculations. Therefore they proposed a set of "practical equations".

The new approach of sect. 2.10, taking into account all terms of order H^2 due to the wave motion or the variation of the mean water level, yields the formal solution, exact to the order H^2, based on the conservative equations. This solution is easily applicable for direct calculations as shown in sect. 2.10.9. A comparison with the solution based on the practical equations of Jonsson et al. (1970) reveals some discrepancy only for what concerns the variation of the mean water level (see fig. 2.19). This discrepancy becomes significant only for large positive Q where the exact solution predicts a wave set-down (increasing quickly as Q approaches Q_{max}) whereas the solution based on the practical equations predicts a small wave set-up.

References

Burnside W., 1915 On the modification of a train of waves as it advances into shallow water. *Proc. London Math. Soc.* 14, 131-133.

Griswold G. M., 1963 Numerical calculation of wave refraction. *J. Geophys. Res.* 68, 1715-1723.

Jonsson I. G., Skougaard C. and Wang J. D., 1970 Interaction between waves and currents. *Proc.* 12th *Conf. Coastal Eng. ASCE*, 489-507.

Jonsson I. G., Brink-Kiaer O. and Thomas G.P., 1978 Wave action and set-down for waves on a shear current. *J. Fluid Mech.* 87, 401-416.

Jonsson I. G. and Arneborg L., 1995 Energy properties and shoaling of higher-order Stokes waves on a current. *Ocean Eng.* 22, 819-857.

Longuet-Higgins M. S. and Stewart R. W., 1960 Changes in the form of short gravity waves on long waves and tidal currents. *J. Fluid Mech.* 8, 565-583;

1961 The changes in amplitude of short gravity waves on steady, non-uniform currents. *J. Fluid Mech.* 10, 529-549;

1962 Radiation stress and mass transport in the gravity waves. *J. Fluid Mech.* 13, 481-504.

Miche R., 1944 Mouvements ondulatoire de la mer en profondeur constante ou dècroissante. *Annales Ponts et Chausses*, 25-78, 131-164, 270-292, 369-406.

Miles J. W., 1962 Transient gravity wave response to an oscillating pressure. *J. Fluid Mech.* 13, 145-150.

Whitham G. B., 1962 Mass, momentum and energy flux in water waves. *J. Fluid Mech.* 12, 135-147.

Chapter 3

WAVE EFFECTS ON COASTS

3.1 The control volume from the breaker line to the beach

The control volume of fig. 3.1 extends from the breaker line to the beach. It is bounded by a vertical plane at the breaker line and by two vertical planes piercing the shoreline. Let us assume that

(i) the flow is ideal up to the breaker line,

(ii) the contour lines are x-parallel.

From the linear momentum equation for this control volume we have

$$< F_{fx} > = \int_{x_1}^{x_2} -R_{yx}(x, y_b)\,\mathrm{d}x + \int_{y_b}^{\infty} -R_{xx}(x_1, y) + R_{xx}(x_2, y)\,\mathrm{d}y, \qquad (3.1a)$$

$$< F_{fy} > + < T_y(x_1) > - < T_y(x_2) > = \int_{x_1}^{x_2} -R_{yy}(x, y_b)\,\mathrm{d}x + \int_{y_b}^{\infty} -R_{xy}(x_1, y) + R_{xy}(x_2, y)\,\mathrm{d}y, \qquad (3.1b)$$

where

(i) the subscript b denotes breaking;

(ii) $\mathbf{T}$ is the shear stress on vertical planes parallel to the y-axis, from the breaker line to the beach;

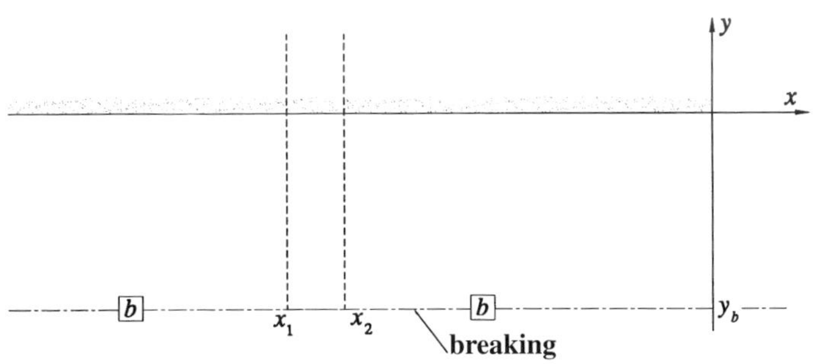

Fig. 3.1 The control volume within the surf zone.

(iii) the y-upper limit of the integrals is ∞ since the control volume extends well over the shoreline. Of course, the contribution to these integrals is different from zero only up to the extreme y where the water arrives on the beach.
The novelty of (3.1*a*-*b*) with respect to (2.22*a*-*b*) consists in the shear stress T_y due to the real-fluid flow within the surf zone.

The expressions of the radiation stresses obtained in sect. 2.4 retain their validity only up to the breaker line. Luckily, the symmetry of the problem (straight contour lines) implies that even in the surf zone, that is for $y > y_b$, $\mathbf{R}$, $<\mathbf{T}>$ and $<\mathbf{f}_f>$ (the mean force per unit length of shoreline) are constant with respect to x. Therefore equations (3.1*a*-*b*) reduce themselves to

$$< f_{fx} > = -R_{yx}(y_b),$$

$$< f_{fy} > = -R_{yy}(y_b).$$

Note, as the contour lines are x-parallel, f_{fx} is a shear force and is nonzero because of the real-fluid flow.

Bearing in mind that the water depth at breaking is $d_b - S_D$, from (2.29) of R_{yx} and (2.33) of R_{yy} we obtain the equations

$$< f_{fx} > = -\frac{1}{16}\rho g H_b^2 \left[1 + \frac{2k_b d_b}{\sinh(2k_b d_b)}\right] \sin\alpha_b \cos\alpha_b, \tag{3.2a}$$

$$< f_{fy} > = -\frac{1}{2}\rho g d_b^2 + \rho g d_b S_D - \frac{1}{16}\rho g H_b^2 \left[\frac{2k_b d_b}{\sinh(2k_b d_b)}(1+\sin^2\alpha_b) + \sin^2\alpha_b\right]. \tag{3.2b}$$

Given that R_{yx} keeps constant from deep water to the breaker line, $< f_{fx} >$ can also be expressed in the alternative forms

$$< f_{fx} > = -\frac{1}{16}\rho g H^2 \left[1 + \frac{2kd}{\sinh(2kd)}\right] \sin\alpha \cos\alpha, \tag{3.3}$$

where H and α are wave height and angle on any fixed water depth d, and

$$< f_{fx} > = -\frac{1}{16}\rho g H_0^2 \sin\alpha_0 \cos\alpha_0. \tag{3.4}$$

3.2 The run-up

3.2.1 *Spilling breaker type*

Let us assume that

(i) the seabed is flat and makes an angle λ with the horizontal,

(ii) the y-component of the average shear stress exerted by the seabed is null.

The assumption (ii) is commonly done [cf. Longuet-Higgins (1971)] for the so called spilling breaker. The usual pattern given of this breaker type is: starting from the breaker line the ratio between the wave height and the water depth keeps constant. (We shall see shortly when this breaker type takes place.)

Under assumptions (i) and (ii), we have

$$< f_{fy} > = -\frac{1}{2}\rho g d_b^2 - \tan\lambda \rho g < \Delta W > , \tag{3.5}$$

where $\rho < \Delta W >$ is the mean variation of the water mass per unit length of the volume of fig. 3.1.

Equating the right-hand sides of (3.2*b*) and (3.5) we obtain

$$< \Delta W > = \cot\lambda \left\{ -d_b S_D + \frac{1}{16} H_b^2 \left[\frac{2k_b d_b}{\sinh(2k_b d_b)} (1 + \sin^2 \alpha_b) + \sin^2 \alpha_b \right] \right\} .$$

The $< \Delta W >$ can be expressed also in terms of set-up S_U [which is the superelevation of the point of intersection between the mean water surface and the shore]. Referring to fig. 3.2, we have

$$< \Delta W > = \frac{1}{2}(d_b - S_D)(d_b + S_U)\cot\lambda - \frac{1}{2} d_b^2 \cot\lambda .$$

The last two equalities yield

$$\frac{S_U}{H_b} = \frac{H_0/H_b}{d_b/L_0} \left\{ -\frac{d_b}{L_0}\frac{S_D}{H_0} + \frac{1}{8}\left(\frac{H_b}{H_0}\right)^2 \frac{H_0}{L_0} \left[\frac{2k_b d_b}{\sinh(2k_b d_b)} (1 + \sin^2 \alpha_b) + \sin^2 \alpha_b \right] \right\} , \tag{3.6}$$

where all terms of the r.h.s. can be obtained through the procedure of sect. 2.6, for given values of H_0/L_0 and α_0.

In the spilling breaker the set-up coincides with the run-up (R_U) that is the highest level where the water arrives on the beach (the wave fades away till its height becomes zero). Therefore equation (3.6) gives also the run-up of the spilling breaker.

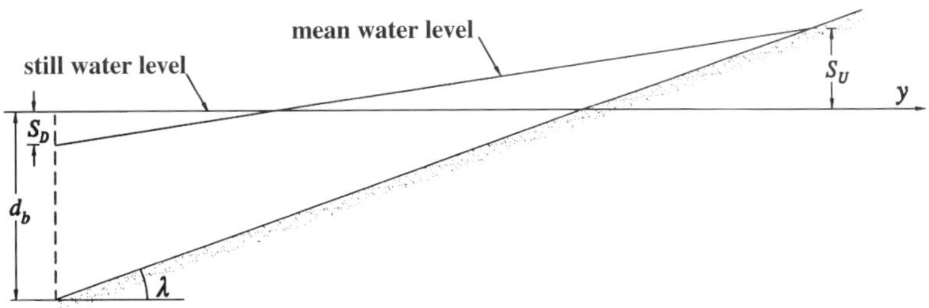

Fig. 3.2 The set-up (S_U) is the superelevation of the point of intersection between the mean water level and the shore. In the spilling breaker type the set-up coincides with the run-up (R_U).

Using (3.6) we have obtained fig. 3.3 which shows the quotient R_U/H_b as a function of H_b/L_0 for the case of an orthogonal wave attack ($\alpha_0 = \pi/2$). The figure shows a basic property: the larger the steepness of the wave, the smaller the run-up. For the rest, we see that the run-up ranges roughly between $0.1H_b$ and $0.3H_b$. This is a small run-up peculiar to the spilling breaker. A greater run-up takes place with the new type of breaker which we will deal with in the next section.

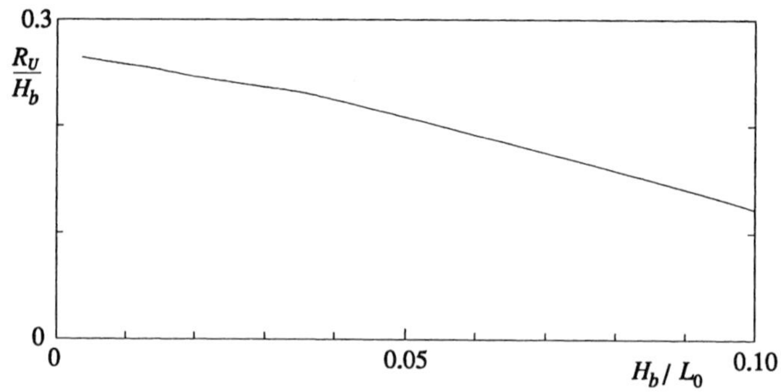

Fig. 3.3 Spilling breaker type: quotient R_U/H_b as a function of H_b/L_0 for an orthogonal attack ($\alpha_0 = \pi/2$). For obtaining this function: fix H_0/L_0, compute d_b/L_0, H_b/H_0 and S_D/H_0 [by means of the equations of sect. 2.6] and then use (3.6).

3.2.2 *Plunging breaker type*

The spilling breaker [fig. 3.4*a*] occurs if the distance from the breaker line to the shore is large with respect to the wavelength (we shall see shortly how large). Otherwise the wave overturns and produces a plunging breaker [fig. 3.4*b*], which is a phenomenon with some striking features. Peregrine et al. (1980) found that the water rising up the front of the wave into the jet is subject to accelerations well greater than g!

As we have already pointed out, in the spilling breaker there is always some water beyond the shoreline. In the plunging breaker, the water withdraws from the

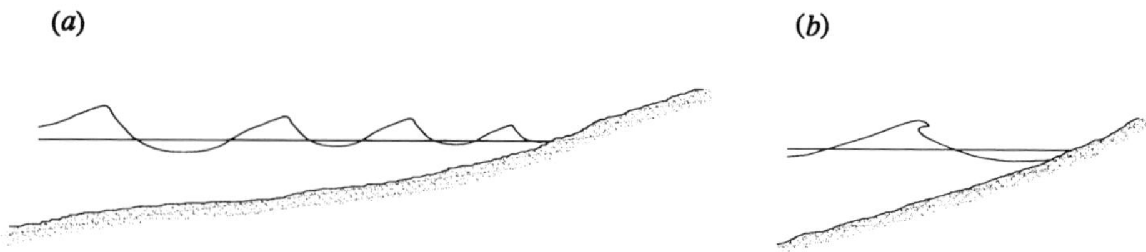

Fig. 3.4 Breaker types: (*a*) spilling; (*b*) plunging.

shoreline during the wave cycle. This run-down is then counterbalanced by a run-up larger than in the spilling breaker.

The prediction of the breaker type is usually done from the value of

$$\mathcal{I} \equiv \tan\lambda \,/\, \sqrt{H_b/L_0}$$

called Irribarren's number, which is related to the quotient $|y_b|/L_b$:

$$\mathcal{I} \cong \sqrt{\frac{1}{(H_b/d_b)2\pi}} \,/\, \frac{|y_b|}{L_b}\,.$$

[To prove this, use the relation $L_b \cong \sqrt{2\pi L_0 d_b}$ which is exact as $d_b/L_0 \to 0$.]

Hence, the greater $|y_b|/L_b$ is, the smaller is $\mathcal{I}$. That is: spilling breaker $\Leftrightarrow$ large $|y_b|/L_b \Leftrightarrow$ small $\mathcal{I}$; plunging breaker $\Leftrightarrow$ small $|y_b|/L_b \Leftrightarrow$ large $\mathcal{I}$. To judge from the work of Battjes (1974) the spilling breaker should occur for $\mathcal{I} < 0.2$ that, with the characteristic ratio $H_b/d_b = 0.8$, is equivalent to $|y_b|/L_b > 2 \div 2.5$.

In the range of the plunging breaker, R_U/H_b grows generally with $\mathcal{I}$. Looking at the data of the U.S. Army Corps of Engineers (1984), we see that R_U/H_b is rather close to $\mathcal{I}$, for $\mathcal{I}$ smaller than some threshold which depends on the bottom slope. In particular

$$R_U/H_b \cong \mathcal{I} \quad \text{for} \quad \mathcal{I} < 0.7 \quad \text{and} \quad \tan\lambda = 0.10\,.$$

For $\mathcal{I}$ exceeding this threshold, the rate of grow of R_U/H_b gets smaller and smaller.

The basic matter is thus confirmed also by the plunging breaker: the larger the wave steepness, the smaller the run-up. An interesting example of this is sometimes visible from the long beach which stretches from Cape Spartivento (eastern side) to Cape dell'Armi (western side) at the extreme south of the Italian peninsula (see fig. 4.14). We happen to see the waves approaching the beach from the right-hand side (south-west) and the foam rushing up the beach from the left-hand side. The waves from the right-hand side are generated by a local wind blowing from Sicily. The run-up is due to the swells from the south-east which are less steep than the wind waves from the south-west. Under roughly the same wave height, the eye is caught by the steeper waves and usually the swells are not noticed. The presence of the swells is only revealed by the run-up because the run-up is greater if the wave steepness is smaller.

3.2.3 *Other breaker types*

Spilling and plunging are the more common breaker types of the wind waves on natural beaches. For very large values of $\mathcal{I}$, say $\mathcal{I}$ in excess of about 3, some new breaker types (collapsing or surging) occur.

A tsunami (see sect. 14.1) can produce a huge run-up. If the greater water depths only are considered, tsunami's run-up should be typically nearly equal to H_b. If we include the shallow flooding, this run-up can be even much greater than H_b (see Camfield, 1980). A laboratory run-up of about 1.7 H_b (including the shallow flooding) on a beach with a slope of 1/20 is described in detail by Lin et al. (1999).

3.3 The longshore transport

3.3.1 *The formula for the longshore transport rate*

For Newton's third law, the shear force exerted by the water on the beach is

$$< f_{ax} > = - < f_{fx} >,$$

and hence it is given by any of the three expressions (3.2*a*), (3.3) and (3.4) multiplied by -1.

The immersed weight P^* (per unit length) of the mobile sand is related to the shear force $< f_{ax} >$ by

$$P^* = < f_{ax} > / \mu,$$

where μ is the coefficient of friction between the mobile sand and the beach. P^* is then related to the cross section A of the mobile layer by

$$P^* = (\gamma_s - \gamma_a)(1 - \mathscr{V})\, A,$$

where γ_s is the specific weight of the sand and $\mathscr{V}$ is the sediment porosity.

The bulk longshore sediment transport rate is

$$Q_s = A v_s,$$

where v_s is a mean longshore sand velocity, which is expressed as

$$v_s = K^* \sqrt{g d_b},$$

with K^* being dependent on the size of the sand.

The last four equations taken together yield

$$Q_s = \widetilde{K} \frac{1}{\gamma_a} < f_{ax} > \sqrt{g d_b}, \tag{3.7}$$

with the following definitions

$$\widetilde{K} \equiv \frac{K}{(\gamma_s/\gamma_a - 1)(1 - \mathscr{V})}, \tag{3.8}$$

$$K \equiv K^*/\mu.$$

A typical value for K suggested by Komar and Inman (1970) is 0.77. The scatter of this factor has been dealt with by Bodge and Kraus (1991).

With the formulae for $< f_{fx} >$, that is with (3.2*a*), or (3.3) or (3.4), formula (3.7) for Q_s becomes

$$Q_s = \frac{1}{32} \widetilde{K} H_b^2 \left[1 + \frac{2 k_b d_b}{\sinh(2 k_b d_b)} \right] \sin(2\alpha_b) \sqrt{g d_b}, \tag{3.9a}$$

or

$$Q_s = \frac{1}{32}\widetilde{K}H^2\left[1+\frac{2kd}{\sinh(2kd)}\right]\sin(2\alpha)\sqrt{gd_b} \tag{3.9b}$$

(where H and α are wave height and angle on any fixed depth d), or

$$Q_s = \frac{1}{32}\widetilde{K}H_0^2\sin(2\alpha_0)\sqrt{gd_b}\,. \tag{3.9c}$$

3.3.2 *The equation of sediment conservation*

So far, we have dealt with x-parallel contour lines. Now let us examine the case of a coast with generally curved contour lines. We shall call ξ the offshore distance of the shoreline. ξ is parallel to the y-axis, but it has the opposite direction: y is landward oriented, ξ is seaward oriented .

Generally, the beach profile is assumed to slide along a horizontal base located at a closure depth d_{cl}. Referring to fig. 3.5, we have

$$(d_{cl}+B)\left(\frac{\partial\xi}{\partial t}\,\mathrm{d}t\right)\mathrm{d}x = -\frac{\partial Q_s}{\partial x}\,\mathrm{d}x\,\mathrm{d}t\,.$$

The l.h.s. of this equation gives the variation in the small time $\mathrm{d}t$ of the volume of sediment in a stretch of coast of length $\mathrm{d}x$. The r.h.s. gives the difference between the volume of sediment entering and the volume leaving the stretch of coast of length $\mathrm{d}x$ in the small time interval $\mathrm{d}t$. Clearly, some terms of this equation cancel out and it can be rewritten in the form

$$\frac{\partial\xi}{\partial t} = -\frac{1}{d_{cl}+B}\frac{\partial Q_s}{\partial x} \tag{3.10}$$

which is the well-known form of the equation of sediment conservation.

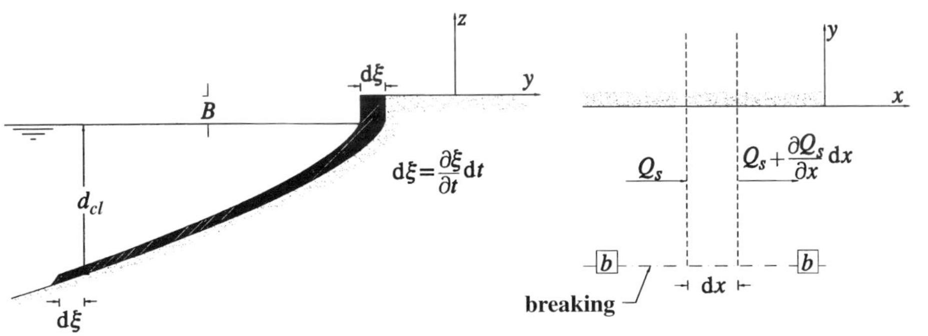

Fig. 3.5 A variation $\mathrm{d}\xi$ of the shoreline planform is due to a nonzero $\partial Q_s/\partial x$.

3.4 The analytical approach to the problem of beach planform evolution

The analytical approach to the problem of beach planform evolution was originally started by Pelnard-Considère (1956). The treatment was then developed thanks to the radiation stress tensor and the control volume from the breaker line to the shore. A very clear picture of the state of the art was given by Dean (1992) in a few lectures.

It is generally assumed that the deviation of the shoreline planform from the general shoreline alignment (x-axis) approaches zero:

$$\frac{\partial \xi}{\partial x} \to 0, \tag{3.11}$$

and in addition it is assumed that

(i) the water depth at breaking is constant along the x-axis and equal to d_b;

(ii) the wave height at breaking is constant and equal to H_b;

(iii) the angle between the direction of wave advance and the x-axis, at breaking, is constant and equal to α_b;

(iv) the contour lines, at any given cross-section, are straight and parallel to a tangent vector to the shoreline;

where d_b, H_b and α_b are, respectively, the water depth, the wave height and the wave angle at breaking *in the case of x-parallel contour lines.*

Under these assumptions, the bulk sediment transport rate can be evaluated by means of (3.9*a*), with $\alpha_b + \partial\xi/\partial x$ as the angle between the direction of wave advance and the contour lines at breaking:

$$Q_s(x) = \frac{\widetilde{K}}{32} H_b^2 \left[1 + \frac{2k_b d_b}{\sinh(2k_b d_b)}\right] \sqrt{g d_b} \sin\left[2\left(\alpha_b + \frac{\partial \xi}{\partial x}\right)\right].$$

Hence, differentiating with respect to x we obtain

$$\frac{\partial Q_s}{\partial x} = \frac{\widetilde{K}}{16} H_b^2 \left[1 + \frac{2k_b d_b}{\sinh(2k_b d_b)}\right] \sqrt{g d_b} \cos(2\alpha_b) \frac{\partial^2 \xi}{\partial x^2}, \tag{3.12}$$

where we have used the assumption (3.11).

From (3.10) and (3.12) we arrive at the equation describing the planform evolution of a shoreline

$$\frac{\partial \xi}{\partial t} = G \frac{\partial^2 \xi}{\partial x^2}, \tag{3.13}$$

where

$$G \equiv -\frac{\widetilde{K}}{16} \frac{H_b^2}{d_{cl} + B} \sqrt{g d_b} \left[1 + \frac{2k_b d_b}{\sinh(2k_b d_b)}\right] \cos(2\alpha_b). \tag{3.14}$$

We shall see in sect. 3.9 that equation (3.13) can be solved analytically for given initial conditions.

Equation (3.13) was obtained originally by Pelnard-Considère himself (1956) but the relation between longshore diffusivity G and wave characteristics was not fully specified. Also the relation (3.14), which is still accepted nowadays, is not yet exact. Indeed, if the shoreline is not straight at least a part of the contour lines will not be straight, and thus at breaking the water depth, the wave height and the angle with the x-axis will vary with x. It is true that these variations are infinitesimal thanks to assumption (3.11) on $\partial\xi/\partial x$; but it is also true that, for the same assumption, $\partial Q_s/\partial x$ is infinitesimal, and thus we cannot conclude that the effect of the aforesaid variations is negligible. On the contrary, we shall see this effect to be not negligible. Moreover the effect of the curvature of the contour lines can be neglected only under some precise assumptions on the higher order derivatives of ξ with respect to x.

3.5 Problem of beach planform evolution: the case of contour lines parallel up to deep water

3.5.1 *Assumptions*

Typically, the contour lines have roughly the same shape as the shoreline. Therefore, we assume the offshore distance $\xi_{(d)}$ of the contour line of depth d to be related to the offshore distance ξ of the shoreline by

$$\xi_{(d)}(x) = \Delta y + \xi(x),$$

where Δy is constant along x and depends only on the fixed depth d of the contour line. Let us call Δy_0 the Δy of the contour line of depth $d_0 = \frac{1}{2}L_0$ (deep water). Thus Δy_0 is the distance between the contour line of depth d_0 and the shoreline.

In what follows we shall make the usual assumption (3.11) and in addition we shall assume that

$$|\Delta y_0| \left|\frac{\partial^{n+1}\xi}{\partial x^{n+1}}\right|_{\max} << \left|\frac{\partial^{n}\xi}{\partial x^{n}}\right|_{\max} \qquad \text{for } n = 1, 2, 3, ..., \tag{3.15}$$

where max denotes the maximum with respect to x.

The assumptions (3.11) and (3.15) are fulfilled by a shoreline of the type

$$\xi(x) = a \sin\left(\frac{2\pi}{\mathcal{N}\Delta y_0}x + \varepsilon\right), \tag{3.16}$$

provided that $\mathcal{N}$ is very large (a representing here the amplitude of the shoreline undulation). Of course, assumptions (3.11) and (3.15) are satisfied also if $\xi(x)$ is a sum of functions like (3.16):

$$\xi(x) = \sum_{i=1}^{N} a_i \sin\left(\frac{2\pi}{\mathcal{N}_i \Delta y_0} x + \varepsilon_i\right), \tag{3.17}$$

which is useful since we can represent a very large variety of planforms of the shoreline through series with the form (3.17).

3.5.2 *The formula for Q_s*

Let us consider any cross-section of the beach, as an example cross-section ① in fig. 3.6. To obtain the bulk sediment transport rate we shall assume that the contour lines are straight and parallel to a tangent vector to the shoreline at this cross-section. In doing so, we shall make an error since the contour lines have a nonzero curvature. But, if condition (3.15) is fulfilled this error has some negligible consequences, as we shall see later.

The direction of wave advance on deep water makes an angle α_0 with x-axis. Consequently, it makes an angle $\alpha_0 + \delta\alpha_0$ with the local direction of the contour lines:

$$\delta\alpha_0 = \frac{\partial \xi}{\partial x}. \tag{3.18}$$

The variation $\delta\alpha_0$ of the deep water angle between the direction of wave advance and the contour lines causes some variations in the breaking conditions. We shall have

$$\text{water depth at breaking} = d_b + \delta d_b,$$

$$\text{wave height at breaking} = H_b + \delta H_b,$$

angle between direction of wave advance and contour lines at breaking $\alpha_b + \delta\alpha_b$,

where d_b, H_b and α_b are the breaking characteristics if the contour lines are x-parallel.

Using (3.9*a*) we have

$$Q_s = \frac{1}{32}\widetilde{K}(H_b + \delta H_b)^2 \left[1 + \frac{2kd}{\sinh(2kd)}\right]_{d=d_b+\delta d_b} \sin[2(\alpha_b + \delta\alpha_b)]\sqrt{g(d_b + \delta d_b)}.$$

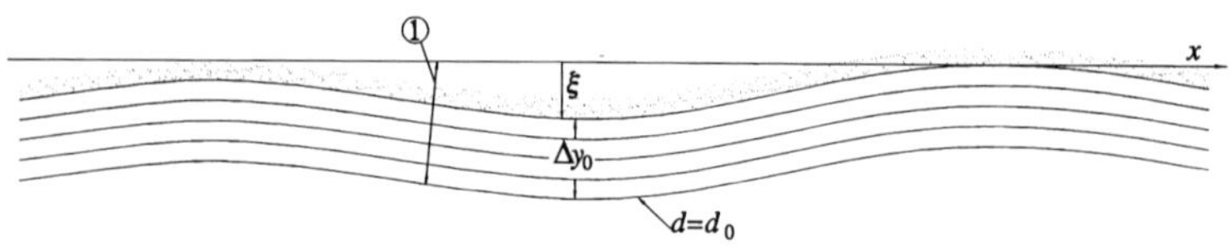

Fig. 3.6 Contour lines with a small curvature. At any cross-section like ①, we assume the contour lines to be straight.

This formula is rather difficult to handle because δH_b, δd_b, and $\delta\alpha_b$ depend on $\partial\xi/\partial x$ in a complex way. Certainly, the formula that proceeds from (3.9c) is more simple:

$$Q_s = \frac{1}{32}\,\widetilde{K}\,H_0^2 \sin[2(\alpha_0+\delta\alpha_0)]\sqrt{g(d_b+\delta d_b)}\,. \tag{3.19}$$

Indeed $\delta\alpha_0$ is simply equal to $\partial\xi/\partial x$, so that only δd_b must be sought. What follows is devoted to this operation.

3.5.3 *The variation δd_b of the water depth at breaking*

The problem is in the following terms. We know: if

(i) the contour lines are straight,

(ii) the wave period is T,

(iii) the wave height on deep water is H_0,

(iv) the angle between the direction of wave advance on deep water and the contour lines is α_0,

then the water depth and the wave height at breaking are respectively d_b and H_b. We wonder: how does it change the water depth at breaking if the direction of wave advance on deep water changes from α_0 into $\alpha_0+\delta\alpha_0$?

Assuming that the quotient between the wave height and the water depth at breaking is constant and equal to 0.8, we have

$$\frac{H_b+\delta H_b}{d_b+\delta d_b} = \frac{H_b}{d_b} = 0.8 \tag{3.20}$$

which implies

$$\delta d_b = \delta H_b/0.8\,. \tag{3.21}$$

To get δH_b let us consider $H(x_1,x_2,d)$ where

$x_1 \equiv$ wave height on deep water,

$x_2 \equiv$ angle between the direction of wave advance on deep water and contour lines.

$H(x_1,x_2,d)$ gives the wave height as a function of water depth d and of wave characteristics x_1 and x_2 on deep water (for a fixed wave period T), under the assumption of straight contour lines. In terms of this function, we have

$$\delta H_b = \left(\frac{\partial H}{\partial x_2}\right)_{x_1=H_0,\,x_2=\alpha_0,\,d=d_b}\delta\alpha_0 + \left(\frac{\partial H}{\partial d}\right)_{x_1=H_0,\,x_2=\alpha_0,\,d=d_b}\delta d_b\,,$$

which together with (3.21) leads to

$$\delta d_b = K_1 H_0\,\delta\alpha_0\,, \tag{3.22}$$

where

$$K_1 \equiv \frac{\dfrac{1}{H_0}\left(\dfrac{\partial H}{\partial x_2}\right)_{x_1=H_0,\,x_2=\alpha_0,\,d=d_b}}{0.8-\left(\dfrac{\partial H}{\partial d}\right)_{x_1=H_0,\,x_2=\alpha_0,\,d=d_b}}\,. \tag{3.23}$$

Function $H(x_1, x_2, d)$ proceeds straightforwardly from equation (2.45) of shoaling-refraction:

$$H(x_1, x_2, d) = x_1 \sqrt{\frac{\sinh(2kd)}{\tanh(kd)[\sinh(2kd) + 2kd]}} \quad \sqrt[4]{\frac{1 - \cos^2 x_2}{1 - \tanh^2(kd)\cos^2 x_2}}.$$

Differentiating this function, we obtain

$$\left(\frac{\partial H}{\partial x_2}\right)_{x_1 = H_0,\, x_2 = \alpha_0,\, d = d_b} = \frac{1}{4} H_0 \left[\frac{S2}{T1(S2 + B2)}\right]^{\frac{1}{2}} \left[\frac{1 - C0^2}{1 - T1^2 C0^2}\right]^{-\frac{3}{4}} \left[\frac{1 - T1^2}{(1 - T1^2 C0^2)^2}\right] \sin(2\alpha_0), \tag{3.24a}$$

$$\left(\frac{\partial H}{\partial d}\right)_{x_1 = H_0,\, x_2 = \alpha_0,\, d = d_b} = 2\pi \frac{H_0}{L_0} \left[\frac{2C1^2}{S2 + B2}\right] \left\{ \left[\frac{S2}{T1(S2 + B2)}\right]^{-\frac{1}{2}} \left[\frac{B2(C2 - 1) - 2S2}{T1(S2 + B2)^2}\right] \cdot \left[\frac{1 - C0^2}{1 - T1^2 C0^2}\right]^{\frac{1}{4}} + \frac{1}{2} \left[\frac{S2}{T1(S2 + B2)}\right]^{\frac{1}{2}} \left[\frac{1 - C0^2}{1 - T1^2 C0^2}\right]^{-\frac{3}{4}} \left[\frac{C0^2(1 - C0^2)T1/C1^2}{(1 - T1^2 C0^2)^2}\right] \right\}, \tag{3.24b}$$

where

$$C1 \equiv \cosh(k_b d_b), \quad T1 \equiv \tanh(k_b d_b), \quad B2 \equiv 2k_b d_b, \quad S2 \equiv \sinh(2k_b d_b),$$

$$C2 \equiv \cosh(2k_b d_b), \quad C0 \equiv \cos\alpha_0.$$

Equation (3.22) with (3.23) and (3.24*a-b*) represent the solution to the problem of δd_b.

To prove the step from (3.20) to (3.21), bear in mind that

$$f(x_o + \delta x) = f(x_o) + \dot{f}(x_o)\delta x,$$

provided that the derivative $\dot{f}(x_o)$ exists. Thus, for $f(x) \equiv 1/x$, $x_o \equiv d_b$, and $\delta x \equiv \delta d_b$, we have

$$\frac{1}{d_b + \delta d_b} = \frac{1}{d_b} - \frac{1}{d_b^2}\delta d_b.$$

Hence, the proof proceeds straightforwardly.

3.5.4 *The relation between Q_s and $\partial\xi/\partial x$*

Equations (3.18), (3.19) and (3.22) taken together yield

$$Q_s = \boxed{\frac{1}{32}\tilde{K} H_0^2 \sin(2\alpha_0)\sqrt{gd_b}} + \frac{1}{32}\tilde{K} H_0^2 \sqrt{gd_b} \left[2\cos(2\alpha_0) + \frac{1}{2} K_1 \sin(2\alpha_0) \frac{H_0}{d_b}\right] \frac{\partial \xi}{\partial x}. \tag{3.25}$$

The framed term on the r.h.s. of this formula is the sediment transport rate for the case of x-parallel contour lines. The additional term is the variation of the sediment transport rate due to the deviation of the shoreline planform from the x-axis.

To prove the step from (3.19) to (3.25), use the relations

$$\sin[2(\alpha_0 + \delta\alpha_0)] = \sin(2\alpha_0) + 2\cos(2\alpha_0)\delta\alpha_0,$$

$$[g(d_b + \delta d_b)]^{\frac{1}{2}} = (gd_b)^{\frac{1}{2}} + \frac{1}{2}\left(\frac{g}{d_b}\right)^{\frac{1}{2}}\delta d_b.$$

3.5.5 *The equation of beach planform evolution*

Combining equations (3.10) and (3.25), we obtain a single equation describing the planform evolution of a shoreline:

$$\frac{\partial \xi}{\partial t} = G_1 \frac{\partial^2 \xi}{\partial x^2}, \tag{3.26}$$

where

$$G_1 \equiv -\frac{\widetilde{K}}{32}\frac{H_0^2}{d_{cl} + B}\sqrt{gd_b}\left[2\cos(2\alpha_0) + \frac{1}{2}K_1 \sin(2\alpha_0)\frac{H_0}{d_b}\right]. \tag{3.27}$$

We see that the exact equation has the same framework of the traditional (3.13), but the longshore diffusivity is markedly different. It suffices to note that G of the traditional equation is nearly always positive (usually α_b is rather close to $\pi/2$, so that $\cos(2\alpha_b)$ is smaller than zero); while G_1 of the exact solution can well be smaller than zero for a small α_0. We shall see in sect. 3.7 that a negative G_1 has a clear physical meaning. The consequence of the difference between the exact and the traditional solution should be already apparent: the fact that in a few cases the longshore diffusivity is negative rather than positive implies an opposite evolution of the beach planform, that is erosion instead of accretion or vice versa.

3.5.6 *The effect of the curvature of the contour lines on the sediment transport*

Equation (3.25) of Q_s has been obtained under the assumption that the contour lines at any fixed cross-section are straight. If we counted the effect of contour lines' curvature, the expression of Q_s would include some additional terms in

$$\Delta y_0 \frac{\partial^2 \xi}{\partial x^2} \quad \text{and} \quad \Delta y_0^2 \frac{\partial^3 \xi}{\partial x^3}.$$

These additional terms can be neglected because of the assumption (3.15).

3.6 Problem of beach planform evolution: the case of contour lines parallel only within a certain distance from the shoreline

3.6.1 *An idealized configuration of a nourished beach*

The original (unnourished) beach of fig. 3.7 consists of x-parallel contour lines, with a berm at a depth d_n [see fig. 3.7]. A restoration program modifies the beach profile up to the berm. The new beach contour lines are parallel to the new shoreline [see again fig. 3.7]. Therefore, from offshore to coast, we find first the x-parallel contour lines of the original topography, then the berm, and then the curved contour lines of the nourished beach.

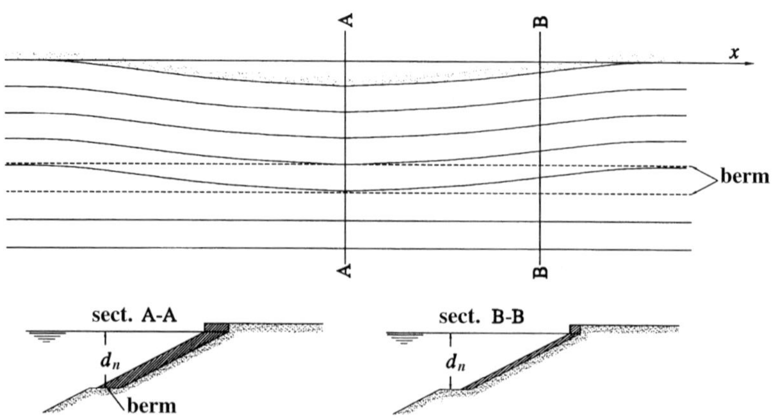

Fig. 3.7 Our pattern of a nourished beach: the initially straight contour lines are modified up to depth d_n because of the added sand.

Also in this case the evolution of the shoreline planform is analytically predictable. To this end, we remake assumptions (3.11) and (3.15) on the derivatives of the shoreline. Moreover, for getting the transport rate Q_s we assume that the contour lines of the nourished beach at any fixed cross-section are straight. Doing so, that is neglecting the contour lines' curvature, we shall make a negligible error thanks to assumption (3.15).

3.6.2 *The formula for Q_s*

On water depth d_n the wave characteristics are the same that would be there without the nourishment; indeed, the original seabed topography with the x-parallel contour lines still remains on water depths larger than d_n. Hence, on water depth d_n, the wave height takes on a constant value H_n and the direction of wave advance makes a constant angle α_n with the x-axis; and H_n and α_n are related to H_0 and α_0 by equations (2.43) and (2.45) of shoaling-refraction:

$$\cos\alpha_n = \cos\alpha_0 \tanh(k_n d_n), \tag{3.28a}$$

$$H_n\{\tanh(k_n d_n)[\sinh(2k_n d_n) + 2k_n d_n]/\sinh(2k_n d_n)\}^{\frac{1}{2}} = H_0\left[\frac{1-\cos^2\alpha_0}{1-\cos^2\alpha_n}\right]^{\frac{1}{4}}, \tag{3.28b}$$

where k_n is the wave number on water depth d_n.

The sediment transport rate Q_s at any fixed cross-shore section is given by (3.9a) or by (3.9b) (this only for $d \le d_n$); equation (3.9c) is no longer valid because the contour lines are parallel to the shoreline only up to water depth d_n. It is convenient to express Q_s in the form (3.9b) with $d = d_n$ because on water depth d_n the wave height takes on the constant value H_n and the angle between the direction of wave advance and the contour lines of the nourished beach is $\alpha_n + \delta\alpha_n$ with

$$\delta\alpha_n = \frac{\partial\xi}{\partial x},$$

and α_n constant. Therefore we have

$$Q_s = \frac{1}{32}\widetilde{K}\,H_n^2\left[1 + \frac{2k_n d_n}{\sinh(2k_n d_n)}\right]\sin[2(\alpha_n + \delta\alpha_n)]\,\sqrt{g(d_b + \delta d_b)}, \tag{3.29}$$

where $d_b + \delta d_b$ is the new water depth at breaking and d_b was the water depth at breaking when the contour lines were x-parallel.

3.6.3 *The variation δd_b of the water depth at breaking*

The reasoning is essentially the same as in sect. 3.5.3, with the only difference that δH_b now is the variation of the wave height at breaking for a variation $\delta\alpha_n$ of the wave direction on depth d_n and *no variation of the wave height on depth d_n*.

Retracing the steps of sect. 3.5.3 we arrive at

$$\delta d_b = K_2 H_0 \delta\alpha_n, \tag{3.30}$$

$$K_2 \equiv \frac{\dfrac{1}{H_0}\left(\dfrac{\partial H}{\partial x_4}\right)_{x_3 = H_n,\ x_4 = \alpha_n,\ d = d_b}}{0.8 - \left(\dfrac{\partial H}{\partial d}\right)_{x_3 = H_n,\ x_4 = \alpha_n,\ d = d_b}}, \tag{3.31}$$

where

$x_3 \equiv$ wave height on water depth d_n,

$x_4 \equiv$ angle between the direction of wave advance and the contour lines on water depth d_n.

$H(x_3, x_4, d)$ gives the wave height as a function of water depth d and of wave characteristics x_3 and x_4 on water depth d_n (for a fixed wave period T) under the assumption of straight contour lines. This function proceeds from the usual equations (2.43) and (2.45) of shoaling-refraction:

$$H(x_3, x_4, d) = x_3 \left\{ \frac{\tanh(k_n d_n)[\sinh(2k_n d_n) + 2k_n d_n]\sinh(2kd)}{\tanh(kd)[\sinh(2kd) + 2kd]\sinh(2k_n d_n)} \right\}^{\frac{1}{2}} \cdot$$

$$\cdot \left\{ \frac{1 - \cos^2 x_4}{1 - [\tanh(kd)/\tanh(k_n d_n)]^2 \cos^2 x_4} \right\}^{\frac{1}{4}}.$$

Differentiating it, we get

$$\left(\frac{\partial H}{\partial x_4}\right)_{x_3 = H_n,\, x_4 = \alpha_n,\, d = d_b} = H_0 \frac{1}{2} \left[\frac{S2}{T1\,(S2 + B2)}\right]^{\frac{1}{2}} \left[\frac{1 - C0^2}{1 - T0^2 C0^2}\right]^{\frac{1}{4}} C0\,T0\,\sqrt{1 - T0^2 C0^2} \cdot$$

$$\cdot \left[\frac{1 - T1^2/T0^2}{(1 - T1^2 C0^2)^2}\right] \left[\frac{1 - T0^2 C0^2}{1 - T1^2 C0^2}\right]^{-\frac{3}{4}},$$

$$\left(\frac{\partial H}{\partial d}\right)_{x_3 = H_n,\, x_4 = \alpha_n,\, d = d_b} = \text{r.h.s. of } (3.24b),$$

where we have used (3.28*a*-*b*) and the compact symbols

$$C1 \equiv \cosh(k_b d_b), \quad T1 \equiv \tanh(k_b d_b), \quad B2 \equiv 2k_b d_b, \quad S2 \equiv \sinh(2k_b d_b),$$

$$C2 \equiv \cosh(2k_b d_b), \quad C0 \equiv \cos\alpha_0, \quad T0 \equiv \tanh(k_n d_n).$$

3.6.4 *The equation of planform evolution of the nourished beach*

Bearing in mind that $\delta\alpha_n = \partial\xi/\partial x$, from (3.29) and (3.30) we obtain

$$Q_s = \frac{1}{32} \widetilde{K} H_n^2 \left[1 + \frac{2k_n d_n}{\sinh(2k_n d_n)}\right] \left[\sin(2\alpha_n)\sqrt{g\,d_b} + 2\cos(2\alpha_n)\sqrt{g d_b}\,\frac{\partial\xi}{\partial x} + \right.$$

$$\left. + \frac{1}{2}\sin(2\alpha_n)\sqrt{\frac{g}{d_b}}\,H_0 K_2 \frac{\partial\xi}{\partial x}\right].$$

Hence, with equation (3.10) of the sediment conservation, we arrive at

$$\frac{\partial\xi}{\partial t} = G_2 \frac{\partial^2\xi}{\partial x^2}, \tag{3.32}$$

where

$$G_2 \equiv -\frac{\widetilde{K}}{32}\frac{H_0^2}{d_{cl} + B}\sqrt{gd_b}\,\frac{1}{T0}\sqrt{\frac{1 - C0^2}{1 - T0^2 C0^2}}\left[2(2T0^2 C0^2 - 1) + K_2\sqrt{1 - T0^2 C0^2}\,T0\,C0\,\frac{H_0}{d_b}\right]. \tag{3.33}$$

Use (3.28*a-b*) to check this expression of G_2, and prove that it correctly coincides with the expression (3.27) of G_1 for $T0 = 1$.

3.7 Planform evolution of a natural shoreline

3.7.1 *Analysis of the longshore diffusivity* G_1

Let us consider the longshore diffusivity G_1 given by (3.27). G_1 thought of as a function of α_0 proves to be symmetrical with respect to $\alpha_0 = \pi/2$:

$$G_1\left(\frac{\pi}{2} + \Delta\alpha_0\right) = G_1\left(\frac{\pi}{2} - \Delta\alpha_0\right).$$

[In examining (3.27) of G_1, bear in mind that K_1 depends on $\sin(2\alpha_0)$, so that G_1 is the sum of a term depending on $\cos(2\alpha_0)$ and of a term depending on $\sin^2(2\alpha_0)$.] Therefore we limit ourselves to analyse G_1 on the domain

$$0 < \alpha_0 \le \frac{\pi}{2},$$

and, to make this analysis easier, we define

$$G_1 \equiv G_1' + G_1'',$$

where

$$G_1' \equiv -\left[\frac{\widetilde{K}}{32}\frac{H_0^2}{d_{cl} + B}2\sqrt{gd_b}\right]\cos(2\alpha_0),$$

$$G_1'' \equiv -\left[\frac{\widetilde{K}}{32}\frac{H_0^2}{d_{cl} + B}\frac{1}{2}\sqrt{gd_b}\frac{H_0}{d_b}\right]K_1\sin(2\alpha_0).$$

Here $\widetilde{K}$ is positive. As to K_1 [definition (3.23)], it is given by the product of a positive term and $\sin(2\alpha_0)$. Therefore, it follows that

$$G_1'\begin{cases} < 0 & \text{for} \quad 0 < \alpha_0 < \dfrac{\pi}{4}, \\ = 0 & \text{for} \quad \alpha_0 = \dfrac{\pi}{4}, \\ > 0 & \text{for} \quad \dfrac{\pi}{4} < \alpha_0 \le \dfrac{\pi}{2}, \end{cases}$$

$$G_1''\begin{cases} < 0 & \text{for} \quad 0 < \alpha_0 < \dfrac{\pi}{2}, \\ = 0 & \text{for} \quad \alpha_0 = \dfrac{\pi}{2}. \end{cases}$$

The conclusion being that G_1 is smaller than zero if $0 < \alpha_0 \le \pi/4$, and is greater than zero if $\alpha_0 = \pi/2$. Hence, given that $G_1(\alpha_0)$ is strictly increasing on $[\pi/4, \pi/2]$, it will exist

$$\alpha_{0\text{crit}} > \frac{\pi}{4},$$

such that

$$G_1 \begin{cases} < 0 & \text{for} \quad \alpha_0 < \alpha_{0\text{crit}}, \\ > 0 & \text{for} \quad \alpha_0 > \alpha_{0\text{crit}}. \end{cases} \tag{3.34}$$

3.7.2 *Whether a foreland undergoes erosion or accretion*

$G_1 > 0$ implies that a foreland is eroded, and, on the contrary, $G_1 < 0$ implies that the foreland grows. This is a consequence of equation (3.26) and of the fact that $\partial^2 \xi / \partial x^2$ of a foreland is negative. Therefore, basing ourselves on (3.34) we conclude the foreland's evolution will be: erosion if $\alpha_0 > \alpha_{0\,\text{crit}}$, accretion if $\alpha_0 < \alpha_{0\,\text{crit}}$.

As discussed in sect. 3.5, the difference with respect to the traditional solution is remarkable. Indeed the longshore diffusivity G in the traditional equation (3.13) is nearly always positive because α_b is usually close to $\pi/2$ as a consequence of refraction. Hence, the traditional solution foresees that the foreland is eroded whatever the wave direction on deep water.

3.7.3 *Why the foreland grows or is eroded according to whether the wave direction on deep water is very inclined or nearly orthogonal to the shoreline*

Fig. 3.8*a* illustrates the case of waves attacking orthogonally a foreland. The sediment transport rate Q_s is positive at point ② and negative at point ①, because of the opposite angle of the wave direction at breaking. As a consequence, the sediments are removed from the foreland that gradually gets smaller.

Fig. 3.8*b* illustrates the case of an inclined wave attack. Here the sign of $Q_{s①}$ is the same as the sign of $Q_{s②}$, and, as for the absolute value, two contrasting matters are involved: for the first one $Q_{s②}$ should prevail over $Q_{s①}$; and, for the second one, $Q_{s①}$ should prevail over $Q_{s②}$. The first fact is that the wave direction at breaking is more inclined at location ② than at location ①. The second fact is that the wave height at breaking is larger at ① than at ②, since ② is on the lee side of the foreland. For $\alpha_0 > \alpha_{0\text{crit}}$ it is the first fact to prevail, so that $Q_{s②}$ is larger than $Q_{s①}$ and the foreland is eroded. On the contrary, for $\alpha_0 < \alpha_{0\text{crit}}$ it is the second fact to prevail, so that $Q_{s①}$ is larger than $Q_{s②}$ and the foreland grows.

3.8 Stability of a nourished beach

3.8.1 *Analysis of the longshore diffusivity G_2*

Here we analyse (3.33) to understand how does G_2 vary with α_0 and d_n/L_0 (d_n being the depth of the nourished beach). It should be noted that G_2 depends on α_0 for the term $C0 \equiv \cos\alpha_0$ and depends on d_n/L_0 for the term $T0 \equiv \tanh(k_n d_n)$.

Specifically, let us analyse the quotient

$$G_2(\alpha_0, d_n/L_0)/G_2(\pi/2, \infty),$$

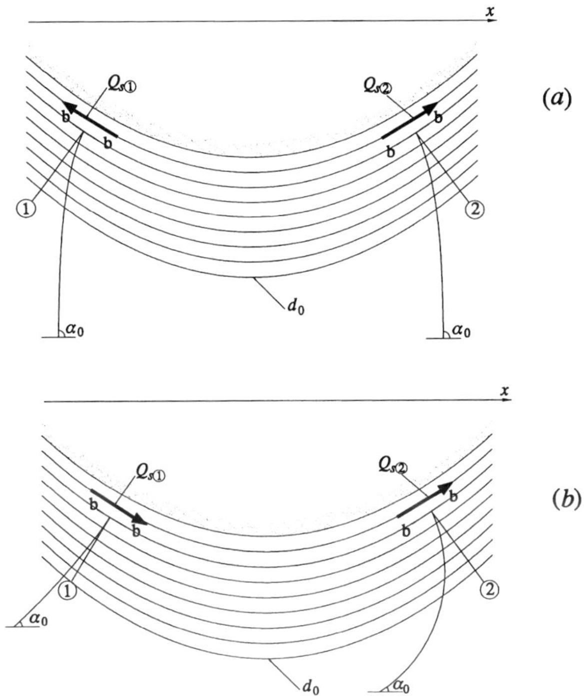

Fig. 3.8 (*a*) A foreland attacked orthogonally is eroded because both $Q_{s①}$ and $Q_{s②}$ are directed outward. (*b*) If the foreland is subjected to an inclined attack, erosion will occur if $Q_{s①}$ is smaller than $Q_{s②}$, while accretion will occur if $Q_{s①}$ is greater than $Q_{s②}$.

where $G_2(\pi/2, \infty)$ is the value of G_2 for the basic case of orthogonal wave attack ($\alpha_0 = \pi/2$) and deep nourishment $\left(d_n/L_0 > \dfrac{1}{2}\right)$. This quotient has been represented in fig. 3.9 as a function of α_0, for a few different values of d_n/L_0, and any fixed H_0 and $d_{cl} + B$, and $H_0/L_0 = 0.05$ (here note that the sensitivity of the graph to a variation of H_0/L_0 is very small).

For $d_n/L_0 > \dfrac{1}{2}$, G_2 tends to coincide with G_1 and thus we see what was illustrated in sect. 3.7.1: G_2 smaller than zero for α_0 within a $\alpha_{0\text{crit}}$. For d_n/L_0 smaller and smaller, the domain on which G_2 is negative shrinks gradually, and in addition the largest positive value of G_2 grows.

3.8.2 *Different degree of stability of a nourished beach according to its depth*

To understand the physical consequences of the analysis on G_2, we must bear in mind equation (3.32) and fig. 3.10. This shows that $\partial^2 \xi/\partial x^2$ is negative on the

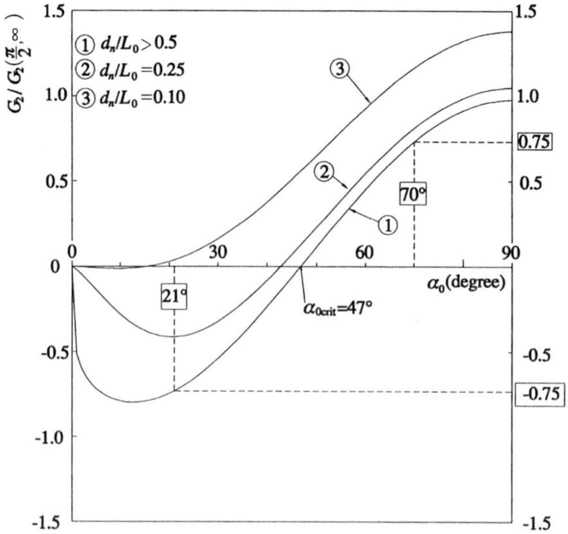

Fig. 3.9 The longshore diffusivity G_2 as a function of the wave direction on deep water, for a few depths of the nourished beach. The nourished beach proves to be the more stable, the greater its depth is and the more inclined the waves are (bearing in mind that $\alpha_0 = \pi/2$ means orthogonal wave attack).

central part of the nourished beach while $\partial^2\xi/\partial x^2$ is positive on the two sides. Therefore, if G_2 is greater than zero, the central part of the nourished beach is eroded and the two sides grow. This means that the nourished beach melts, since the nourished material spreads over the coast. On the contrary, if G_2 is smaller than zero the central part of the nourished beach grows.

The conclusion is that the deeper the toe of the nourished beach, the slower the erosion is. Under the same depth of the nourished beach, the erosion is stronger if the wave direction on deep water is orthogonal to the general alignment of the shoreline. In the case of a deep nourished beach, the waves whose direction on deep water is very inclined with respect to the shoreline restore (rather than destroy) the nourished beach.

Kamphuis et al. (1986) suggest that $\tilde{K}$ in formulae (3.9) for Q_s is proportional to H_b/D (D being the grain size). With their expression for $\tilde{K}$, one can obtain the exact forms of G_1 and G_2, using the formal solutions for δd_b given in sects. 3.5.3 and 3.6.3. As a result, you will find that the range on which G_2 is smaller than zero gets larger than in fig. 3.9, which means smaller beach erosion.

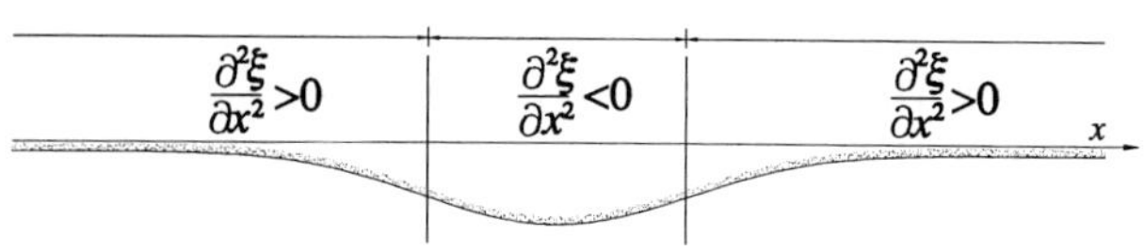

Fig. 3.10 Curvature of a nourished beach planform.

3.8.3 *Different degree of stability of a nourished beach according to its location*

Under the same depth d_n and wave characteristics on deep water, the nourished beach of fig. 3.11*a* on a long straight coast should be more stable than the nourished beach of fig. 3.11*b* between two promontories. Indeed the very inclined waves in case (*b*) cannot reach the nourished beach that is sheltered by the two promontories.

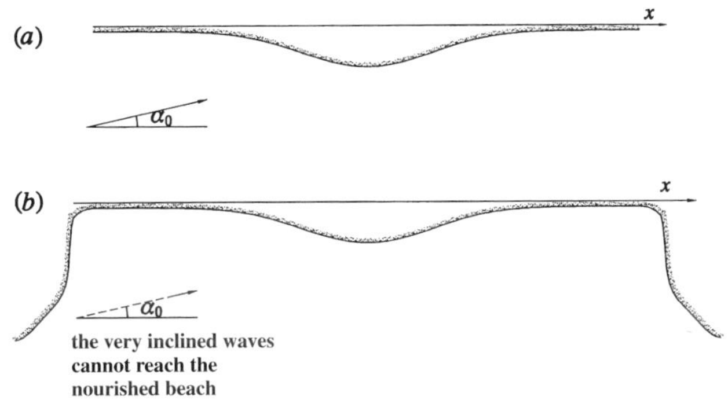

Fig. 3.11 The nourished beach (*b*) is more easily eroded than (*a*), because it cannot be restored by the very inclined waves.

The difference between the case of fig. 3.11*a* and the case of fig. 3.11*b* gets relevant only if the nourished beach is deep, since only in this case do the very inclined waves fully exert their beneficial action. In such a case, it may even be that the nourished beach (*a*), thanks to the action of the very inclined waves, keeps stable. For this to occur, it suffices that the waves with angles smaller than $\alpha_{0\text{crit}}$ and the waves with angles greater than $\alpha_{0\text{crit}}$ have nearly the same frequency of occurrence and the same height on deep water. We shall examine this point in the next section.

Of course the stability of a beach nourishment project is widely affected also by factors such as grain size-distribution and fill placement techniques (e.g. Kana and Mohan, 1998), which go beyond the scopes of our basic illustration of the wave action on a nourished beach.

3.9 Planform evolution of beach nourishment projects

3.9.1 *The solution for the initial shoreline of rectangular planform*

Equations (3.32-33) are exact whatever the time t provided that d_{cl} is equal to d_n, or else they are valid only at time $t = 0$. Henceforth we shall assume

$$d_{cl} = d_n$$

and we shall analyse the evolution of an initial shoreline planform.

Equation (3.32) with the initial condition

$$\xi(x,0)\begin{cases} = a & \text{if} \quad -\dfrac{1}{2}b < x < \dfrac{1}{2}b, \\ = 0 & \text{if} \quad x < -\dfrac{1}{2}b \text{ or } x > \dfrac{1}{2}b, \end{cases}$$

with a and b any pair of positive values, admits the solution

$$\xi(x,t) = \frac{a}{2}\left\{\operatorname{erf}\left[\mathscr{A}_1(x,t)\right] - \operatorname{erf}\left[\mathscr{A}_2(x,t)\right]\right\}, \tag{3.35}$$

with

$$\mathscr{A}_1(x,t) \equiv \frac{1}{2}\left(x + \frac{b}{2}\right) \Big/ \sqrt{G_2 t}, \tag{3.36a}$$

$$\mathscr{A}_2(x,t) \equiv \frac{1}{2}\left(x - \frac{b}{2}\right) \Big/ \sqrt{G_2 t}, \tag{3.36b}$$

where it is assumed that $G_2 > 0$ and $t > 0$.

Here one has simply to obtain the partial derivatives of function (3.35) and verify that these derivatives satisfy equation (3.32). To this end it needs only to recall the definition

$$\operatorname{erf}(w) \equiv \frac{2}{\sqrt{\pi}} \int_0^w e^{-u^2}\, du. \tag{3.37}$$

As to the initial condition, it suffices to note that

$$\lim_{t\to 0} \mathscr{A}_1(x,t) \begin{cases} = +\infty & \text{for every fixed} \quad x > -b/2, \\ = -\infty & \text{for every fixed} \quad x < -b/2, \end{cases}$$

$$\lim_{t\to 0} \mathscr{A}_2(x,t) \begin{cases} = +\infty & \text{for every fixed} \quad x > b/2, \\ = -\infty & \text{for every fixed} \quad x < b/2, \end{cases}$$

$$\lim_{w\to\infty} \operatorname{erf}(w) = 1,$$

$$\lim_{w\to -\infty} \operatorname{erf}(w) = -1,$$

from which it follows that, as $t \to 0$, $\xi(x,t)$ approaches zero for any fixed x greater than $b/2$ or smaller than $-b/2$ and approaches a for any fixed x between $-b/2$ and $b/2$. Of course, ξ is a continuous function of x also as $t \to 0$, and indeed it goes gradually from 0 to a in a neighbourhood of order $\sqrt{G_2 t}$ of $x = -b/2$, and it goes gradually from a to 0 in a neighbourhood of order $\sqrt{G_2 t}$ of $x = b/2$. In simple words: as $t \to 0$, (3.35) is the equation of a rectangle of base b and height a, with round corners.

Fig. 3.12 shows $\xi(x)$ at a few time instants. Function ξ, as we can easily verify from its expression (3.35), is always positive, for every x and t. Moreover the area between $\xi(x,t)$ (the shoreline of the nourished beach) and the x-axis (the original shoreline) keeps constant in time. Indeed

$$\int_{-\infty}^{+\infty} \xi(x,t)\,\mathrm{d}x = ab, \tag{3.38}$$

for every t.

3.9.2 *A solution for a more realistic initial condition*

Function (3.35-36) with G in place of G_2 is used [Dean (1992)] to foresee the planform evolution of a beach nourishment project (G being given a simplified form valid on shallow water).

Really function (3.35) satisfies the differential equation, but fails to satisfy condition (3.11). Indeed, as $t \to 0$, the derivative $\partial\xi/\partial x$ at $x = -b/2$ and $x = b/2$, far from approaching zero, tends to infinity. To obviate this discrepancy, and also to get a more realistic initial condition, Walton (1994) has obtained the analytical solution for beach fills with tapers at the end. Alternatively, we can resort again to function (3.35), but with the following new forms of $\mathscr{A}_1(x,t)$ and $\mathscr{A}_2(x,t)$:

$$\mathscr{A}_1(x,t) \equiv \frac{1}{2}\left(x + \frac{b}{2}\right) \Big/ \sqrt{G_2 t_o + G_2 t}, \tag{3.39a}$$

$$\mathscr{A}_2(x,t) \equiv \frac{1}{2}\left(x - \frac{b}{2}\right) \Big/ \sqrt{G_2 t_o + G_2 t}, \tag{3.39b}$$

where it is assumed that $G_2 > 0$ and $t \geq 0$. With these forms of $\mathscr{A}_1(x,t)$ and $\mathscr{A}_2(x,t)$, the initial planform of the shoreline is the one that originally (i.e. with formulae 3.36*a*-*b* for $\mathscr{A}_1$ and $\mathscr{A}_2$) occurred at instant t_o. The value of t_o will be defined after a few attempts, so to fit the initial design planform of the nourished beach. In summary, we shall have to fix two parameters: t_o, and a or b (bearing in mind that $a \cdot b$ is the total area of the gained beach, which obviously is a design input).

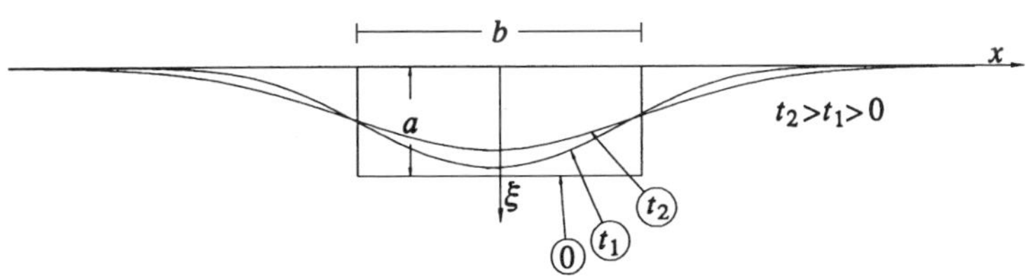

Fig. 3.12 Evolution of an initially rectangular beach planform.

3.9.3 *The solution valid for unsteady wave conditions*

Solution (3.35) with $\mathscr{A}_1, \mathscr{A}_2$ given by (3.39*a*-*b*) enables us to foresee the beach planform evolution provided that the waves keep constant in time, so that G_2 also keeps constant. Let us now allow the possibility that the wave height and direction are not constant in time. In this case, G_2 is no longer a constant, that is we have a $G_2(t)$ in place of the constant G_2. Nevertheless, function (3.35) satisfies the differential equation (3.32) provided that

$$\mathscr{A}_1(x,t) \equiv \frac{1}{2}\left(x + \frac{b}{2}\right) \bigg/ \sqrt{\text{const} + \int_0^t G_2(t')\,\mathrm{d}t'},$$

$$\mathscr{A}_2(x,t) \equiv \frac{1}{2}\left(x - \frac{b}{2}\right) \bigg/ \sqrt{\text{const} + \int_0^t G_2(t')\,\mathrm{d}t'},$$

where const is a positive constant and G_2 may be positive or negative. Naturally, the constant const (whose dimension is square length) will be obtained after a few attempts, so that the initial condition fits the design planform of the nourished beach. Also in this case the parameters to be determined are only two, i.e. const and a or b, given that $a \cdot b$ is the known area of the gained beach.

If

$$\int_0^t G_2(t')\,\mathrm{d}t' > 0,$$

we shall have something like fig. 3.13*a*: erosion of the nourished beach. On the contrary, if

$$\int_0^t G_2(t')\,\mathrm{d}t' < 0,$$

we shall have something like fig. 3.13*b*: accretion of the nourished beach.

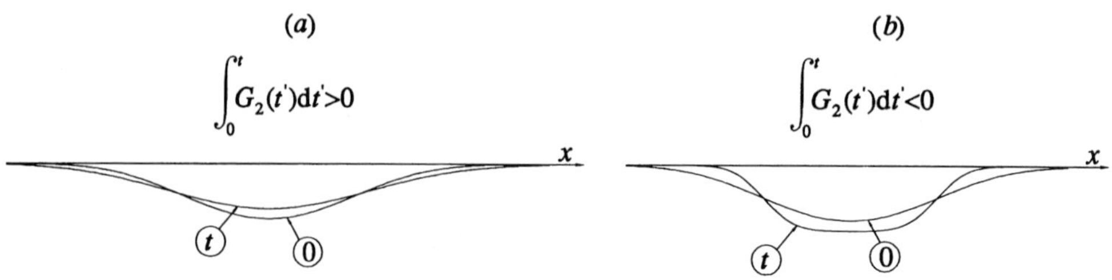

Fig. 3.13 (*a*) The integral of $G_2(t)$ is positive: erosion of the nourished beach. (*b*) The integral of $G_2(t)$ is negative: accretion of the nourished beach.

3.9.4 *Example: planform evolution of the nourished beach under unsteady wave conditions*

Let us imagine that a nourished beach is subjected to one hundred hours of waves with given H_0 and L_0, and α_0 equal to 70°; subsequently it is subjected to one hundred hours of waves with the same H_0 and L_0, and α_0 equal to 20°. Let us assume also that $d_n/L_0 \geq \frac{1}{2}$. As we see from fig. 3.9, the G_2 of $\alpha_0 = 20°$ is nearly opposite to the G_2 of $\alpha_0 = 70°$. Hence

$$\int_0^t G_2(t')\,\mathrm{d}t'$$

is zero for $t = 200$ hours, and consequently the offshore distance $\xi(x)$ after the two hundred hours is unchanged with respect to the initial shoreline.

Note, fig. 3.9 is based on the assumption of constant H_0 and L_0, and hence it is suitable for the aim of this example. Whereas, for comparing the G_2 of two wave trains with different values of H_0 and/or L_0, one has to resort to equation (3.33).

3.10 A useful simplification

3.10.1 *The core of the equation of $\partial\xi/\partial t$*

Since Q_s depends on $< f_{ax} >$ and on $\sqrt{gd_b}$, $\partial\xi/\partial t$ depends on the longshore variations of $< f_{ax} >$ and $\sqrt{gd_b}$. Here we shall prove that the basic dependence is on $< f_{ax} >$. To achieve this we shall restart the treatment from the beginning, using the following simplified forms of Q_s and $\partial\xi/\partial t$:

$$Q_s \cong \text{const} < f_{ax} > = -\,\text{const} < f_{fx} >, \tag{3.40a}$$

$$\frac{\partial\xi}{\partial t} \cong -\,\text{const}\,\frac{\partial Q_s}{\partial x}, \tag{3.40b}$$

where the constants are positive.

3.10.2 *Planform evolution of a nourished beach: an approximate equation*

From (3.40*a*) and (3.3) we have

$$Q_s \cong \text{const}\,H^2\left[1 + \frac{2kd}{\sinh(2kd)}\right]\sin(2\alpha), \tag{3.41}$$

where H and α are the wave height and the angle of the direction of wave advance on an arbitrary water depth d (provided that the contour lines are parallel to each other within this water depth).

Let us now consider the idealized configuration of a nourished beach described in sect. 3.6.1 and fig. 3.7. In this case, it is convenient to apply (3.41) with $d = d_n$, since the wave height on water depth d_n takes on the constant value H_n, and the angle between the direction of wave advance and the beach contour lines is equal to $\alpha_n + \partial\xi/\partial x$ with α_n constant. Therefore

$$Q_s \cong \text{const}\, H_n^2 \left[1 + \frac{2k_n d_n}{\sinh(2k_n d_n)}\right]\left[\sin(2\alpha_n) + 2\cos(2\alpha_n)\frac{\partial\xi}{\partial x}\right].$$

Then with (3.28*a-b*) relating H_n, α_n to H_0, α_0, it follows that

$$Q_s \cong \text{const}\, H_0^2 \frac{1}{T0}\sqrt{\frac{1 - C0^2}{1 - T0^2 C0^2}}\left[2\sqrt{1 - T0^2 C0^2}\; T0\; C0 + 2(2T0^2 C0^2 - 1)\frac{\partial\xi}{\partial x}\right],$$

where *T0* stands for $\tanh(k_n d_n)$ and *C0* for $\cos\alpha_0$.

Finally, with this formula for Q_s and equation (3.40*b*) of $\partial\xi/\partial t$, we arrive at

$$\frac{\partial\xi}{\partial t} \cong G_2^* \frac{\partial^2\xi}{\partial x^2}, \tag{3.42}$$

where G_2^* (that stands for *approximate* G_2) is given by

$$G_2^* \equiv -\text{const}\, H_0^2 \frac{1}{T0}\sqrt{\frac{1 - C0^2}{1 - T0^2 C0^2}}\,(2T0^2 C0^2 - 1).$$

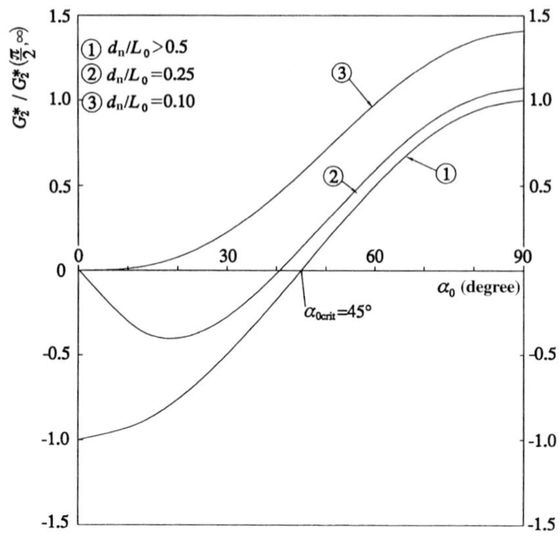

Fig. 3.14 The longshore diffusivity obtained by means of the simplified equations (3.40*a-b*). The exact form is given in fig. 3.9.

3.10.3 *The prediction of beach planform evolution by means of the approximate equation*

The approximate longshore diffusivity G_2^* is shown in fig. 3.14 as a function of α_0 and d_n/L_0. We see that it re-proposes all the essential features of the exact solution that was given in fig. 3.9. In particular G_2^*, like G_2, is greater than zero only for α_0 greater than a threshold $\alpha_{0\text{crit}}$ which is nearly the same for G_2^* and G_2. Moreover G_2^*, like G_2, grows as d_n/L_0 decreases.

Thus, even from the approximate solution, we arrive at the two essential concepts: the waves nearly orthogonal to the shoreline destroy the nourishment, while the very inclined waves restore the nourishment; the demolition gets less intense and the restoration gets more intense as the depth of the nourished beach grows.

Conclusion: the simplified forms (3.40*a-b*) are effective to understand the main trend of the beach planform evolution; and, thanks to these simplified forms, also the prediction of the beach planform evolution caused by coastal structures gets easier, as we shall see in the following section.

3.11 Beach planform evolution caused by structures

3.11.1 *A general equation for the evolution trend of a beach near to coastal structures*

In order to predict the beach planform evolution caused by a structure we shall model the beach as an ideal *absorber* which does not alter the wave motion in front of it. It does not reflect nor transmit energy. The absorber is located at the original shoreline (x-axis) as is shown in figs. 3.15 and 3.17 which illustrate, respectively, the case of a groin and of a detached breakwater.

Let us consider any small control volume of width δx such as the one of fig. 3.15*b*. Let us compare this control volume with the same control volume without the absorber [cf. fig. 3.15*c*]. The difference, for what concerns the x-component of the linear momentum equation, is that in the control volume of fig. 3.15*b* there is the force $< f_{fx}(x) > \delta x$ exerted by the absorber while in the control volume of fig. 3.15*c* there is the efflux of linear momentum $R_{yx}(x,0)\ \delta x$. The other terms are the same in the two control volumes, given that the ideal absorber does not alter the wave motion before it. Therefore, the x-component of the linear momentum equation for the control volume of fig. 3.15*b* is equal to the x-component of the linear momentum equation for the control volume of fig. 3.15*c*, apart from two differences: the presence on the l.h.s. of the force exerted by the absorber; and the absence on the r.h.s. of the linear momentum efflux from the upper x-parallel side of the control volume. Since both the small control volumes, the one of fig. 3.15*b* and the one of fig. 3.15*c*, must satisfy the linear momentum equation, it follows that

$$< f_{fx}(x) > = -R_{yx}(x,0)\,.$$

Hence, by means of (3.40*a*-*b*), we get

$$\frac{\partial \xi}{\partial t} \cong -\mathrm{const}\ \frac{\partial R_{yx}(x,0)}{\partial x}, \tag{3.43}$$

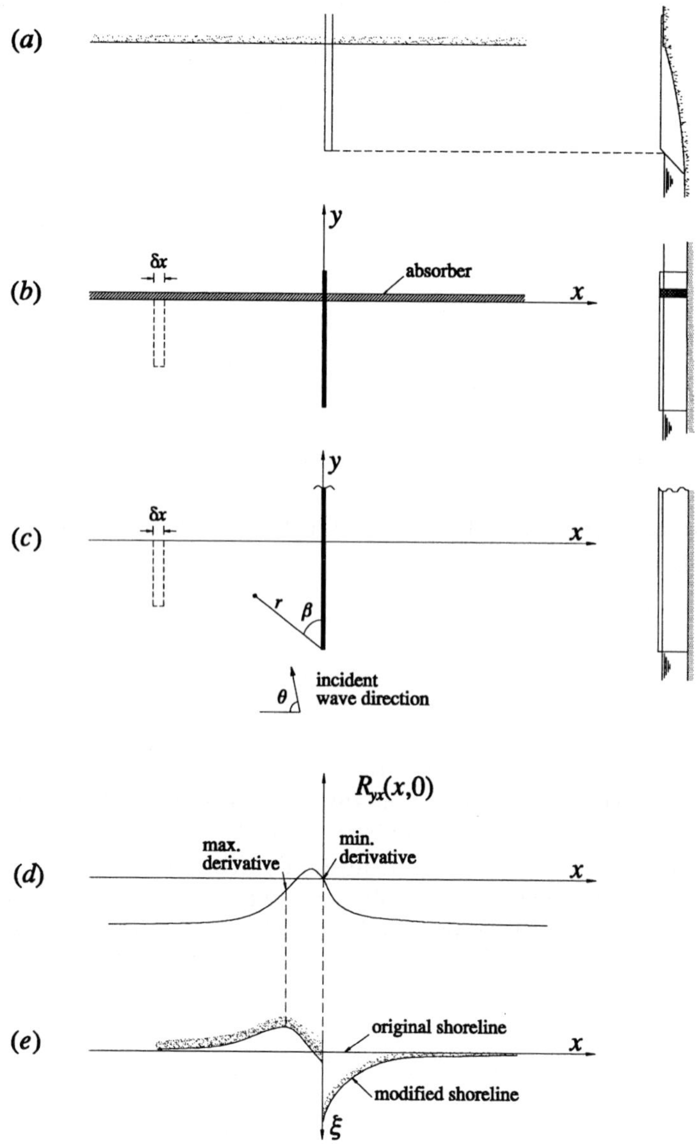

Fig. 3.15 (*a*) Groin. (*b*) Beach thought of as an ideal absorber. (*c*) Absorber removed: comparing (*c*) to (*b*) we realize that $< f_{fx}(x) >= -R_{yx}(x, 0)$. (*d*) Function $R_{yx}(x, 0)$. (*e*) Evolution trend of the shoreline planform obtained by means of (3.43). The R_{yx} of the diffracted waves was contributed by P. Filianoti.

where, as said, const is a positive constant. The derivative $\partial\xi/\partial t$ gives the *evolution trend*: if $\partial\xi/\partial t$ is positive the dry beach grows; if $\partial\xi/\partial t$ is negative the dry beach gets narrower; the larger the absolute value of $\partial\xi/\partial t$, the larger is the deformation of the dry beach.

Equation (3.43) is a general tool to predict the evolution trend of a beach in consequence of the building of some coastal structure. One has to estimate the R_{yx} at the original shoreline (assumed to coincide with the x-axis), as if the beach was not there. Then the evolution trend is simply given by the opposite of the derivative $\partial R_{yx}/\partial x$.

To a first approximation, one assumes a constant water depth (the mean water depth within the shoreline and the groin's tip or detached breakwater). As for the radiation stress R_{yx}, it can be obtained analytically for a few configurations of basic interest (we shall see two examples in the next section).

With this approach, we can estimate the diffraction effect which is the principal responsible for the beach deformation near coastal structures.

3.11.2 *The cases of a groin and of a detached breakwater*

Groin

For the case of the groin (fig. 3.15), we obtain R_{yx} from the formula (1.59*b*) for the velocity potential of the waves interacting with a semi-infinite reflecting wall (the groin). In (1.59*b*), ϕ is given in terms of the polar coordinates r, β. Thus we must resort to the conversion rules

$$\frac{\partial\phi}{\partial x}\,\mathrm{d}x = \frac{\partial\phi}{\partial r}\,(-\mathrm{d}x\sin\beta) + \frac{\partial\phi}{\partial\beta}\left(-\frac{\mathrm{d}x\cos\beta}{r}\right), \tag{3.44a}$$

$$\frac{\partial\phi}{\partial y}\,\mathrm{d}y = \frac{\partial\phi}{\partial r}\,\mathrm{d}y\cos\beta + \frac{\partial\phi}{\partial\beta}\left(-\frac{\mathrm{d}y\sin\beta}{r}\right) \tag{3.44b}$$

which have been written in the coarse form (without cancelling dx and dy) in order to make easier their verification by means of fig. 3.16.

Fig. 3.15*d* shows $R_{yx}(x,0)$, and fig. 3.15*e* shows the *evolution trend* of the beach. R_{yx} is zero at the groin where v_x is zero. The derivative $\partial R_{yx}/\partial x$ has its negative maximum at the wave beaten side of the groin, and its positive maximum at a certain distance from the lee side. Therefore the maximum beach accretion is at the wave beaten side of the groin, and the maximum erosion at a certain distance from the lee side.

Note, in equation (1.59) θ is the angle between the orthogonal to the wall and the direction of wave advance. Accordingly, in fig. 3.15, θ denotes the angle between the orthogonal to the groin and the direction of wave advance.

Note also that the $R_{yx}(x,0)$ shown in the figure has been obtained through the more realistic theory of the wind waves (chap. 8) with a characteristic directional spectrum and an inclined dominant direction. Resorting to the periodic wave pattern, one would find a sequence of local maxima and minima of $R_{yx}(x,0)$ for $x>0$. As a consequence, the disturbance of the groin would affect the shoreline for a great distance.

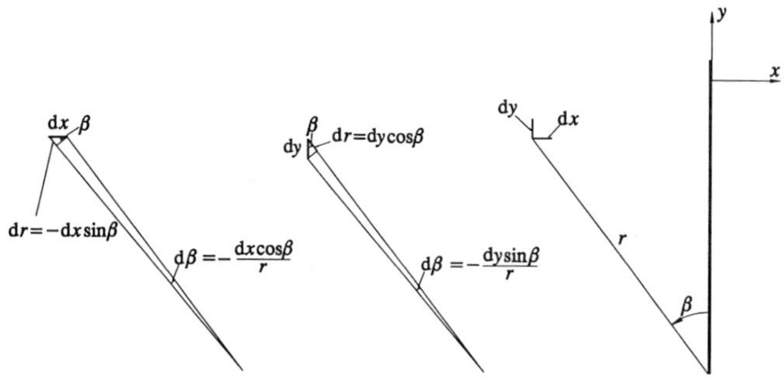

Fig. 3.16 Aid for checking equations (3.44*a*-*b*).

Detached breakwater

In the case of the detached breakwater (fig. 3.17), the radiation stress R_{yx} approaches zero as $|x|$ tends to infinity, since we have assumed the wave attack to be orthogonal to the original shoreline. Moreover, for symmetry, R_{yx} is zero at $x = 0$ (breakwater centre). Before $x = 0$, R_{yx} has a local maximum since the wave diffracted by the left tip of the breakwater attacks the beach from the left side. On the contrary, after $x = 0$, R_{yx} has a minimum (negative maximum) because the wave diffracted by the right tip of the breakwater attacks the beach from the right side. The aforesaid local maximum and minimum of R_{yx} are located roughly opposite the two tips of the detached breakwater. Thus, in the strip between these two tips, $\partial R_{yx}/\partial x$ is smaller than zero, which results in beach accretion. Outside this strip, $\partial R_{yx}/\partial x$ is greater than zero, which results in beach erosion.

Conclusive note

The solution for the set-up is due to the works of Longuet-Higgins and Stewart (1963 and 1964) and of Bowen et al. (1968). The solution for the longshore sediment transport is due to the works of Longuet-Higgins (1970-1971) and Komar and Inman (1970).

As previously said, the traditional solution for the evolution of a beach planform is due to Pelnard-Considère (1956). Dean and Yoo (1992), Work and Dean (1995), and Work and Rogers (1997) pointed out that the longshore variation in the wave direction and height modifies the rate of evolution of a nourished beach, and they gave some improved solutions. However, these solutions are still approximate in that some terms being linearly dependent on $\partial\xi / \partial x$ are neglected or assumed to be constant. The solutions given in sects. 3.5-8 are novelties. They are exact as $\partial\xi / \partial x \to 0$ under the assumption of sect. 3.5.1, at the initial time instant (the solution for the ideal nourished beach is exact whatever the time instant, provided $d_n = d_{cl}$).

The pattern of sect. 3.11, based on the radiation stress tensor of the diffracted waves, is a new proposal. Indeed, so far the equation (3.13), which is valid if the beach deformation is due to wave *refraction*, is applied also to beaches with groins or detached breakwaters, where the deformation is essentially due to wave *diffraction*. For example, Larson et al. (1997) apply equation (3.13) with two different values of G: one for the lee-shore and the other one for the rest of the shoreline.

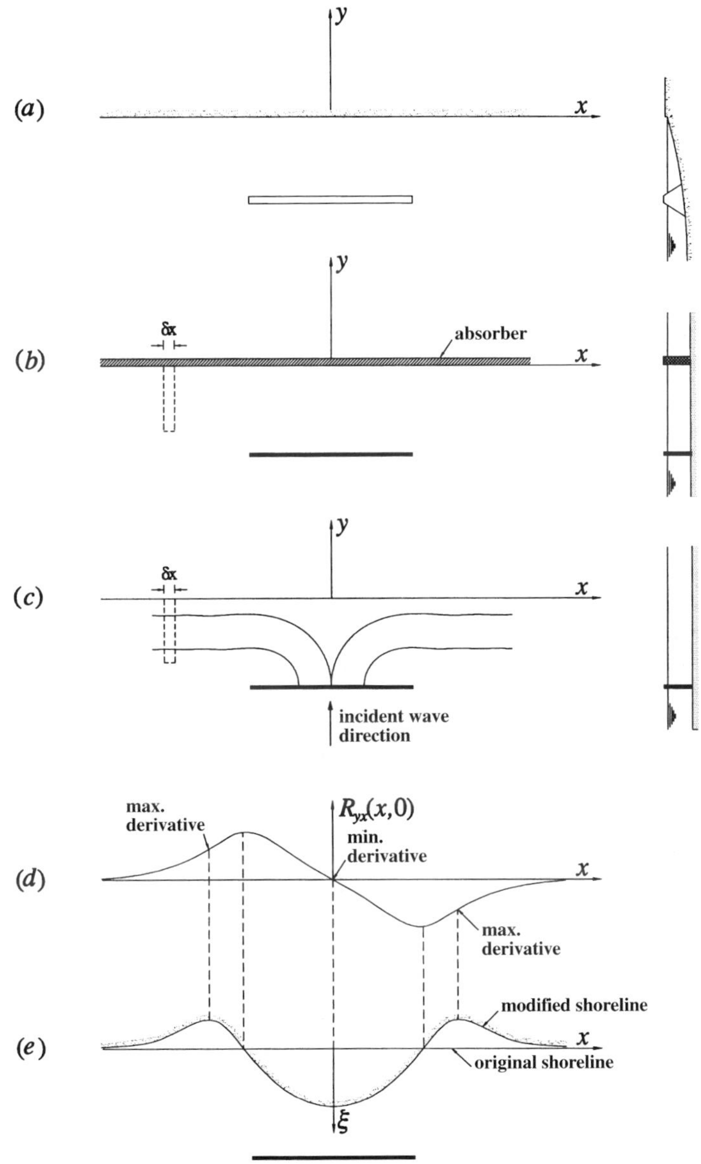

Fig. 3.17 (*a*) Detached breakwater. (*b*), (*c*), (*d*), (*e*) [see caption of fig. 3.15].

References

Battjes J. A., 1974 Computation of set-up, longshore currents, run-up and overtopping due to wind-generated waves. *Dept. Civil Eng. Delft Univ. of Technology*, Report 74.

Bodge K.R. and Kraus N.C., 1991 Critical examination of longshore sediment transport rate magnitude. *Proc. Coastal Sediments '91 ASCE*, 139-155.

Bowen A. J., Inman D. L. and Simmons V. P., 1968 Wave set-down and wave set-up. *J. Geophys. Res.* 73, 2569-2577.

Camfield F.E., 1980 Tsunami engineering. *U.S. Army, Coastal Engineering Research Center, Report* 6, 1-222.

Dean R.G., 1992 Beach nourishment: Design Principles. *Proc. Short Course attached to the* 23rd *Conf. Coastal Eng.*, 301-349.

Dean R.G. and Yoo C.-H., 1992 Beach-nourishment performance predictions. *J. Waterway, Port, Coastal, and Ocean Eng.* 118, 567-586.

Kamphuis J.W., Davies M.H., Nairn R.B. and Sayao O.J., 1986 Calculation of littoral sand transport rate. *Coastal Eng.* 10, 1-21.

Kana T.W. and Mohan R.K., 1998 Analysis of nourished profile stability following the fifth Hunting Island (SC) beach nourishment project, *Coastal Eng.* 33, 117-136.

Komar P. D. and Inman D. L., 1970 Longshore sand transport on the beaches. *J. Geophys. Res.* 75, 5914-5927.

Larson M., Hanson H. and Kraus N.C., 1997 Analytical solutions of one-line model for shoreline change near coastal structures. *J. Waterway, Port, Coastal, and Ocean Eng.* 123, 180-191.

Lin P., Chang K.-A. and Liu P.L.-F., 1999 Runup and rundown of solitary waves on sloping beaches. *J. Waterway, Port, Coastal, and Ocean Eng.* 125, 247-255.

Longuet-Higgins M. S., 1970 Longshore currents generated by oblique incident waves, 1. *J. Geophys. Res.* 75, 6778-6789;
1971 Recent progress in the study of longshore currents, in Waves on beaches and resulting sediment transport, *Academic Press*, New York.

Longuet-Higgins M. S. and Stewart R. W., 1963 A note on wave set-up. *J. Marine Res.* 21, 4-10;
1964 Radiation Stresses in water waves; a physical discussion with application. *Deep-Sea Res.* 11, 529-562.

Pelnard-Considère R., 1956 Essai de thèorie de l'évolution des formes de rivages en plages de sable et de galets. *Quatrième Journées de l'Hydraulique, Les Energies de la Mer, Question* 3.

Peregrine D.H., Cokelet E.D. and McIver P., 1980 The fluid mechanics of the waves approaching breaking. *Proc. 17th Conf. Coastal Eng., ASCE*, 512-528.

U. S. Army Corps of Engineers, CERC, 1984 Shore Protection Manual. *U.S. Government Printing Office*, Washington, D.C.

Walton T.L., 1994 Shoreline solution for tapered beach fill. *J. Waterway, Port, Coastal, and Ocean Eng.* 120, 651-655.

Work P.A. and Dean R.G., 1995 Assessment and prediction of beach-nourishment evolution. *J. Waterway, Port, Coastal, and Ocean Eng.* 121, 182-189.

Work P.A. and Rogers W.E., 1997 Wave transformation for beach nourishment projects. *Coastal Eng.* 32, 1-18.

Chapter 4
WIND GENERATED WAVES: BASIC CONCEPTS

4.1 The sea state

Fig. 4.1 shows a record of the surface displacement η at a fixed location at sea. As with the periodic waves, $\eta(t)$ represents the free surface elevation above the mean water level. A single wave is a piece of $\eta(t)$ between two consecutive zero up-crossings; the wave period is the interval between the two extreme zero up-crossings; the wave crest is the highest local maximum and the wave trough is the lowest local minimum of the wave; the wave height is measured from trough to crest. We see from the figure that the waves generally have different sizes and shapes.

By *ideal sea state* we mean an infinitely long stationary time series of wind generated waves. To understand this definition, let us imagine we gather a number of sets of N consecutive waves and we estimate the mean height and period of each of these sets: $\overline{H}_1$, $\overline{T}_1$ will be respectively the mean wave height and the mean wave period of the first set, $\overline{H}_2$, $\overline{T}_2$ will be the mean wave height and the mean wave period of the second set, and so on. For a small N, say $N=5$, the pairs $(\overline{H}_1, \overline{T}_1)$, $(\overline{H}_2, \overline{T}_2)$, ... will generally be very different from one another. However, as N grows, the differences between these pairs will tend to vanish, and as $N \to \infty$ all the pairs will become equal to each other. This is a simple way to introduce the ideal sea state whose mathematical description will be given in the next section.

By *real sea state* we mean a sequence of a few hundred wind-generated waves (typically $100 \div 300$ waves). Such a sequence is sufficiently short to be nearly

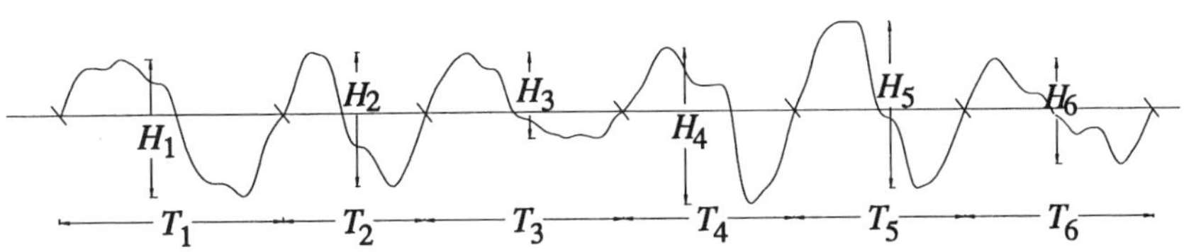

Fig. 4.1 Record of sea waves at a fixed point.

stationary, and it is long enough for its statistical properties to be meaningful. In other words, it can be thought of as a sequence drawn from an ideal sea state, and we can assume that its mean wave height and period are very close to the mean wave height and period of this ideal sea state.

If we drew a sequence of a few waves from an ideal sea state (i.e. from a stationary random process), the mean wave height and period of this sequence could be very different from the mean wave height and period of the ideal sea state. On the contrary, if we recorded a sequence of a few thousand sea waves, generally it could not be thought of as a sequence from an ideal sea state, in that the assumption of stationary time series might be grossly unsatisfied. For example, the first half of the sequence might belong to a stage of nearly calm sea, while the second half might include a sea storm.

The best duration of a real sea state is that leading to the smallest difference between the first half and the second half. To seek the best duration of a sea state, we have resorted to the time series of experiment RC 1990 in the Straits of Messina off Reggio Calabria.

From the wave records we have obtained a number of sequences, each consisting of $N/2$ consecutive waves. The first sequence goes from wave number 1 (the first recorded wave) to wave number $N/2$; the second sequence goes from wave number $N/2+1$ to wave number N, and so on. Then we have evaluated the root mean square surface displacement of each sequence: σ_1 is the r.m.s. of the first sequence, σ_2 the r.m.s. of the second sequence, and so on. Finally, we have defined the random variable

$$V_i \equiv \frac{\sigma_{i+1}}{\sigma_i} - 1\,,$$

and we have obtained the variance of this variable:

$$\overline{V_i^2} = \frac{1}{(n-1)} \sum_{i=1}^{n-1} V_i^2\,,$$

where n is the number of sequences of $N/2$ waves.

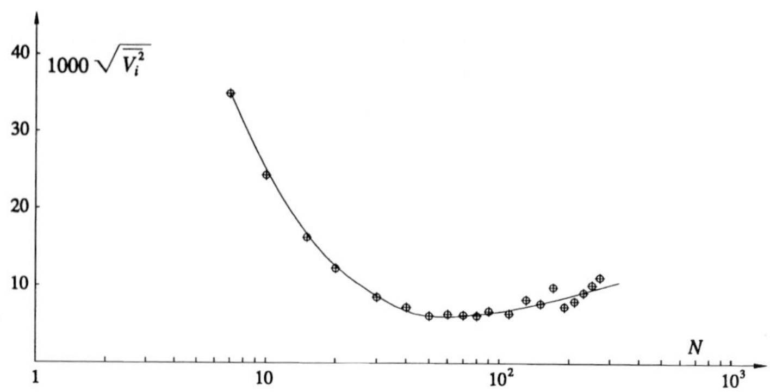

Fig. 4.2 Degree of difference between the first half and the second half in a sequence of N waves (from experiment RC 1990).

This operation has been repeated with different values of number N (of course N is an even number). As a result we have obtained the function

$$\sqrt{\overline{V_i^2}} = f(N)$$

that gives the degree of difference between the first half and the second half of the sequence of N waves. This function is represented by fig. 4.2 which shows that the degree of difference is minimum for N of about one hundred.

4.2 The theory of the sea states

4.2.1 *General overview*

Let us choose a point at sea and to make the ideas more specific let us think of the point as being at the centre of the Tyrrhenian Sea. At a certain time instant waves begin forming at this point. These waves may be under the influence of the wind in a generating area, or may also be waves out of their generating area. In the first case they are called *wind waves* and in the second case they are called *swells*.

Let us record $\eta(t)$ at the fixed point after some time from the beginning of the wave motion, say from instant $t_{\rm inf}$ to instant $t_{\rm sup}$, with $t_{\rm sup} - t_{\rm inf}$ = duration of a real sea state, that is a few hundred consecutive waves. Let us now suppose that the same storm is repeated many times, each time with the same speed, direction and duration of the wind all over the Thyrrhenian Sea. Then let us imagine we record the surface displacement $\eta(t)$ at the fixed point, each time starting at instant $t_{\rm inf}$ and ending at instant $t_{\rm sup}$. We call $\eta_1(t)$, $\eta_2(t)$, ..., $\eta_n(t)$ the records so obtained.

According to the theory of the sea states to the first order in a Stokes' expansion, each of the n time series $\eta_1(t)$, $\eta_2(t)$, ..., $\eta_n(t)$ is a piece of a new realization of a stationary Gaussian process. Each realization of this process has an infinite

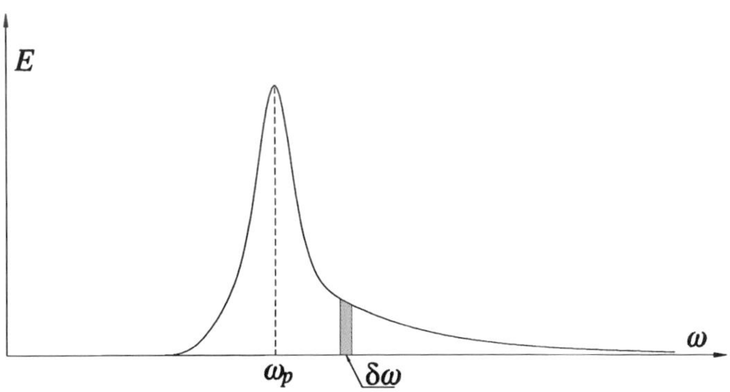

Fig. 4.3 Definition of spectrum: the small dark area is equal to the partial sum $\sum_i \frac{1}{2} a_i^2$ for i such that ω_i belongs to the small interval of amplitude $\delta\omega$.

duration, and thus it represents the ideal sea state introduced in sect. 4.1. The analytical form of the process is

$$\eta(t) = \sum_{i=1}^{N} a_i \cos(\omega_i t + \varepsilon_i), \tag{4.1}$$

where it is assumed that frequencies ω_i are different from each other, number N is infinitely large, phase angles ε_i are uniformly distributed on $(0,2\pi)$ and are stochastically independent of each other, and all amplitudes a_i are of the same order. Finally, the frequency spectrum $E(\omega)$, which is defined as

$$E(\omega)\,\delta\omega \equiv \sum_i \frac{1}{2} a_i^2 \quad \text{for } i \text{ such that } \omega - \delta\omega/2 < \omega_i < \omega + \delta\omega/2 \tag{4.2}$$

[see fig. 4.3], is assumed to be continuous and to be the same in each realization. Under these hypotheses, (4.1) represents a stationary Gaussian random process. We shall prove this statement in sect. 5.1. The fact that (4.1) is exact to the first order in a Stokes' expansion will be explained in sect. 8.2, where the theory will be developed on the space-time. At the present stage and as far as chap. 7, we deal only with the waves on the time domain at some fixed point.

4.2.2 *Numerical simulation of a sea state*

For a deeper insight into the theory it is useful to see how a sea state can be numerically simulated. To this end we must fix the spectrum $E(\omega)$ (we shall see later some characteristic shapes of the spectrum). Then let us fix a very large number N, say $N = 10^9$. Let us now subdivide the ω-axis into 10^9 small intervals, so that the integral of the spectrum over each of these small intervals is equal to $10^{-9} m_0$, where

$$m_0 \equiv \int_0^\infty E(\omega)\,\mathrm{d}\omega. \tag{4.3}$$

Let us call ω_1 the centroid's abscissa of the first small strip of the spectrum (*small* because its area is one billionth of the total area of the spectrum), ω_2 the centroid's abscissa of the second small strip, and so on as far as ω_{10^9}. Let us fix the amplitudes a_i all equal to $\sqrt{2m_0/N}$.

This is the simplest way to fulfil the assumptions on a_i and ω_i. These a_i and ω_i will be left unchanged in the whole set of realizations of the random process. At this stage only the set of ε_i remains to be fixed. To this end, let us write the integers from 1 to 10^9, each of them on a new sheet of paper. Then let us put these sheets of paper into a box. Let us shuffle and draw the first number. Let us multiply this number by $(2\pi/10^9)$, so obtaining ε_1. Let us put again the drawn number into the box, shuffle again and draw a new number from which we get ε_2 (being understood that the drawn number must be multiplied by $2\pi/10^9$). Let us go on as far as draw No 10^9 from which we shall obtain ε_{10^9}.

The ones so obtained will be the phase angles ε_i of the first realization of the process. For obtaining the phase angles ε_i of the second realization, we have to repeat the whole procedure: to draw the first number; to multiply it by $(2\pi/10^9)$; to put the sheet again into the box; to shuffle; to draw the second number, and so on. Clearly, we are speaking of an ideal procedure!

4.3 Some basic relations in the theory of the sea states

4.3.1 *Definitions*

The standard deviation

$$\sigma \equiv \sqrt{<\eta^2(t)>}, \tag{4.4}$$

with the temporal mean covering the whole duration of the sea state, gives the most direct idea of the strength of the wave motion. The larger the σ, the higher the waves.

The parameter most commonly used is probably the *significant wave height* being defined as

$$H_s \equiv 4\sigma. \tag{4.5}$$

The first theories (in the 50's) gave H_s as equal to $H_{1/3}$ that is the average height of the highest one-third of all the waves of a particular sea state. To better understand the meaning of $H_{1/3}$, let us think of ordering the wave heights of a sea state: H_1 is the highest wave height, H_2 the second highest wave height, and so on as far as H_n that is the smallest wave height (n being the number of the waves forming the sea state). Then $H_{1/3}$ proceeds from the following operation

$$\frac{H_1 + H_2 + ... + H_{n/3}}{n/3}.$$

In other words, the researchers who started the modern ocean engineering aimed to introduce a significant average being representative of the greatest wave heights occurring in a sea state. Later the term *significant average wave height* was shortened to the form *significant wave height*. As to fix H_s equal to 4σ, now we know that $H_{1/3}$ is systematically smaller (of $5 \div 10\%$) with respect to 4σ. Nevertheless, H_s is still used for the strength of the wave motion, like some old units of length are used in place of the metre.

The peak frequency (symbol ω_p) is the frequency of the highest peak of the spectrum. The peak period is the wave period associated with the peak frequency:

$$T_p \equiv 2\pi/\omega_p.$$

L_p denotes the wavelength relevant to the peak period, and L_{p0} is the wavelength on deep water:

$$L_{p0} \equiv \frac{gT_p^2}{2\pi}.$$

The symbol m_j denotes the jth order moment of the spectrum

$$m_j \equiv \int_0^\infty \omega^j E(\omega)\, \mathrm{d}\omega. \tag{4.6}$$

The zeroth moment, that is the integral of the spectrum over $(0, \infty)$, has been already defined in sect. 4.2.2 [see definition (4.3)].

The definition of *autocovariance* is

$$\psi(T) \equiv < \eta(t)\, \eta(t+T) > , \tag{4.7}$$

and thus it is the mean value of the product of the surface displacement at time t and the surface displacement at the later time $t+T$. Naturally, this mean value depends on T, that is the autocovariance is a function of T.

The definition of autocovariance may look abstract like some useless mathematical exercise. On the contrary, we shall see in chap. 9 that the autocovariance takes on a central role in the light of the quasi-determinism theory. Moreover, we shall also see in this chapter how the autocovariance can be applied to make a quick and clear "diagnosis" of a sea state; in short, the use of autocovariance is like the use of X-rays in medicine. We shall also see in chap. 5 that the basic statistical properties of the waves in a sea state depend on the characteristics of the absolute minimum of the autocovariance.

4.3.2 *The relation between variance of the surface displacement and spectrum*

Let us prove that

$$< \eta^2(t) > = \int_0^\infty E(\omega)\, \mathrm{d}\omega. \tag{4.8}$$

With (4.1) of $\eta(t)$ we get

$$< \eta^2(t) > = < \sum_{i=1}^{N} \sum_{j=1}^{N} a_i a_j \cos(\omega_i t + \varepsilon_i) \cos(\omega_j t + \varepsilon_j) > ,$$

and hence, interchanging the order of temporal mean and summation, we write

$$< \eta^2(t) > = \sum_{i=1}^{N} \sum_{j=1}^{N} a_i a_j < \cos(\omega_i t + \varepsilon_i) \cos(\omega_j t + \varepsilon_j) > . \tag{4.9}$$

Here we have

$$< \cos(\omega_i t + \varepsilon_i) \cos(\omega_j t + \varepsilon_j) > \begin{cases} = \dfrac{1}{2} & \text{if } i = j, \\ = 0 & \text{if } i \neq j, \end{cases} \tag{4.10}$$

given that $\omega_i \neq \omega_j$ if $i \neq j$. Therefore (4.9) is reduced to

$$< \eta^2(t) > = \sum_{i=1}^{N} \frac{1}{2} a_i^2 \tag{4.11}$$

which completes the proof, in that definition (4.2) of spectrum implies

$$\sum_{i=1}^{N} \frac{1}{2} a_i^2 = \int_0^\infty E(\omega)\, \mathrm{d}\omega .$$

4.3.3 *The relation between autocovariance and spectrum*

Let us prove that

$$\psi(T) = \int_0^\infty E(\omega) \cos(\omega T)\, \mathrm{d}\omega . \tag{4.12}$$

To arrive at this relation we have to start from the definition (4.7) of autocovariance, and use (4.1) of $\eta(t)$. The result is

$$\psi(T) = \sum_{i=1}^{N} \sum_{j=1}^{N} a_i a_j < \cos(\omega_i t + \varepsilon_i) \cos[\omega_j (t+T) + \varepsilon_j] > ,$$

where the order of temporal mean and summation has been interchanged. Then applying the addition formula to the second cosine we obtain

$$\psi(T) = \sum_{i=1}^{N} \sum_{j=1}^{N} a_i a_j [\cos(\omega_j T) < \cos(\omega_i t + \varepsilon_i) \cos(\omega_j t + \varepsilon_j) > +$$

$$- \sin(\omega_j T) < \cos(\omega_i t + \varepsilon_i) \sin(\omega_j t + \varepsilon_j) >].$$

Here the first temporal mean is given by (4.10), and the second temporal mean is zero whichever the i and j, and therefore

$$\psi(T) = \sum_{i=1}^{N} \frac{1}{2} a_i^2 \cos(\omega_i T) . \tag{4.13}$$

Let us consider the summation on the r.h.s. of this equation. Since

$$\left\{ \text{contribution to} \sum_{i=1}^{N} \frac{1}{2} a_i^2 \cos(\omega_i T) \text{ from the terms whose frequency } \omega_i \text{ is between } \omega - \delta\omega/2 \text{ and } \omega + \delta\omega/2 \right\} =$$

$$= \left\{ \cos(\omega T) \sum_i \frac{1}{2} a_i^2 \text{ for } i \text{ such that } \omega - \delta\omega/2 < \omega_i < \omega + \delta\omega/2 \right\} = \cos(\omega T) E(\omega) \delta\omega ,$$

we have

$$\sum_{i=1}^{N} \frac{1}{2} a_i^2 \cos(\omega_i T) = \int_0^\infty E(\omega) \cos(\omega T)\, \mathrm{d}\omega$$

that together with (4.13) completes the proof of (4.12).

4.3.4 *The three basic properties of the autocovariance*

The main properties of the autocovariance are:

1st property: $\psi(-T) = \psi(T)$,
2nd property: $|\psi(T)| < \psi(0)$ for $T \neq 0$,
3rd property: $\lim_{|T|\to\infty} \psi(T) = 0$.

The first property proceeds straightforwardly from relation (4.12) between autocovariance and spectrum. The second property can be readily proved from (4.12), given that the spectrum is positive and continuous. The third property can be deduced as follows.

If we fix a very large $|T|$ the function $\cos(\omega T)$ (function of ω) has a very small period $\delta\omega = 2\pi/|T|$, so that $E(\omega)$ keeps practically constant while $\cos(\omega T)$ makes a full cycle. Therefore the contribution to the integral

$$\int_0^\infty E(\omega)\cos(\omega T)\,\mathrm{d}\omega$$

from the small interval $\delta\omega$ approaches zero. The same conclusion holds for each small interval $\delta\omega$, and consequently the integral approaches zero as $|T| \to \infty$.

4.3.5 *A few alternative ways to express the variance of the surface displacement*

Gathering

(i) the definition (4.4) of σ,
(ii) the definition (4.5) of H_s,
(iii) the definition (4.6) of m_j,
(iv) the relation (4.8) between $<\eta^2(t)>$ and the spectrum,
(v) the relation (4.12) between $\psi(T)$ and the spectrum,

we obtain the following equalities

$$<\eta^2(t)> = \sigma^2 = \frac{H_s^2}{16} = m_0 = \int_0^\infty E(\omega)\,\mathrm{d}\omega = \psi(0)\,.$$

4.4 How to obtain the input data of the theory

4.4.1 *A few parameters that can be readily obtained*

The surface displacement of a sea state is measured at a fixed sampling rate. For a good description of the waves, Δt_{samp} [the time interval between successive measurements] should not exceed 1/15 of T_p.

The elevation of the free surface is measured above a fixed level generally different from the mean water level. Nevertheless, the surface displacement η can be deduced easily (one has simply to estimate the average of the measured elevations and to subtract this average from these elevations).

Thus we obtain the data $(\eta_1, \eta_2, ..., \eta_n)$ of a real sea state. Then from these data, we can straightforwardly compute the standard deviation:

$$\sigma = \sqrt{\frac{\eta_1^2 + \eta_2^2 + ... + \eta_n^2}{n}}.$$

Hence, multiplying σ by 4, we obtain the significant wave height.

Even the autocovariance can be obtained quite simply. For example, let us imagine we have a wave record of 10 minutes with a sampling rate of 10 per second. At the end we shall have obtained 6000 sampled data $(\eta_1, \eta_2, ..., \eta_{6000})$ taken at regular intervals of 0.1 s. Thus to evaluate, say, $\psi(T)$ for $T = 1$ s, it will suffice to perform the following operation:

$$\psi(1\,\mathrm{s}) = \frac{\eta_1\eta_{11} + \eta_2\eta_{12} + ... + \eta_{5990}\eta_{6000}}{5990},$$

which gives the mean value of the product of the surface displacements at time t and the surface displacement at time $t + 1$ s.

Obtaining the spectrum is not as simple, and hence the rest of this section is devoted to this operation.

4.4.2 *The Fourier series and the line spectrum*

Given an odd number n of data $(\eta_1, \eta_2, ..., \eta_n)$ recorded with a sampling interval Δt_{samp}, that is to say recorded at instants

$$t_1 = 0,\ t_2 = \Delta t_{\mathrm{samp}}, ...,\ t_n = (n-1)\Delta t_{\mathrm{samp}},$$

the function

$$\eta_F(t) = \sum_{i=1}^{N} a_i' \cos(\omega_i t) + a_i'' \sin(\omega_i t)$$

with

$$N = (n-1)/2,\quad \omega_i = \frac{2\pi}{\Delta t_{\mathrm{samp}}}\frac{i}{n},\quad a_i' = \frac{2}{n}\sum_{j=1}^{n} \eta_j \cos(\omega_i t_j),\quad a_i'' = \frac{2}{n}\sum_{j=1}^{n} \eta_j \sin(\omega_i t_j)$$

is such that

$$\eta_F(t_1) = \eta_1,\ \eta_F(t_2) = \eta_2, ...,\ \eta_F(t_n) = \eta_n.$$

$\eta_F(t)$ is the Fourier series which is periodic of period $T_F = n\Delta t_{\mathrm{samp}}$:

$$\eta_F(t + T_F) = \eta_F(t). \tag{4.14}$$

Such a series can also be given the form

$$\eta_F(t) = \sum_{i=1}^{N} a_i \cos(\omega_i t + \varepsilon_i),$$

where a_i, ε_i are related to a_i', a_i''. [Write $\cos(\omega_i t + \varepsilon_i)$ in the form $\cos(\omega_i t)\cos\varepsilon_i$ – $\sin(\omega_i t)\sin\varepsilon_i$ to obtain the relation between a_i, ε_i and a_i', a_i''.]

The spectrum $E_F(\omega)$ of function $\eta_F(t)$ consists of a sequence of delta functions:

$$E_F(\omega) = \sum_{i=1}^{N} \frac{1}{2} a_i^2 \delta(\omega - \omega_i).$$

It is the line spectrum whose graph is a sequence of vertical arrows with heights $\frac{1}{2} a_i^2$ at frequencies ω_i.

The fact that $\eta_F(t)$ is periodic with a period T_F implies that also the autocovariance $\psi_F(T)$ is periodic with a period T_F. Indeed

$$\psi_F(T) = < \eta_F(t)\,\eta_F(t+T) > = < \eta_F(t)\,\eta_F(t+T+T_F) > = \psi_F(T+T_F).$$

Autocovariance $\psi_F(T)$ is related to amplitudes a_i and frequencies ω_i of the Fourier series by

$$\psi_F(T) = \sum_{i=1}^{N} \frac{1}{2} a_i^2 \cos(\omega_i T), \tag{4.15}$$

and, for a wide interval after the origin $T = 0$, it proves to be nearly coincident with $\psi(T)$. Some discrepancies between the two functions arise only for large T, which is understandable if one thinks that $\psi(T)$ is defined for T smaller than the duration of the sea state, whereas $\psi_F(T)$ is defined even for arbitrarily large T.

4.4.3 *Theoretical foundations to obtain the continuous spectrum*

Let us define the function

$$\widetilde{E}_F(\omega) \equiv \frac{2}{\pi} \int_0^{\mathscr{T}} \psi_F(T) \cos(\omega T)\, \mathrm{d}T$$

that, thanks to (4.15), is rewritten in the form

$$\widetilde{E}_F(\omega) = \frac{2}{\pi} \sum_{i=1}^{N} \frac{1}{2} a_i^2 \int_0^{\mathscr{T}} \cos(\omega_i T) \cos(\omega T)\, \mathrm{d}T. \tag{4.16}$$

This function satisfies the following property:

$$\int_{\omega_j - \tilde{x}2\pi/\mathscr{T}}^{\omega_j + \tilde{x}2\pi/\mathscr{T}} \widetilde{E}_F(\omega)\, \mathrm{d}\omega = \frac{1}{2} a_j^2 \int_{-\tilde{x}}^{\tilde{x}} \frac{1}{\pi} \frac{\sin(2\pi x)}{x}\, \mathrm{d}x \quad \text{for any fixed } \tilde{x}, \text{ as } \mathscr{T} \to \infty \tag{4.17}$$

(where it is understood that, for $x = 0$, the quotient $\sin(2\pi x)/x$ has to be substituted by its limit as $x \to 0$). The integral on the r.h.s. rapidly converges on 1 if we fix larger and larger values of $\tilde{x}$. This implies that $\widetilde{E}_F(\omega)$ (the integrand on the l.h.s.) has an infinitely high and narrow peak at ω_j, and the area under this peak is $\frac{1}{2}a_j^2$.

The conclusion is that both $E_F(\omega)$ and $\widetilde{E}_F(\omega)$ as $\mathscr{T} \to \infty$ consist of the same sequence of infinitely high and narrow peaks on frequencies $\omega_1, \omega_2, \ldots \omega_N$.

For verifying (4.17), use the definition (4.16) of $\widetilde{E}_F(\omega)$. Pay attention to the fact that index j is fixed, while index i varies from 1 to N. Replace variable ω with

$$x \equiv (\omega - \omega_j)\frac{\mathscr{T}}{2\pi}.$$

Apply the addition formula to $\cos(\omega_j T + 2\pi x T/\mathscr{T})$. Note that all the terms of the summation are negligible with respect to the one for $i = j$. Note that $\cos^2(\omega_j T)$ and $\cos(2\pi x T/\mathscr{T})$ are functions like the ones of fig. 4.4, so that you can use the equality

$$\int_0^{\mathscr{T}} \cos^2(\omega_j T)\cos\left(\frac{2\pi}{\mathscr{T}}xT\right)\mathrm{d}T = \frac{1}{2}\int_0^{\mathscr{T}} \cos\left(\frac{2\pi}{\mathscr{T}}xT\right)\mathrm{d}T \quad \text{as } \mathscr{T} \to \infty,$$

where $\frac{1}{2}$ on the r.h.s. is the average of $\cos^2$. Then you will be able to prove equation (4.17) in a few simple steps.

4.4.4 *The problem of the continuous spectrum*

Given that the continuous spectrum of the ideal sea state must of necessity be similar to the line spectrum of the real sea state, we seek a continuous spectrum that is *equivalent* to the line spectrum. We shall judge the equivalence through two main features:

(i) the cumulative function

$$\mathscr{F}(\omega) \equiv \frac{\int_0^{\omega} E(\omega')\,\mathrm{d}\omega'}{\int_0^{\infty} E(\omega')\,\mathrm{d}\omega'},$$

(ii) the autocovariance

$$\psi(T) = \int_0^{\infty} E(\omega)\cos(\omega T)\,\mathrm{d}\omega.$$

Clearly, the autocovariance of the spectrum $E_F(\omega)$ is $\psi_F(T)$; and the cumulative function is

$$\mathscr{F}_F(\omega) \equiv \frac{\int_0^{\omega} E_F(\omega')\,\mathrm{d}\omega'}{\int_0^{\infty} E_F(\omega')\,\mathrm{d}\omega'} = \frac{\sum_j \frac{1}{2}a_j^2}{\sum_{i=1}^{N} \frac{1}{2}a_i^2} \quad \text{for } j \text{ such that } \omega_j < \omega.$$

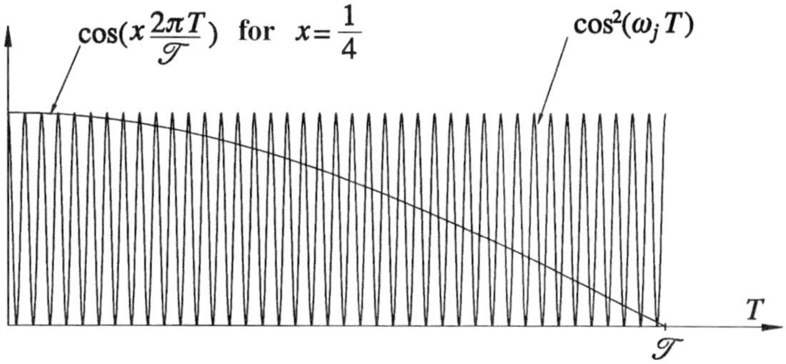

Fig. 4.4 $\mathscr{T}$ is very large, ω_j and x are fixed, and T is the independent variable. Therefore the function $\cos[(2\pi/\mathscr{T})xT]$ varies much more gradually than $\cos^2(\omega_j T)$.

4.4.5 *How to obtain the continuous spectrum*

The gist of sect. 4.4.3 is: evaluate

$$I(\omega) \equiv \int_0^{\mathscr{T}} \psi_F(T)\cos(\omega T)\,\mathrm{d}T, \tag{4.18}$$

for increasing values of the upper limit $\mathscr{T}$, and you will obtain a continuous function that, apart from a constant, approximates the line spectrum better and better. An example is given by figs. 4.5-6-7. We see a line spectrum, the same in each figure, and function (4.18) for a few different $\mathscr{T}$: $\mathscr{T} \cong 2.5T_p$ in fig. 4.5, $\mathscr{T} \cong 5T_p$ in fig. 4.6, $\mathscr{T} \cong 10T_p$ in fig. 4.7. (The graphs are given in a normalized form, and this is why the spectrum's peak is equal to 1 in each picture.)

We see the result of integral (4.18) fits the line spectrum better as $\mathscr{T}$ grows. We start from a small $\mathscr{T}$ with a sketch, and then we tend to the exact picture as $\mathscr{T} \to \infty$. The interesting point is that even the sketches accurately reflect the essential features of the line spectrum. In order to realize this crucial point, let us take one of these sketches as the continuous spectrum. Specifically, let us assume

$$E(\omega)\begin{cases} = \text{const}\, I(\omega) & \text{if} \quad I(\omega) \geq 0, \\ = 0 & \text{if} \quad I(\omega) < 0, \end{cases} \tag{4.19}$$

with const such as to satisfy condition (4.8). Then, let us compare $\mathscr{F}(\omega)$ and $\psi(T)$ from this spectrum with $\mathscr{F}_F(\omega)$ and $\psi_F(T)$ from the line spectrum. We see that the agreement between $\mathscr{F}(\omega)$ and $\mathscr{F}_F(\omega)$ and $\psi(T)$ and $\psi_F(T)$ is rather good even for $\mathscr{T} \cong 2.5T_p$ (fig. 4.5), and becomes nearly perfect for $\mathscr{T} \cong 5T_p$ (fig. 4.6) and $\mathscr{T} \cong 10T_p$ (fig. 4.7).

Naturally, for large $|T|$, $\psi(T)$ will deviate from $\psi_F(T)$, since $\psi(T) \to 0$ as $|T| \to \infty$, while $\psi_F(T)$ is periodic. But what is important is that the two autocovariances are coincident on a large interval after the origin. Indeed, we shall see in chap. 5 that the basic statistical properties of a sea state depend only on the first wave of the autocovariance (that is the wave including the origin $T = 0$); moreover the

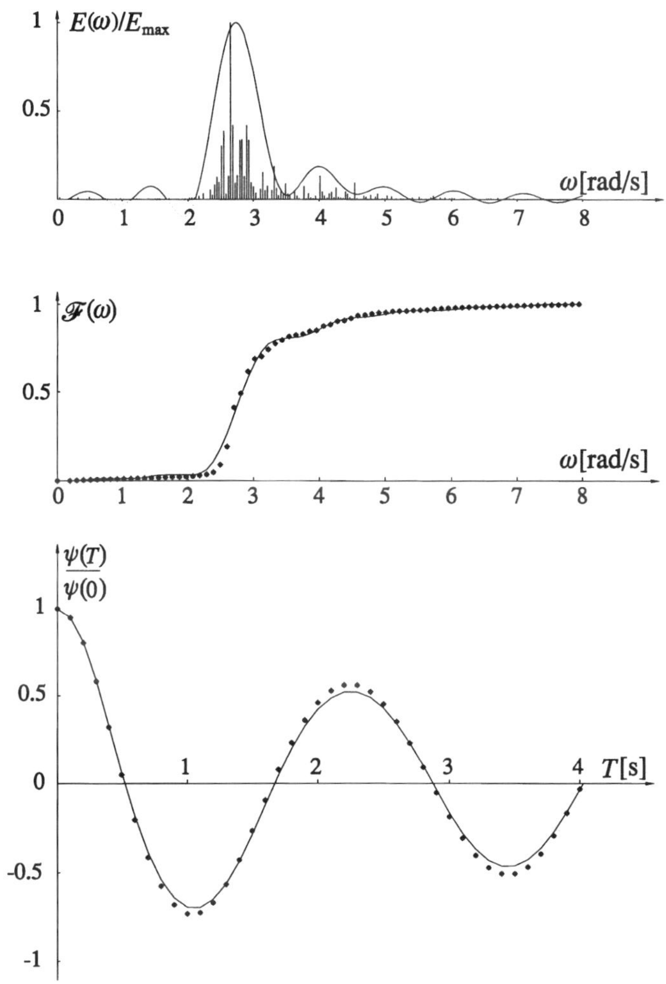

Fig. 4.5 A typical line spectrum from a record of wind waves. The continuous spectrum has been obtained by means of (4.18-19) with $\mathcal{T} \cong 2.5T_p$. The two lower panels show the function $\mathcal{F}(\omega)$ and the autocovariance $\psi(T)$ (points: line spectrum, continuous line: continuous spectrum).

configuration of the wave groups depends only on the first two or three waves of the autocovariance, that we shall call *core of the autocovariance*.

The conclusion is that a number of continuous spectra, sometimes appearing very different from each other, are equally equivalent to the line spectrum. Therefore the continuous spectrum is indeterminate!

In practice this indeterminacy is not an obstacle, because two sea states with two different spectra like the ones of fig. 4.6 and fig. 4.7 have the same basic statistical properties. Hence, we can assume at our choice the one or the other of the two

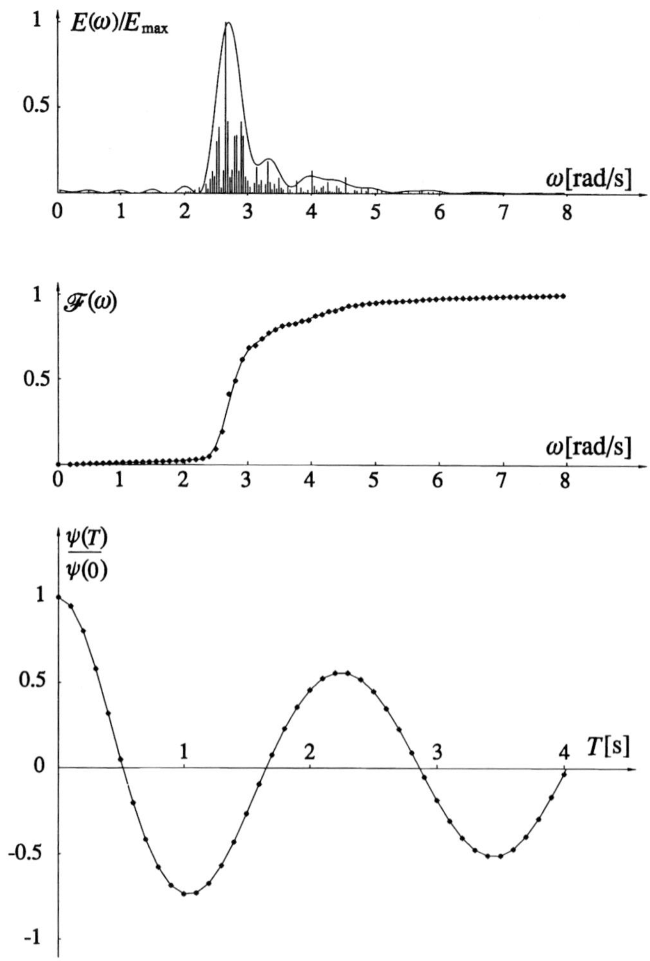

Fig. 4.6 The line spectrum is the same as in fig. 4.5; the continuous spectrum has been obtained with $\mathscr{T} \cong 5T_p$ in (4.18). We see that now the $\mathscr{F}(\omega)$ and $\psi(T)$ of the continuous spectrum coincide perfectly with the $\mathscr{F}(\omega)$ and $\psi(T)$ of the line spectrum.

spectra. For convenience, the best choice is to take the smallest $\mathscr{T}$ giving the full equivalence between the continuous spectrum and the line spectrum. This is because the spectrum shape gets simpler. Typically, $\mathscr{T} = 4 \div 5T_p$ is large enough.

For $T < 5T_p$, $\psi_F(T)$ is usually very close to the $\psi(T)$ obtained from the time series data of $\eta(t)$, so we can use this $\psi(T)$ in place of $\psi_F(T)$. Before we obtain the spectrum, we do not know T_p, and therefore the question arises, how to estimate $4 \div 5T_p$? A good preliminary estimate can be done on taking T_p equal to the average period of the two highest waves of function $\psi(T)$.

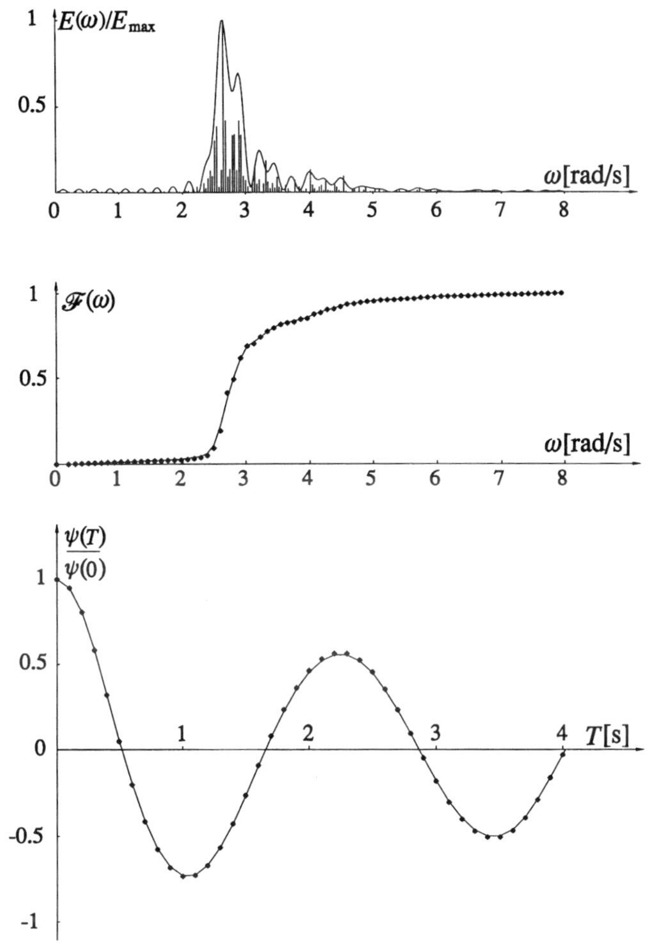

Fig. 4.7 The line spectrum is the same as in the two previous figures; $\mathscr{T}$ has been raised up to $10T_p$. The result is a more indented spectrum. The agreement between the continuous spectrum and the line spectrum for what concerns $\mathscr{F}(\omega)$ and $\psi(T)$ keeps on being excellent.

To make the study of this section easier, it will be useful to bear in mind that three autocovariances are involved:

(i) the one obtained straightforwardly from the time series data of $\eta(t)$, through the definition itself of autocovariance;

(ii) the one obtained from the continuous spectrum;

(iii) the one obtained from the line spectrum.

The first two have been denoted by the symbol $\psi(T)$, while the third one has been denoted by $\psi_F(T)$. The three functions prove to be coincident with one another on a wide interval after the origin. For large $|T|$ they become different from one another: the first function is

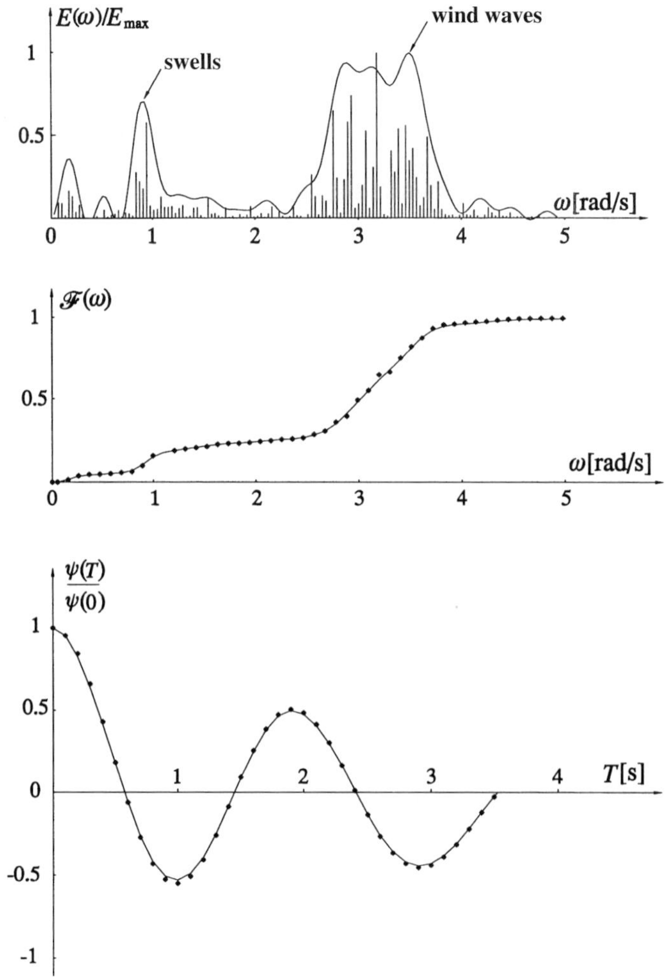

Fig. 4.8 A sea state consisting of wind waves and swells: the continuous spectrum has been obtained by means of (4.18-19).

defined only for $|T|$ smaller than the duration of the wave record; the second function approaches zero as $|T| \to \infty$; the third function is periodic with a period equal to the length of the wave record.

4.4.6 *The case of wind waves superimposed on swells*

The one shown in the previous section is a typical spectrum of the wind waves but the method is efficient even for the swells. An example is given in fig. 4.8 which shows the spectrum of wind waves of 2s period on swells of 6.5s. The highest peak

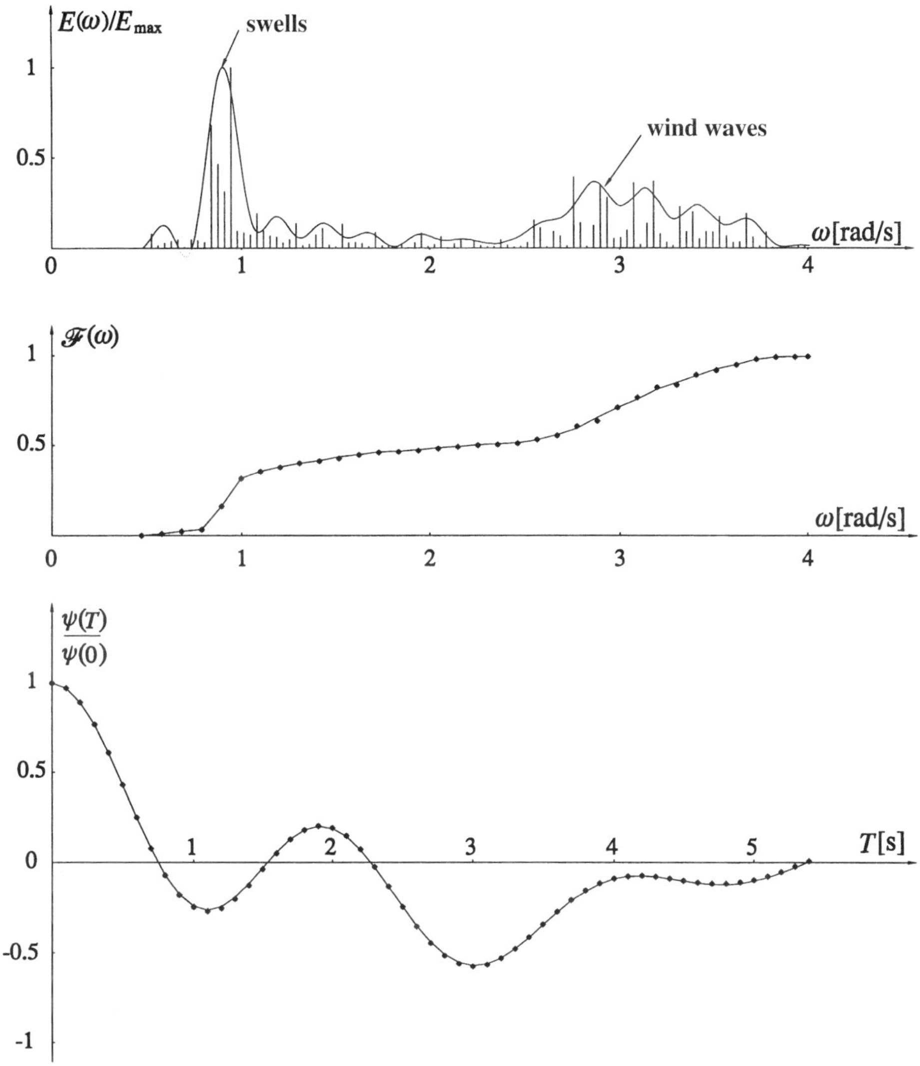

Fig. 4.9 A very broad spectrum: about a half of the energy content belongs to wind waves and a half to swells with a frequency nearly three times smaller. Here too the continuous spectrum has been obtained by means of (4.18-19).

of the spectrum is that of the wind waves, the second peak is that of the swells. The two peaks of $E(\omega)$ give rise to the two steps of $\mathscr{F}(\omega)$; of course the two domains where $\mathscr{F}(\omega)$ has a high rate of growth are those of the two peaks of $E(\omega)$. We see that the 25% of the spectrum's energy belongs to the swells and the remaining 75% to the wind waves.

This is the picture emerging both from $\mathscr{F}(\omega)$ and from $\mathscr{F}_F(\omega)$, which are nearly coincident with each other. $\mathscr{T}$ has been taken of 20 s, that is about ten times the T_p of the wind waves and three times the T_p of the swells. Generally, as we noted, the best value of $\mathscr{T}$ is nearly $5T_p$ but in this case the spectrum consists of two stumps with some very different T_p, so that we must make a compromise, taking $\mathscr{T}$ greater than five times the smaller T_p and smaller than five times the larger T_p. For the rest, the continuous spectrum $E(\omega)$ is deeply equivalent to the line spectrum, as we can judge on comparing $\mathscr{F}(\omega)$ to $\mathscr{F}_F(\omega)$ and $\psi(T)$ to $\psi_F(T)$.

Lastly, let us consider an extremely chaotic sea state, where some wind waves are superimposed on swells of nearly the same height. This case is illustrated in fig. 4.9 where $\mathscr{T}$ has been taken of 25 s, that is somewhat less than four times the T_p of the swells, which is equal to 6.7 s. Also in this case the equivalence between the continuous spectrum and the line spectrum clearly emerges if we compare $\mathscr{F}(\omega)$ to $\mathscr{F}_F(\omega)$ and $\psi(T)$ to $\psi_F(T)$.

Really those of figs. 4.8-9, rather than spectra of surface waves, are the spectra of some pressure head waves beneath the water surface. However, for the aims of this section, it does not matter whether the spectrum is of the surface waves or of the pressure head waves. Resorting to the pressure head waves is useful for didactic purposes, since a spectrum with a fixed ratio between swells' energy and wind waves' energy can be found more easily in these waves.

4.5 A mathematical form of the wind wave spectrum

4.5.1 *The JONSWAP spectrum*

As previously mentioned, when a wind generates waves and hence waves and wind have nearly the same direction, we speak of wind waves. Such waves typically have a spectrum like that of fig. 4.6 where $E(\omega)$ approaches rapidly zero on the left side and approaches zero more gradually on the right side. A mathematical form suggested for describing this characteristic spectrum shape is

$$E(\omega) = A g^2 \omega^{-5} \exp\left[-\frac{5}{4}\left(\frac{\omega_p}{\omega}\right)^4\right] \exp\left\{\ln\chi_1 \exp\left[-\frac{(\omega-\omega_p)^2}{2\chi_2^2\omega_p^2}\right]\right\}. \tag{4.20}$$

This is the so called JONSWAP spectrum (Hasselmann et al., 1973) which is effective for deep water:

$$d > \frac{1}{2} L_{p0}.$$

The variations of the spectrum on shallow water will be dealt with in chap. 8.

The JONSWAP spectrum was the final result of a work developed in the 50's and 60's. Phillips was the first to observe in the 50's that the spectrum approaches zero, for large ω, as ω^{-5}. The term $A\ g^2\omega^{-5}$ in the formula (4.20) is due to him, and indeed A is called the *Phillips parameter*. The form

$$E(\omega) = A\,g^2\omega^{-5}\exp\left[-\frac{5}{4}\left(\frac{\omega_p}{\omega}\right)^4\right] \tag{4.21}$$

was introduced by Pierson and Moskowitz in the 60's. The last improvement, that is the introduction of the second exponential function, was due to the JONSWAP project in the early 70's.

According to the researchers of the JONSWAP project, the more characteristic values of the shape parameters χ_1 and χ_2 are

$$\chi_1 = 3.3, \quad \chi_2 \begin{cases} = 0.07 \text{ if } \omega \leq \omega_p, \\ = 0.09 \text{ if } \omega > \omega_p, \end{cases} \tag{4.22}$$

but we can as well assume

$$\chi_1 = 3, \quad \chi_2 = 0.08\,,$$

with some negligible consequences. The spectrum with values (4.22) of parameters χ_1 and χ_2 is called the mean JONSWAP. It is shown by fig. 4.10 in the nondimensional form $E(\omega)/E(\omega_p)$ as a function of ω/ω_p. It should be noted that such a nondimensional form depends only on the values of χ_1 and χ_2. Therefore, fig. 4.10 represents the mean JONSWAP spectrum, whatever the peak frequency ω_p and Phillips' parameter.

As to parameter A, it depends on the characteristics of the wave generation: the smaller is fetch $\mathscr{Y}$ and the greater is wind speed u, the larger is A. The researchers of the JONSWAP project suggested the relation

$$A = 0.076\left(\frac{g\,\mathscr{Y}}{u^2}\right)^{-0.22}.$$

As an exercise, the reader could verify that the peak of the Pierson and Moskowitz spectrum occurs at $\omega = \omega_p$. To this end, it suffices to differentiate (4.21) with respect to ω and

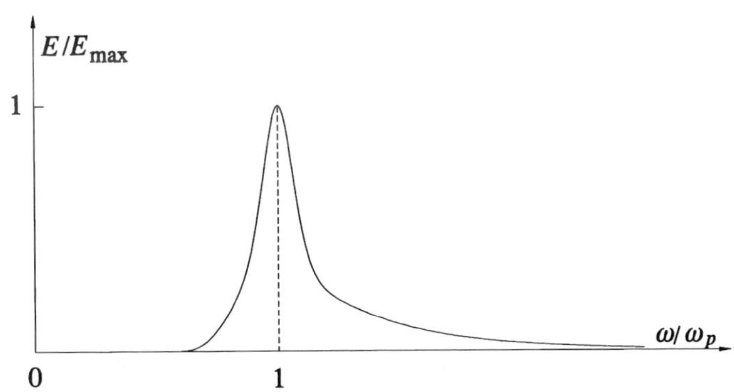

Fig. 4.10 The mean JONSWAP spectrum.

verify that the derivative is zero only for $\omega = \omega_p$. Then for verifying that also the JONSWAP spectrum [function (4.20)] has its maximum at $\omega = \omega_p$, it will suffice to note that

(i) the second exponential function in (4.20) is equal to χ_1 at $\omega = \omega_p$ and approaches 1 as $|\omega - \omega_p|$ grows;

(ii) χ_1 takes on values greater than 1.

Therefore the second exponential function in (4.20) makes even more prominent the peak at ω_p.

4.5.2 *The relation between T_p and H_s*

With the JONSWAP spectrum in (4.3), we get

$$m_0 = Ag^2 \int_0^\infty \omega^{-5} \exp\left[-\frac{5}{4}\left(\frac{\omega_p}{\omega}\right)^4\right] \exp\left\{\ln \chi_1 \exp\left[-\frac{(\omega - \omega_p)^2}{2\chi_2^2 \omega_p^2}\right]\right\} \mathrm{d}\omega. \tag{4.23}$$

Here it is convenient to replace the dimensional variable ω with the nondimensional variable

$$w \equiv \omega/\omega_p,$$

and it is convenient to define the nondimensional spectrum

$$\mathscr{E}(w) \equiv w^{-5} \exp\left(-\frac{5}{4} w^{-4}\right) \exp\left\{\ln \chi_1 \exp\left[-\frac{(w-1)^2}{2\chi_2^2}\right]\right\} \tag{4.24}$$

that will be repeatedly used in what follows. Substituting w for ω in (4.23), we have

$$m_0 = Ag^2 \omega_p^{-4} \int_0^\infty \mathscr{E}(w)\, \mathrm{d}w.$$

The integral of $\mathscr{E}(w)$ is evaluated numerically and, with values (4.22) of χ_1 and χ_2, the result is 0.305. Therefore

$$m_0 = 0.305 A g^2 \omega_p^{-4}. \tag{4.25}$$

Hence, bearing in mind that $m_0 \equiv H_s^2/16$ and $T_p \equiv 2\pi/\omega_p$, we arrive at the following relation:

$$T_p = \sqrt[4]{\frac{1}{0.305A}}\, 2\pi \sqrt{\frac{H_s}{4g}}. \tag{4.26}$$

As previously mentioned, the value of A depends on the characteristics of the wave generation. A typical value for design conditions is

$$A = 0.01,$$

with which, the relation between T_p and H_s becomes

$$T_p = 8.5\pi \sqrt{\frac{H_s}{4g}}. \tag{4.27}$$

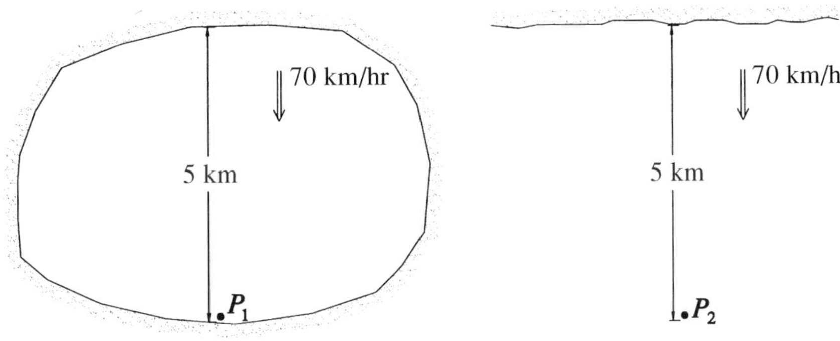

Fig. 4.11 The points P_1 in a lake and P_2 off a coast have the same relatively small $g\mathscr{Y}/u^2$. Consequently, the waves at these points will be very steep.

Note that, even a variation of 20% on A leads to a variation of only 5% on the numerical coefficient 8.5 in (4.27).

Relation (4.26) shows that the peak period gets smaller under the same significant wave height for a larger A. Since we saw in sect. 4.5.1 that the smaller the fetch and the stronger the wind the larger is A, we can see a characteristic feature of the wind waves. The reader may have observed a relatively small fetch lashed by a strong wind (two possible situations are shown in fig. 4.11 just to fix the ideas). In that case it will have been noted that the waves break even on deep water. The phenomenon is due to the fact that the waves are too steep owing to the large A. Indeed from (4.26) it follows that the characteristic wave steepness is

$$\frac{H_s}{L_{p0}} = \frac{2}{\pi}\sqrt{0.305A}\,.$$

4.5.3 *Dynamic similarity between stormy seas and slight seas*

Let us consider a sea state consisting of small wind waves generated by a weak wind, say waves of

$$H_s = 0.30\,\mathrm{m}\,.$$

Relation (4.26) with the characteristic value $A = 0.01$ of Phillips' parameter gives a peak period

$$T_p = 2.33\,\mathrm{s},$$

for this sea state. Let us consider now a stormy sea state, say one with

$$H_s = 9.0\,\mathrm{m}\,.$$

Here, relation (4.26) with $A = 0.01$ gives

$$T_p = 12.79\,\mathrm{s}\,.$$

Therefore the first sea state is the small scale model (1:30) of the second sea state, with the same Froude number. Indeed the quotient between the wave periods of the two sea states is equal to the square root of the quotient between the wave heights. The dynamic similarity is full, in that the normalized autocovariance is the same for the model sea state and the prototype sea state. Which means that the same statistical properties exist in the model sea state as in the prototype sea state (this statement will be proved in chap 5); in particular the probability of a wave height greater than a threshold $\tilde{\alpha} H_s$ ($\tilde{\alpha}$ being arbitrarily fixed) is the same for the model sea state and the prototype sea state.

Let us prove that the normalized autocovariance is the same for the model sea state and the prototype sea state. From relation (4.12) between the autocovariance and the spectrum, and formula (4.20) of the JONSWAP spectrum, we have

$$\psi(T) = \int_0^\infty Ag^2\omega^{-5}\exp\left[-\frac{5}{4}\left(\frac{\omega_p}{\omega}\right)^4\right]\exp\left\{\ln\chi_1\exp\left[-\frac{(\omega-\omega_p)^2}{2\chi_2^2\omega_p^2}\right]\right\}\cos(\omega T)\,\mathrm{d}\omega.$$

Then, replacing ω with $w \equiv \omega/\omega_p$ and using definition (4.24) of $\mathscr{E}(w)$, we arrive at

$$\frac{\psi(T)}{\psi(0)} = \frac{\displaystyle\int_0^\infty \mathscr{E}(w)\cos\left(2\pi w\frac{T}{T_p}\right)\mathrm{d}w}{\displaystyle\int_0^\infty \mathscr{E}(w)\,\mathrm{d}w}. \tag{4.28}$$

Hence, given T/T_p we can obtain $\psi(T)/\psi(0)$ leaving H_s and ω_p out of consideration. This implies that $\psi(T)/\psi(0)$ as a function of T/T_p does not change from the prototype sea state to the model sea state.

4.6 Possibility of testing small scale models in sea or lakes

4.6.1 *The main obstacle*

We shall see in sect. 14.2 that the proper dynamic similarity for ocean structures requires the constancy of the Froude number. Therefore the conclusion of the previous section should open the road for natural small scale models: models of ocean structures being tested directly in sea or lakes rather than in the laboratory tanks. In practice this is difficult, because it is not at all easy to find pure wind waves of a size suitable for small scale models. Indeed the waves of this size usually include a significant swell component. The swell preserves the period of the generating area but has a smaller height. Therefore, relation (4.26) often results in a large underestimate of the T_p of the swells.

The trouble is that even the presence of some small swells can spoil the dynamic similarity. To realize this point it suffices to look at fig. 4.12 which shows the spectrum of the surface waves and the spectra of the pressure head waves at a few

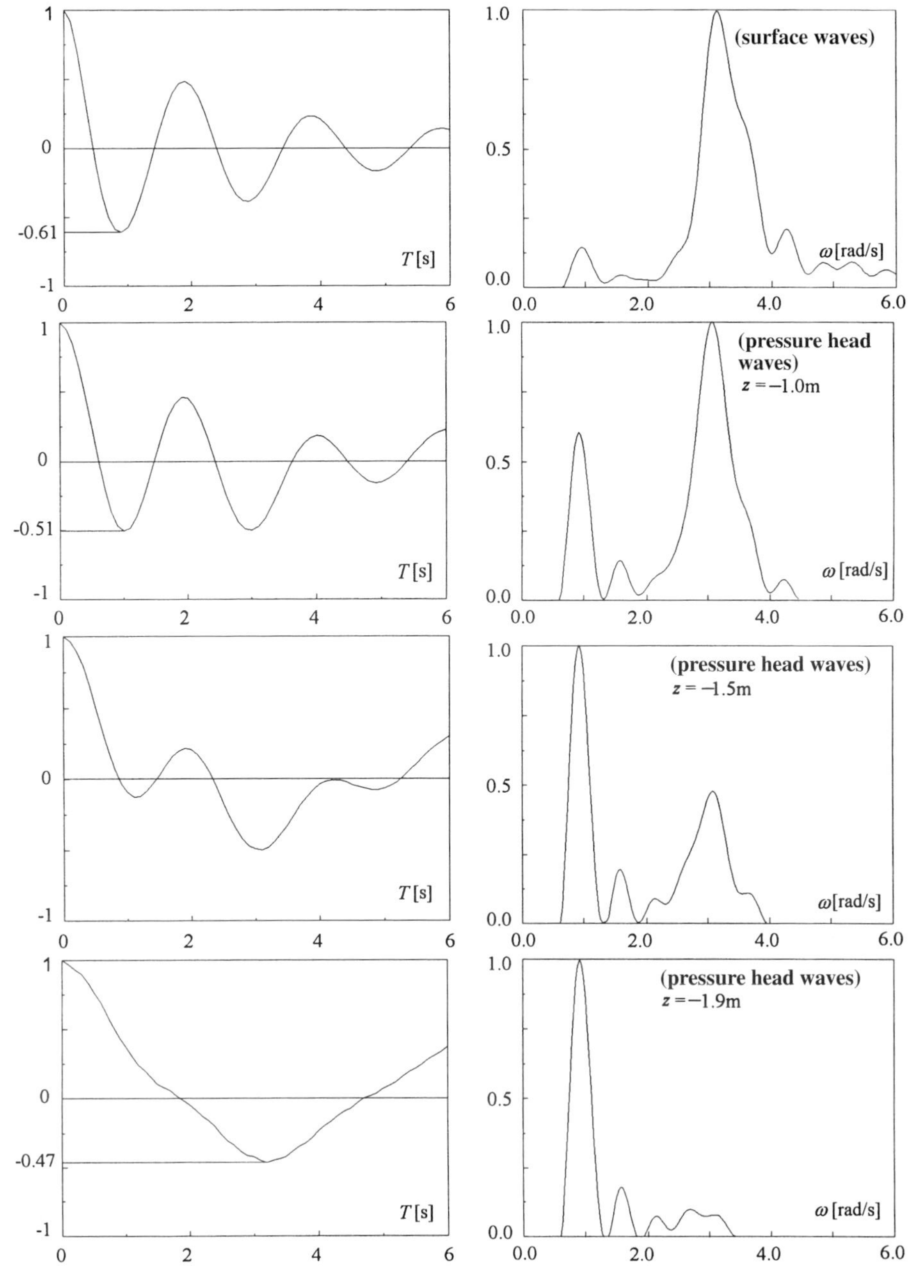

Fig. 4.12 Autocovariances and spectra of surface waves and of pressure head waves beneath the water surface. We see that the ratio between swell amplitude and wind-wave amplitude grows as the depth increases. The reason is that the swell is less attenuated, owing to its greater period. The spectra are given in a normalized form, and this is why the peak has the same height in each picture.

growing depths beneath the water surface (note that the spectra are normalized and this is why they all have the same height). To judge from the surface waves we should conclude that a sea state with this spectrum is a good small scale model of a sea storm. Indeed the significant wave height is of 0.23 m and the peak period is of 2.00 s, which, in a scale of 1:35, lead to a prototype with $H_s = 8.0$ m and $T_p = 11.8$ s, which are realistic values of a severe storm in the Mediterranean Sea.

However, the dynamic similarity is spoiled by some small swells whose presence is revealed by the small bump on the angular frequency of 0.95 rad/s. Indeed the swells, even if markedly smaller than the wind waves, are able to upset the flow field beneath the water surface. Note in particular in fig. 4.12 the spectrum of the pressure head waves 1.5 m beneath the mean water level: the swell's peak has overtaken the wind wave's peak. This upset takes place since the swells undergo a lower attenuation with depth, owing to their greater period. Therefore, the ratio between the swell amplitude and the wind wave amplitude grows from the water surface to the seabed.

Clearly, a situation like that of fig. 4.12 is quite unreal for the prototype, indeed the peak period of the pressure head waves would be of

$$(2\pi/0.95)\,\sqrt{35} = 39\,\text{s}!$$

Here is the problem: the waves of a size suitable for small scale models often include a swell component with a period several times greater than the period of the wind waves, whereas in a severe storm this is impossible. Keep in mind that a spectrum having a wind wave peak at a frequency ω_1 and a swell peak at a frequency ω_2 implies that some time before, in some far-off area, there were wind waves of frequency ω_2.

4.6.2 *An exceptionally favourable site*

Several small scale models were recently executed off the beach at Reggio Calabria (east coast of the Straits of Messina). The model tests of a few different structures were successful thanks to some exceptionally favourable environmental conditions. The most important is that the sea states often consist of pure wind waves with the typical size of a big laboratory tank ($0.20\,\text{m} < H_s < 0.40\,\text{m}$, $1.8\,\text{s} < T_p < 2.6\,\text{s}$). An example of these pure wind waves with a laboratory size is given by fig. 4.13. We see the surface wave spectrum has no peak on low frequencies, which is confirmed by the spectra of the pressure head waves beneath the water surface.

It is not easy to find sea waves with an H_s smaller than 0.50 m and spectra like the ones of fig. 4.13. Off Reggio Calabria this is possible for a few days in a month thanks to:

(i) the high stability of a local NNW wind, the so called *wind of the Straits* blowing from Messina towards Reggio Calabria;

(ii) the orientation of the coast. The coast being oriented from SW to NE it is naturally sheltered from the swells travelling towards the north;

(iii) the relatively short fetch (about 10 km).

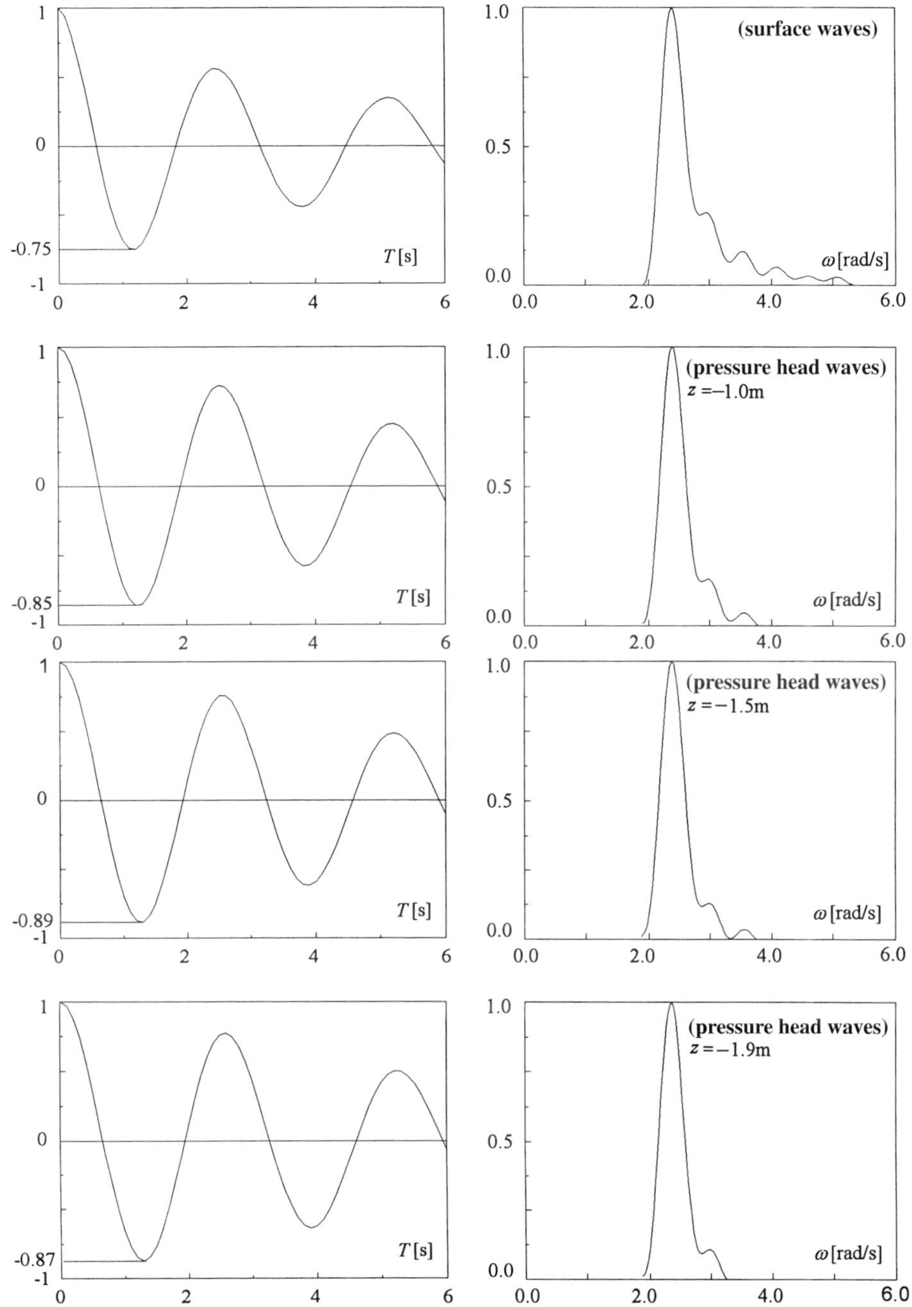

Fig. 4.13 Normalized autocovariances and spectra of surface waves and of pressure head waves at growing depths beneath the water surface. They are pure wind waves with a size suitable for small scale models.

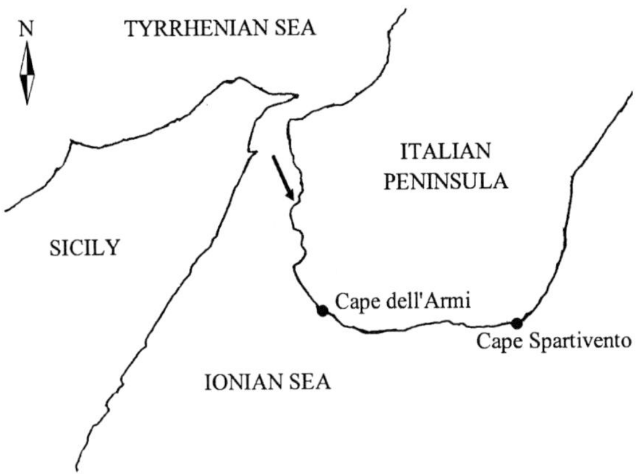

Fig. 4.14 Straits of Messina. The arrow points to the location of the natural laboratory of Reggio Calabria.

Probably similar conditions occur in some lakes, and we have to look at the lakes rather than at the sea for developing the natural models. However the laboratory of Reggio Calabria possesses an additional series of favourable conditions making it special.

First: the high stability of the local wind, which we have already pointed out as a cause of the near absence of swells (this wind blowing from the north opposes the swells from the south). Clearly, the wind stability is also useful for operating. Indeed, the wind often keeps steady from morning to evening, which provides many suitable hours to test the models.

Second: the small tide amplitude (typically within 0.10 m). The small tide amplitude is a requisite for the small scale models. The tide amplitude of Reggio Calabria has not yet raised any difficulty.

Third: the clearness of the water, which is really exceptional to be found in front of a city. The water is extremely clear because of the Straits' current that flows twice a day. It is quite obvious that the clearness of the water is another requisite for a natural laboratory. Indeed, in the sea or in the lake the small scale model of an ocean structure requires a lot of underwater work.

In fixing the model scale one has to bear in mind that the most typical H_s at Reggio Calabria is about 0.30 m. Therefore, if one wishes to test an offshore platform under 15 m of significant wave height, the model scale should be 1:50 or, if one wishes to test a submerged floating tunnel under $H_s \cong 10$ m, the model scale should be 1:30. The 1:50 scale model of an offshore platform (height 3 m, base diameter 2 m, water depth 2.5 m) and the 1:30 scale model of a floating tunnel (diameter 0.90 m, depth of the centre 1.5 m, water depth 3 m) were tested in the natural laboratory of Reggio Calabria as we shall see in chap. 11.

4.7 Inferring the nature of waves from the bandwidth

4.7.1 *The concept of infinitely narrow spectrum*

The knowledge of the bandwidth is useful to understand the nature of a sea state. To realize the relation between bandwidth and wave characteristics it is convenient to start from the concept of *infinitely narrow spectrum.*

Let us refer to the theory of the sea states and let us take the spectrum of fig. 4.15, a very high triangle with a very small base $\delta\omega$, and a finite area m_0. The surface displacement at a fixed time t_o can be written in the form

$$\eta(t_o) = \sum_{i=1}^{N} a_i \cos(\omega_p t_o + \delta\omega_i t_o + \varepsilon_i), \tag{4.29}$$

where

$$\delta\omega_i \equiv \omega_i - \omega_p .$$

The surface displacement at the instant $t_o + nT_p$, with n an arbitrary integer, can be written as

$$\eta(t_o + nT_p) = \sum_{i=1}^{N} a_i \cos\left[(\omega_p t_o + \delta\omega_i t_o + \varepsilon_i) + n2\pi + n\frac{\delta\omega_i}{\omega_p} 2\pi\right] . \tag{4.30}$$

Hence

$$\frac{\eta(t_o + n\,T_p)}{\eta(t_o)} \cong 1, \tag{4.31}$$

provided that

$$n << \frac{\omega_p}{\delta\omega} .$$

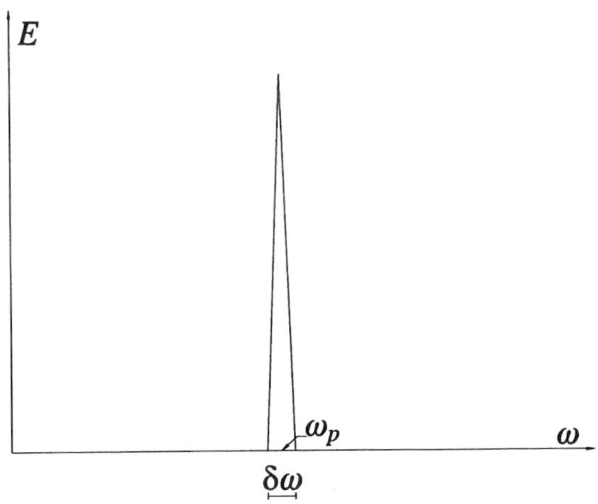

Fig. 4.15 Infinitely narrow spectrum.

Equation (4.31) proceeds straightforwardly from (4.29) and (4.30); it suffices to note that the argument of the ith cosine in (4.30) differs from the argument of the ith cosine in (4.29) of $n2\pi$, plus a term smaller than $n(\delta\omega/2\omega_p)2\pi$.

As an example, let us assume

$$\omega_p/\delta\omega = 10^6, \quad n = 10.$$

Then, the difference between the ith cosine of the summation (4.30) and the ith cosine of the summation (4.29) will be smaller than $\pi/10^5$ (as can be easily verified). This means that $\eta(t_o + 10T_p)$ will be nearly coincident with $\eta(t_o)$. Naturally, even more so, the conclusion holds for $\eta(t_o + T_p)$, $\eta(t_o + 2T_p)$, ..., $\eta(t_o + 9T_p)$. Through the same reasoning, we can also realize that $\eta(t_o + 10^6 T_p)$ will be generally different from $\eta(t_o)$.

Being t_o arbitrary, we conclude that $\eta(t)$ will locally approach a periodic function of period T_p. Then we can easily prove this to be a sinusoidal function. The difference with respect to a pure sinusoidal function is that each wave shows a small random variation from the preceding one, and the sum of these small variations, after a very large number of waves, can give rise to some large variation of the wave height.

In conclusion, a sea state with an infinitely narrow spectrum would be similar to a sequence of periodic waves. However there would be a substantial difference as the wave height would vary largely though very gradually. Clearly, the narrower the spectrum, the closer the waves would be to this ideal condition. Vice versa, the wider the spectrum, the more irregular the waves, that is, the greater the differences among consecutive waves.

4.7.2 *Bandwidth parameters ε and v*

It is helpful to express a comment on a sea state with only a few key parameters. Among these parameters we should include an index of how much the waves deviate from the ideal condition of the infinitely narrow spectrum. Such an index is called *bandwidth parameter*, and is generally 0 for the infinitely narrow spectrum and approaches 1 as the bandwidth grows.

Cartwright and Longuet-Higgins (1956) used the bandwidth parameter

$$\varepsilon \equiv \sqrt{1 - \frac{m_2^2}{m_0 m_4}}$$

which really is not always efficient, in that it is too sensitive to the high frequency noise. Let us see why with an example.

The moments of the spectrum of fig. 4.16 are

$$m_0 = m_{01} + m_{01}/n = m_{01}(1 + n^{-1}),$$

$$m_2 = m_{01}\omega_p^2 + (m_{01}/n)n^2\omega_p^2 = m_{01}\omega_p^2(1 + n),$$

$$m_4 = m_{01}\omega_p^4 + (m_{01}/n)n^4\omega_p^4 = m_{01}\omega_p^4(1+n^3),$$

and hence

$$\varepsilon = \sqrt{1 - \frac{(1+n)^2}{(1+n^{-1})(1+n^3)}}.$$

Thus, ε approaches 1 as $n \to \infty$, which suggests the idea of an extreme difference with respect to the case of the infinitely narrow spectrum. Really, as $n \to \infty$, the waves with this spectrum are practically the same as the waves with the infinitely narrow spectrum. With the naked eye these waves look exactly like the waves of the infinitely narrow spectrum, and only if we used a magnifier should we find that their surface is pitted with a lot of very small ripples.

Thus, in this case ε misses the mark. The fact is that resorting to ε is equivalent to judging by the number of the local maxima (or minima) being present in each wave; the greater this number, the greater the difference from the infinitely narrow spectrum. With the spectrum of fig. 4.16, each wave has an infinitely large number of local maxima due to the very small noise on the wave surface, and this is why ε gets the upper limit.

For the same reason, ε is equal to 1 also for the JONSWAP spectrum. Indeed also with the JONSWAP spectrum, the waves are affected by a high frequency noise of very small amplitude. The noise is due to the high frequency tail of the spectrum which approaches zero as ω^{-5} as $\omega \to \infty$. Consequently m_4 tends to infinity and ε approaches 1, just as in the above example.

In 1975 Longuet-Higgins used a new bandwidth parameter:

$$\nu \equiv \sqrt{\frac{m_0 m_2}{m_1^2} - 1}$$

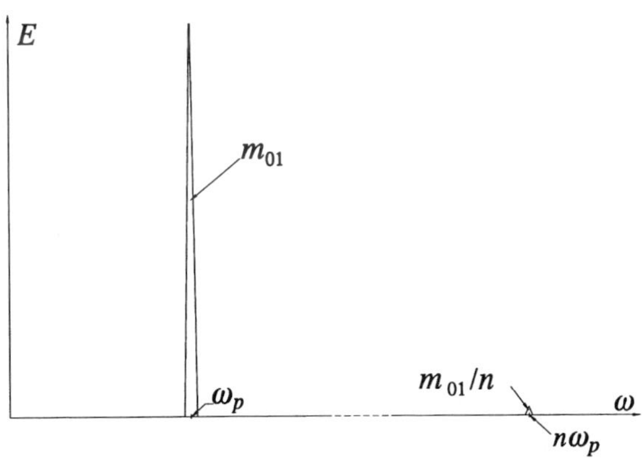

Fig. 4.16 The spectrum used for the analysis of the bandwidth parameters.

whose range is $(0, \infty)$. This new parameter is more effective than ε. However, for the spectrum of fig. 4.16 it tends to infinity.

4.7.3 *Narrow bandedness parameter* ψ^*

Let us define

$\psi^* \equiv$ absolute value of the quotient between the absolute minimum and the absolute maximum of the autocovariance,

that is

$$\psi^* \equiv |\psi(T^*)/\psi(0)|,$$

where

$T^* \equiv$ abscissa of the absolute minimum of the autocovariance function.

If the spectrum is infinitely narrow, the autocovariance approaches a cosine, and thus ψ^* approaches 1. As the bandwidth grows, ψ^* gets smaller and smaller approaching 0. Therefore ψ^* is a narrow bandedness parameter, i.e. it is the one's complement of a bandwidth parameter.

It is a natural parameter in that it appeared from the solution for the wave height probability. Indeed, as we shall see later, the distribution of the wave heights is governed by this parameter; and two sea states having the same ψ^* and H_s have also the same distribution of the wave heights. The smaller the ψ^*, the larger are the discrepancies of this distribution from that pertaining to the infinitely narrow spectrum. This is because ψ^* succeeds in giving a clear idea of how much the waves deviate from the condition of the infinitely narrow spectrum.

Using ψ^* we shall not be muddled by the high frequency noise. For example the autocovariance of the spectrum of fig. 4.16 is

$$\psi(T) = m_{01}\cos(\omega_p T) + \frac{1}{n} m_{01}\cos(n\omega_p T).$$

As $n \to \infty$, this $\psi(T)$ is a cosine affected by a very small noise. Therefore ψ^* approaches 1 as in the case of the infinitely narrow spectrum; that is ψ^* correctly classifies the waves with this spectrum as very close to the waves with the infinitely narrow spectrum.

Really, ψ^* is efficient if the absolute minimum of the autocovariance is also the first minimum of this function. Indeed the above mentioned solution for the wave height distribution requires that the absolute minimum of $\psi(T)$ is also the first minimum after the origin. The fact (rather rare) that this condition is not satisfied is itself important to understand the nature of the sea state, as we shall see in sect. 4.7.5.

4.7.4 *Understanding the nature of the waves from* ψ^*

The quotient $\psi(T)/\psi(0)$ as a function of T/T_p [function (4.28)] is shown in fig. 4.17, for the mean JONSWAP spectrum ($\chi_1 = 3.3$; $\chi_2 = 0.08$). From this figure we see that

$$\psi^* = 0.73 \quad \text{for the mean JONSWAP spectrum.}$$

The dependence of ψ^* on the spectrum shape proves to be rather weak. Letting χ_1 range from 1 to 7, which according to the researchers of the JONSWAP project is the whole range of variability of this parameter, ψ^* grows from 0.65 to 0.75. The variability of ψ^* due to the second shape parameter χ_2 is even smaller. In summary, based on the JONSWAP spectrum, we conclude that

$$0.65 < \psi^* < 0.75$$

is the typical domain of the wind waves.

The experience, based on a few thousand sea states recorded in the natural laboratory of Reggio Calabria, confirms (0.65, 0.75) as the typical domain of ψ^* for the wind waves. If ψ^* falls below 0.60, we are probably dealing with wind waves superimposed on swells. Indeed the presence of wind waves and swells leads to a wider spectrum, and consequently to a smaller ψ^*.

An efficient way to properly understand the nature of a sea state is to compare the ψ^* of the surface waves with the ψ^* of the pressure head waves beneath the water surface. Two emblematic examples are those of figs. 4.12 and 4.13 which were discussed in sect. 4.6. In the case of the wind waves (fig. 4.13), the spectrum shrinks gradually from the water surface to the seabed, since it sheds its high frequency tail; accordingly ψ^* grows from the surface waves to the pressure head waves. The opposite occurs if the wind waves are superimposed on some smaller swells (fig.

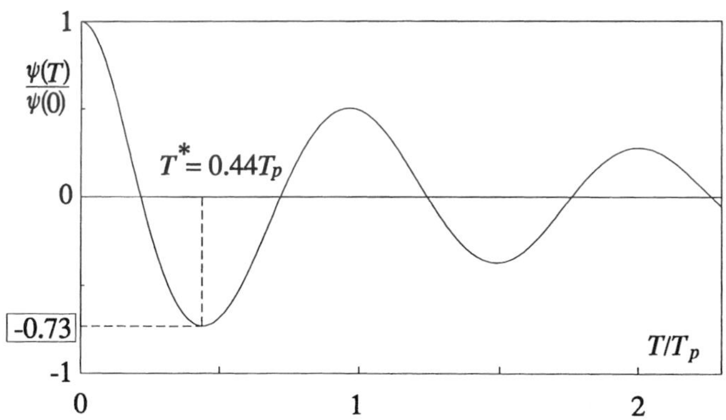

Fig. 4.17 Normalized autocovariance obtained from the mean JONSWAP spectrum.

4.12). Here, as the depth grows the high frequency peak (the one of the wind waves) gets smaller more rapidly than the low frequency peak (the one of the swells). Therefore, generally there is a *critical depth* where the two peaks of the spectrum take on the same height. At this critical depth the bandwidth reaches its maximum, and hence ψ^* has a minimum.

4.7.5 *The nature of the waves if the first local minimum of the autocovariance is not the absolute minimum*

Looking at fig. 4.12, we note that the absolute minimum of the autocovariance occurs at $T^* \cong 1$ s as far as the critical depth, while the absolute minimum occurs at $T^* \cong 3$ s for larger depths. Now, the T^* of 1 s is of the wind waves, while the T^* of 3 s is of the swells. Indeed, our wind waves have a peak period of about 2 s, so that the crest-trough interval is typically about 1 s; while our swells have a peak period of somewhat more than 6 s, so that the crest-trough interval is typically about 3 s. Therefore, at a certain depth beneath the water surface, T^* passes from the wind wave's value to the swell's value.

Close to the depth where swells overcome wind waves, the first local minimum of the autocovariance ceases to be the absolute minimum. At this depth the autocovariance is mainly affected by the swells, but the wind waves still have a considerable weight so that they leave a clear track of themselves in the local minimum of the autocovariance at $T \cong 1$ s. Clearly, at greater depths, the weight of the wind waves becomes negligible, the local minimum of the autocovariance at T of 1 s disappears, and the absolute minimum of $\psi(T)$ comes back to coincide with the first local minimum, with the only difference being that this time it occurs at T of 3 s.

We therefore realize that the case of the first local minimum not being the absolute minimum of the autocovariance is the special case of wind waves superimposed on somewhat higher swells. In the sea state of fig. 4.12, this phenomenon occurred at some depth beneath the water surface, affecting only the pressure head waves.

Conclusive note

The link between sea state and Gaussian random process was first noticed by Longuet-Higgins (1952). The theory of the sea states then ripened in the 60's, thanks to the contribution of several authors [see the references at the end of chap. 8 which gives the theory on the space-time].

The central role we have attributed to the autocovariance proceeds from the quasi-determinism theory (see chap. 9). In particular the fact that the main statistical properties of a sea state are coded in the first two waves of the autocovariance (the *core* of the autocovariance) is a consequence of the aforesaid theory. Also the narrow bandedness parameter ψ^* proceeds from a corollary of this theory, as we shall see in sect. 9.10.

The continuos spectrum is usually estimated following the Blackman-Tukey method (1958), or the method based on the Fourier Transform of the time series data (Bendat and Piersol, 1986). With the first method, $E(\omega)$ is obtained through the relation

$$E(\omega) = \frac{2}{\pi} \int_0^\infty \psi(T) \cos(\omega T)\, \mathrm{d}T$$

which proceeds from the Wiener-Khinchine theorem. In the applications, $\psi(T)$ is the autocovariance obtained from the time series data through definition (4.7). Given that $\psi(T)$ gets less and less reliable as T approaches the length of the record, it is common practice to multiply $\psi(T)$ by some deterministic window function which is close to 1 near the origin and approaches 0 as T grows.

With the second method the record is divided into a number of segments. The line spectrum is obtained for each segment, and these line spectra are averaged to obtain the continuous spectrum. A time window is usually used to eliminate the discontinuities at the beginning and end of each segment.

Here we have followed an alternative approach which removes the degree of subjectivity inherent in both the above cited methods: choice of the window in the first method (Moskowitz, 1964) and choice of the number of segments in the second method (Borgman, 1972 and Harris, 1974). We have used $\psi_F(T)$ (the periodic autocovariance associated with the line spectrum) rather than $\psi(T)$, and we have obtained the continuous spectrum through (4.18-19). The advantage is that for $\mathcal{T} > 5T_p$ the solution converges, in that the core of the autocovariance associated with the continuous spectrum keeps constant.

References

Bendat J. S. and Piersol A. G., 1986 Random data: analysis and measurement procedures. *J. Wiley & Sons*, 1-566.

Blackman R.B. and Tukey J.W., 1958 The measurement of power spectra. *Dover Publications Inc.*, 1-190.

Borgman L. E., 1972 Confidence intervals for ocean wave spectra. *Proc. 13th Conf. Coastal Eng. ASCE*, 237-250.

Cartwright D. E. and Longuet-Higgins M. S., 1956 The statistical distribution of the maxima of a random function. *Proc. Roy. Soc. London* A 237, 213-232.

Harris D. L., 1974 Finite spectrum analyses of wave records. *Proc. Int. Symp. On Wave Measurement and Analysis ASCE*, 107-124.

Hasselmann K., Barnett T. P., Bouws E. et al., 1973 Measurements of wind wave growth and swell decay during the Joint North Sea Wave Project (JONSWAP). *Deut. Hydrogr. Zeit.* A8, 1-95.

Longuet-Higgins M. S., 1952 On the statistical distribution of the heights of sea waves. *J. Mar. Res.* 11, 245-266;
1975 On the joint distribution of the periods and amplitudes of sea waves. *J. Geophys. Res.* 80, 2688-2694.

Moskowitz L., 1964 Estimates of power spectrums for fully developed seas for wind speeds of 20 to 40 knots. *J. Geophys. Res.* 69, 5161-5179.

Phillips O. M., 1958 The equilibrium range in the spectrum of wind-generated waves. *J. Fluid Mech.* 4, 426-434.

Pierson W. J. and Moskowitz L., 1964 A proposed spectral form for fully developed waves based on the similarity theory of S. A. Kitaigorodskii. *J. Geophys. Res.* 69, 5181-5190.

Chapter 5

ANALYSIS OF THE SEA STATES: THE TIME DOMAIN

5.1 Why the surface displacement represents a stationary Gaussian process

5.1.1 *Outline of the proof*

The random process (4.1) with the assumptions we have made on N, a_i, ω_i and ε_i is stationary and Gaussian. This means that the probability that $\eta(t)$ of a given realization falls within a fixed small interval $(w,\ w+\mathrm{d}w)$ is equal to the probability that $\eta(t_o)$ at any fixed time t_o, in a realization taken at random, falls in the fixed small interval $(w,\ w+\mathrm{d}w)$; and this probability is given by

$$p(\eta=w)\,\mathrm{d}w=\frac{1}{\sqrt{2\pi m_0}}\exp\left(-\frac{w^2}{2m_0}\right)\mathrm{d}w. \tag{5.1}$$

Before proving this property let us get a deeper insight into the meaning of the two above mentioned probabilities. The first one [probability that $\eta(t)$ of a given realization falls within the fixed small interval] is equal to the quotient between the time in which η is between w and $w+\mathrm{d}w$ and the total time [see fig. 5.1]. The second one [probability that $\eta(t_o)$ at a fixed t_o in a realization chosen at random falls within the fixed small interval] is equal to the quotient between the number of realizations

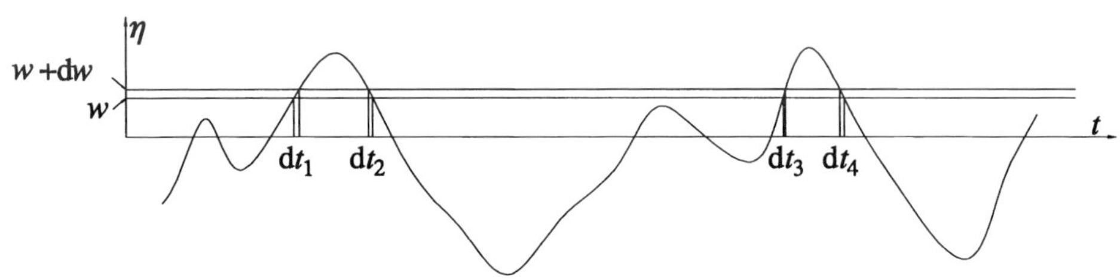

Fig. 5.1 $p(\eta=w)\,\mathrm{d}w$ is the probability that the surface displacement falls in the small interval $w,\ w+\mathrm{d}w$. It is equal to the quotient between the summation $(\mathrm{d}t_1+\mathrm{d}t_2+\ldots)$ and the total time.

in which the surface displacement η at time t_o is within $(w, w+\mathrm{d}w)$ and the total number of realizations.

Now let us see how equation (5.1) is achieved. First, let us consider two arbitrary random variables V_1 and V_2. If

$$\overline{V_1^n} = \overline{V_2^n} \qquad \forall n \tag{5.2}$$

that reads "if the mean value of the nth power of V_1 is equal to the mean value of the nth power of V_2, whichever the n", then the two variables have the same probability density function, that is

$$p(V_1 = w) = p(V_2 = w). \tag{5.3}$$

This rather intuitive property which proceeds formally from the theorem of moments will enable us to prove (5.1).

Before the proof of (5.1), it is also worth specifying that we shall adopt two different symbols for the mean value: one for the temporal mean, the other for the ensemble average. Specifically, $<\eta^n(t)>$ will denote the mean value of the nth power of $\eta(t)$ in a given realization of the process; and $\overline{\eta^n(t_o)}$ will denote the mean value of the nth power of η at the fixed time t_o.

5.1.2 *Proof relevant to any given realization*

From equation (4.1) of $\eta(t)$ we have

$$<\eta^4(t)>= \lim_{\mathcal{T}\to\infty} \frac{1}{\mathcal{T}} \int_0^{\mathcal{T}} \left[\sum_{i=1}^{N} a_i \cos(\omega_i t + \varepsilon_i)\right]^4 \mathrm{d}t = \sum_{i=1}^{N}\sum_{j=1}^{N}\sum_{m=1}^{N}\sum_{n=1}^{N} a_i a_j a_m a_n \cdot$$

$$\cdot \lim_{\mathcal{T}\to\infty} \frac{1}{\mathcal{T}} \int_0^{\mathcal{T}} \cos(\omega_i t + \varepsilon_i)\cos(\omega_j t + \varepsilon_j)\cos(\omega_m t + \varepsilon_m)\cos(\omega_n t + \varepsilon_n)\,\mathrm{d}t. \tag{5.4}$$

Besides the four assumptions of sect. 4.2, we assume $\omega_i \neq \omega_j + \omega_m + \omega_n$ (this being only a sufficient condition for the proof) and hence we have

$$\lim_{\mathcal{T}\to\infty} \frac{1}{\mathcal{T}} \int_0^{\mathcal{T}} \cos(\omega_i t + \varepsilon_i)\cos(\omega_j t + \varepsilon_j)\cos(\omega_m t + \varepsilon_m)\cos(\omega_n t + \varepsilon_n)\,\mathrm{d}t =$$

$$= \begin{cases} = \dfrac{1}{4} \text{ if } i = j \neq m = n \ \text{ or } \ i = m \neq j = n \ \text{ or } \ i = n \neq j = m, \\ = \dfrac{3}{8} \text{ if } i = j = m = n, \\ = 0 \text{ otherwise,} \end{cases}$$

and (5.4) reduces itself to

$$<\eta^4(t)>= 3\sum_{i=1}^{N}\sum_{j=1\,(j\neq i)}^{N} \frac{1}{4}\, a_i^2 a_j^2 + \sum_{i=1}^{N} \frac{3}{8}\, a_i^4. \tag{5.5}$$

Here the assumptions of sect. 4.2 on N and a_i come into play (N being infinitely large, a_i being of the same order of one another). Under these hypotheses, (5.5) can be rewritten in the form

$$< \eta^4(t) >= 3\sum_{i=1}^{N}\sum_{j=1}^{N}\frac{1}{4}a_i^2 a_j^2 \tag{5.6}$$

which taken together with (4.11) yields

$$< \eta^4(t) >= 3\,[< \eta^2(t) >]^2 = 3m_0^2.$$

Now, assuming that (5.1) is actually the probability of the surface displacement, we get the same value of $< \eta^4(t) >$:

$$< \eta^4(t) >= \int_{-\infty}^{+\infty} w^4 p(\eta = w)\,\mathrm{d}w = \frac{1}{\sqrt{2\pi m_0}}\int_{-\infty}^{+\infty} w^4 \exp\left(-\frac{w^2}{2m_0}\right)\mathrm{d}w = 3m_0^2.$$

By the same way of reasoning we can prove that, whatever the n, $< \eta^n(t) >$ takes on the same value if evaluated from equation (4.1) of $\eta(t)$ or from equation (5.1) of the probability of $\eta(t)$. If n is odd the proof is trivial, indeed it can be readily verified that $< \eta^n(t) >$ is zero whether one proceeds from (4.1) or from (5.1). The proof is trivial also for $n = 2$: from (4.1) we obtain $< \eta^2(t) >= m_0$, and the same result proceeds from (5.1). The first non-trivial proof, for which the assumption $N \to \infty$ is needed, is the one concerning $< \eta^4(t) >$ which we have given in this section. The proofs pertaining to $< \eta^6(t) >$, $< \eta^8(t) >$ and so on are like the proof for $< \eta^4(t) >$.

The fact that

$$< \eta^n(t) > \text{ obtained from (4.1)} = < \eta^n(t) > \text{ obtained from (5.1)} \quad \forall n$$

implies that (5.1) is actually the probability of $\eta(t)$.

Let us verify the step from (5.5) to (5.6) for the particular case in which all the a_i are equal to each other, that is

$$a_i = a \quad \forall i.$$

In this case, the r.h.s of (5.5) is equal to $\frac{3}{4}(N^2 - N)a^4 + \frac{3}{8}Na^4$, and the r.h.s. of (5.6) is equal to $\frac{3}{4}N^2a^4$, so that their quotient approaches 1 as $N \to \infty$.

5.1.3 *Proof relevant to the ensemble at a fixed time instant*

Fixing any time t_o, from (4.1) we have

$$\overline{\eta^4(t_o)} = \overline{\left[\sum_{i=1}^{N} a_i \cos(\omega_i t_o + \varepsilon_i)\right]^4}. \tag{5.7}$$

Here it is convenient to define

$$\tilde{\varepsilon}_i \equiv \omega_i t_o + \varepsilon_i - 2\pi \,\mathrm{int}\left[(\omega_i t_o + \varepsilon_i)/2\pi\right],$$

where

$$\mathrm{int}\,(x) \equiv \text{truncated value of } x,$$

and rewrite (5.7) in the form

$$\overline{\eta^4(t_o)} = \sum_{i=1}^{N}\sum_{j=1}^{N}\sum_{m=1}^{N}\sum_{n=1}^{N} a_i a_j a_m a_n \,\overline{\cos\tilde{\varepsilon}_i \cos\tilde{\varepsilon}_j \cos\tilde{\varepsilon}_m \cos\tilde{\varepsilon}_n}\,. \tag{5.8}$$

Given that $\tilde{\varepsilon}_i$, like ε_i, are uniformly distributed in $(0,2\pi)$ and are stochastically independent of one another, we have

$$\overline{\cos\tilde{\varepsilon}_i \cos\tilde{\varepsilon}_j \cos\tilde{\varepsilon}_m \cos\tilde{\varepsilon}_n} \begin{cases} = \dfrac{1}{4} & \text{if } i=j\neq m=n \text{ or } i=m\neq j=n \text{ or } i=n\neq j=m, \\ = \dfrac{3}{8} & \text{if } i=j=m=n, \\ = 0 & \text{otherwise}, \end{cases} \tag{5.9}$$

and consequently the equation (5.8) of $\overline{\eta^4(t_o)}$ reduces itself to

$$\overline{\eta^4(t_o)} = 3\sum_{i=1}^{N}\sum_{j=1\,(j\neq i)}^{N} \frac{1}{4}\, a_i^2 a_j^2 + \sum_{i=1}^{N} \frac{3}{8}\, a_i^4 .$$

Comparing this with equation (5.5) of $<\eta^4(t)>$, we see that

$$\overline{\eta^4(t_o)} = <\eta^4(t)> .$$

Similarly, we can verify the equality

$$\overline{\eta^n(t_o)} = <\eta^n(t)> \quad \forall n$$

which implies that the probability of $\eta(t_o)$ (relevant to the ensemble at a fixed time instant) is equal to the probability of $\eta(t)$ relevant to any given realization.

Check (5.9). If $i\neq j\neq m\neq n$, you obtain

$$\overline{\cos\tilde{\varepsilon}_i \cos\tilde{\varepsilon}_j \cos\tilde{\varepsilon}_m \cos\tilde{\varepsilon}_n} =$$

$$= \int_0^{2\pi}\int_0^{2\pi}\int_0^{2\pi}\int_0^{2\pi} \frac{1}{(2\pi)^4} \cos w_1 \cos w_2 \cos w_3 \cos w_4 \, dw_4\, dw_3\, dw_2\, dw_1 = 0,$$

where $1/(2\pi)^4$ is the joint probability density function $p\,(\tilde{\varepsilon}_i = w_1,\, \tilde{\varepsilon}_j = w_2,\, \tilde{\varepsilon}_m = w_3,\, \tilde{\varepsilon}_m = w_4)$. Then you can easily verify that also the terms with $i=j\neq m\neq n$ and $i=j=m\neq n$ are zero.

As to the terms with $i = j \neq m = n$ they give

$$\overline{\cos\tilde{\varepsilon}_i \cos\tilde{\varepsilon}_j \cos\tilde{\varepsilon}_m \cos\tilde{\varepsilon}_n} = \overline{\cos^2\tilde{\varepsilon}_i \cos^2\tilde{\varepsilon}_m} =$$
$$= \int_0^{2\pi}\int_0^{2\pi} \frac{1}{(2\pi)^2} \cos^2 w_1 \cos^2 w_2 \, \mathrm{d}w_2 \, \mathrm{d}w_1 = \frac{1}{4},$$

where $1/(2\pi)^2$ is the joint probability density function $p(\tilde{\varepsilon}_i = w_1, \tilde{\varepsilon}_j = w_2)$. Finally, the terms with $i = j = m = n$ give

$$\overline{\cos\tilde{\varepsilon}_i \cos\tilde{\varepsilon}_j \cos\tilde{\varepsilon}_m \cos\tilde{\varepsilon}_n} = \overline{\cos^4\tilde{\varepsilon}_i} = \int_0^{2\pi} \frac{1}{2\pi} \cos^4 w_1 \, \mathrm{d}w_1 = \frac{3}{8},$$

where $1/2\pi$ is the probability density function $p(\tilde{\varepsilon}_i = w_1)$.

5.1.4 *The process is ergodic*

Really, $\eta(t)$ of a sea state is a stationary *ergodic* Gaussian random process. The fact is that generally we omit the attribute *ergodic*, both because the attributes here are a good deal, and because the term ergodic is not very familiar. In simple words, the process is stationary Gaussian since $p[\eta(t_o) = w]$ at any fixed time t_o is given by (5.1), and is ergodic Gaussian since $p[\eta(t) = w]$ in any realization is given by (5.1). To better understand this concept, we give an example of a stationary Gaussian process being not ergodic, and an example of an ergodic Gaussian process being not stationary.

Let us consider the process

$$\eta_i(t) = V_i,$$

where the ith realization is a constant function of time, with the value of the constant being generally different for each realization. If the V_i are distributed according to (5.1), that is if

$$p(V_i = w) = \frac{1}{\sqrt{2\pi m_0}} \exp\left(-\frac{w^2}{2m_0}\right),$$

the random process is stationary Gaussian but not ergodic.

Let us consider now a stationary ergodic Gaussian process and let us imagine to bring the origin of each realization to coincide with a zero of η. Thus, our process keeps ergodic Gaussian, but it is no longer stationary since $p[\eta(0) = w]$ is given by a delta function rather than by (5.1).

5.2 Joint probability of surface displacements

5.2.1 *The mathematical form*

Let us define n random variables $V_1, V_2, \ldots, V_n$ each of them representing the surface displacement η or a derivative of any order of η taken at instants generally different from one another.

For example

$$V_1 \equiv \eta(t_o), \quad V_2 \equiv \dot{\eta}(t_o + T), \; ..., \quad V_n \equiv \ddot{\eta}(t_o + T'),$$

where the dot denotes the derivative, and t_o is any fixed time instant and T, T' are fixed time lags. The product

$$p(V_1 = w_1, \; V_2 = w_2, \; ..., \; V_n = w_n) \, \mathrm{d}w_1 \, \mathrm{d}w_2 \ldots \mathrm{d}w_n$$

represents the probability that V_1 falls in a fixed small interval $\mathrm{d}w_1$ including w_1; V_2 falls in a fixed small interval $\mathrm{d}w_2$ including w_2, and so on.

In sect. 5.1 we have proved that $p[\eta(t_o) = w]$ is a Gaussian (normal) probability density function. Expanding the reasoning from the probability density of a single variable to the joint probability density of a set of random variables, we can prove that $p(V_1 = w_1, \; V_2 = w_2, \; ..., \; V_n = w_n)$ is multivariate Gaussian, that is to say

$$p(V_1 = w_1, \; V_2 = w_2, ..., V_n = w_n) = \frac{1}{(2\pi)^{n/2}\sqrt{M}} \exp\left[-\frac{1}{2M}\sum_{i=1}^{n}\sum_{j=1}^{n} M_{ij} w_i w_j\right], \tag{5.10}$$

where M_{ij} and M are, respectively, the i, j cofactor and the determinant of the covariance matrix of $V_1, V_2, ..., V_n$:

$$\begin{array}{l} M_{ij} \equiv i, j \text{ cofactor}, \\ M \equiv \text{determinant of} \end{array} \begin{pmatrix} \overline{V_1^2} & \overline{V_1 V_2} & \cdots & \overline{V_1 V_n} \\ \overline{V_2 V_1} & \overline{V_2^2} & \cdots & \overline{V_2 V_n} \\ \vdots & \vdots & & \vdots \\ \vdots & \vdots & & \vdots \\ \overline{V_n V_1} & \overline{V_n V_2} & \cdots & \overline{V_n^2} \end{pmatrix}.$$

The elements of this matrix are ensemble averages like $\overline{\eta^4(t_o)}$ obtained in sect. 5.1. Since the ensemble averages are equal to the temporal means, the elements of the covariance matrix can be obtained also from temporal means. This approach is advisable.

5.2.2 *How to get a covariance matrix*

We seek the covariance matrix of $\eta(t_o)$, $\dot{\eta}(t_o)$ which will be applied later (as usual, t_o is any fixed time). The result is

$$\begin{pmatrix} \overline{\eta^2(t_o)} & \overline{\eta(t_o)\,\dot{\eta}(t_o)} \\ \overline{\dot{\eta}(t_o)\eta(t_o)} & \overline{\dot{\eta}^2(t_o)} \end{pmatrix} = \begin{pmatrix} m_0 & 0 \\ 0 & m_2 \end{pmatrix}. \tag{5.11}$$

Here, apart from the 1,1 entry that is simply the variance of the surface displacement, the other elements proceed from the following operations

$$\overline{\eta(t_o)\dot{\eta}(t_o)} = < \eta(t)\,\dot{\eta}(t) > = \frac{1}{\mathcal{T}}\int_0^{\mathcal{T}} \left[\sum_{i=1}^N a_i \cos(\omega_i t + \varepsilon_i)\right]\left[-\sum_{j=1}^N a_j \omega_j \sin(\omega_j t + \varepsilon_j)\right] \mathrm{d}t =$$

$$= -\sum_{i=1}^N \sum_{j=1}^N a_i a_j \omega_j \frac{1}{\mathcal{T}} \int_0^{\mathcal{T}} \cos(\omega_i t + \varepsilon_i)\sin(\omega_j t + \varepsilon_j)\,\mathrm{d}t = 0;$$

$$\overline{\dot{\eta}^2(t_o)} = < \dot{\eta}^2(t) > = \frac{1}{\mathcal{T}}\int_0^{\mathcal{T}} \left[-\sum_{i=1}^N a_i \omega_i \sin(\omega_i t + \varepsilon_i)\right]^2 \mathrm{d}t =$$

$$= \sum_{i=1}^N \sum_{j=1}^N a_i a_j \omega_i \omega_j \frac{1}{\mathcal{T}} \int_0^{\mathcal{T}} \sin(\omega_i t + \varepsilon_i)\sin(\omega_j t + \varepsilon_j)\,\mathrm{d}t = \sum_{i=1}^N \frac{1}{2} a_i^2 \omega_i^2 = m_2. \tag{5.12}$$

In these equations, we have understood that $\mathcal{T}$ tends to infinity; we have used the fact that the mean value of $\sin(\omega_i t + \varepsilon_i)\ \sin(\omega_j t + \varepsilon_j)$ is $\frac{1}{2}$ if $i = j$ or else is 0, and the fact that the mean value of $\cos(\omega_i t + \varepsilon_i)\ \sin(\omega_j t + \varepsilon_j)$ is always 0.

Check the rightmost equality (5.12). Hint: start from the question "which is the contribution to the summation from the small waves whose frequency ω_i is within a fixed small interval $\omega, \omega + \delta\omega$?". Then, use the definition (4.2) of spectrum and the definition (4.6) of moments of the spectrum. The reasoning is essentially the same we did for the summation on the r.h.s. of (4.13).

5.3 Rice's problem

We shall call b up-crossing (symbol b_+) the up-crossing of a fixed threshold b from the random function $\eta(t)$ [see fig. 5.2]. Our goal is $\mathcal{N}_+(b;\mathcal{T})$, *the number of* b_+ *in a given very large interval* $\mathcal{T}$. To this end we start subdividing the time axis into very small intervals $\mathrm{d}t$, and define

$p_+(b)\mathrm{d}t \equiv$ probability that any fixed $\mathrm{d}t$, say $(-\mathrm{d}t/2, \mathrm{d}t/2)$, includes a b_+.

$\mathcal{N}_+(b;\mathcal{T})$ is equal to the probability $p_+(b)\mathrm{d}t$ multiplied by the number $(\mathcal{T}/\mathrm{d}t)$ of small intervals in $\mathcal{T}$:

$$\mathcal{N}_+(b;\mathcal{T}) = p_+(b)\mathcal{T}. \tag{5.13}$$

Let us define now $p_+(b, w)\,\mathrm{d}t\,\mathrm{d}w$, the probability that

(i) the small interval $(-\mathrm{d}t/2, \mathrm{d}t/2)$ includes a b_+;

(ii) the derivative of η at the time of this b_+ falls within a fixed small interval $w, w + \mathrm{d}w$.

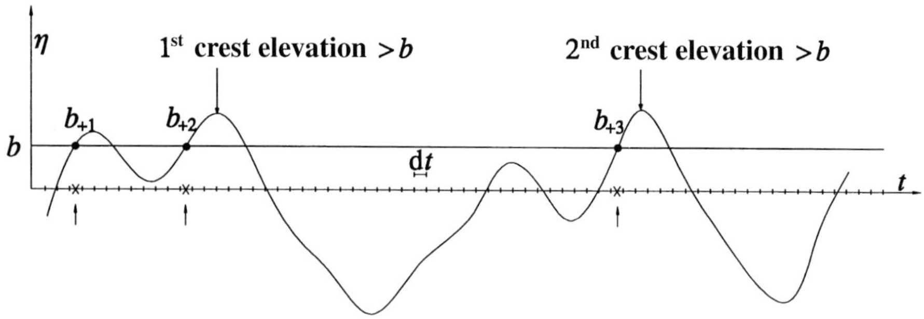

Fig. 5.2 A b_+ is an up-crossing of some fixed threshold b.

The relation between $p_+(b)$ and $p_+(b, w)$ proceeds straightforwardly from their definitions:

$$p_+(b) = \int_0^\infty p_+(b, w)\, \mathrm{d}w. \tag{5.14}$$

The joint probability of (i) and (ii) is equal to the probability that

(iii) $\eta(0)$ falls within $\left(b - \frac{1}{2}\,\mathrm{d}t\, w,\ b + \frac{1}{2}\,\mathrm{d}t\, w\right)$;

(iv) $\dot{\eta}(0)$ falls within $(w,\ w + \mathrm{d}w)$.

Fig. 5.3 is helpful to understand this equality, that is to say it is helpful to understand that

$$\text{probability of (i) and (ii)} = \text{probability of (iii) and (iv)}. \tag{5.15}$$

In an analytical form this equality is

$$p_+(b, w)\ \mathrm{d}t\, \mathrm{d}w = p[\eta(0) = b,\ \dot{\eta}(0) = w]w\, \mathrm{d}t\ \mathrm{d}w.$$

Here the joint p.d.f. on the r.h.s. is multivariate Gaussian and the relevant covariance matrix is (5.11), so that we get

$$p_+(b, w) = \frac{1}{2\pi\sqrt{m_0 m_2}} \exp\left[-\frac{1}{2m_0 m_2}\,(m_2 b^2 + m_0 w^2)\right] w. \tag{5.16}$$

Finally, equations (5.13), (5.14), and (5.16) taken together yield

$$\mathscr{N}_+(b; \mathscr{T}) = \frac{1}{2\pi}\sqrt{\frac{m_2}{m_0}} \exp\left(-\frac{b^2}{2m_0}\right)\mathscr{T}. \tag{5.17}$$

This is a result with some important consequences, as we shall see in the rest of this chapter. But before we see these consequences, it is worth making some remarks on the crucial step of our reasoning, that is to say, on equality (5.15).

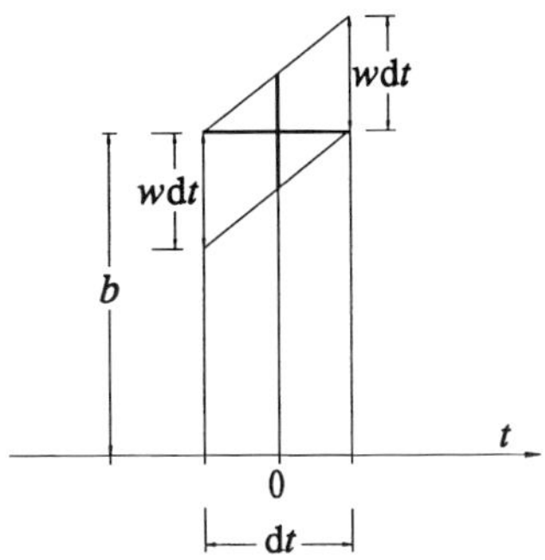

Fig. 5.3 The probability that the small interval $(-\mathrm{d}t/2, \mathrm{d}t/2)$ contains a b_+ of the surface displacement with a derivative between w and $w + \mathrm{d}w$ is equal to the probability that, at $t = 0$, the surface displacement falls between $b - \frac{1}{2} w\,\mathrm{d}t$ and $b + \frac{1}{2} w\,\mathrm{d}t$ and its derivative falls between w and $w + \mathrm{d}w$.

5.4 Rice's logic

Equation (5.15) is crucial in that it relates an unknown *horizontal probability* to a known *vertical probability*. Here by *horizontal probability* we mean the probability that a given value of the random function occurs on a fixed small interval dt: *horizontal* because dt belongs to the horizontal coordinate. By *vertical probability* we mean the probability that the random function falls within a fixed small interval dη at a fixed time instant: *vertical* because dη belongs to the vertical coordinate. The logic leading to (5.15) is general and enables us to solve an entire class of problems relevant to the differentiable random processes. This is why it is worth examining this logic closely.

Let us suppose that the events (i) and (ii) [cf. the definition of sect. 5.3] jointly occur. Specifically, a b_+ occurs at time δt with

$$-\mathrm{d}t/2 < \delta t < \mathrm{d}t/2, \tag{5.18}$$

and the derivative of the random function η at this instant falls between w and $w + \mathrm{d}w$, with $w > 0$. It follows that

$$\eta(0) = b - w\delta t, \tag{5.19a}$$

$$w - \ddot{\eta}(0)\,\delta t < \dot{\eta}(0) < w - \ddot{\eta}(0)\,\delta t + \mathrm{d}w. \tag{5.19b}$$

The two pairs of inequalities (5.18) and (5.19*b*) and the equality (5.19*a*) imply

$$b - w\,\mathrm{d}t/2 < \eta(0) < b + w\,\mathrm{d}t/2, \tag{5.20a}$$

$$w - |\ddot{\eta}(0)|\,\mathrm{d}t/2 < \dot{\eta}(0) < w + |\ddot{\eta}(0)|\,\mathrm{d}t/2 + \mathrm{d}w. \tag{5.20b}$$

Since the small intervals dt and dw are independent of each other, we can take dt so small that

$$\frac{|\ddot{\eta}(0)|\,\mathrm{d}t}{\mathrm{d}w} \to 0, \tag{5.21}$$

so that the inequality (5.20b) reduces itself to

$$w < \dot{\eta}(0) < w + \mathrm{d}w. \tag{5.22}$$

Conclusion: having assumed the occurrence of both event (i) and event (ii) we have proved that the events (iii) and (iv) occur too. Indeed the pair of inequalities (5.20a) describe just the event (iii) and the pair of inequalities (5.22) describe just the event (iv). The proof is valid under the assumption (5.21) on $\mathrm{d}t$ and $\mathrm{d}w$.

Through a similar reasoning we can also verify the converse: if the events (iii) and (iv) occur, then the events (i) and (ii) also occur, provided that $\mathrm{d}t$ and $\mathrm{d}w$ satisfy condition (5.21). Hence, the relation (5.15) between the probability of (i) and (ii) and the probability of (iii) and (iv) is proved under the assumption that $\mathrm{d}t$ and $\mathrm{d}w$ satisfy condition (5.21).

This proof is sufficient. Indeed, the probability of (i) and (ii) is proportional to the product $\mathrm{d}t\,\mathrm{d}w$, and also the probability of (iii) and (iv) is proportional to $\mathrm{d}t\,\mathrm{d}w$. Therefore if one succeeds in proving (as we have done) that for a particular pair $\mathrm{d}t, \mathrm{d}w$ the probability of (i) and (ii) is equal to the probability of (iii) and (iv), then it follows that the two probabilities are equal to each other whichever the pair $\mathrm{d}t$, $\mathrm{d}w$.

It is as if we had proved that a prism with a small base $\mathrm{d}A$ and a relatively large height $\mathrm{d}h$ of a homogeneous material ① has the same weight of a prism of the same base $\mathrm{d}A$ and height $\mathrm{d}h$ of a homogeneous material ②, and hence we should have concluded that, given any volume with a shape generally different from the prism, the weight of this volume of material ① and the weight of this volume of material ② are equal to each other. Clearly, in this comparison, the volume of small base $\mathrm{d}A$ and relatively large height $\mathrm{d}h$ stands for $\mathrm{d}t\,\mathrm{d}w$, with $\mathrm{d}t$ and $\mathrm{d}w$ satisfying condition (5.21); the weight of the volume of material ① stands for the probability of (i) and (ii); the weight of the volume of material ② stands for the probability of (iii) and (iv).

5.5 Corollaries of Rice's problem

5.5.1 *The mean wave period*

The mean wave period is given by the quotient between the very large time interval $\mathscr{T}$ and the number $\mathscr{N}_+(0;\mathscr{T})$ of zero up-crossings in this interval:

$$\overline{T} = \mathscr{T}/\mathscr{N}_+(0;\mathscr{T}) \tag{5.23}$$

(bearing in mind that the number of waves is equal to the number of zero up-crossings). With (5.17) of $\mathscr{N}_+(b;\mathscr{T})$, (5.23) becomes

$$\overline{T} = 2\pi\sqrt{\frac{m_0}{m_2}} \tag{5.24}$$

that is the formula of the mean period.

With the JONSWAP spectrum we have

$$\frac{m_0}{m_2} = \frac{A g^2 \omega_p^{-4} \int_0^\infty \mathscr{E}(w)\, \mathrm{d}w}{A g^2 \omega_p^{-2} \int_0^\infty w^2 \mathscr{E}(w)\, \mathrm{d}w},$$

where $\mathscr{E}(w)$ is defined by (4.24), and hence

$$\bar{T} = T_p \sqrt{\frac{\int_0^\infty \mathscr{E}(w)\, \mathrm{d}w}{\int_0^\infty w^2 \mathscr{E}(w)\, \mathrm{d}w}}.$$

The two integrals can be numerically evaluated for given values of the shape parameters χ_1 and χ_2 in the expression of $\mathscr{E}(w)$, and with the values of the mean JONSWAP spectrum ($\chi_1 = 3.3$, $\chi_2 = 0.08$) the result is

$$\bar{T} = 0.78\, T_p. \tag{5.25}$$

Hence, with the formula (4.27) of T_p we obtain

$$\bar{T} = 6.6\pi \sqrt{\frac{H_s}{4g}}. \tag{5.26}$$

5.5.2 *The general inequality of the crest elevation probability*

If a wave has a crest higher than a fixed threshold b then it contains *at least* a b_+ [cf. fig. 5.2]. Hence, the number of waves whose crest exceeds a fixed threshold b is smaller than or equal to the number of b_+. Only if the spectrum is infinitely narrow the waves approach the sinusoidal function so that there is only one local maximum per wave, which implies: the number of wave crests higher than a fixed threshold b is equal to the number of b_+, for every b. Summarizing, we write

$$\mathscr{N}_{cr}(b;\mathscr{T}) \begin{cases} = \mathscr{N}_+(b;\mathscr{T}) & \text{if the spectrum is infinitely narrow,} \\ \leq \mathscr{N}_+(b;\mathscr{T}) & \text{in general;} \end{cases} \tag{5.27}$$

where $\mathscr{N}_{cr}(b;\mathscr{T})$ is the number of waves whose crest exceeds a fixed threshold b, in a very large time $\mathscr{T}$, and $\mathscr{N}_+(b;\mathscr{T})$ is the number of b_+ in the interval $\mathscr{T}$, which has been already defined.

The probability $P(\mathscr{C} > b)$ of a wave with a crest elevation exceeding a fixed threshold b is equal to the quotient between $\mathscr{N}_{cr}(b;\mathscr{T})$ and $\mathscr{N}_+(0;\mathscr{T})$ (this being equal to the number of waves), that is

$$P(\mathscr{C} > b) = \mathscr{N}_{cr}(b;\mathscr{T}) / \mathscr{N}_+(0;\mathscr{T}). \tag{5.28}$$

From (5.27) and (5.28) it follows that

$$P(\mathscr{C} > b)\begin{cases} = \mathscr{N}_+(b;\mathscr{T})/\mathscr{N}_+(0;\mathscr{T}) & \text{if the spectrum is infinitely narrow,}\\ \leq \mathscr{N}_+(b;\mathscr{T})/\mathscr{N}_+(0;\mathscr{T}) & \text{in general;}\end{cases}$$

Hence, using the formula (5.17) for $\mathscr{N}_+(b;\mathscr{T})$, we obtain

$$P(\mathscr{C} > b)\begin{cases} = \exp(-b^2/2m_0) & \text{if the spectrum is infinitely narrow,} \quad (5.29a)\\ \leq \exp(-b^2/2m_0) & \text{in general.} \quad (5.29b)\end{cases}$$

The nondimensional form

$$P(\beta) = \exp(-\beta^2/2) \tag{5.30}$$

proceeding from (5.29*a*) gives the probability that a wave crest in a stationary Gaussian process with an infinitely narrow spectrum is higher than β times the standard deviation of this process (with β being arbitrarily fixed). It will be used later in the analysis of the wave forces on ocean structures.

5.5.3 *The wave height probability under the assumption of infinitely narrow spectrum*

We have just recalled that a single wave approaches a sinusoidal function if the spectrum is infinitely narrow. Another consequence of this is that the wave height is twice the crest elevation. Therefore

$$P(\text{wave height} > H) = P\left(\mathscr{C} > \frac{H}{2}\right) \quad \text{if the spectrum is infinitely narrow,}$$

which yields

$$P(\text{wave height} > H) = \exp\left(-\frac{H^2}{8m_0}\right) \quad \text{if the spectrum is infinitely narrow.} \tag{5.31}$$

This is one of the more commonly applied expressions in ocean engineering; the so called Rayleigh distribution. It gives the probability of exceedance of the wave heights under the assumption of infinitely narrow spectrum.

In what follows we shall express (5.31) in the alternative form

$$P(H; H_s = h) = \exp\left[-2\left(\frac{H}{h}\right)^2\right] \text{if the spectrum is infinitely narrow,} \tag{5.32}$$

where $P(H; H_s = h)$ is the probability of a wave height exceeding a fixed threshold H, in a sea state with a given significant wave height. For the step from (5.31) to (5.32) it suffices to use the definition $H_s \equiv 4\sqrt{m_0}$.

Besides the probability of exceedance, we shall use also the probability density function $p(H; H_s = h)$ that is related to the probability of exceedance by

$$p(H; H_s = h) = -\frac{\mathrm{d}P(H; H_s = h)}{\mathrm{d}H}. \tag{5.33}$$

Clearly, the product $p(H; H_s = h)\,\mathrm{d}H$ gives the probability of a wave height within the fixed small interval $H, H + \mathrm{d}H$, in a sea state with a given significant wave height. Equation (5.32) and (5.33) yield

$$p(H; H_s = h) = 4\frac{H}{h^2}\exp\left[-2\left(\frac{H}{h}\right)^2\right] \quad \text{if the spectrum is infinitely narrow.} \tag{5.34}$$

Finally, for the experimental verification of the wave height distribution the nondimensional form

$$P(\alpha) = \exp\left(-\frac{\alpha^2}{8}\right) \quad \text{if the spectrum is infinitely narrow} \tag{5.35}$$

is commonly used. It gives the probability that the quotient between the wave height and the root mean square surface displacement of the sea state exceeds any fixed threshold α. Here we use the simplest symbol $[P(\alpha)]$ because this is the probability most commonly used in dealing with field measurements.

5.5.4 *The general inequality of the mean wave height*

The mean crest elevation

The mean value of a random nonnegative variable is equal to the integral over $(0, \infty)$ of the probability of exceedance [see sect. 5.10.3]. In particular the mean crest elevation is given by

$$\overline{\mathscr{C}} = \int_0^\infty P(\mathscr{C} > b)\,\mathrm{d}b,$$

so that using (5.29*a*-*b*) we obtain

$$\overline{\mathscr{C}} \begin{cases} = \sqrt{\dfrac{\pi}{2}}\sqrt{m_0} & \text{if the spectrum is infinitely narrow,} \\[2ex] \leq \sqrt{\dfrac{\pi}{2}}\sqrt{m_0} & \text{in general.} \end{cases} \tag{5.36}$$

The mean trough depth

The stationary Gaussian process is statistically symmetrical, which means in particular the mean trough depth to be equal to the mean crest elevation:

$$\overline{\mathscr{C}^{(-)}} = \overline{\mathscr{C}}.$$

For verifying this equality is suffices to note that the process $-\eta(t)$ is also stationary Gaussian and that the troughs of $\eta(t)$ are the crests of $-\eta(t)$ (from which the symbol $\mathscr{C}^{(-)}$ that stands for crest elevation of the process $-\eta(t)$).

The mean wave height

For getting the mean wave height we should perform the following operation

$$\overline{H} = \frac{H_1 + H_2 + \ldots + H_n}{n}, \tag{5.37}$$

where H_1 is the height of the first wave of the sea state, H_2 the height of the second wave, and so on as far as H_n being the height of the last wave. For the wave definition, (5.37) can be rewritten in the form

$$\overline{H} = \frac{(\mathscr{C}_1 + \mathscr{C}_1^{(-)}) + (\mathscr{C}_2 + \mathscr{C}_2^{(-)}) + \ldots + (\mathscr{C}_n + \mathscr{C}_n^{(-)})}{n} = \overline{\mathscr{C}} + \overline{\mathscr{C}^{(-)}}, \tag{5.38}$$

indeed the wave height is the sum of the crest elevation and the trough depth.

From (5.36) and (5.38) it follows that

$$\overline{H} \begin{cases} = \sqrt{2\pi}\sqrt{m_0} \text{ if the spectrum is infinitely narrow,} & (5.39a) \\ \leq \sqrt{2\pi}\sqrt{m_0} \text{ in general.} & (5.39b) \end{cases}$$

Put into words, this means that the mean wave height is generally smaller than $\sqrt{2\pi}\,\sqrt{m_0}$ and becomes equal to $\sqrt{2\pi}\,\sqrt{m_0}$ in the limit as the bandwidth approaches zero.

5.6 Solved and still unsolved problems

How to know whether or not a problem relevant to the stationary Gaussian process is formally solvable for an arbitrary bandwidth? One should try to split the problem into the estimates of one or more probabilities that some given events occur in a finite number of given infinitesimal time intervals. If one succeeds in doing this, the problem is formally solvable. However, if the problem needs the estimate of only one probability that a given event occurs in a given interval of *finite extent*, then the closed solution, at present, is not achievable.

The first class of problems should be dedicated to S.O. Rice, since he gave the general solution to these problems (see references). A classic problem of this class is that of the mean wave period, which calls for the estimate of the probability that a zero up-crossing occurs in a given infinitesimal interval dt [cf. sects. 5.3 and 5.5.1].

The closed solution for the probability that a given interval (t_1, t_2) of finite extent contains at least one zero up-crossing has not yet been obtained. If this solution was obtained, we should be able to get, as a simple corollary, the distribution of the random times between successive zero up-crossings, that is the wave period distribution. Indeed, the distribution of the random times between successive points of a random point process is related to the probability that at least one point occurs in a fixed interval by a general equation that will be deduced in sect. 7.3.3.

For the probability of a wave height within a fixed small interval H, $H + \mathrm{d}H$ we should estimate the probability that the following events jointly occur:

(i) a fixed small interval $(t_o, t_o + \mathrm{d}t)$ contains a 0_+;

(ii) a fixed small interval $(t_o + T, t_o + T + \mathrm{d}T)$ contains a 0_+;

(iii) a fixed small interval $(t_o + T', t_o + T' + \mathrm{d}T')$, with $T' < T$, contains a local maximum with an elevation between b and $b + \mathrm{d}b$ (b and $\mathrm{d}b$ fixed, $b < H$);

(iv) a fixed small interval $(t_o + T'', t_o + T'' + \mathrm{d}T'')$, with $T' < T'' < T$, contains a local minimum with an elevation between $b - H - \mathrm{d}H$ and $b - H$;

(v) no other 0_+ occurs on $(t_o, t_o + T)$;

(vi) no other local maximum with an elevation greater than b occurs on $(t_o, t_o + T)$;

(vii) no other local minimum with an elevation smaller than $b - H$ occurs on $(t_o, t_o + T)$.

Dividing the joint probability of the seven events (i) ... (vii) by the probability of (i) we should obtain the probability of a wave with a height within $(H, H + \mathrm{d}H)$, period within $(T, T + \mathrm{d}T)$, crest elevation within $(b, b + \mathrm{d}b)$, interval between the first zero up-crossing and the crest within $(T', T' + \mathrm{d}T')$, interval between the first zero up-crossing and the trough within $(T'', T'' + \mathrm{d}T'')$. Then integrating this probability over

$$\{(T, T', T'', b) | T \in (0, \infty), T' \in (0, T), T'' \in (T', T), \text{and } b \in (0, H)\}$$

we should arrive at the sought probability of a wave whose height falls within the fixed small interval $(H, H + \mathrm{d}H)$.

It will have been understood that the general solution for the wave height probability is beyond our present skill. Indeed, apart from the overall complexity, we should estimate all of three probabilities that some given events [the events (v), (vi) and (vii)] occur on a fixed time interval of finite extent.

The formal solution for the wave height probability in the limit $H/\sigma \to \infty$ was obtained in the early 80's as a corollary of the theory of quasi-determinism. It was the first closed solution to a problem outside the Rice's class.

5.7 The period of a very high wave and the wave height probability under general bandwidth assumptions

5.7.1 *Elements of the quasi-determinism theory*

Let us consider the set of the waves with a given height H, say $H = 3\sigma$ in a stationary Gaussian process. (Really, the probability to find a wave height being *exactly* equal to 3σ is zero, and thus it is understood that we fix a small interval $\mathrm{d}H$ and take the set of the waves whose height is between H and $H + \mathrm{d}H$, specifically between 3σ and $3\sigma + \mathrm{d}H$.) The waves of this set will be different, even very different, from one another.

If we fixed a larger H, say $H = 8\sigma$, we should find that the waves contained in the set differ much less from one another and, in the limit as $H/\sigma \to \infty$, all waves of the set, apart from a negligible share, would prove to be equal to one another. More specifically, each wave of the set would occupy the centre of a well defined group that is the sum of a deterministic framework and a residual random noise of a smaller order. The form of the deterministic component is

$$\overline{\eta}(T) = \frac{\psi(T) - \psi(T - T^*)}{\psi(0) - \psi(T^*)} \frac{H}{2}, \tag{5.40}$$

where T^* is the abscissa of the absolute minimum, being assumed to be also the first minimum, of the autocovariance.

With the JONSWAP spectrum, the deterministic group (5.40) becomes

$$\overline{\eta}(T) = \frac{\int_0^\infty \mathscr{E}(w)\left\{\cos\left(2\pi w \frac{T}{T_p}\right) - \cos\left[2\pi w\left(\frac{T}{T_p} - \frac{T^*}{T_p}\right)\right]\right\} \mathrm{d}w}{\int_0^\infty \mathscr{E}(w)\left\{1 - \cos\left(2\pi w \frac{T^*}{T_p}\right)\right\} \mathrm{d}w} \frac{H}{2}.$$

This is shown in fig. 5.4 for the case of the mean JONSWAP spectrum ($\chi_1 = 3.3$, $\chi_2 = 0.08$), for which T^* proves to be equal to $0.44T_p$.

This is a short summary of the quasi-determinism theory. This theory deals with the *mechanics* of the wind generated waves, and hence it will be shown later (chapter 9). But the quasi-determinism theory has also two important consequences on wave statistics, that is on the subject of this chapter.

5.7.2 *Period T_h of a very high wave*

The first consequence of the quasi-determinism theory is that a wave of given height H has a well defined period, with a probability approaching 1, as $H/\sigma \to \infty$. This is

$$T_h = \text{period of the central wave of the group (5.40)}, \tag{5.41}$$

where the subscript h stands for *high waves*. With the mean JONSWAP spectrum, T_h can be simply taken from fig. 5.4:

$$T_h = 0.92 T_p \quad \text{for the mean JONSWAP spectrum}. \tag{5.42}$$

Clearly, the limit as $H/\sigma \to \infty$ of the regression $\overline{T}(H)$ of wave periods versus wave heights coincides with T_h:

$$\lim_{H/\sigma \to \infty} \overline{T}(H) = T_h,$$

and hence it can be calculated by means of (5.41) for *any given spectrum*. This is very useful since the numerical simulations of the stationary Gaussian process re-

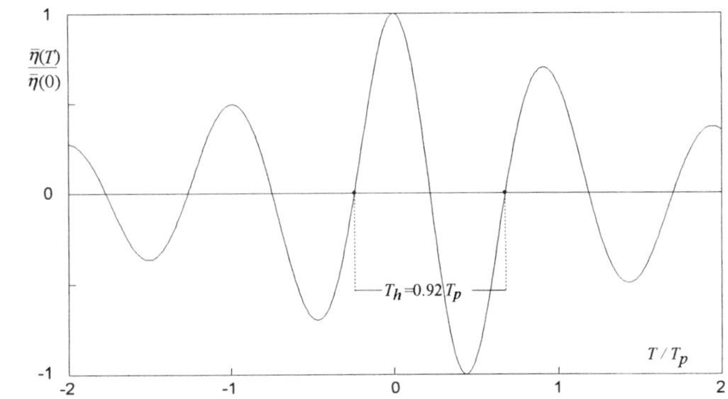

Fig. 5.4 The function (5.40) obtained from the mean JONSWAP spectrum.

veal that $\overline{T}(H)$ converges on its limit for $H/\sigma > 4 \div 5$, and the periods of the waves with $H/\sigma > 7 \div 8$ are typically very close to this limit.

5.7.3 *The wave height probability under general bandwidth assumptions*

The second consequence of the quasi-determinism theory, in the field of wave statistics, deals with the very wave height probability. Let us see how quasi-determinism theory and wave height probability are related to each other.

A wave of given height H necessarily has crest elevation and trough depth between 0 and H. Fig. 5.5*a* shows two possibilities. These possibilities are ∞^2 because the crest elevation can take on any value between 0 and H and the time interval between the crest and the trough can take on any positive value. The situation is more suitably represented in the plane τ-ξ, where τ is the crest-trough lag, and ξ is the quotient between the crest elevation and the crest-to-trough wave height. In particular the two waves ① and ② of fig. 5.5*a* are represented by two distinct points in the plane τ-ξ [see fig. 5.5*b*].

Let us imagine we examine a very large time interval $\mathcal{T}$, we gather all waves whose height is within a fixed small interval H, $H + \mathrm{d}H$, and we mark the points representative of these waves in the plane τ-ξ. For a finite H/σ the marked points would spread over the plane τ-ξ. [For example, the number of points falling in rectangles like A or B (see fig. 5.5*b*) in general would not be negligible.] On the contrary, as $H/\sigma \to \infty$ we would look at a great concentration: all the points but a negligible share would fall in an open 2-ball with centre at $T^*, \dfrac{1}{2}$ and radius of order $(H/\sigma)^{-1}$. Moreover all the points in this small 2-ball would be associated with waves whose profile is very close to the deterministic profile (5.40). In partial confirmation of this, the wave of given height H very large, according to (5.40), has crest elevation $H/2$ and crest-trough lag T^*, and thus it is represented by the point $T^*, \dfrac{1}{2}$ in the plane τ-ξ.

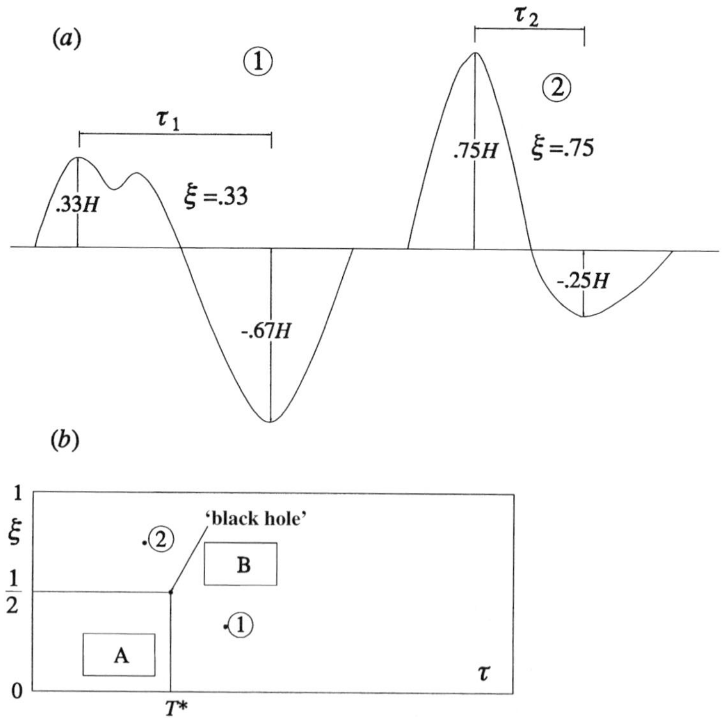

Fig. 5.5 (a) The waves with a fixed height H generally show a large variety of ξ and τ. (b) Plotting ξ vs τ, generally we get a wide cloud of points. Only in the limit as $H/\sigma \to \infty$, all points would gather at a special location, something like a black hole. This special location is at $\tau = T^*$, $\xi = \dfrac{1}{2}$.

If we associated a mass with each point in the plane τ-ξ (the same mass for each point), in the case of a finite H/σ we should find the mass rather uniformly distributed over the plane τ-ξ. While as $H/\sigma \to \infty$, the whole mass but a negligible share would pile up at point T^*, $\dfrac{1}{2}$, which would become something like a black hole. The mass density would vary widely in the black hole, despite its radius being very small, in the order of $(H/\sigma)^{-1}$. Specifically, the mass density would approach zero from the centre to the outskirts of the black hole.

All this will be formally proved in chap. 9. Moreover in sect. 9.10.1 we shall count all the points being present in the black hole. That is to say, we shall integrate the number of points over a small 2-ball with a radius of order $(H/\sigma)^{-1}$ and centre at point T^*, $\dfrac{1}{2}$. Clearly, once we shall have obtained the number of points, that is the number of waves whose height falls within the fixed small interval H, $H + \mathrm{d}H$, during the very large time interval $\mathscr{T}$, we shall arrive at once at the wave height probability. The result is

$$P(\text{wave height} > H) = K\exp\left[-\frac{1}{4(1+\psi^*)}\frac{H^2}{m_0}\right], \tag{5.43}$$

where K depends on the quotient between the absolute minimum and the absolute maximum of $\psi(T)$, and on the quotient of the curvatures of these maximum and minimum. Its value is equal to 1.21 for the Pierson and Moskowitz spectrum, 1.15 for the mean JONSWAP spectrum, and naturally is 1 for the infinitely narrow spectrum.

The form (5.43) has been compared with the data of numerical simulations of stationary Gaussian processes, and for the processes with the characteristic spectra of sea waves, (5.43) proves to be efficient for H/σ greater than $3 \div 4$ (we shall see this in more details in sect. 9.10.4). For $H/\sigma > 6$, which is the range of practical interest for ocean engineering, the effect of K proves to be negligible. This means that for $H/\sigma > 6$ we can assume $K = 1$, that is

$$P(\text{wave height} > H) = \exp\left[-\frac{1}{4(1+\psi^*)}\frac{H^2}{m_0}\right]. \tag{5.44}$$

This, without any appreciable consequence, in the sense that for a given P we get practically the same H whether using (5.43) or (5.44) (provided, of course, that $H/\sigma > 6$). Therefore we can use (5.44) for the practical applications. We can use it in the alternative form

$$P(H; H_s = h) = \exp\left[-\frac{4}{1+\psi^*}\left(\frac{H}{h}\right)^2\right], \tag{5.45}$$

or in the form

$$P(\alpha) = \exp\left[-\frac{\alpha^2}{4(1+\psi^*)}\right] \tag{5.46}$$

which gives the probability that the quotient between a wave height and the root mean square surface displacement of the sea state exceeds any fixed threshold α.

Another useful formula is that of the probability density function, which proceeds from the probability of exceedance through (5.33), and proves to be

$$p(H; H_s = h) = \frac{8}{1+\psi^*}\frac{H}{h^2}\exp\left[-\frac{4}{1+\psi^*}\left(\frac{H}{h}\right)^2\right]. \tag{5.47}$$

It should be noted that the forms (5.44-45-46) of the probability of exceedance reduce themselves correctly to (5.31-32-35), and the form (5.47) of the probability density function reduces itself to (5.34), if the spectrum is infinitely narrow so that $\psi^* = 1$.

Finally, it is useful to get the expressions of the probability of exceedance and of the probability density function, for the case of the mean JONSWAP spectrum. We have already seen in sect. 4.7.4 that the ψ^* of this spectrum is equal to 0.73, and hence:

$$\text{mean JONSWAP spectrum}\begin{cases} P(H; H_s = h) = \exp\left[-2.31\left(\dfrac{H}{h}\right)^2\right], & (5.48a) \\ p(H; H_s = h) = 4.62\dfrac{H}{h^2}\exp\left[-2.31\left(\dfrac{H}{h}\right)^2\right]. & (5.48b) \end{cases}$$

5.8 Experimental verification

5.8.1 *The two basic graphs*

A typical verification of the wave statistics calls for the measurement of the surface displacement $\eta(t)$ of a sea state and the following operations:

(i) to calculate the standard deviation σ of the time series data of $\eta(t)$;

(ii) to evaluate the autocovariance $\psi(T)$ from the time series data of $\eta(t)$;

(iii) to obtain T^* from $\psi(T)$, and $\psi^* = |\psi(T^*)/\psi(0)|$;

(iv) to compute the wave group (5.40) from $\psi(T)$ (we mean the nondimensional wave group $\overline{\eta}(T)/H$);

(v) to obtain T_h from the wave group (5.40) (T_h is the period of the wave whose crest falls at $T = 0$) ;

(vi) to single out the waves from the time series data of $\eta(t)$, and to measure their heights H_1, H_2, ..., H_n and periods T_1, T_2, ..., T_n;

(vii) to divide the height of each single wave by σ, and to divide the period by T_h, so to get a set of pairs α_1, $\widetilde{T}_1$; α_2, $\widetilde{T}_2$; ...; α_n, $\widetilde{T}_n$; where

$$\alpha_i \equiv \frac{H_i}{\sigma}\,,\; \widetilde{T}_i \equiv \frac{T_i}{T_h}\,.$$

The operations (i) ... (vii) can be repeated for an arbitrarily large number of wave records, and the sets of pairs α_i, $\widetilde{T}_i$ of the different sea states can be gathered so as to make up a single big set. The sea states can be recorded in different areas, and under environmental conditions which are even very different from one another. Only one must take care to discard the sea states where the absolute minimum of $\psi(T)$ is not also the first local minimum on the positive domain of this function (it is something that may happen if the wind waves are superimposed on some slightly higher swells as was shown in sect. 4.7.5).

In order to verify the theory we have to draw two graphs. The first one is the representation on the plane α-$\widetilde{T}$ of all the pairs α_i, $\widetilde{T}_i$. The result, as evident, will be a point cloud. For the second graph we have to work only on the nondimensional values α_i, from which we get the experimental $P(\alpha)$. Let us imagine, for example, we have 10^5 values of α_i and that $2 \cdot 10^3$ of these α_i exceed the threshold $\alpha = 5$, then

$$P(5) = \frac{2 \cdot 10^3}{10^5} = 2 \cdot 10^{-2}.$$

It is convenient to resort to a semilogarithmic coordinate paper where the functions (5.35) and (5.46) are represented by two parabolas. Indeed from these two functions we have respectively

$$\ln\frac{1}{P}=\frac{\alpha^2}{8}\quad \text{if the spectrum is infinitely narrow,}$$

$$\ln\frac{1}{P}=\frac{\alpha^2}{4(1+\psi^*)}\quad \text{in general.}$$

We shall represent the experimental $P(\alpha)$ (obtained from the set of the α_i) and the two functions (5.35) and (5.46). As to ψ^* (which is necessary for 5.46), we shall take the average of the ψ^* in the set of wave records.

5.8.2 *Graphs obtained from field data*

Basing ourselves on the theory of sect. 5.7 we expect that the ordinates of the rightmost points α, $\widetilde{T}$ of the first graph are close to 1. (Indeed the rightmost points are those with the largest $\alpha = H/\sigma$, and in sect. 5.7.2 we have seen that a wave with a very large H/σ must have a period T very close to T_h, and hence a nondimensional period $\widetilde{T}$ very close to 1.) We also expect that the experimental $P(\alpha)$ approaches the function (5.46) as α grows. An apparent confirmation of both the predictions is given by figs. 5.6-7 being obtained from the data of the experiment RC 1990.

Typically, the point cloud α, $\widetilde{T}$ appears as in fig. 5.6, with the maximum width roughly on $\alpha = 2$. As to the $P(\alpha)$, we see the data points for small α to be practically coincident with the function (5.35). Then, for $\alpha > 3$, the data points deviate gradu-

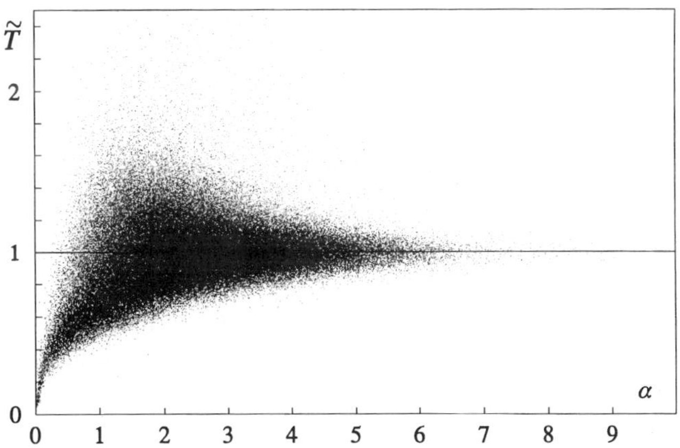

Fig. 5.6 Data points $\widetilde{T}_i \equiv T_i/T_h$, $\alpha_i \equiv H_i/\sigma$ obtained from the experiment RC 1990. We see that the $\widetilde{T}$ of the highest waves are very close to 1.

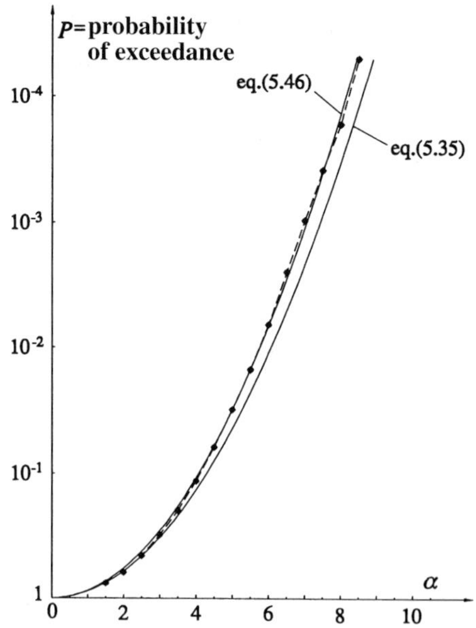

Fig. 5.7 Probability $P(\alpha)$ obtained from the experiment RC 1990. We see that $P(\alpha)$ approaches the form (5.46) for increasing α.

ally from (5.35) and move towards (5.46). Starting from $\alpha = 4$, the data points are practically coincident with the function (5.46). This implies that the experimental $P(\alpha)$ all over its domain is smaller than or equal to (5.35), which confirms also the inequality (5.39*b*) on the mean wave height (bearing in mind that the mean value of a nonnegative random variable is equal to the integral from 0 to ∞ of its probability of exceedance).

5.8.3 *Non-linearity effects*

In conclusion, the theory of the sea states to the first order in a Stokes' expansion enables us to foresee the basic things on wave height and wave period variability. This does not mean that a sea state is wholly a stationary Gaussian process, as predicted by the aforesaid theory. Indeed this theory is not able to predict the profile distortion due to the non-linearity effects. Going into detail, in an actual sea state, the probability of exceedance of the crest elevation is generally different from the probability of exceedance of the trough depth; for the surface waves, the probability of exceedance of the crest elevation is greater than the probability of exceedance of the trough depth; vice-versa for the pressure head waves beneath the water surface. Hence, there is an apparent difference with respect to the stationary Gaussian processes which are statistically symmetrical.

On the other hand, as we have seen, there is an excellent agreement between sea states and Gaussian process for what concerns the wave height probability. The reason for this agreement is that the non-linearity affects the crest elevation and the trough depth, but does not affect the crest-to-trough wave height. Similarly, the non-linearity shortens the wave crest and lengthens the wave trough, but does not affect the wave period.

5.8.4 *A firm landmark*

Sometimes the Italians dream about setting their watches to train departures. Well, something like this does actually occur at sea; the wave height probability is usually so regular that we can use it to check the wave gauges.

In the natural laboratory of Reggio Calabria, we resort to graphs like that of fig. 5.7 for a general check of the wave gauges. An emblematic case occurred in 1990 with a few capacitance wave staffs. The experimental $P(\alpha)$ looked somewhat different from that of fig. 5.7, in the sense that the data points slightly deviated rightward with respect to the function (5.35). This made us think of a gauge's flaw, and in fact a few laboratory tests revealed that these gauges had some small departures from linearity.

5.9 Characteristic wave heights

5.9.1 *The formal solution*

Let us fix a number $\mathscr{F}$ between 0 and 1:

$$0 < \mathscr{F} < 1,$$

and let us define

$H^*_{\mathscr{F}} \equiv$ the threshold being exceeded by $\mathscr{F} \cdot N$ wave heights in a sea state (N being the total number of waves).

This definition is simply expressed by the equality

$$P(H^*_{\mathscr{F}}; H_s = h) = \mathscr{F},$$

that, with the formula (5.32) for the probability of exceedance, yields

$$H^*_{\mathscr{F}} = h\sqrt{\frac{1}{2}\ln\frac{1}{\mathscr{F}}}.$$

Then let us define

$H_{\mathscr{F}} \equiv$ the average wave height of the $\mathscr{F} \cdot N$ highest waves of a sea state.

The definitions of $H^*_{\mathscr{F}}$ and $H_{\mathscr{F}}$ yield

$$H_{\mathscr{F}} \equiv \text{the average height of the waves higher than } H^*_{\mathscr{F}},$$

and consequently

$$H_{\mathscr{F}} = \frac{\displaystyle\int_{H^*_{\mathscr{F}}}^{\infty} H p(H; H_s = h)\,\mathrm{d}H}{\displaystyle\int_{H^*_{\mathscr{F}}}^{\infty} p(H; H_s = h)\,\mathrm{d}H} = h\left[\sqrt{\frac{1}{2}\ln\frac{1}{\mathscr{F}}} + \frac{1}{\mathscr{F}}\sqrt{\frac{\pi}{8}}\left(1 - \operatorname{erf}\sqrt{\ln\frac{1}{\mathscr{F}}}\right)\right], \quad (5.49)$$

where the erf function has been already used in chap. 3 [cf. the definition (3.37)].

Check the rightmost equality (5.49). Use (5.34) of $p(H; H_s = h)$, and note that the integral on the denominator is equal to $\mathscr{F}$, and the integral on the numerator can be evaluated by parts.

5.9.2 *Analytical exercise: find a well approximate formula for the direct evaluation of $H_{\mathscr{F}}$*

Equation (5.49) gives the exact solution for $H_{\mathscr{F}}$. But it has the inconvenience of containing the erf function that requires a numerical integration. Thus let us seek an approximate formula of quick use.

The function $1 - \operatorname{erf}\sqrt{x}$ appearing in (5.49) satisfies the following pair of inequalities

$$S_1(x) < 1 - \operatorname{erf}\sqrt{x} < S_2(x), \quad (5.50)$$

with

$$S_1(x) \equiv \frac{1}{\sqrt{\pi}} \mathrm{e}^{-x}\left(x^{-\frac{1}{2}} - \frac{1}{2}x^{-\frac{3}{2}}\right),$$

$$S_2(x) \equiv \frac{1}{\sqrt{\pi}} \mathrm{e}^{-x}\left(x^{-\frac{1}{2}} - \frac{1}{2}x^{-\frac{3}{2}} + \frac{3}{4}x^{-\frac{5}{2}}\right).$$

For verifying the inequalities (5.50) it suffices to note that

$$\lim_{x\to\infty}\begin{cases} 1 - \operatorname{erf}\sqrt{x} = 0, \\ S_1(x) = 0, \\ S_2(x) = 0, \end{cases} \quad (5.51)$$

and that

$$\left\{\frac{\mathrm{d}}{\mathrm{d}x}S_2(x) = -\frac{1}{\sqrt{\pi}}\mathrm{e}^{-x}\left(x^{-\frac{1}{2}} + \frac{15}{8}x^{-\frac{7}{2}}\right)\right\} < \left\{\frac{\mathrm{d}}{\mathrm{d}x}(1 - \operatorname{erf}\sqrt{x}) = -\frac{1}{\sqrt{\pi}}\mathrm{e}^{-x}x^{-\frac{1}{2}}\right\} <$$

$$< \left\{\frac{\mathrm{d}}{\mathrm{d}x}S_1(x) = -\frac{1}{\sqrt{\pi}}\mathrm{e}^{-x}\left(x^{-\frac{1}{2}} - \frac{3}{4}x^{-\frac{5}{2}}\right)\right\}. \quad (5.52)$$

The equalities (5.51) and (5.52) imply also that

$$\lim_{x\to\infty}\frac{1-\operatorname{erf}\sqrt{x}}{S_1(x)}=1\,,\qquad \lim_{x\to\infty}\frac{1-\operatorname{erf}\sqrt{x}}{S_2(x)}=1\,.$$

As x takes on smaller and smaller values, the difference between $1-\operatorname{erf}\sqrt{x}$ and $S_1(x)$ or $S_2(x)$ progressively grows. However for $x>1$, the function $1-\operatorname{erf}\sqrt{x}$ can be very well fitted by an intermediate function between $S_1(x)$ and $S_2(x)$. Specifically, the form

$$1-\operatorname{erf}\sqrt{x}\cong\frac{1}{\sqrt{\pi}}\,\mathrm{e}^{-x}\left(x^{-\frac{1}{2}}-\frac{1}{2}x^{-\frac{3}{2}}+\frac{1}{4}x^{-\frac{5}{2}}\right)\quad\text{for } x>1$$

proves to be efficient. This approximate form enables us to rewrite (5.49) as

$$H_{\mathscr{F}}\cong h\left[\sqrt{\frac{1}{2}x}+\frac{1}{\sqrt{8}}\left(x^{-\frac{1}{2}}-\frac{1}{2}x^{-\frac{3}{2}}+\frac{1}{4}x^{-\frac{5}{2}}\right)\right]\quad\left(x\equiv\ln\frac{1}{\mathscr{F}}\right)\quad\text{for } \mathscr{F}<0.35.\tag{5.53}$$

5.9.3 *The bandwidth effect*

Equation (5.49) and its approximate form (5.53) are valid provided the spectrum is infinitely narrow, since we have used formula (5.34) for $p(H;H_s=h)$. If we use formula (5.47) valid for an arbitrary bandwidth, we obtain again (5.49) (and 5.53) multiplied by the factor $\sqrt{(1+\psi^*)/2}$.

Thus the characteristic height $H_{\mathscr{F}}$ is generally smaller than it is predicted by (5.49) or (5.53): the broader the spectrum the smaller the characteristic height. In particular, with the mean JONSWAP spectrum, $H_{\mathscr{F}}$ proves to be 7% smaller than is predicted by (5.49) or (5.53). But $H_{\mathscr{F}}$ has come into use under the assumption of infinitely narrow spectrum, and whenever some design rules prescribe $H_{1/3}$, $H_{1/10}$, $H_{1/100}$ it is understood that these are $H_{1/3}$, $H_{1/10}$, $H_{1/100}$ for the infinitely narrow spectrum.

5.10 The maximum expected wave height in a sea state of given duration

5.10.1 *The stochastic independence of the wave heights and its consequences on the result*

Let us consider N consecutive waves of a sea state of given significant height. The probability that the largest wave height of this set of N waves is smaller than a given threshold H is equal to the probability that all N wave heights are smaller than H. (Indeed if the maximum wave height $H_{\max}$ is smaller than H, then all wave heights are smaller than H; and if all wave heights are smaller than H, then also $H_{\max}$ is smaller than H.) Thus we can write

$$\mathscr{P}(H_{\max}<H;H_s=h)=\mathscr{P}(H_1<H,H_2<H,\ldots,H_N<H;H_s=h)\,.\tag{5.54}$$

In literature the probability on the r.h.s. of this equation is expressed in the form

$$\mathscr{P}(H_1 < H, H_2 < H, \dots, H_N < H; H_s = h) = [1 - P(H; H_s = h)]^N \tag{5.55}$$

that would be exact if the wave heights were stochastically independent of one another. Indeed (5.55) says that the joint probability that H_1 is smaller than H, H_2 is smaller than H, and so on is equal to the probability that H_1 is smaller than H multiplied by the probability that H_2 is smaller than H, and so on. Really, the wave heights are *not* stochastically independent of one another, and hence it is necessary to reflect upon (5.55).

It is convenient to consider first the condition of infinitely narrow spectrum. In this case, if N is finite, the N consecutive waves have only some infinitesimal differences from one another [cf. sect. 4.7.1]. Therefore the probability that all N waves are smaller than H is equal to the probability that one single wave is smaller than H:

$$\mathscr{P}(H_1 < H, H_2 < H, \dots, H_N < H; H_s = h) = 1 - P(H; H_s = h)$$
$$\text{if the spectrum is infinitely narrow.}$$

As a consequence, if the spectrum was infinitely narrow, using (5.55) we would largely underestimate the probability that all the wave heights are smaller than H, that is we would overestimate largely the probability that H_{max} is greater than H, which means that our estimate would be conservative.

If the spectrum has a finite bandwidth like that of the sea wave spectra, using (5.55) we make an error of the same kind, i.e. a conservative one, but much smaller. The quasi-determinism theory helps us to understand why. Indeed it shows that the characteristic wave groups consist of three or four consecutive waves [see the pictures of these groups in chap. 10]. Before and after groups, we can find everything: a sequence of small random waves or even a new group. In other words the three or four waves forming the group are statistically independent from the preceding and the following ones. On the contrary, the three or four waves of the group are obviously not stochastically independent of one another. In practice, this implies that given an exceptionally high wave (centre of a group) we can expect that one or two waves before and one or two waves after will also be higher than the mean wave height.

To summarize, with the characteristic spectra of the sea waves we have a correlation affecting sequences of three or four consecutive waves, and this is not even a full correlation in the sense that the three or four waves do not have the same height; with the infinitely narrow spectrum, the correlation is full and affects the whole set of N waves. Thus we understand that the greater the N the more efficient is the assumption of stochastic independence, and for N of a few hundred this assumption will lead to only a small conservative error. In conclusion, equation (5.55), assuming the stochastic independence of the wave heights, is efficient for engineering purposes, since it is simple and slightly conservative.

5.10.2 *The formula for the maximum expected wave height*

Equations (5.54) and (5.55) yield

$$P(H_{\max} > H; H_s = h) = 1 - [1 - P(H; H_s = h)]^N \tag{5.56}$$

which leads us at once to the maximum expected wave height $\overline{H_{\max}}$ in a sequence of N waves. For a physical interpretation of $\overline{H_{\max}}$, let us suppose to take n sets each of N consecutive waves from the same sea state. The first set will have a maximum wave height $H_{\max 1}$, the second set will have a maximum wave height $H_{\max 2}$ generally different from $H_{\max 1}$, and so on as far as the nth set whose maximum wave height will be $H_{\max n}$. The maximum expected wave height $\overline{H_{\max}}$ is the average of $H_{\max 1}, H_{\max 2}, \ldots, H_{\max n}$.

Since the mean value of a nonnegative random variable like $H_{\max}$ is equal to the integral over $(0, \infty)$ of its probability of exceedance, from (5.56) we obtain

$$\overline{H_{\max}} = \int_0^\infty P(H_{\max} > H; H_s = h)\,\mathrm{d}H = \int_0^\infty 1 - \left\{1 - \exp\left[-\frac{4}{1+\psi^*}\left(\frac{H}{h}\right)^2\right]\right\}^N \mathrm{d}H \tag{5.57}$$

which can be given the nondimensional form

$$\overline{\tilde{\alpha}_{\max}} = \int_0^\infty 1 - \left[1 - \exp\left(-\frac{4}{1+\psi^*}\tilde{\alpha}^2\right)\right]^N \mathrm{d}\tilde{\alpha}\,, \tag{5.58}$$

where $\tilde{\alpha} \equiv H/h$. The integrand on the r.h.s. of this equation is shown in fig. 5.8 for $N = 200$ and $\psi^* = 0.73$ which is the value pertaining to the mean JONSWAP spectrum. The integral is equal to 1.586 which implies the maximum expected wave height in a sequence of 200 waves to be equal to 1.586 times the significant wave height.

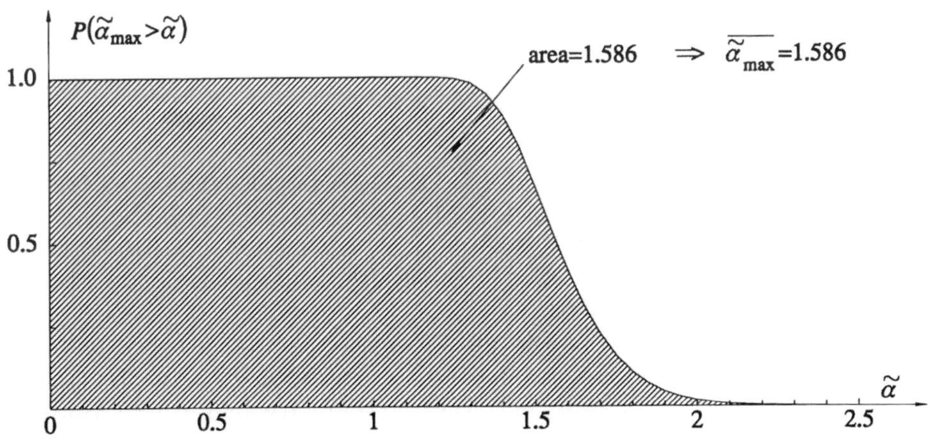

Fig. 5.8 $\overline{\tilde{\alpha}_{\max}}(N)$ proceeds from the integral of $P(\tilde{\alpha}_{\max} > \tilde{\alpha})$ over $(0, \infty)$. Here N has been taken equal to 200.

Usually we have to evaluate the maximum expected wave height in a sea state of given duration Δt. To this end equation (5.58) is used with

$$N = \frac{\Delta t}{\overline{T}(h)}, \tag{5.59}$$

and with the expression of $\overline{T}(h)$ obtained in sect. 5.5.1. Indeed $\Delta t/\overline{T}(h)$ is the expected number of waves in an interval Δt, in the given sea state.

5.10.3 *Appendix: the relation between mean value and probability of exceedance of a nonnegative random variable*

Let us consider a nonnegative random variable V. By $p(V = x)\,\mathrm{d}x$ we mean the probability that V takes on a value within a fixed small interval $x, x + \mathrm{d}x$, and by $P(V > x)$ we mean the probability that V takes on a value greater than a fixed x, so that

$$p(V = x) = -\frac{\mathrm{d}P(V > x)}{\mathrm{d}x}.$$

Now, probably everybody knows that the mean value of V can be obtained through the relation

$$\overline{V} = \int_0^\infty x\,p(V = x)\,\mathrm{d}x \tag{5.60}$$

which is well documented in every book on probability and statistics, and is also quite easy to verify. Perhaps, it is less well-known that the mean value of a nonnegative random variable can be readily evaluated from the probability of exceedance. Indeed, evaluating (by parts) the integral on the r.h.s. of (5.60), we obtain

$$\overline{V} = -x\,P(V > x)\Big|_0^\infty + \int_0^\infty P(V > x)\,\mathrm{d}x,$$

from which, bearing in mind that $\lim_{x\to\infty} xP(V > x) = 0$ (this being a necessary condition for the existence of the mean value), it follows that

$$\overline{V} = \int_0^\infty P(V > x)\,\mathrm{d}x.$$

Conclusive note

The classic approach of Cartwright and Longuet-Higgins (1956) proceeds from Rice's solution (1944-45) for the distribution of the elevations of the local maxima of a stationary Gaussian process. Equation (5.32) is obtained as a corollary of this solution. Then a new solution by Rice (1944-45), the one for the expected number per unit time of zero up-crossings, gives (5.24).

In this chapter we have followed an alternative approach suggested by the author (1982), which proceeds from Rice's solution (1958) for the expected number per unit time of b up-crossings, where b is any fixed threshold. Both basic equations (5.24) and (5.32) are obtained from this solution, and furthermore also inequalities (5.29*b*) and (5.39*b*) are obtained which are valid for an arbitrary bandwidth.

The very typical discrepancy between data points and the Rayleigh distribution of wave heights was illustrated by Haring et al. (1976) and Forristall (1978) and was definitively pointed out, and classified as a bandwidth effect, by Longuet-Higgins (1980).

References

Boccotti P., 1982 Relations between characteristic sea wave parameters. *J. Geophys. Res.* 87, 4267-4268.

Cartwright D. E. and Longuet-Higgins M. S., 1956 The statistical distribution of the maxima of a random function. *Proc. Roy. Soc.* A 237, 213-232.

Forristall G.Z., 1978 On the statistical distribution of wave heights in a storm. *J. Geophys. Res.* 83, 2353-2358.

Haring R.E., Osborne A.R. and Spencer L.P. 1976 Extreme wave parameters based on continental shelf storm wave records. *Proc. 15th Conf. Coastal Eng. ASCE.*

Longuet-Higgins M.S., 1980 On the distribution of the heights of sea waves: some effects of nonlinearity and finite bandwidth. *J. Geophys. Res.* 85, 1519-1523.

Rice S. O., 1944 Mathematical analysis of random noise. *Bell Syst. Tech. J.* 23, 282-332;
1945 Mathematical analysis of random noise. *Bell Syst. Tech. J.* 24, 46-156;
1958 Distribution of the duration of fades in radio transmission. *Bell Syst. Tech. J.* 37, 581-635.

Chapter 6
THE WAVE CLIMATE

6.1 The function $H_s(t)$

6.1.1 *Measurements of H_s*

Let us imagine we take a continuous record of the surface displacement $\eta(t)$ at a fixed site. We shall call $H_s(t)$ the significant wave height of the sea state lasting from $t-\Delta t/2$ to $t+\Delta t/2$:

$$H_s(t) = 4\sqrt{\frac{1}{\Delta t}\int_{t-\Delta t/2}^{t+\Delta t/2} \eta^2(t')\,\mathrm{d}t'},$$

where Δt is the duration of a sea state. $H_s(t)$ is a random continuous function that is gradually variable and has a mean value which depends on the site.

The random function $H_s(t)$ is not stationary because of a seasonal component. Moreover, it has a strong statistical asymmetry with respect to the mean. Some numerical simulations of this function have been recently presented by Cunha and Guedes Soares (1999).

Usually the records of $H_s(t)$ are taken by means of instrumented buoys with a Δt of 20 minutes. This is a rather large duration for a weak sea state (for example, with an H_s of 0.5 m, we expect about 500 wind waves in 20 minutes). But we aim to get with more accuracy the H_s of the heavier sea states. Thus the choice of fixing a Δt of 20 minutes is justified and convenient.

Usually, the records are intermittent. A typical way of working is: record of $\eta(t)$ for a continuous duration of 20 minutes in three hours or in one hour, which means sampling $H_s(t)$ at a rate of 1/3 per hour or 1 per hour. The data points of $H_s(t)$ are then interpolated (e. g. resorting to a Fourier series) and the interpolation proves to be efficient because the function varies gradually.

6.1.2 *Definition of sea storm*

From function $H_s(t)$ we can single out the sea storms. Let us define a sea storm as "a sequence of sea states in which $H_s(t)$ exceeds a fixed threshold h_{crit}, and does not fall below this threshold for a continuous time interval greater than 12 hours". As for h_{crit} we suggest that we assume it is equal to $1.5 < H_s(t) >$, that is 1.5 times the mean annual significant wave height. In practice, the one in fig. 6.1*a* is a single storm, and that of fig. 6.1*b* is also a single storm, while those of fig. 6.1*c* are two distinct storms.

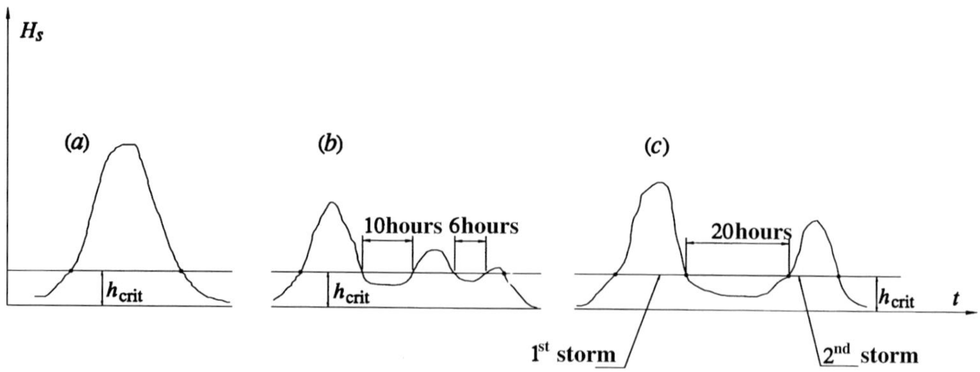

Fig. 6.1 (*a*) A sea storm with a single peak. (*b*) A sea storm with two peaks. (*c*) Two distinct sea storms (H_s keeps below the threshold h_{crit} for a span of time greater than 12 hours).

6.1.3 *The probability of exceedance* $P(H_s > h)$

We shall see later that, thanks to a general property of the sea storms, the design wave can be obtained from $P(H_s > h)$. This is the probability of exceedance of $H_s(t)$, that is the share of time in which the significant wave height keeps above any fixed threshold h at the location under examination.

Referring to fig. 6.2, we have

$$P(H_s > h) = \frac{1}{\mathscr{T}} \sum_i \Delta t_i(h),$$

where $\Delta t_i(h)$ are the time intervals in which H_s exceeds h at the fixed location, and $\mathscr{T}$ is the total time. Hence, for estimating $P(H_s > h)$ it is not necessary to analyse each single storm. It simply suffices to count how many times the significant wave height has exceeded the fixed threshold h, and to divide this number by the total number of measurements.

6.1.4 *The probability density function* $p(H_s = h)$

Besides $P(H_s > h)$ we shall use $p(H_s = h)\,\mathrm{d}h$ that is the probability that H_s falls within a fixed small interval $(h, h + \mathrm{d}h)$. It is equal to the share of time in which

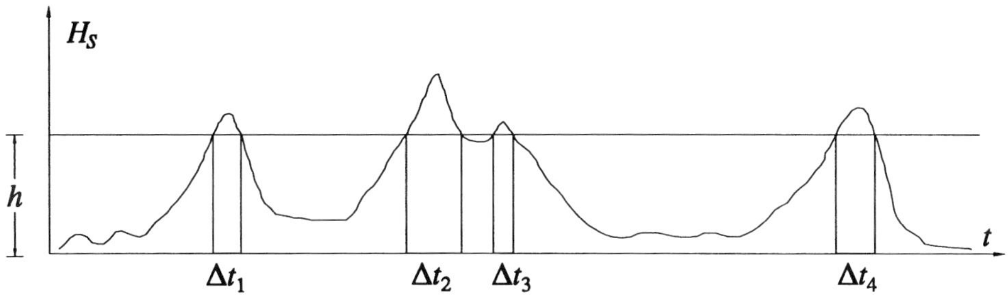

Fig. 6.2 $P(H_s > h)$ is the quotient between summation $(\Delta t_1 + \Delta t_2 + ...)$ and total time. Clearly it can strongly change from a location to another.

$H_s(t)$ falls within $(h, h + \mathrm{d}h)$, or in other words, it is equal to the quotient between time in which $h < H_s < h + \mathrm{d}h$ and total time. Referring to fig. 6.3, we have

$$p(H_s = h)\,\mathrm{d}h = \frac{1}{\mathscr{T}} \sum_i \mathrm{d}t_i(h; \mathrm{d}h),$$

where $\mathrm{d}t_i(h; \mathrm{d}h)$ is the ith time interval in which $h < H_s < h + \mathrm{d}h$.

The functions $P(H_s > h)$ and $p(H_s = h)$ are related to each other by

$$P(H_s > h) = \int_h^{\infty} p(H_s = h')\,\mathrm{d}h',$$

$$p(H_s = h) = -\frac{\mathrm{d}P(H_s > h)}{\mathrm{d}h} \qquad (6.1)$$

which proceed from the definitions of these two functions.

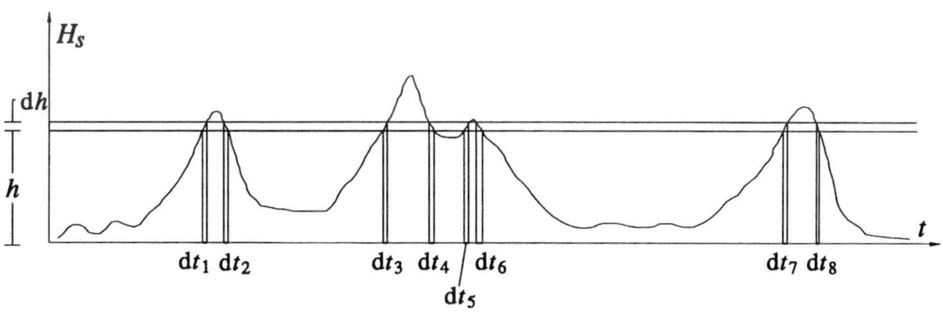

Fig. 6.3 $p(H_s = h)\,\mathrm{d}h$ is the quotient between summation $(\mathrm{d}t_1 + \mathrm{d}t_2 + ...)$ and total time.

6.2 The probability of the significant wave height

6.2.1 *The form of* $P(H_s > h)$, *and the definition of ancillary variables* X, Y

On the domain of h greater than h_{crit}, the probability $P(H_s > h)$ often can be fitted by

$$P(H_s > h) = \exp\left[-\left(\frac{h}{w}\right)^u\right]. \tag{6.2}$$

This is a Weibull distribution whose parameters u and w depend on the special location under examination and are obtained as described here below.

Let us imagine we have 20000 H_s data uniformly distributed over the seasons of the year, and let us imagine 2000 of these 20000 H_s data exceed 2.5 m. Then we shall have

$$P(H_s > 2.5\,\text{m}) = \frac{2000}{20000} = 0.10.$$

Similarly, we shall obtain $P(H_s > 3\,\text{m})$, $P(H_s > 3.5\,\text{m})$, and so on.

For plotting the data it is convenient to resort to two ancillary variables. We shall use

$$X \equiv 100\ln(2.5h) \qquad \text{with } h \text{ in metres,} \qquad Y \equiv 100\ln\ln\frac{1}{P}. \tag{6.3}$$

With this choice one works typically on X ranging from 100 to 350 and Y ranging from 50 to 250.

If the probability is actually given by (6.2), the data points should be on a straight line

$$Y = a + bX, \tag{6.4}$$

apart from some natural scatter that can be described in terms of confidence intervals [see sect. 6.2.4]. After having obtained the constants a and b of the Cartesian equation for this line, we obtain the two parameters u and w of $P(H_s > h)$, by means of

$$u = b, \tag{6.5a}$$

$$w = \frac{1}{2.5}\exp\left(-\frac{a}{100b}\right) \text{ with } w \text{ in metres.} \tag{6.5b}$$

The verification of these relations is left as an exercise. To this end

(i) replace in (6.4) X and Y with their expressions (6.3), so obtaining a relation between $\ln\ln(1/P)$ and $\ln(h)$, with a and b;

(ii) from (6.2) obtain a new relation between $\ln\ln(1/P)$ and $\ln(h)$, with u and w;

(iii) compare these two relations between $\ln\ln(1/P)$ and $\ln(h)$.

Note, the wave recorders having to work continuously for long periods are subject to interruptions due to various causes. If there is no correlation between these interruptions and the wave height there is no consequence on the estimate of $P(H_s > h)$. Indeed $P(H_s > h)$ is equal to the quotient between the number of records where H_s exceeds the fixed threshold h and the total number of records. This, whether or not the times between successive records are constant, provided these times are stochastically independent of the significant wave height.

6.2.2 *A physical interpretation of the parameters u and w of $P(H_s > h)$*

Let us consider the inverse function

$$h(P) = w\left(\ln\frac{1}{P}\right)^{\frac{1}{u}} \tag{6.6}$$

which gives the threshold h having a probability P to be exceeded by the significant wave height. Equation (6.6) shows that the significant wave height is proportional to w. This implies that, if w of a location A is twice the w of a location B and both these locations have the same u, the wave heights at A will be twice greater than at B. For example, the threshold of significant wave height that is exceeded for one hour, on average, in 100 years at A will be twice greater than at B; or the maximum expected wave height in 100 years at A will be twice greater than at B (really this last statement also requires that the duration of the storms is equal at the two locations, as we shall see in chap. 7).

As to the parameter u, it determines the rate of growth of the significant wave height as P decreases. To realize this item, let us consider two probability levels, say, $P = 1:1000$ and $P = 1:10$. From (6.6) we have

$$\frac{h\left(\frac{1}{1000}\right)}{h\left(\frac{1}{10}\right)} = \left(\frac{\ln 1000}{\ln 10}\right)^{\frac{1}{u}},$$

from which it appears that the quotient between the significant wave height of probability 1:1000 and the significant wave height of probability 1:10 is larger the smaller u is.

In the Oceans both w and u are usually greater than in the Mediterranean Sea. The fact that w of the Oceans is greater than w of the Mediterranean Sea implies that the significant wave height in the Oceans is greater than in the Mediterranean Sea. The fact that u of the Oceans is greater than u of the Mediterranean Sea implies that the difference between Oceans and Mediterranean Sea, as to wave height, gets smaller at low levels of the probability of exceedance. This phenomenon appears from fig. 6.4 showing a $P(H_s > h)$ of the Mediterranean Sea and a $P(H_s > h)$ of the North Sea. [The graph of the two probabilities is given in coordinates (X,Y) defined by (6.3).] Look at the ratio between the significant

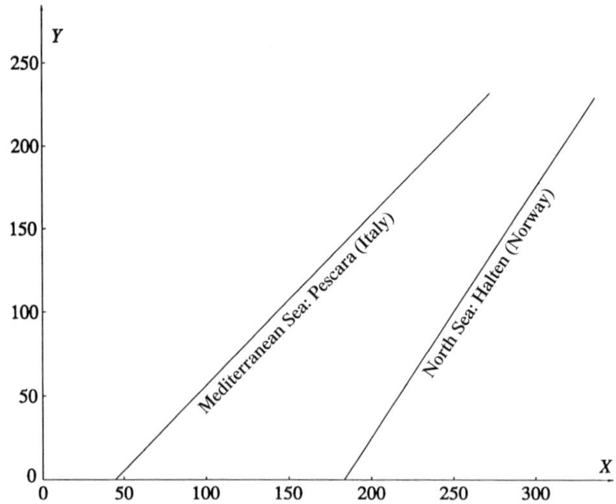

Fig. 6.4 A $P(H_s > h)$ of the Mediterranean Sea and one of the North Sea. [Table 4 of Krogstad (1985) was the data source for Halten.]

wave heights of the two locations at a fixed level of the probability P (that is of the ordinate Y). As P gets smaller also this ratio gets smaller. (Bear in mind that the aforesaid ratio would be constant were the two lines parallel.) Note also that the $h(P)$ of the North Sea is substantially greater than the $h(P)$ of the Mediterranean Sea up to the lowest P of any practical interest.

6.2.3 *The expression and the main property of the probability density function* $p(H_s = h)$

The probability density function $p(H_s = h)$, which is related to $P(H_s > h)$ by (6.1), takes on the form

$$p(H_s = h) = \frac{u}{w}\left(\frac{h}{w}\right)^{u-1} \exp\left[-\left(\frac{h}{w}\right)^{u}\right]. \tag{6.7}$$

The derivative

$$\frac{\mathrm{d}p(H_s = h)}{\mathrm{d}h} = \left[\frac{u(u-1)}{w^2}\left(\frac{h}{w}\right)^{u-2} - \frac{u^2}{w^2}\left(\frac{h}{w}\right)^{2u-2}\right]\exp\left[-\left(\frac{h}{w}\right)^{u}\right]$$

has one zero on the domain $h > 0$ only if $u > 1$. The value of h where the derivative is zero is

$$h_m = w\left(1 - \frac{1}{u}\right)^{\frac{1}{u}},$$

so that it is smaller than w, and consequently it is also smaller than h_{crit} (the threshold of the sea storms).

The consequence being that

$$\frac{\mathrm{d}p(H_s = h)}{\mathrm{d}h} < 0 \quad \text{for} \quad h > h_{\text{crit}},$$

which will be used in sect. 7.1.

6.2.4 *Example of a probability estimate*

Fig. 6.5 shows the $P(H_s > h)$ of Mazara del Vallo in the Straits of Sicily. Data: 14650 values of H_s taken with a sampling rate of 1/3 per hour from July 1989 to December 1994.

We see the fitting line pass at points $X = 160$, $Y = 84.5$ and $X = 260$, $Y = 212$, and hence its Cartesian equation is

$$Y = -119.5 + 1.275X.$$

Therefore, from the equations (6.5*a-b*) of u and w, we have

$$u = 1.275, \qquad w = 1.021\,\text{m},$$

that is

$$P(H_s > h) = \exp\left[-\left(\frac{h}{1.021}\right)^{1.275}\right] \text{ with } h \text{ in metres.} \tag{6.8}$$

For many localities the data points exhibit a marked linear trend (on the plane X, Y) starting from some Y between 0 and 100, as we see in fig. 6.5. In the case under examination, having about 15000 data we attain the probability level of

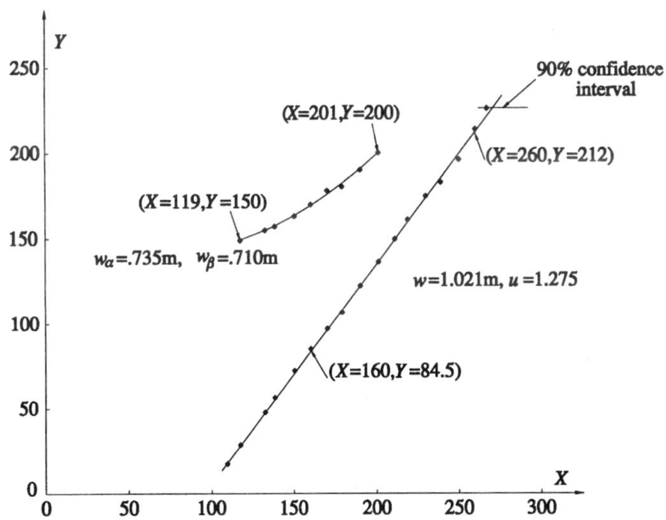

Fig. 6. 5 Example of sects. 6.2.4 and 6.3.4: how to obtain $P(H_s > h)$ from the data of a waverider.

$P = 1:15000$ which corresponds to $Y = 225$. Really, the extreme data have some wide confidence intervals (as we shall see shortly) so that, in this case, our knowledge is effective roughly up to $Y \sim 200$.

In conclusion, a straight line fits the data points rather well in the range from $Y = 50$ to $Y = 200$. Clearly this line is then extrapolated. How much? Well, the extreme Y of practical interest lies between $Y = 250$ and $Y = 300$. Here it should be noted that the maximum return period for design purposes is 2000 years, and the threshold of significant wave height that is exceeded for one hour, on average, every 2000 years has a probability

$$P = \frac{1}{2000 \cdot 365 \cdot 24} = 5.7 \cdot 10^{-8}$$

which is equivalent to $Y = 280$.

As to the scatter of the data, it is worth making a few remarks. Indeed this subject is partly beyond the classic treatment of the confidence intervals.

With the classic treatment we assume that our 14650 data are stochastically independent of one another. This implies the probability that all 14650 data are smaller than a fixed threshold h to be equal to

$$\left[1 - P(H_s > h)\right]^{14650}.$$

We call $h_{0.05}$ the h making this probability equal to 0.05. Hence, we have

$$\left[1 - P(H_s > h_{0.05})\right]^{14650} = 0.05 \Rightarrow h_{0.05} = 1.021 \left\{ \ln\left[\frac{1}{1 - 0.05^{1/14650}}\right]\right\}^{\frac{1}{1.275}} = 5.47\,\mathrm{m}.$$

Similarly, we can evaluate $h_{0.95}$ (the probability that all 14650 data are smaller than $h_{0.95}$ being equal to 0.95), with the result

$$h_{0.95} = 7.43\,\mathrm{m}.$$

Conclusion: if $P(H_s > h)$ is given by (6.8), then the maximum of a set of 14650 outcomes of the random variable H_s is smaller than 5.47 m five times in one hundred, and is smaller than 7.43 m ninety five times in one hundred. If the maximum is 5.47 m, the extreme datum point is

$$X = 262, \quad Y = 226.$$

If the maximum is 7.43 m, the extreme datum point becomes

$$X = 292, \quad Y = 226.$$

The horizontal segment whose end points are $X = 262$, $Y = 226$ and $X = 292, Y = 226$ is shown in fig. 6.5. It gives the confidence interval within which the extreme datum is expected to be contained with a probability of 90%. Clearly the amplitude of the 90% confidence interval decreases with Y. For example the 90% confidence interval of the second datum, the one of ordinate

$$Y = 218 \Leftrightarrow P = 2/14650,$$

is 2/3 of the 90% confidence interval of the first datum.

What we have described is the classic approach to the problem of the confidence intervals. But, as we have said, the classic approach is not quite satisfactory. The reason is simply that the values of H_s sampled at a rate of 1/3 per hour, or even more so at a rate of 1 per hour, are not stochastically independent of one another. To realize this, note that the probability of H_s exceeding 5 m at noon, given that H_s is 7 m at 9 a. m., is surely greater than the probability of H_s exceeding 5 m at noon given that H_s is 0.1 m at 9 a. m. Whereas the stochastic independence involves these two probabilities being equal to each other.

The assumption of stochastic independence would be effective were $\Delta t_{\rm samp}$ (the sampling interval) of the order of one day; and it would become perfect as $\Delta t_{\rm samp} \to \infty$. To realize what is the consequence of taking a relatively small $\Delta t_{\rm samp}$, it is helpful to consider what would happen if we worked in the limit as $\Delta t_{\rm samp} \to 0$. In this case, all 14650 data would be equal to one another because they all would represent the H_s at the same time. Therefore, in this case, the probability that all 14650 data are smaller than a fixed threshold h would be equal to the probability that one single datum is smaller than h, which is $[1 - P(H_s > h)]$. Therefore we should have

$$1 - \exp\left[-\left(\frac{h_{0.05}}{1.021}\right)^{1.275}\right] = 0.05 \Rightarrow h_{0.05} = 0.10\,\text{m} \Leftrightarrow X_{0.05} = -139,$$

$$1 - \exp\left[-\left(\frac{h_{0.95}}{1.021}\right)^{1.275}\right] = 0.95 \Rightarrow h_{0.95} = 2.42\,\text{m} \Leftrightarrow X_{0.95} = 180.$$

As we can see, the 90% confidence interval should widen and shift leftward: from $(X_{0.05} = 262, X_{0.95} = 292)$ as $\Delta t_{\rm samp} \to \infty$, to $(X_{0.05} = -139,\ X_{0.95} = 180)$ as $\Delta t_{\rm samp} \to 0$.

The same trend, even if in a much subdued way will arise in passing from $\Delta t_{\rm samp} \to \infty$ to $\Delta t_{\rm samp} = 3$ hours. Therefore, in our condition ($\Delta t_{\rm samp} = 3$ hours) we expect the confidence interval to be somewhat shifted leftward and to be somewhat wider than the classic confidence interval based on the assumption of stochastic independence.

6.3 The probability of the significant wave height for a given direction of wave advance

6.3.1 *The form of $P(H_s > h)$ for a given direction of wave advance*

Let us consider the probability $P(H_s > h;\ \theta_1 < \theta < \theta_2)$ that the significant wave height at a given location exceeds a fixed threshold h, with the dominant direction being within a fixed sector (θ_1, θ_2). Here too it is convenient resorting to coordinates (X,Y) defined by (6.3). The graph of $P(H_s > h;\ \theta_1 < \theta < \theta_2)$ in these coordinates is a line with a positive curvature we shall call *sector line*. Moreover we shall call *all-directions line*, the straight line $Y = a + bX$ which represents $P(H_s > h)$ regardless of the wave direction. The sector line necessarily lies on the left of the all-directions line. Indeed from the very definitions it follows that

$$P(H_s > h;\ \ \theta_1 < \theta < \theta_2) \le P(H_s > h). \tag{6.9}$$

As has been said, the slope of the sector line grows with X, but over the data range, usually, it looks smaller than the slope of the all-directions line (see fig. 6.5).

However, the asymptote of the sector line as X and/or $Y \to \infty$ cannot take on a slope smaller than the all-directions line's (otherwise the two lines would intersect each other, and the inequality (6.9) would no longer be satisfied). In the light of these remarks it seems reasonable to assume the asymptote of the sector line as X and/or $Y \to \infty$ to be parallel to the all-directions line.

A form that satisfies the assumption on the asymptote and usually fits the data points above a certain threshold (of course, provided 6.2 is effective) is

$$P(H_s > h;\, \theta_1 < \theta < \theta_2) = \exp\left[-\left(\frac{h}{w_\alpha}\right)^u\right] - \exp\left[-\left(\frac{h}{w_\beta}\right)^u\right]. \tag{6.10}$$

This is the difference between two Weibull forms, where the parameters w_α and w_β generally vary from one sector to another, while u takes on the same value as in the probability $P(H_s > h)$.

Parameters w_α and w_β must be nonnegative and w_α must be greater than w_β, otherwise the probability would take on some negative values. Finally, w_α must be smaller than or equal to w [the parameter of $P(H_s > h)$], otherwise condition (6.9) would no longer be satisfied.

Since $w_\beta < w_\alpha$, the asymptotic form of (6.10) as $h \to \infty$ is

$$P(H_s > h;\, \theta_1 < \theta < \theta_2) = \exp\left[-\left(\frac{h}{w_\alpha}\right)^u\right] \quad \text{as} \quad h \to \infty. \tag{6.11}$$

The graph of (6.11) in coordinates (X, Y) is a straight line parallel to the all-directions line. This means that the asymptote of the sector line, as X and/or $Y \to \infty$, is parallel to the all-directions line, as it was required.

6.3.2 *How to obtain the parameters w_α and w_β*

For the aim of evaluating w_α and w_β, we do as follows. First looking at the data, we fix two points X_1, Y_1 and X_2, Y_2 which should belong to the sector line. Since h is a function of X, and P is a function of Y, a single pair h_1, P_1 corresponds to X_1, Y_1 and a single pair h_2, P_2 corresponds to X_2, Y_2. Let us assume

$$h_1 < h_2 \tag{6.12}$$

that implies

$$P_1 > P_2. \tag{6.13}$$

From (6.10) we have

$$P_1 = \exp\left(-\frac{1}{w_\alpha^u} h_1^u\right) - \exp\left(-\frac{1}{w_\beta^u} h_1^u\right), \tag{6.14a}$$

$$P_2 = \exp\left(-\frac{1}{w_\alpha^u} h_2^u\right) - \exp\left(-\frac{1}{w_\beta^u} h_2^u\right). \tag{6.14b}$$

Let us define

$$\tilde{x} \equiv \frac{w^u}{w_\alpha^u}, \quad \tilde{y} \equiv \frac{w^u}{w_\beta^u}. \tag{6.15}$$

Then, the equations (6.14*a-b*) yield a nonlinear system in the two unknowns $\tilde{x}$ and $\tilde{y}$. Clearing $\tilde{y}$ from this system, we arrive at

$$\exp(-h_1^u\tilde{x}/w^u) = P_1 + \left[\exp(-h_2^u\tilde{x}/w^u) - P_2\right]^{(h_1/h_2)^u}. \tag{6.16}$$

Thus $\tilde{x}$ (provided that it exists) is that giving the equality of the two following functions:

$$f_1(x) \equiv \exp(-h_1^u x/w^u),\ x \geq 0, \tag{6.17a}$$

$$f_2(x) \equiv P_1 + \left[\exp(-h_2^u x/w^u) - P_2\right]^{(h_1/h_2)^u},\ 0 \leq x \leq x_{\mathrm{sup}}, \tag{6.17b}$$

where x_{sup} is the value of x for which $\exp(-h_2^u x/w^u) - P_2$ is equal to zero:

$$x_{\mathrm{sup}} = \frac{w^u}{h_2^u}\ln\left(\frac{1}{P_2}\right). \tag{6.18}$$

After having evaluated $\tilde{x}$, we obtain w_α using the definition (6.15) of $\tilde{x}$:

$$w_\alpha = w\left(\frac{1}{\tilde{x}}\right)^{1/u}. \tag{6.19}$$

Then, from w_α, we obtain also w_β by means of (6.14*a*) or, alternatively, of (6.14*b*):

$$w_\beta = h_1\left\{\ln\left[\exp\left(-\frac{h_1^u}{w_\alpha^u}\right) - P_1\right]^{-1}\right\}^{-\frac{1}{u}}. \tag{6.20}$$

6.3.3 *The necessary and sufficient condition for the existence of the solution*

Given that

(i) $f_2(0) > f_1(0)$, as a consequence of the inequalities (6.12-13),
(ii) $f_1(x)$ and $f_2(x)$ are strictly decreasing functions,
(iii) $f_2(x)$ is defined on $[0, x_{\mathrm{sup}}]$,

it follows that $\tilde{x}$ exists and is the only solution to (6.16) if

$$f_2(x_{\mathrm{sup}}) < f_1(x_{\mathrm{sup}}).$$

Since

$$f_1(x_{\mathrm{sup}}) = P_2^{(h_1/h_2)^u},\quad f_2(x_{\mathrm{sup}}) = P_1,$$

the existence and uniqueness of $\tilde{x}$ requires that

$$P_1 < P_2^{(h_1/h_2)^u}. \tag{6.21}$$

What does it mean? We have fixed two points X_1, Y_1 and X_2, Y_2. Generally, there is not an equation of the form (6.10) with the given u, whose graph in coordinates

(X,Y) contains the two fixed points. The existence and uniqueness of such an equation is guaranteed only if the inequality (6.21) is satisfied.

For a geometrical interpretation of condition (6.21) let us substitute in it P_1, P_2, h_1 and h_2 respectively by Y_1, Y_2, X_1 and X_2, using the definitions (6.3) relating Y to P and X to h. The result is

$$\frac{Y_2 - Y_1}{X_2 - X_1} < u. \tag{6.22}$$

This shows that the necessary and sufficient condition for the existence and uniqueness of an equation of the form (6.10) (with the fixed u) whose graph in coordinates (X,Y) contains the two points X_1, Y_1 and X_2, Y_2 is that the slope of the straight line containing these two points be smaller than u.

6.3.4 *Example of a probability estimate*

Fig. 6.5 shows also the data of

$$P(H_s > h; 135^\circ - 11.25^\circ < \theta < 135^\circ + 11.25^\circ),$$

where θ is the azimuth of the direction of wave advance. The location is naturally the same for which in sect. 6.2.4 we got the probability $P(H_s > h)$ regardless of the wave direction.

The points marked by the two arrows are those we suppose the sector line should contain. Naturally, this is a tentative pair of points. Then, possibly, we shall have to fix a new pair of points till we shall find the line that best fits the data. The two points are

$$X_1 = 119, \quad Y_1 = 150,$$
$$X_2 = 201, \quad Y_2 = 200.$$

Since $u = 1.275$ (this was evaluated in sect. 6.2.4) and

$$\frac{Y_2 - Y_1}{X_2 - X_1} = 0.61,$$

the condition (6.22) is fulfilled.

The values of h_1, P_1 and h_2, P_2 corresponding, respectively, to X_1, Y_1 and X_2, Y_2 are obtained through definitions (6.3) relating X to h and Y to P. The result is

$$h_1 = 1.315\,\text{m}, \quad P_1 = 1.131 \cdot 10^{-2},$$
$$h_2 = 2.985\,\text{m}, \quad P_2 = 6.180 \cdot 10^{-4}.$$

The maximum element ($x_{\sup}$) of the domain of $f_2(x)$ proceeds from (6.18):

$$x_{\sup} = \frac{(1.021)^{1.275}}{(2.985)^{1.275}} \ln\left(\frac{1}{6.180 \cdot 10^{-4}}\right) = 1.882$$

(1.021 m being the value of w which we have obtained in sect. 6.2.4).

The functions $f_1(x)$ and $f_2(x)$ [defined by (6.17*a*-*b*)] take on the forms

$$f_1(x) = \exp(-1.381x),$$

$$f_2(x) = 1.131 \cdot 10^{-2} + \left[\exp(-3.927x) - 6.18 \cdot 10^{-4}\right]^{0.352}.$$

The abscissa $\tilde{x}$ is that for which the difference $f_2(x) - f_1(x)$ is zero. Fig. 6.6 shows that

$$\tilde{x} = 1.519,$$

and hence from the equations (6.19), (6.20) we obtain

$$w_\alpha = 1.021\left(\frac{1}{1.519}\right)^{1/1.275} = 0.735\,\mathrm{m},$$

$$w_\beta = 1.315\left\{\ln\left\{\exp\left[-\left(\frac{1.315}{0.735}\right)^{1.275}\right] - 1.131 \cdot 10^{-2}\right\}^{-1}\right\}^{-1/1.275} = 0.710\,\mathrm{m}.$$

At this stage, w_α and w_β being known, the sector line can be computed. Let us compute an ordinate of this line, for example $Y(150)$. We have

$$X = 150 \Rightarrow h = 1.793\,\mathrm{m} \Rightarrow P = \exp\left[-\left(\frac{1.793}{0.735}\right)^{1.275}\right] - \exp\left[-\left(\frac{1.793}{0.710}\right)^{1.275}\right] =$$

$$= 5.808 \cdot 10^{-3} \Rightarrow Y = 163.9.$$

Similarly, we compute the other values of $Y(X)$, and hence we can plot the sector line $Y(X)$.

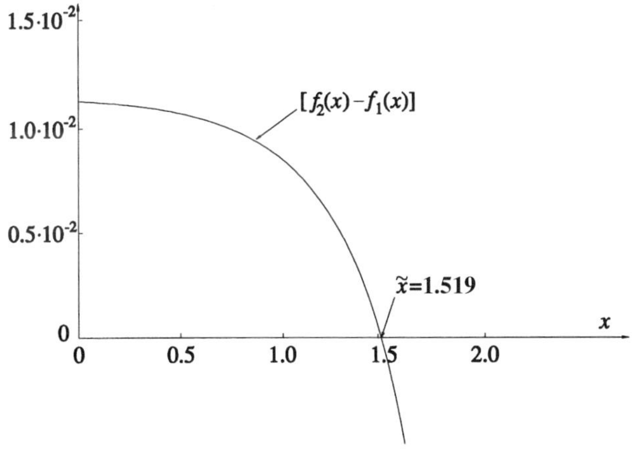

Fig. 6.6 Example of sect. 6.3.4: solution to equation (6.16).

As said, the calculations have to be repeated till one finds the sector line $Y(X)$ that best fits the data. In doing so, we should bear in mind that an equation of the form (6.10) generally is able to fit the data only above some lower bound $Y_{\min}$ (typically $Y_{\min} \approx 150$). Naturally, here too the extreme data points show a scatter that may be due also to some uncertainty in the estimate of the dominant direction.

6.4 Probabilities of the significant wave height for a few areas of the globe

Table 6.1 gives the parameters u and w of the probability $P(H_s > h)$, which we have evaluated for a few locations. The H_s data are those of the NDBC buoys, up to the year 1997. The H_s data for Halten (North Sea) have been taken from table 4 of the paper of Krogstad (1985). The Atlantic is not covered by table (6.1) and will be dealt with separately in Appendix A.2.3, because of some special features.

As we noticed in sect. 6.2.4, the data points of $P(H_s > h)$ on the plane X-Y, at many localities, exhibit a linear trend starting from $Y = 50 \div 100$, and the u, w values of table 6.1 represent this linear trend. The graphs of $P(H_s > h)$ on the plane X-Y for the three locations of the North Sea, the Bering Sea and the North-Eastern Pacific exhibit a linear trend up to the extreme datum point. Whilst, the higher data points ($Y > 200$) of the two locations of Hawaii and the Gulf of Alaska deviate somewhat rightward (something like in fig. A.2*a* of Appendix A, even if in a more subdued form). The extreme datum point falls at the upper limit of the 95% confidence interval (Hawaii) and at the upper limit of the 80% confidence interval (Gulf of Alaska). Therefore some caution should be exercised in dealing with these two areas, because they could be affected, to some extent, by the phenomenon of the anomalous extreme storms, we shall describe in sect. A.2.3 of Appendix A.

Table 6.1 *Parameters of* $P(H_s > h)$ *for a few areas of the globe.*

LOCATION	NDBC buoy	w[m]	u
Halten, Norway - North Sea	–	2.083	1.333
56°5′N, 177°48′W - Bering Sea	46035	2.566	1.382
56°17′N, 148°10′W - Gulf of Alaska	46001	2.685	1.555
40°50′N, 137°29′W - North-Eastern Pacific	46006	2.877	1.631
23°24′N, 162°15′W - Hawaii	51001	2.093	1.550

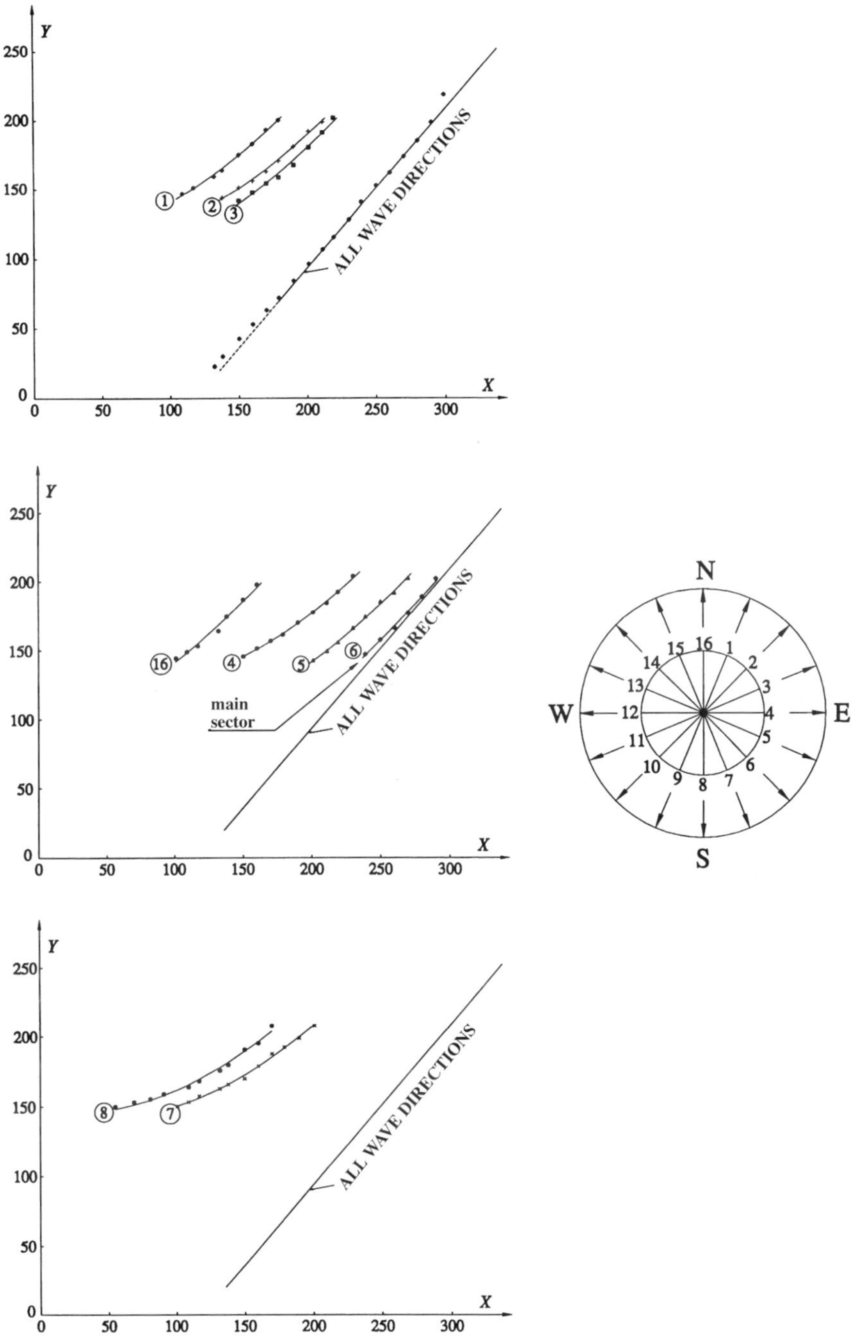

Fig. 6.7 Alghero (West Sardinia): $P(H_s > h)$ regardless of the wave direction, and $P(H_s > h)$ for some given directions of wave advance.

Table 6.2 gives the parameters u, w of $P(H_s > h)$, and w_α, w_β of $P(H_s > h; \theta_1 < \theta < \theta_2)$, which we have obtained for a few locations of the Mediterranean Sea. The data of H_s and θ used for this table are those of the Italian Hydrographic Service, for the period 1989-1994. From this table we can see which is the sector with the highest waves of each location. It suffices to read the values

Table 6.2 *Mediterranean Sea: $P(H_s > h)$ for given directions of wave advance (*)*

LOCATION	w [m]	u
La Spezia (SP)	0.731	1.023
Alghero (AL)	1.280	1.137
Ponza (PO)	0.874	1.200
Mazara del Vallo (MZ)	1.021	1.275
Crotone (CR)	0.637	0.958
Monopoli (MN)	0.645	1.143
Pescara (PE)	0.625	1.018

	SECTOR → / LOCATION ↓	①	②	③	④	⑤	⑥	⑦	⑧	⑨	⑩	⑪	⑫	⑬	⑭	⑮	⑯
w_α [m]	La Spezia	.450	.669	.731	.437	.332										.307	.331
	Alghero	.444	.613	.650	.750	1.040	1.236	.556	.442								.367
	Ponza	.718	.812	.727	.865	.653	.468							.544	.587	.477	.639
	Mazara del Vallo	.666	.600	.808	.978	1.003	.735	.440						.759	.756	.514	.555
	Crotone								.472	.338	.360	.388	.538	.637	.637	.441	.286
	Monopoli						.592	.512	.645	.512	.590	.549	.429				
	Pescara						.404	.486	.605	.611	.487	.378	.375	.378			

	LOCATION	①	②	③	④	⑤	⑥	⑦	⑧	⑨	⑩	⑪	⑫	⑬	⑭	⑮	⑯
w_β [m]	La Spezia	.421	.333	.660	.423	.324										.265	.291
	Alghero	.404	.557	.565	.699	.905	.942	.534	.427								.311
	Ponza	.688	.779	.688	.754	.633	.458							.525	.566	.460	.609
	Mazara del Vallo	.655	.591	.795	.895	.907	.710	.431						.740	.628	.479	.533
	Crotone								.342	.231	.337	.376	.521	.605	.576	.395	.158
	Monopoli						.553	.453	.601	.484	.565	.522	.410				
	Pescara						.384	.374	.561	.552	.470	.365	.340	.353			

(*) See fig. 6.7 for sectors' numbers.

of w_α: the larger is w_α, the higher the waves. Indeed we have seen the sector line $Y(X)$ tends to an asymptote that moves rightward more, the larger w_α gets. As to w_β, it has a minor effect: the larger the w_β, the slower is the convergence of the sector line on its asymptote. This means that, under the same w_α, the larger the w_β, the smaller the waves. For example, at Crotone the maximum w_α belongs to two distinct sectors: sector ⑬ ($\overline{\theta} = 292.5^\circ$) and sector ⑭ ($\overline{\theta} = 315^\circ$). Thus, to decide which of the two sectors has the highest waves, we must resort to w_β. Since w_β of sector ⑭ is smaller than w_β of sector ⑬, the highest waves are those of sector ⑭.

From table 6.2 note that the sea of West Sardinia (Alghero) is by far the strongest near the Italian coasts. This is also confirmed by measurements from satellite (which we shall deal with in sect. 14.3.3).

6.5 The maximum expected wave height in a storm with a given history

The probability that all waves of the ith sea state of a storm are smaller than a fixed threshold H is given by (5.55) with $N = \Delta t/\overline{T}(h_i)$ (Δt being the duration of the sea state, h_i the significant wave height of the ith sea state and $\overline{T}(h_i)$ its mean period).

The probability that the maximum wave height of a storm is smaller than a fixed threshold H is equal to the probability that all wave heights of the storm are smaller than H. This in turn is equal to the product of [probability that all wave heights of the *first* sea state are smaller than H] $\times$ [probability that all wave heights of the *second* sea state are smaller than H] $\times$..., that is to say

$$\mathscr{P}(H_{\max} < H) = \prod_{i=1}^{\mathscr{N}} \left[1 - P\left(H; H_s = h_i\right)\right]^{\frac{\Delta t}{\overline{T}(h_i)}},$$

where $\mathscr{N}$ is the total number of sea states forming the storm.

The probability of exceedance of the maximum wave height in the storm is the complement (relative to 1) of $\mathscr{P}(H_{\max} < H)$:

$$P(H_{\max} > H) = 1 - \prod_{i=1}^{\mathscr{N}} \left[1 - P\left(H; H_s = h_i\right)\right]^{\frac{\Delta t}{\overline{T}(h_i)}}$$

that is more conveniently rewritten in the form

$$P(H_{\max} > H) = 1 - \exp\left\{\sum_{i=1}^{\mathscr{N}} \frac{\Delta t}{\overline{T}(h_i)} \ln[1 - P(H; H_s = h_i)]\right\}. \tag{6.23}$$

This formula enables us to evaluate $P(H_{\max} > H)$ as a function of H for given $H_s(t)$ [the time history of the storm]. To this end we have to work on a step function like the one of fig. 6.8, where each step is a new sea state. The result of the calculation

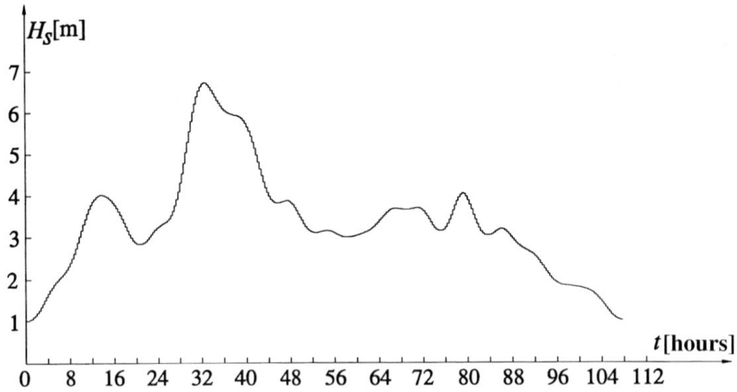

Fig. 6.8 Sea storm and its sea states.

does not change if we let Δt approach zero. This is useful from a mathematical point of view, since it enables us to rewrite (6.23) in the form

$$P(H_{\max} > H) = 1 - \exp\left\{\int_0^{\mathscr{D}} \frac{1}{\overline{T}[h(t)]} \ln[1 - P(H; H_s = h(t))]\, \mathrm{d}t\right\}, \qquad (6.24)$$

where $\mathscr{D}$ is the duration of the storm.

After having obtained $P(H_{\max} > H)$ (as a function of H), we can compute the maximum expected wave height of the storm. It suffices to recall that the mean value of a nonnegative random variable, like $H_{\max}$, is equal to the integral over $(0, \infty)$ of its probability of exceedance:

$$\overline{H_{\max}} = \int_0^{\infty} 1 - \exp\left\{\int_0^{\mathscr{D}} \frac{1}{\overline{T}[h(t)]} \ln[1 - P(H; H_s = h(t))]\, \mathrm{d}t\right\} \mathrm{d}H. \qquad (6.25)$$

The physical interpretation of $\overline{H_{\max}}$ is as follows. Let us consider n storms with the same time history. The maximum wave height of the first storm will be $H_{\max 1}$, the maximum wave height of the second storm will be $H_{\max 2}$, and so on. $\overline{H_{\max}}$ is the average of $H_{\max 1}$, $H_{\max 2}$, and so on.

6.6 The concept of "equivalent triangular storm"

6.6.1 *Definition*

We can associate with each actual sea storm an equivalent triangular storm so defined

(i) the height a of the triangle is equal to the maximum significant wave height in the actual storm;

(ii) the base b of the triangle (that is the duration of the equivalent triangular storm) is such that the maximum expected wave height of the triangular storm is equal to the maximum expected wave height of the actual storm.

According to this definition, the height of the triangle is immediately obtained, while the base will be obtained after a few attempts. It is convenient to fix a small base (duration of the triangular storm). Thus we shall find the maximum expected wave height of the triangular storm to be smaller than the maximum expected wave height of the actual storm. Then we shall gradually widen the base till the maximum expected wave height of the triangular storm will be equal to the maximum expected wave height of the actual storm.

The computation of $\overline{H_{max}}$ for the triangular storm is done more quickly by means of a specific formula. Indeed we have

$$\int_0^{\mathscr{D}} \frac{1}{\overline{T}[h(t)]} \ln[1 - P(H; H_s = h(t))]\,\mathrm{d}t = \frac{b}{a} \int_0^a \frac{1}{\overline{T}(h)} \ln[1 - P(H; H_s = h)]\,\mathrm{d}h,$$

which enables us to rewrite the expressions of $P(H_{max} > H)$ and $\overline{H_{max}}$ in the forms

$$P(H_{max} > H) = 1 - \exp\left\{\frac{b}{a} \int_0^a \frac{1}{\overline{T}(h)} \ln[1 - P(H; H_s = h)]\,\mathrm{d}h\right\}, \qquad (6.26a)$$

$$\overline{H_{max}} = \int_0^{\infty} 1 - \exp\left\{\frac{b}{a} \int_0^a \frac{1}{\overline{T}(h)} \ln[1 - P(H; H_s = h)]\,\mathrm{d}h\right\}\mathrm{d}H. \qquad (6.26b)$$

6.6.2 *The basic property*

We should expect that $P(H_{max} > H)$ of the equivalent triangular storm is generally different from $P(H_{max} > H)$ of the actual storm. That is we should expect

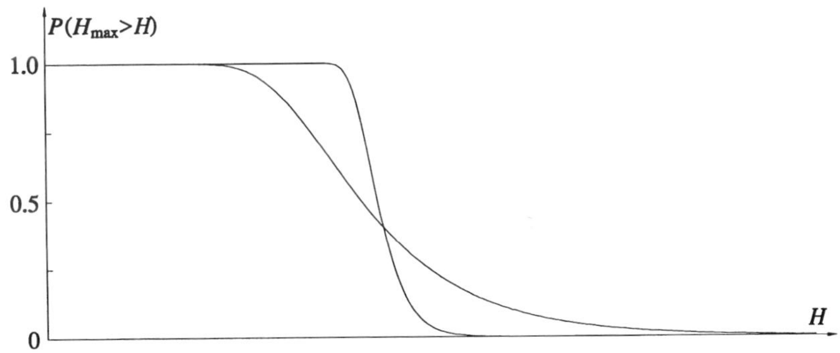

Fig. 6.9 Having required only that the $\overline{H_{max}}$ of the e.t.s. was equal to the $\overline{H_{max}}$ of the actual storm, the $P(H_{max} > H)$ of the e.t.s. and of the actual storm were expected to be generally different from each other, similar to that shown in the figure. Surprisingly enough it was found that the two $P(H_{max} > H)$ were practically coincident, which proved to be very useful for applications.

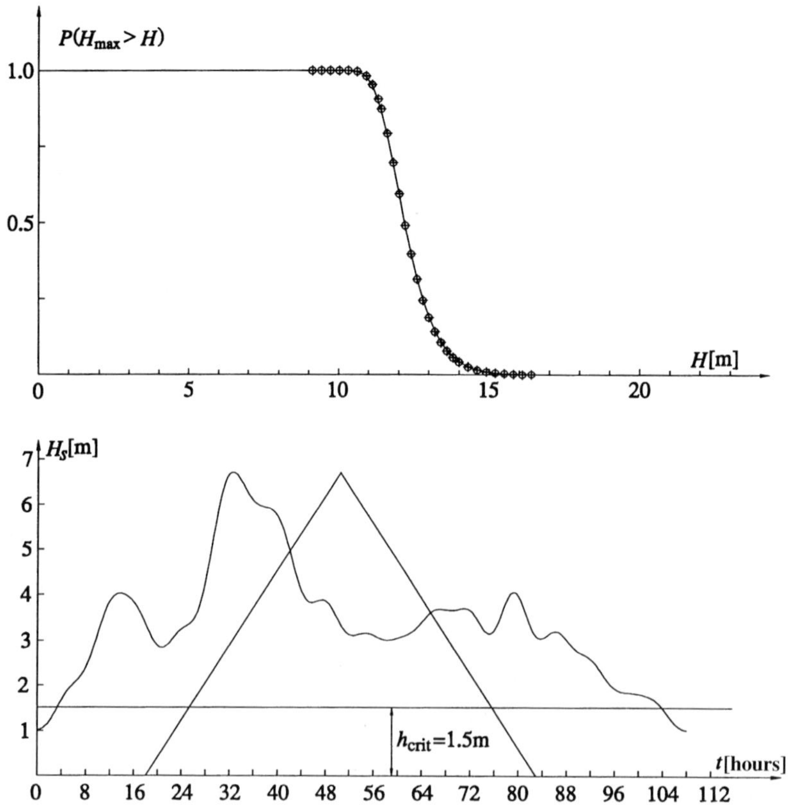

Fig. 6.10 A sea storm recorded at La Spezia (Tyrrhenian Sea) and its e.t.s. We see the $P(H_{\max} > H)$ of the e.t.s. (points) coincides with the $P(H_{\max} > H)$ of the actual storm (continuous line).

a condition like that of fig. 6.9: the two $P(H_{\max} > H)$ are different from each other and only their integrals over $(0, \infty)$ are equal (since these integrals represent $\overline{H_{\max}}$). On the contrary, the two $P(H_{\max} > H)$ prove to be nearly coincident. This is a general property, that was found on comparing the two $P(H_{\max} > H)$, the one of the triangular equivalent storm and the one of the actual storm, for several scores of locations [see an example in fig. 6.10].

In conclusion, the actual storms have some irregular histories generally different from one another. However, an equivalent triangular storm can be associated with each actual storm. The equivalence between this triangular storm and the actual storm is full, since they have the same maximum value of the significant wave height and have the same probability that the maximum wave height exceeds *any* fixed threshold H. Clearly, dealing with the equivalent triangular storms rather than with the actual storms very much simplifies the mathematical treatment.

6.6.3 *The "equivalent sea"*

We shall call *equivalent* sea the sequence of the e.t.s. (equivalent triangular storms). Thus the equivalent sea consists of the same number of storms as the actual sea, each of them with the same maximum H_s. As a consequence the return period of a storm whose significant wave height exceeds a fixed threshold h is the same in the actual sea and in the equivalent sea. Moreover, we have seen in sect. 6.6.2 the probability that the maximum wave height exceeds a fixed threshold H to be the same in the actual storm as in its e.t.s., which implies that the maximum expected wave height in any given time interval will be the same in the actual sea as in the equivalent sea.

Only one topic remains to be considered: what is the relationship between $P(H_s > h)$ of the equivalent sea and $P(H_s > h)$ of the actual sea? $P(H_s > h)$ of the equivalent sea is estimated through a simple quotient:

$$P(H_s > h) \equiv \text{quotient between time in which } H_s > h \text{ and total time.}$$

For example, to get $P(H_s > 4\,\text{m})$ it suffices to take the intervals $\Delta t(4\,\text{m})$ [see fig. 6.11] from the e.t.s. whose $H_{s\max} > 4\,\text{m}$. The probability will then be obtained through the following operation:

$$P(H_s > 4\,\text{m}) = \frac{1}{\mathscr{T}} \sum_i \Delta t_i(4\,\text{m}),$$

where $\mathscr{T}$ is the time interval (typically a few years) under examination.

Comparisons between the $P(H_s > h)$ of the actual sea and the $P(H_s > h)$ of the equivalent sea show some minor discrepancies for $h > h_{\text{crit}}$, and these discrepancies disappear as h (or X) grows. (An example is given in fig. 6.12.) Hence we can assume for the equivalent sea the same $P(H_s > h)$ of the actual sea.

Conclusion: in the next chapter, which is devoted to the design wave estimate, we shall work on the equivalent sea rather than on the actual sea, and in doing so we shall very much simplify the treatment, and

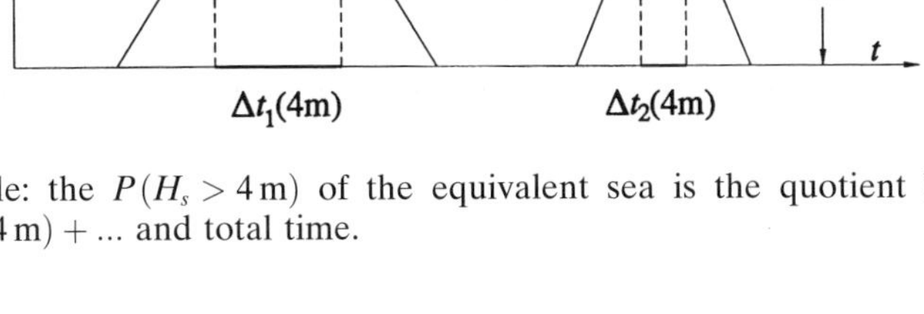

Fig. 6.11 Example: the $P(H_s > 4\,\text{m})$ of the equivalent sea is the quotient between the sum $\Delta t_1(4\,\text{m}) + \Delta t_2(4\,\text{m}) + \ldots$ and total time.

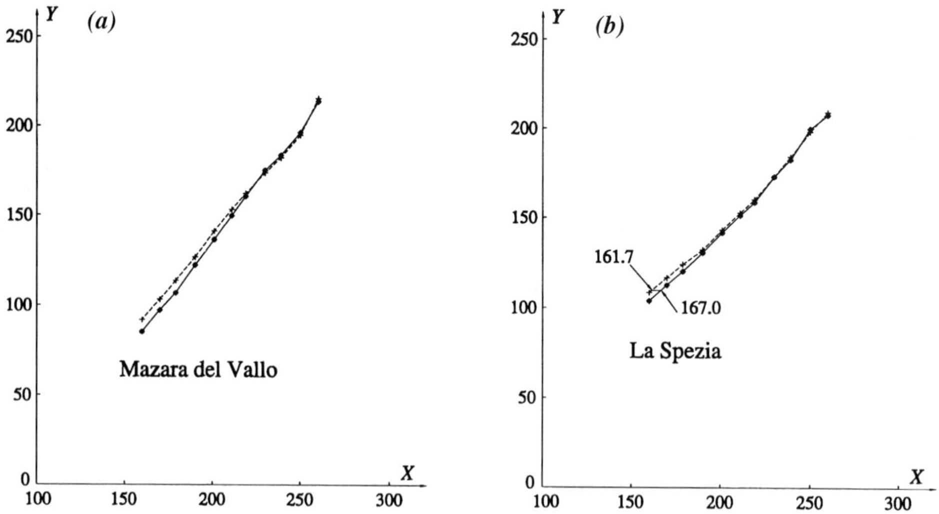

Fig. 6.12 The actual $P(H_s > h)$ (continuous line) and the $P(H_s > h)$ of the equivalent sea (dashed line): (*a*) Mazara del Vallo (Straits of Sicily); (*b*) La Spezia (Tyrrhenian Sea).

(i) we shall not introduce any approximation, because the equivalent sea has the same return periods of storms of given characteristics;

(ii) we shall have no additional work because, on the range of practical interest, $P(H_s > h)$ estimated for the actual sea is good also for the equivalent sea.

6.7 Storm durations

Being essentially interested in the heavier sea storms, we shall consider the set of the N heavier e.t.s. at a given location in a number of years (n_{years}). In particular we fix

$$N = 10\, n_{\text{years}}.$$

This means that having a record of, say, five years we shall consider the 50 more severe storms in these five years. We call a_{10} and b_{10} respectively the mean height and the mean base of the N e.t.s. belonging to this set.

Fig. 6.13 shows the pairs $\tilde{a} \equiv a/a_{10}$, $\tilde{b} \equiv b/b_{10}$, for three locations of the North-Western Atlantic. The line represents the function

$$K_1 \exp(K_2 \tilde{a})$$

which best fits the regression $\overline{\overline{\tilde{b}}}(\tilde{a})$. In the dimensional form, it yields

$$\overline{b}(a) = K_1 b_{10} \exp\left(K_2 \frac{a}{a_{10}} \right). \tag{6.27}$$

Table 6.3 *Parameters of equation (6.27) for $\overline{b}(a)$*

AREA	LOCATION	a_{10}[m]	b_{10} [hours]	K_1	K_2
North-Western Atlantic	NDBC 41001 34°40′N, 72°38′W	6.9	52	1.40	– 0.46
	NDBC 44004 38°27′N, 70°41′W	7.2	55		
	NDBC 44008 40°30′N, 69°25′W	6.1	54		
North-Eastern Pacific	NDBC 46002 42°32′N, 130°16′W	7.7	61	1.80	– 0.59
	NDBC 46005 46°05′N, 131°00′W	7.9	61		
	NDBC 46006 40°51′N, 137°29′W	8.6	63		
Central Mediterranean Sea	La Spezia 43°56′N, 9°49′E	3.6	74	1.12	– 0.115
	Alghero 40°33′N, 8°6′E	5.2	88		
	Ponza 40°52′N, 12°57′E	3.4	69		
	Mazara del Vallo 37°31′N, 12°31′E	3.6	73		
	Crotone 39°1′N, 17°13′E	3.2	74		
	Monopoli 40°58′N, 17°23′E	2.9	70		
	Pescara 42°24′N, 14°32′E	3.3	65		

Table 6.3 gives the values of the parameters of (6.27), which we have obtained for a few areas of the globe.

$\overline{b}(a)$ is decreasing (K_2 negative) because the highest peaks of the random process $H_s(t)$ are usually sharp and consequently the bases (durations) of their equivalent triangular storms prove to be rather short. The choice of the

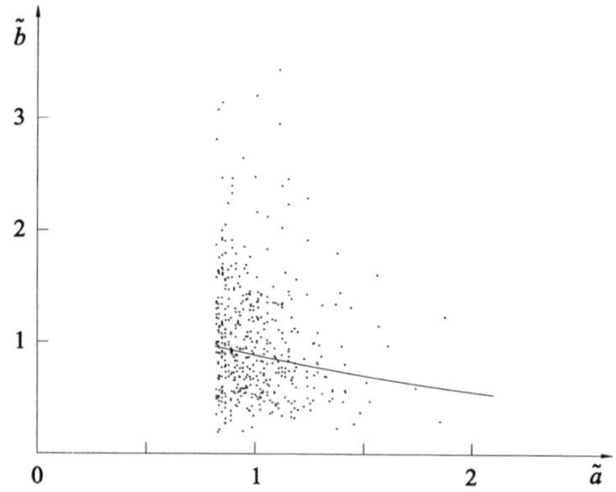

Fig. 6.13 North-Western Atlantic: normalized bases of the equivalent triangular storms versus normalized heights.

mathematical form that fits the regression $\bar{b}(a)$ generally proves to have some very small effects on the calculation of the design wave height, which is dealt with in the next chapter.

Conclusive note

The Weibull model for the probability $P(H_s > h)$ is widely cited in literature, and the first systematic use of it was made by Battjes (1970). Some alternative models were used, among the others, by Ochi and Whalen (1980) and Ferreira and Guedes Soares (1999). The formula (6.10) for the $P(H_s > h)$ relevant to a given wave direction is a new proposal. The formula (6.24) for the probability of exceedance of the maximum wave height in a storm is due to Borgman (1970 and 1973). The concept of equivalent triangular storm was introduced by the author in the 80's, and was used to get the solution for return periods of storms with some given characteristics [see references at the end of chap. 7].

References

Battjes J. A., 1970 Long term wave height distribution at seven stations around the British isles. *Report A 44 National Oceanographic Institute*, Wormley U. K.

Borgman L. E., 1970 Maximum wave height probabilities for a random number of random intensity storms. *Proc. 12th Conf. Coastal Eng. ASCE*, 53-64;
1973 Probabilities for highest wave in hurricane. *J. Waterways Harbors and Coastal Eng. Div. ASCE* 99, WW2, 185-207.

Cunha C. and Guedes Soares C., 1999 On the choice of data transformation for modelling time series of significant wave height. *Ocean Eng.* 26, 489-506.

Ferreira J.A. and Guedes Soares C., 1999 Modelling the long-term distribution of significant wave height with the Beta and Gamma models. *Ocean Eng.* 26, 713-725.

Krogstad H.E., 1985 Height and period distribution of extreme waves. *Appl. Ocean Res.* 7, 158-165.

Ochi K. and Whalen E., 1980 Prediction of the severest significant wave height. *Proc. 17th Conf. Coastal Eng. ASCE*, 587-599.

Chapter 7
DESIGN WAVES AND RISK ANALYSIS

7.1 The return period of a sea storm where the significant wave height exceeds a fixed threshold

7.1.1 *Preliminary solution: the probability density function $p_A(a)$ for the heights of the equivalent triangular storms*

The return period of a given event is the average time interval between two consecutive occurrences of this event. In this chapter we shall obtain the formal solutions for a few return periods, starting from the return period $R(H_s > h)$ of a storm whose significant wave height exceeds a fixed threshold h. To this end, we shall refer to the concept of *equivalent triangular storm* (e.t.s.) and of *equivalent sea* introduced in sect. 6.6, and in the proofs we shall emphasize the geometric point of view using terms like triangle, triangle height and triangle base in place of e.t.s., $H_{s\max}$ of the e.t.s. and duration of the e.t.s.

Let us define

$p_A(a) \equiv$ probability density function of the triangle height;

$p_B(b|a) \equiv$ conditional probability density function of the triangle base given the triangle height;

$\mathcal{N}(\mathcal{T}) \equiv$ number of triangles being present in a very large time interval $\mathcal{T}$;

$\delta t(h, \mathrm{d}h, a, b) \equiv$ time in which H_s falls within a fixed small interval $(h, h+\mathrm{d}h)$ in a triangle of height a and base b.

Using these definitions, we write

$$p_A(a)\,\mathrm{d}a\,\mathcal{N}(\mathcal{T}) = \text{number of triangles during } \mathcal{T}, \text{ the height of which belongs to a given interval } (a, a+\mathrm{d}a); \tag{7.1}$$

$$p_A(a)\,\mathrm{d}a\,\mathcal{N}(\mathcal{T})\,p_B(b|a)\mathrm{d}b = \text{number of triangles during } \mathcal{T}, \text{ the height of which belongs to a given interval } (a, a+\mathrm{d}a) \text{ and the base of which belongs to a given interval } (b, b+\mathrm{d}b); \tag{7.2}$$

$$[p_A(a)\,\mathrm{d}a\,\mathcal{N}(\mathcal{T})\,p_B(b|a)\mathrm{d}b]\,\delta t(h, \mathrm{d}h, a, b) = \text{time during } \mathcal{T}, \text{ in which } H_s \text{ belongs to a given interval } (h, h+\mathrm{d}h), \text{ in the triangles the height of which is between } a \text{ and } a+\mathrm{d}a \text{ and the base of which is between } b \text{ and } b+\mathrm{d}b. \tag{7.3}$$

It will have been noted that the l.h.s. of the first equality, that is $p_A(a)\,\mathrm{d}a\,\mathcal{N}(\mathcal{T})$, multiplied by $p_B(b|a)\,\mathrm{d}b$ becomes the l.h.s. of the second equality, which in turn multiplied by $\delta t(h,\mathrm{d}h,a,b)$ becomes the l.h.s. of the third equality. In practice, the equalities (7.1), (7.2), (7.3) make a sequence. This sequence enables us to gradually trace the steps leading to (7.3), which is the true starting point of our analysis. It implies

$$\Delta t(h,\mathrm{d}h,\mathcal{T}) = \int_0^\infty \int_0^\infty p_A(a)\,\mathcal{N}(\mathcal{T})\,p_B(b|a)\,\delta t(h,\mathrm{d}h,a,b)\,\mathrm{d}b\,\mathrm{d}a\,, \tag{7.4}$$

where $\Delta t(h,\mathrm{d}h,\mathcal{T})$ denotes *the time, during* $\mathcal{T}$, *in which* H_s *is within a small fixed interval* $(h, h+\mathrm{d}h)$.

As to $\delta t(h,\mathrm{d}h,a,b)$, its expression derives at once from its definition. It suffices to note that δt is zero for triangles like ① of fig. 7.1 where $a < h$; while δt is equal to $(\mathrm{d}h/a)b$ for triangles like ②. Therefore

$$\delta t(h,\mathrm{d}h,a,b)\begin{cases} = (\mathrm{d}h/a)\,b & \text{if} \quad a > h\,, \\ = 0 & \text{if} \quad a \le h\,. \end{cases} \tag{7.5}$$

With this expression of δt in equation (7.4), we obtain

$$\Delta t(h,\mathrm{d}h,\mathcal{T}) = \int_h^\infty \int_0^\infty p_A(a)\,\mathcal{N}(\mathcal{T})\,p_B(b|a)\,\frac{\mathrm{d}h}{a}\,b\,\mathrm{d}b\,\mathrm{d}a\,. \tag{7.6}$$

Note that the lower limit of the first integral passes from 0 in equation (7.4) to h in (7.6), as a consequence of the fact that $\delta t(h,\mathrm{d}h,a,b) = 0$ for $a \le h$. Clearly, for the linearity property, the terms $\mathcal{N}(\mathcal{T})$ and $\mathrm{d}h$ in (7.6) can be put outside the integral. Then we use the equality

$$\int_0^\infty p_B(b|a)\,b\,\mathrm{d}b = \overline{b}(a)\,,$$

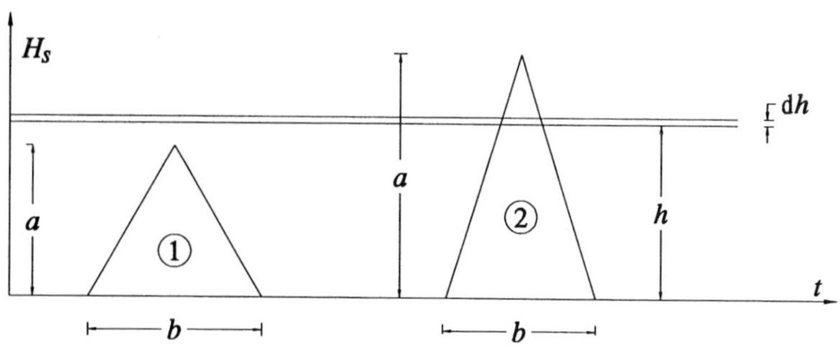

Fig. 7.1 Time duration of H_s within a fixed small interval $(h, h+\mathrm{d}h)$, in an e.t.s. of height a and base b. This duration is equal to $(\mathrm{d}h/a)b$ if $a > h$, while is zero if $a < h$.

where $\overline{b}(a)$, introduced in sect. 6.7, represents the mean base amplitude of the triangles with a given height a. Therefore equation (7.6) is rewritten in the form

$$\Delta t(h, \mathrm{d}h, \mathscr{T}) = \mathscr{N}(\mathscr{T})\mathrm{d}h \int_h^\infty p_A(a)\frac{1}{a}\overline{b}(a)\,\mathrm{d}a. \tag{7.7}$$

More simply $\Delta t(h, \mathrm{d}h, \mathscr{T})$ could have been written at once in the form

$$\Delta t(h, \mathrm{d}h, \mathscr{T}) = p(H_s = h)\,\mathrm{d}h\,\mathscr{T}, \tag{7.8}$$

which proceeds from the definition of $p(H_s = h)$ [cf. sect. 6.1.4.].

Equation (7.7) gives Δt in terms of the probability density function $p_A(a)$, and equation (7.8) gives the same Δt in terms of the probability density function $p(H_s = h)$. Hence equating the right-hand sides of the two equations (7.7) and (7.8) we obtain the relation between the unknown function $p_A(a)$ and the known function $p(H_s = h)$:

$$\mathscr{N}(\mathscr{T}) \int_h^\infty p_A(a)\frac{1}{a}\overline{b}(a)\,\mathrm{d}a = p(H_s = h)\,\mathscr{T}$$

Here differentiating with respect to h both sides of the equation, we get

$$-\mathscr{N}(\mathscr{T})\,p_A(h)\frac{1}{h}\overline{b}(h) = \frac{\mathrm{d}p(H_s = h)}{\mathrm{d}h}\mathscr{T}, \tag{7.9}$$

and hence

$$p_A(a) = -\frac{\mathscr{T}}{\mathscr{N}(\mathscr{T})}\frac{a}{\overline{b}(a)}\frac{\mathrm{d}p(H_s = a)}{\mathrm{d}a}. \tag{7.10}$$

(The change of variable, from h in (7.9) to a in (7.10) is simply formal, h and a playing the role of dummy variables.)

Equation (7.10) calls for $\mathrm{d}p(H_s = h)/\mathrm{d}h \leq 0$ for the equivalent sea, which can be formally proved to be true from the very definition of equivalent sea. [Fix h_1, h_2 and $\mathrm{d}h$ with $h_2 > h_1$. Prove that the duration of time in which H_s is between h_1 and $h_1 + \mathrm{d}h$ is greater than (or equal to) the duration of time in which H_s is between h_2 and $h_2 + \mathrm{d}h$. Hence, the proof follows straightforwardly.]

Note that $p(H_s = h)$ of the equivalent sea is very close to $p(H_s = h)$ of the actual sea only for $h > h_{\mathrm{crit}}$ where $\mathrm{d}p(H_s = h)/\mathrm{d}h$ of the actual sea proves to be always negative (sect. 6.2.3).

7.1.2 *Solution for the return period $R(H_s > h)$*

The number of triangles whose height is greater than a fixed threshold h, in the time interval $\mathscr{T}$, is

$$\mathscr{N}(h, \mathscr{T}) = \mathscr{N}(\mathscr{T}) \int_h^\infty p_A(a)\,\mathrm{d}a.$$

The return period $R(H_s > h)$, that is to say the average time interval between two consecutive triangles with height exceeding h, is related to $\mathcal{N}(h, \mathcal{T})$ by

$$R(H_s > h) = \frac{\mathcal{T}}{\mathcal{N}(h, \mathcal{T})}.$$

The two last equations yield

$$R(H_s > h) = \frac{\mathcal{T}}{\mathcal{N}(\mathcal{T}) \int_h^{\infty} p_A(a) \mathrm{d}a}$$

that, with the formula (7.10) for $p_A(a)$, becomes

$$R(H_s > h) = \frac{1}{-\int_h^{\infty} \frac{a}{\overline{b}(a)} \frac{\mathrm{d}p(H_s = a)}{\mathrm{d}a} \mathrm{d}a}.$$

The convergence of the integral in the last equation is very quick and, on the contrary, $\overline{b}(a)$ is a very gradually varying function. Hence, we achieve an excellent approximate form assuming $\overline{b}(a)$ to be constant on the interval of integration, that is assuming

$$\overline{b}(a) = \overline{b}(h).$$

With this assumption, the expression of $R(H_s > h)$ becomes

$$R(H_s > h) = \frac{\overline{b}(h)}{-\int_h^{\infty} a \frac{\mathrm{d}p(H_s = a)}{\mathrm{d}a} \mathrm{d}a},$$

from which, integrating by parts, we arrive at

$$R(H_s > h) = \frac{\overline{b}(h)}{h p(H_s = h) + P(H_s > h)}. \tag{7.11}$$

Finally, with (6.2) of $P(H_s > h)$ and (6.7) of $p(H_s = h)$, we obtain the following simple formula for the return period:

$$R(H_s > h) = \frac{\overline{b}(h)}{1 + u\left(\frac{h}{w}\right)^u} \exp\left(\frac{h}{w}\right)^u. \tag{7.12}$$

As an example, fig. 7.2 shows $R(H_s > h)$ for Ponza (centre of the Tyrrhenian Sea). The values of u and w are given in table 6.2, the function $\overline{b}(h)$ is given by (6.27) and the values of the parameters in this formula are given in table 6.3.

To check the step leading to (7.11), use the relation (6.1) between $p(H_s = h)$ and $P(H_s > h)$ and the limit

$$\lim_{a \to \infty} a p(H_s = a) = 0$$

which is a necessary condition for the mean value of $H_s(t)$ to exist.

7.2 The significant wave height and its persistence vs the return period

7.2.1 *The significant wave height $h(R)$*

The threshold $h(R)$ being exceeded by the significant wave height with a given return period R can be read at once from graphs like the one of fig. 7.2. It suffices to read the abscissa h associated with a given ordinate R. Alternatively, $h(R)$ can be evaluated by means of the formula we deduce here below.

We seek the solution to the equation

$$\frac{\bar{b}(h)}{R} \exp\left(\frac{h}{w}\right)^u = 1 + u\left(\frac{h}{w}\right)^u$$

that proceeds from (7.12). To this end, it is convenient to define a tentative value of $\bar{b}(h)$, that we shall call B:

$$B \equiv \text{tentative value of } \bar{b}(h).$$

Failing more precise information, we can very well assume $B = b_{10}$. Thus the equation becomes

$$\frac{B}{R} \exp\left(\frac{h}{w}\right)^u = 1 + u\left(\frac{h}{w}\right)^u, \tag{7.13}$$

where R and B (besides u and w) are known and h is the unknown. Defining

$$x \equiv \frac{B}{R} \exp\left(\frac{h}{w}\right)^u, \tag{7.14a}$$

$$A \equiv 1 + u \ln\left(\frac{R}{B}\right), \tag{7.14b}$$

we rewrite (7.13) in the form

$$x = A + u \ln x. \tag{7.15}$$

The x satisfying this equality is the limit of the sequence

$$x_i = A + u \ln x_{i-1} \quad \text{with} \quad x_0 = A, \tag{7.16}$$

which is easily obtained since the sequence converges very quickly. Then, from x, we get $h(R)$ from the definition (7.14a):

$$h(R) = w\left[\ln\left(\frac{R}{B}x\right)\right]^{\frac{1}{u}}. \tag{7.17}$$

Naturally, after having obtained $h(R)$ we can evaluate $\overline{b}(h)$, and hence we take $B = \overline{b}(h)$, and carry out a second iteration. Usually, the third iteration is unnecessary; indeed the difference between h of the second iteration and h of the third iteration is negligible.

Really, equation (7.15) admits two positive solutions, but one can easily prove that the smaller one of these two must be discarded. To this end, prove that the l.h.s. of (7.15) is smaller than the r.h.s. for $x = B/R$ (given that the storm duration B is smaller than the return period R). This implies that the smallest value of x satisfying (7.15) is smaller than B/R. Then, use (7.14a) to prove that $x < B/R$ would imply $(h/w)^u < 0$.

7.2.2 *The persistence $\overline{D}(h)$*

After having evaluated $h(R)$ it is necessary to obtain $\overline{D}(h)$ which is the mean persistence of H_s above the threshold h in the storms where this threshold is exceeded. The expression of $\overline{D}(h)$ proceeds by dividing

$$P(H_s > h)\mathscr{T} = \text{time duration in which } H_s > h\text{, in the long interval } \mathscr{T},$$

by

$$\mathscr{T}/R(H_s > h) = \text{number of storms in which } H_s \text{ goes over the threshold } h\text{, during } \mathscr{T}.$$

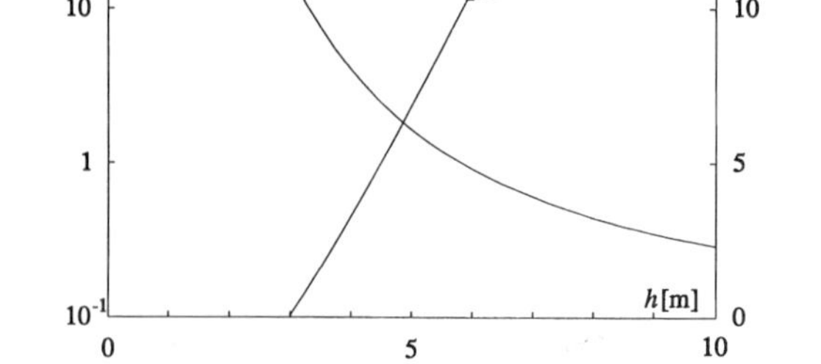

Fig. 7.2 Ponza (Tyrrhenian Sea): return period $R(H_s > h)$ and persistence $\overline{D}(h)$.

The result is

$$\overline{D}(h) = P(H_s > h)\, R(H_s > h), \tag{7.18}$$

which takes on the form

$$\overline{D}(h) = \frac{\overline{b}(h)}{1 + u\left(\dfrac{h}{w}\right)^u}, \tag{7.19}$$

if $P(H_s > h)$ is given by (6.2).

As an example the persistence $\overline{D}(h)$ at Ponza is shown in fig. 7.2.

7.3 The encounter probability

7.3.1 *General definitions*

Fig. 7.3 shows a few points on the time axis. Each of them marks the occurrence time of a natural event of given characteristics. For example, the event could be a storm in which the significant wave height exceeds a fixed threshold h at a fixed location.

The random times T_i (T_1, T_2, ...) between the successive occurrence times are called *interarrival times*. The mean value of the interarrival times is, by definition, the return period:

$$R \equiv \overline{T_i}.$$

The product $p(T_i = \tau)\,\mathrm{d}\tau$ will represent the probability that a T_i is between τ and $\tau + \mathrm{d}\tau$ (τ being arbitrarily fixed). $P(T_i > \tau)$ will be the probability of exceedance, that is the probability that a T_i exceeds a fixed threshold τ. From these definitions it follows that

$$p(T_i = \tau) = -\frac{\mathrm{d}P(T_i > \tau)}{\mathrm{d}\tau}, \tag{7.20}$$

that is

$$P(T_i > \tau) = \int_\tau^\infty p(T_i = \tau')\,\mathrm{d}\tau'. \tag{7.21}$$

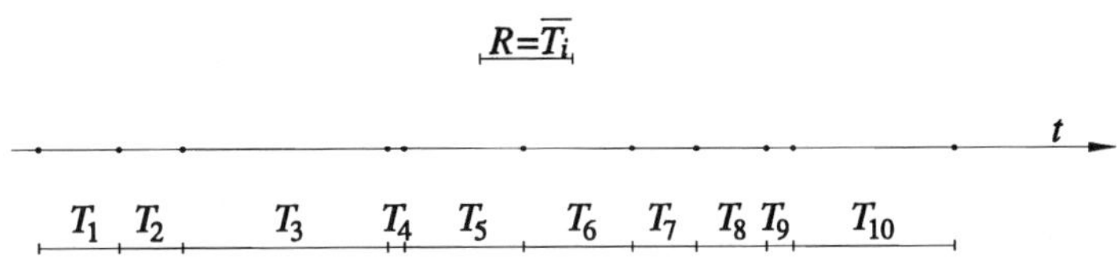

Fig. 7.3 Interarrival times of a point process.

Finally, the symbol $\mathscr{P}(L, R)$ will denote the probability that the event with the given return period R occurs *at least* once in the lifetime of a structure: *encounter probability*.

7.3.2 *The Poisson process*

In general the occurrences of the given event are assumed to form a homogeneous Poisson process. This is a well-known process, but for the convenience of the reader we would propose all the same a simple introduction.

Let us consider a very large time interval $\mathscr{T}$ and subdivide it into a sequence of very small intervals $\mathrm{d}t$, as in fig. 7.4. Let us get $\mathscr{T}/\mathrm{d}t$ white balls, and blacken $\mathscr{T}/R$ of these balls. Then let us put all the balls in a box, shuffle the balls and draw them at random. Each draw is coupled with a new small interval $\mathrm{d}t$: the first draw with the first small interval, the second draw with the second small interval and so on. If the drawn ball is black we mark a point in the small interval, otherwise we do not. At the end, we shall have marked $\mathscr{T}/R$ points, as many points as black balls, and consequently the mean value of the interarrival times T_i will be

$$\overline{T_i} = \frac{\mathscr{T}}{\mathscr{T}/R} = R,$$

according to the definition of sect. 7.3.1.

$P(T_i > \tau)$ is equal to the probability that, given a small interval with a point, the $N = \tau/\mathrm{d}t$ following small intervals do not contain points. Bearing in mind the procedure followed for marking the points, the probability that a given small interval does not contain a point is equal to the quotient between the number of white balls and the total number of balls, that is

$$\frac{\mathscr{T}/\mathrm{d}t - \mathscr{T}/R}{\mathscr{T}/\mathrm{d}t} = 1 - \frac{\mathrm{d}t}{R},$$

and $P(T_i > \tau)$ is equal to the Nth power of this probability with $N = \tau/\mathrm{d}t$:

$$P(T_i > \tau) = \left(1 - \frac{\mathrm{d}t}{R}\right)^{\frac{\tau}{\mathrm{d}t}}.$$

Multiplying and dividing the exponent $\tau/\mathrm{d}t$ by R, and applying the well-known limit

$$\lim_{x \to 0} (1 - x)^{\frac{1}{x}} = e^{-1},$$

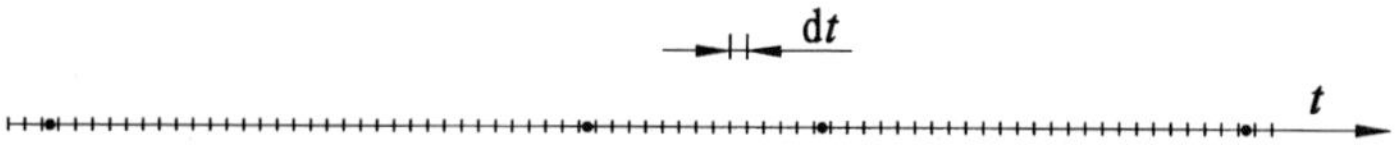

Fig. 7.4 How to generate a Poisson process: subdivide the time axis into a very large number of small intervals $\mathrm{d}t$, and draw at random from a box containing white balls and black balls (one draw for each small interval). If the drawn ball is black, mark a point in the small interval.

we rewrite the expression of $P(T_i > \tau)$ in the form

$$P(T_i > \tau) = \exp\left(-\frac{\tau}{R}\right).$$

Finally, using the relation (7.20) between the probability density function $p(T_i = \tau)$ and the probability of exceedance $P(T_i > \tau)$, we obtain

$$p(T_i = \tau) = \frac{1}{R}\exp\left(-\frac{\tau}{R}\right).$$

As to the probability $\mathscr{P}(L, R)$ that at least one point falls within a fixed time interval L, it is given by

$$1 - \mathscr{P}(L, R) = \left(1 - \frac{\mathrm{d}t}{R}\right)^{\frac{L}{\mathrm{d}t}}. \tag{7.22}$$

Here the l.h.s. represents the probability that the given time interval L does not contain any point, and the r.h.s. represents the Nth power of the probability that a fixed small interval $\mathrm{d}t$ does not contain a point, where N is the number of small intervals $\mathrm{d}t$ in L. Multiplying and dividing the exponent $L/\mathrm{d}t$ by R, and using the same limit as before, we obtain

$$\mathscr{P}(L, R) = 1 - \exp\left(-\frac{L}{R}\right).$$

This would be the situation if the process, as it is usually assumed, were the Poisson process. Really, it is likely that the occurrences of environmental events are not so random as in the Poisson process. In particular sometimes it is pointed out that the occurrences of such events show a tendency to cluster as well as to be recurrent. For this reason, in the next section, we shall consider the relation between $\mathscr{P}$, L and R under general assumptions on the random point process. Indeed we shall allow the process an arbitrary $p(T_i = \tau)$.

7.3.3 *The general solution for the encounter probability*

Let us assume that the random point process of the occurrences of an environmental event is generally different from the Poisson process, and let us imagine we know the probability density function $p(T_i = \tau)$ of the interarrival times T_i. Consequently we also know the return period of the event, which is equal to the mean value of the T_i, and naturally we know the probability of exceedance $P(T_i > \tau)$ that is related to $p(T_i = \tau)$ by (7.21). Without any more information, we seek the probability $\mathscr{P}(L, R)$ that the given event occurs at least once in a fixed time interval L (the lifetime of a structure).

Let us use a thick mark for the part of the time axis consisting of all the $T_i < L$ and of a segment of length L for each $T_i > L$, as we have done in fig. 7.5. Then, without looking at the time axis (imagine fig. 7.5 is covered by a black sheet) let us put a stick of length L over the time axis. $\mathscr{P}(L, R)$ is the probability that, after removing the black sheet, we find the stick covers at least one of the points.

Thus $\mathscr{P}(L, R)$ is equal to the probability that I [the left end of the stick] falls in the thick marked part of the time axis. Indeed if I falls in this part, then the interval L contains at least one point; and, conversely, if the interval L contains at least one point, then I must have fallen in the thick marked part of the time axis. This is instantly obvious on looking at the figure.

The probability that I falls in the thick marked part of the time axis is equal to the quotient between the length $C(\mathscr{T})$ of this part and the total length of the time axis:

$$\mathscr{P}(L, R) = C(\mathscr{T})/\mathscr{T}, \tag{7.23}$$

where

$$C(\mathscr{T}) \equiv \text{thick marked part of the time axis}$$

(here C is for *crassum*, the latin translation of thick).

Let us obtain the expression of $C(\mathscr{T})$ in terms of the probability density function $p(T_i = \tau)$. We have

$$p(T_i = \tau)\, \mathrm{d}\tau \,\widetilde{\mathscr{N}}(\mathscr{T}) = \text{number of } T_i \text{ such that } \tau < T_i < \tau + \mathrm{d}\tau \text{ in } \mathscr{T},$$

$$\tau p(T_i = \tau)\, \mathrm{d}\tau \,\widetilde{\mathscr{N}}(\mathscr{T}) = \text{portion of } \mathscr{T} \text{ occupied by the } T_i \text{ such that } \tau < T_i < \tau + \mathrm{d}\tau,$$

where $\widetilde{\mathscr{N}}(\mathscr{T})$ *is the number of the interarrival times* T_i *in* $\mathscr{T}$. Hence, it follows that

$$\int_L^\infty p(T_i = \tau)\, \mathrm{d}\tau \,\widetilde{\mathscr{N}}(\mathscr{T}) = \text{number of } T_i > L, \text{ in } \mathscr{T}, \tag{7.24a}$$

$$\int_0^L \tau p(T_i = \tau)\, \mathrm{d}\tau \,\widetilde{\mathscr{N}}(\mathscr{T}) = \text{portion of } \mathscr{T} \text{ occupied by the } T_i < L. \tag{7.24b}$$

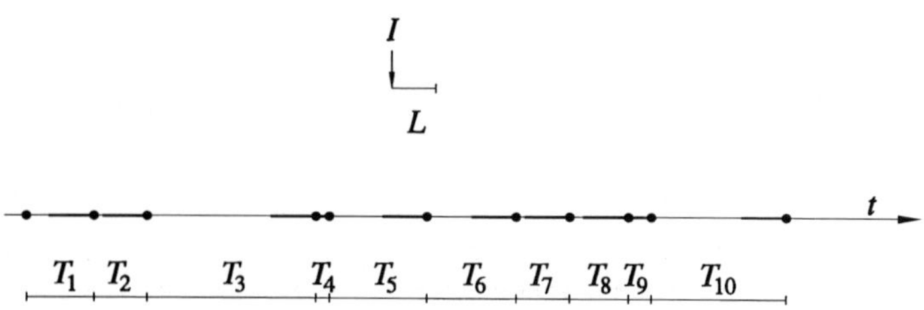

Fig. 7.5 The thick marked part of the time axis consists of all the interarrival times $T_i < L$ and of a span of time L for each interarrival time $T_i > L$.

The length $C(\mathscr{T})$ of the thick marked part of the time axis is equal to the sum of all the $T_i < L$ (given by 7.24*b*) plus L multiplied by the number of $T_i > L$ (this number being given by 7.24*a*). Therefore $C(\mathscr{T})$ is given by

$$C(\mathscr{T}) = \int_0^L \tau p(T_i = \tau)\,\mathrm{d}\tau.\widetilde{\mathscr{N}}^{\cdot}(\mathscr{T}) + L\int_L^\infty p(T_i = \tau)\,\mathrm{d}\tau.\widetilde{\mathscr{N}}^{\cdot}(\mathscr{T}),$$

and (7.23) becomes

$$\mathscr{P}(L,R) = \frac{\widetilde{\mathscr{N}}^{\cdot}(\mathscr{T})}{\mathscr{T}}\left[\int_0^L \tau p(T_i = \tau)\,\mathrm{d}\tau + L\int_L^\infty p(T_i = \tau)\,\mathrm{d}\tau\right], \tag{7.25}$$

which yields

$$\mathscr{P}(L,R) = \frac{1}{R}\int_0^L P(T_i > \tau)\,\mathrm{d}\tau. \tag{7.26}$$

Check the step from (7.25) to (7.26). To this end, use (7.20-21), evaluate by parts the first integral on the r.h.s. of (7.25), and note that

$$R \equiv \overline{T_i} = \mathscr{T} \Big/ \widetilde{\mathscr{N}}^{\cdot}(\mathscr{T}).$$

7.3.4 *A corollary of the general solution*

The mean value of the interarrival times T_i is equal to the integral over $(0,\infty)$ of the probability of exceedance:

$$R \equiv \overline{T_i} = \int_0^\infty P(T_i > \tau)\,\mathrm{d}\tau.$$

Hence, the general formula for $\mathscr{P}(L,R)$ can be rewritten in the form

$$\mathscr{P}(L,R) = \frac{\displaystyle\int_0^L P(T_i > \tau)\,\mathrm{d}\tau}{\displaystyle\int_0^\infty P(T_i > \tau)\,\mathrm{d}\tau},$$

which reveals at once that

$$\mathscr{P}(L,R) \leq 1,$$

as it must be, $\mathscr{P}(L,R)$ being a probability. Moreover, from equation (7.26) we conclude also that

$$\mathscr{P}(L,R) \leq \frac{L}{R},$$

since $P(T_i > \tau)$ is smaller than or equal to 1.

The two last inequalities yield

$$\mathscr{P}(L,R)\begin{cases} \leq \dfrac{L}{R} & \text{if } L < R, \\ \leq 1 & \text{if } L \geq R. \end{cases} \tag{7.27}$$

7.3.5 *The formulae for* $\mathscr{P}(L,R)$ *for a few given forms of* $p(T_i = \tau)$

① *Delta function*: $p(T_i = \tau) = \delta(\tau - R)$, see fig. 7.6.
We have

$$P(T_i > \tau) = \int_{\tau}^{\infty} \delta(\tau' - R)\,\mathrm{d}\tau' = U(\tau; R),$$

where $U(\tau; R)$ is the step function:

$$U(\tau;R)\begin{cases} \equiv 1 & \text{if } \tau \leq R, \\ \equiv 0 & \text{if } \tau > R. \end{cases}$$

With the expression of $P(T_i > \tau)$ in (7.26) we obtain

$$① \quad \mathscr{P}(L,R) = \frac{1}{R}\int_0^L U(\tau;R)\,\mathrm{d}\tau \begin{cases} = \dfrac{L}{R} & \text{if } L < R, \\ = \ 1 & \text{if } L \geq R. \end{cases} \tag{7.28}$$

② *Boxcar*: $p(T_i = \tau) = \dfrac{1}{2R}\,U(\tau; 2R)$, see fig. 7.7.
We have

$$P(T_i > \tau) = \int_{\tau}^{\infty} \frac{1}{2R}\,U(\tau'; 2R)\,\mathrm{d}\tau' \begin{cases} = 1 - \dfrac{\tau}{2R} & \text{if } \tau \leq 2R, \\ = 0 & \text{if } \tau > 2R, \end{cases}$$

and hence

$$② \quad \mathscr{P}(L,R)\begin{cases} = \dfrac{L}{R} - \dfrac{1}{4}\left(\dfrac{L}{R}\right)^2 & \text{if } L \leq 2R, \\ = 1 & \text{if } L > 2R. \end{cases}$$

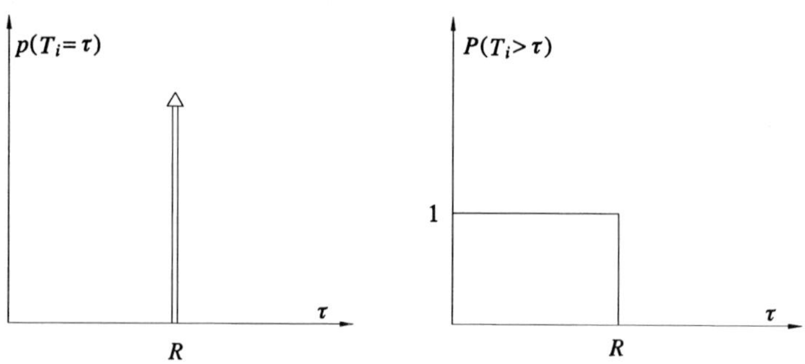

Fig. 7.6 ①: $p(T_i = \tau)$ = delta function.

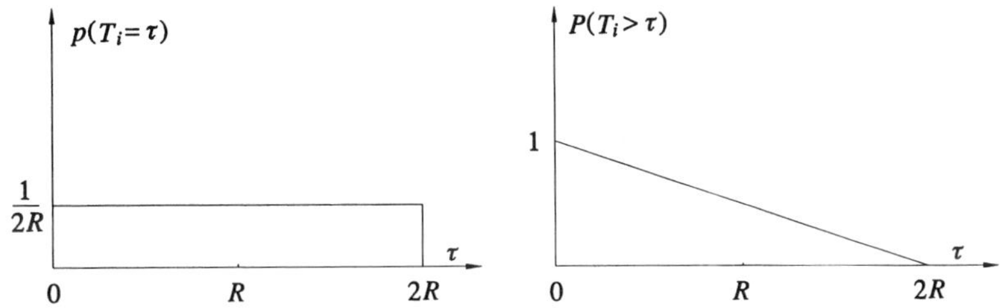

Fig. 7.7 ②: $p(T_i = \tau) =$ boxcar.

③ *Exponential function*: $p(T_i = \tau) = \frac{1}{R}\exp\left(-\frac{\tau}{R}\right)$, see fig. 7.8.
We have

$$P(T_i > \tau) = \int_{\tau}^{\infty} \frac{1}{R}\exp\left(-\frac{\tau'}{R}\right) \mathrm{d}\tau' = \exp\left(-\frac{\tau}{R}\right),$$

and hence

$$③\ \mathscr{P}(L,R) = \frac{1}{R}\int_0^L \exp\left(-\frac{\tau}{R}\right) \mathrm{d}\tau = 1 - \exp\left(-\frac{L}{R}\right).$$

7.3.6 *Comment on the formulae for* $\mathscr{P}(L,R)$

We must make two remarks. First: we see $\mathscr{P}(L,R)$ attains its maximum for given L and R, if $p(T_i = \tau)$ is the delta function (compare formula 7.28 for $\mathscr{P}(L,R)$ with inequality 7.27). In other words, under the same L and R, $\mathscr{P}(L,R)$ is maximum if the interarrival times T_i are all equal to one another, i.e. they are all equal to the return period R.

Second remark: if $p(T_i = \tau)$ is an exponential function, $\mathscr{P}(L,R)$ takes on the same form we obtained in sect. 7.3.2 for the Poisson process. In more detail, in sect. 7.3.2 we proved that $p(T_i = \tau)$ of the Poisson process is an exponential function, and that $\mathscr{P}(L,R)$ of this process takes on the form $1 - \exp(-L/R)$. Here we have shown that $\mathscr{P}(L,R)$ takes on the form $1 - \exp(-L/R)$ if $p(T_i = \tau)$ is an exponential

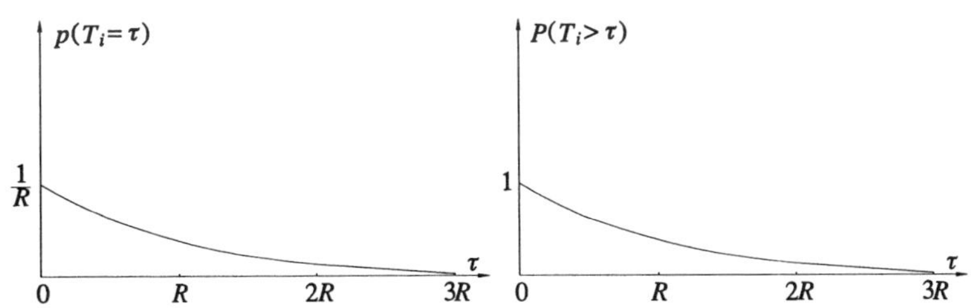

Fig. 7.8 ③: $p(T_i = \tau) =$ exponential function.

function. This is a step forward: indeed the fact that $p(T_i = \tau)$ of the Poisson process is an exponential function does not imply that a process whose $p(T_i = \tau)$ is an exponential function is a Poisson process. To better express this concept, in what follows we suggest the example of a process that in spite of being different from the Poisson process has $p(T_i = \tau)$ given by the exponential function.

Let us imagine that the values of a lot of interarrival times have been written on paper sheets (one value per paper sheet). These values are distributed according to the exponential function so that, for any fixed τ, we have a number $\exp(-\tau/R).\mathcal{N}$ of values exceeding τ ($\mathcal{N}$ being the total number of paper sheets). Let us put the paper sheets in a box, and let us draw: V_1 will be the first value drawn, V_2 the second value drawn, and so on. Let us use the sequence $V_1, V_2, ..., V_{\mathcal{N}}$ to form a process, in the following manner.

We take $T_1 = T_2 = T_3 = V_1$ where T_1 is the time between the first and the second occurrence times, T_2 is the time between the second and the third occurrence times, and T_3 is the time between the third and the fourth occurrence times. Similarly, we take $T_4 = T_5 = T_6 = V_2$, and so on. In doing so, we get a point process where the distribution of the interarrival times T_i is exponential, and indeed the share of the T_i greater than any fixed τ is $\exp(-\tau/R)$. However, it is apparent that this is not at all the Poisson process described in sect. 7.3.2. It is a process that exhibits both recurrence and clustering: recurrence because each time we have four occurrences of the event at regular intervals; clustering because whenever a small V_i is drawn, we have four occurrences in a short time. Nevertheless, $\mathcal{P}(L, R)$ is the same as that of the Poisson process where the T_i are stochastically independent of one another.

Thus we realize that the three expressions ①, ② and ③ of $\mathcal{P}(L, R)$, not only refer to three distributions of the interarrival times T_i very different from one another, but they also cover a variety of ways the process actually is: with or without recurrence, with or without clustering.

7.3.7 *The design calculation: obtaining R for given $\mathcal{P}$ and L*

The design rules usually give the lifetime L and the encounter probability $\mathcal{P}$. The engineer has first to compute R from these given L and $\mathcal{P}$, and then he has to single out the environmental event with this return period. The structure must be able to withstand such an event with a given safety factor.

Using the three different forms ①, ② and ③ of $\mathcal{P}$ (L, R), the relation which gives R from L and $\mathcal{P}$ proves to be

$$① \; R = L/\mathcal{P},$$

$$② \; R = \frac{L}{2\mathcal{P}}\left(1 + \sqrt{1 - \mathcal{P}}\right),$$

$$③ \; R = L/\ln\left(\frac{1}{1 - \mathcal{P}}\right).$$

[Really the equation ② of $\mathscr{P}(L,R)$ has two solutions, but one of the two is discarded because it does not comply with the condition: $R \to \infty$ as $\mathscr{P} \to 0$ for a fixed L.] In the following sect. 7.4 we shall use these three forms of $R(L,\mathscr{P})$ to compute the design wave. Then we shall compare the results obtained, and thus be able to judge whether or not our ignorance on the distribution of the interarrival times T_i has some significant consequence on the structure design.

7.4 The chain: lifetime, encounter probability ⇒ return period ⇒ significant wave height

7.4.1 *Worked example*

Let us assume

$$L = 50 \text{ years}, \qquad \mathscr{P} = 0.1$$

which, as we shall see in sect. 7.5, are realistic inputs for the design of ocean structures. Using formula ③ for $R(L,\mathscr{P})$ obtained in the foregoing section, we have

$$③\ R = 50/\ln\left(\frac{1}{1-0.1}\right) = 475\,\text{years}.$$

Thus, let us evaluate $h(R = 475 \text{ years})$ for Ponza that is characterized by

$$u = 1.200, \quad w = 0.874\,\text{m}, \quad a_{10} = 3.4\,\text{m}, \quad b_{10} = 69 \text{ hours}, \quad K_1 = 1.12, \quad K_2 = -0.115.$$

Following the procedure given in sect. 7.2.1, let us fix a value of B for the first iteration:

$$B = b_{10} = 69 \text{ hours}.$$

Since R = 475 years, from (7.14b) we have

$$A = 1 + 1.200\ln\left(\frac{475 \cdot 365 \cdot 24}{69}\right) = 14.21,$$

and hence the terms of the sequence (7.16) are

$$\begin{aligned}
x_0 &= 14.21,\\
x_1 &= 14.21 + 1.200 \ \ln 14.21 = 17.39,\\
x_2 &= 14.21 + 1.200 \ \ln 17.39 = 17.64,\\
x_3 &= 14.21 + 1.200 \ \ln 17.64 = 17.65,\\
x_4 &= 14.21 + 1.200 \ \ln 17.65 = 17.65.
\end{aligned}$$

We see that the limit of the sequence is

$$x = 17.65.$$

With this value in (7.17) we obtain

$$h(R) = 0.874\left[\ln\left(\frac{475\cdot 365\cdot 24}{69}\,17.65\right)\right]^{\frac{1}{1.200}} = 7.82\,\mathrm{m}.$$

The value of B for the second iteration proceeds from (6.27) of $\overline{b}(h)$:

$$B = \overline{b}(7.82\,\mathrm{m}) = 1.12\cdot 69\exp\left(-0.115\,\frac{7.82}{3.4}\right) = 59\,\text{hours},$$

and, with this new value of B, we obtain

$$2^{\mathrm{nd}}\ \text{iteration}\begin{cases} A = 14.40, \\ x = 17.86, \\ h(R) = 7.90\,\mathrm{m}. \end{cases}$$

The third iteration with $B = \overline{b}(7.90\,\mathrm{m})$ leaves the result unchanged. Therefore, 7.90 m is the threshold of significant wave height associated with the return period of 475 years.

Now let us repeat the procedure from the start, using formula ② for $R(L,\mathscr{P})$. The result is

$$② \ R = \frac{50}{2\cdot 0.1}\,(1+\sqrt{1-0.1}) = 487\,\text{years},$$

$$h(487\ \text{years}) = 7.91\ \mathrm{m}.$$

Finally, with formula ① for $R(L,\mathscr{P})$ we get

$$① \ R = 50/0.1 = 500\,\text{years},$$

$$h(500\ \text{years}) = 7.93\ \mathrm{m}.$$

The summary is

$$① \quad R = 500\ \text{years} \quad\Rightarrow\quad h(R) = 7.93\ \mathrm{m},$$

$$② \quad R = 487\ \text{years} \quad\Rightarrow\quad h(R) = 7.91\ \mathrm{m},$$

$$③ \quad R = 475\ \text{years} \quad\Rightarrow\quad h(R) = 7.90\ \mathrm{m}.$$

7.4.2 *Comment on the results of the worked example*

We have to come back to the conclusion of sect. 7.3. We know practically nothing about the form of the distribution of the interarrival times T_i. This is why we have assumed three distributions very different from one another, from which we have got three alternative formulae for $R(L,\mathscr{P})$. Using these formulae in the worked example, we have found three different values of R. The differences are within 5% (largest difference: $R = 500$ years with formula ① against $R = 475$ years with formula ③). As to $h(R)$, that is the true target, the differences are within only 0.4%.

In reality, the design input ($L = 50$ years, $\mathscr{P} = 0.1$) which we have taken is a rather severe one. For some smaller lifetimes and/or higher encounter probabilities, the differences resulting from the use of the different forms of $R(L, \mathscr{P})$ prove to be somewhat larger. To get an idea of this, let us consider the case of

$$L = 15\,\text{years}, \quad \mathscr{P} = 0.5.$$

These are the smallest lifetime and the highest encounter probability granted by design rules for some special classes of structures. Using the formulae ①, ② and ③ with these values of L and $\mathscr{P}$ we get

① $R = 30.0$ years $\Rightarrow$ $h(R) = 6.45\,\text{m}$,

② $R = 25.6$ years $\Rightarrow$ $h(R) = 6.36\,\text{m}$,

③ $R = 21.6$ years $\Rightarrow$ $h(R) = 6.27\,\text{m}$,

where $h(R)$ is here too for Ponza. As we can see even in this extreme case, the differences between the values of $h(R)$ do not exceed 3%.

In the light of these results, it is apparent that we can use any of the alternative formulae ①, ②, or ③ with practically negligible differences on the structure design. In what follows we shall use formula ③, that is

$$\mathscr{P}(L, R) = 1 - \exp\left(-\frac{L}{R}\right), \tag{7.29a}$$

$$R(L, \mathscr{P}) = L/\ln\left(\frac{1}{1 - \mathscr{P}}\right). \tag{7.29b}$$

The reason for this choice is simply that ③ is valid for the Poisson process, so this formula is usually known. The more we know after the foregoing analysis is that, even in the most probable event that the random point process of the storm occurrences differ greatly from the Poisson process, the use of expression ③ will lead to only negligible errors on the design wave.

7.5 Coastal structures: the design sea state

A design sea state with $h(R)$ as significant wave height is used for the coastal structures. The upright breakwaters must be able to withstand this sea state with some dictated safety factors. The rubble mound breakwaters must be able to withstand the design sea state, suffering at the most damages contained within a prescribed level.

We have seen in the foregoing section that the input data for the calculation of $h(R)$ are lifetime L and encounter probability $\mathscr{P}$. As an example, tables 7.1-2 give the minimum L and the maximum $\mathscr{P}$ prescribed by a few European guidelines on the design of the various classes of maritime structures (cf. Puertos del Estado, 1990).

An additional verification is required for the rubble mound breakwaters. Specifically, the structure must be able to withstand, with practically no damage, a sea state that has a relatively high encounter probability. This relatively high probability is specified in table 7.3.

Table 7.1 *Minimum lifetime L (years)*

Safety level [(1)] → / Structure type ↓	*1*	2	3
General use	25	50	100
Special industrial use	15	25	50

Table 7.2 *Maximum encounter probability* $\mathscr{P}$

Economic consequence in case of failure [(2)]	risk for human life [(3)]	
	small	high
Low	0.20	0.15
Average	0.15	0.10
High	0.10	0.05

Table 7.3 *Maximum encounter probability* $\mathscr{P}$ *(second verification of rubble mound breakwaters)*

Economic consequence in case of failure [(2)]	risk for human life [(3)]	
	small	high
Low	0.50	0.30
Average	0.30	0.20
High	0.25	0.15

[(1)] Level 1: structure of local interest; small risk of loss of human lives or environmental damage in case of failure. Level 2: structure of general interest; moderate risk of loss of human lives or environmental damage in case of failure. Level 3: structure of protection against floods, or structure of international interest; high risk of loss of human lives or environmental damage in case of failure. For more details, see the Italian or Spanish guidelines on the design of maritime structures.

[(2)] The judgement is based on quotient Q between cost of losses and investment:

$$Q \leq 5 \Rightarrow low;\ 5 < Q \leq 20 \Rightarrow medium;\ Q > 20 \Rightarrow high.$$

[(3)] Risk for human life *high* or *small* according to whether human losses are or are not expected in case of failure or damage.

For the design of a few structure components (e.g. the armour of a rubble mound), besides the significant wave height, we have to specify also the duration of the design sea state. To this end, we reason as follows. $h(R)$ is the threshold that is exceeded, on average, once in R years; and the persistence of H_s above this threshold is a random variable whose mean value $\overline{D}(h)$ is given by (7.19). This means that H_s *remains on the threshold* H on average for a span of time of $2\overline{D}(h)$. We mean that h is the mean value of H_s for this span of time. Therefore we assume

$$\mathscr{D} = 2\,\overline{D}(h).$$

Fig. 7.9 is helpful to realize this.

Now we can complete the calculation for $L = 50$ years, $\mathscr{P} = 0.1$ at Ponza. In sect. 7.4 we saw that

$$R = 475\,\text{years},\;\; h(475\,\text{years}) = 7.90\text{ m},\;\; \overline{b}(7.90\,\text{m}) = 59\text{ hours},$$

from which, using (7.19), we obtain

$$\overline{D}(7.90\,\text{m}) = \frac{59}{1 + 1.200\left(\dfrac{7.90}{0.874}\right)^{1.200}} = 3.3\text{ hours},$$

and hence

$$\mathscr{D} = 2 \cdot 3.3 = 6.6\,\text{hours}.$$

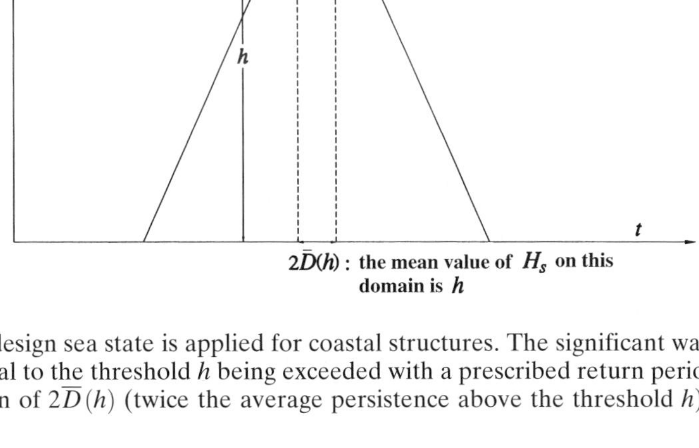

Fig. 7.9 The design sea state is applied for coastal structures. The significant wave height of this sea state is equal to the threshold h being exceeded with a prescribed return period. The duration should be taken of $2\overline{D}(h)$ (twice the average persistence above the threshold h).

7.6 The return period of a wave with a height exceeding a fixed threshold

7.6.1 *Solution for the return period* $R^*(H)$

With this section we start a new reasoning that will lead us, in sect. 7.8, to the design waves for the offshore structures.

Here too we resort to a sequence of equalities, where the l.h.s. of the $i+1$th equality is equal to the l.h.s. of the ith equality multiplied by a new term. Referring to a very large time interval $\mathcal{T}$, we have

$$p(H_s = h)\,\mathrm{d}h\,\mathcal{T} = \text{time in which } h < H_s < h + \mathrm{d}h, \text{ during } \mathcal{T};$$

$$\frac{1}{\overline{T}(h)}\,p(H_s = h)\,\mathrm{d}h\,\mathcal{T} = \text{number of waves in the sea states with } h < H_s < h + \mathrm{d}h, \text{ during } \mathcal{T};$$

$$P(H; H_s = h)\,\frac{1}{\overline{T}(h)}\,p(H_s = h)\,\mathrm{d}h\,\mathcal{T} = \text{number of waves higher than a given threshold } H, \text{ in the sea states with } h < H_s < h + \mathrm{d}h, \text{ during } \mathcal{T}.$$

Integrating the l.h.s. of this last equality over $0 < h < \infty$ we obtain $\mathcal{N}(H, \mathcal{T})$, the number of waves with a height greater than H in $\mathcal{T}$:

$$\mathcal{N}(H, \mathcal{T}) = \int_0^\infty P(H; H_s = h)\,\frac{1}{\overline{T}(h)}\,p(H_s = h)\,\mathrm{d}h\,\mathcal{T}.$$

The return period $R^*(H)$ of a wave with a height exceeding the fixed threshold H is equal to the quotient between $\mathcal{T}$ and the number of waves higher than H in $\mathcal{T}$, that is

$$R^*(H) = \mathcal{T} / \mathcal{N}(H, \mathcal{T}),$$

and hence

$$R^*(H) = \left[\int_0^\infty P(H; H_s = h)\,\frac{1}{\overline{T}(h)}\,p(H_s = h)\,\mathrm{d}h\right]^{-1}. \tag{7.30}$$

As to the functions appearing in this equation, we can use (5.26), (5.48*a*) and (6.7).

The integrand in (7.30) has a maximum. Indeed it approaches zero as $h \to 0$ because of $P(H; H_s = h)$, and approaches zero as $h \to \infty$ because of $p\,(H_s = h)$. The integral can be evaluated numerically even with a relatively large integration step: usually a step of 0.5 m is suitable. Moreover the interval of integration can be reduced to $(0.3H, 2.0H)$ without any appreciable consequence.

7.6.2 *A seeming paradox*

Fig. 7.10 shows a typical $R^*(H)$ for the Mediterranean Sea (Ponza: $w = 0.874$ m, $u = 1.200$), and we suspect it stirs up some doubts. Specifically, the fact that the

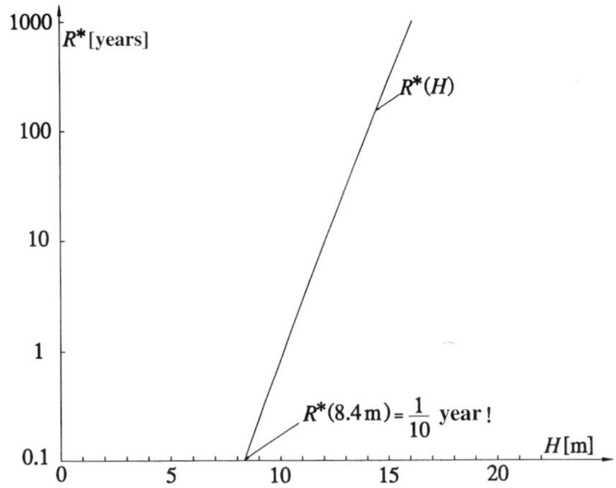

Fig. 7.10 Ponza (Tyrrhenian Sea): return period $R^*(H)$. (Obtained by means of 7.30.)

return period of a wave height exceeding 8.4 m is only 1/10 of a year will probably not be easy to accept. We are afraid that the reader's doubts would change to a refusal if we were to go down to some lower return period. It suffices to say that

$$R^*(5\,\text{m}) = 9\,\text{hours!}$$

Really this result seems to go against common sense, and it suggests something is wrong in the equation (7.30), or in the calculation, or in both. But it is not so. As to the equation, it is based on only a few simple steps, check them little by little and you will realize the solution is exact whatever the H. As to the calculation, anybody can quickly implement a PC program. Anyway, we can do a general verification together.

Let us estimate how many waves higher than 5 m occur on average in one year, during sea states with an H_s between 3.75 m and 4.25 m. The quotient between time in which H_s is within the range (3.75 m, 4.25 m) and total time is

$$P(H_s > 3.75\text{ m}) - P(H_s > 4.25\text{ m}) = \exp\left[-\left(\frac{3.75}{0.874}\right)^{1.200}\right] - \exp\left[-\left(\frac{4.25}{0.874}\right)^{1.200}\right] =$$
$$= 1.94 \cdot 10^{-3}$$

and hence H_s will be between 3.75 m and 4.25 m, on average for

$$365 \cdot 24 \cdot 1.94 \cdot 10^{-3} = 17.0\,\text{hours/year}.$$

The mean wave period in a sea state with an H_s of 4.0 m (4.0 m is the median of the range 3.75 m, 4.25 m) is evaluated by means of (5.26), and proves to be

$$\overline{T} = 6.6\pi\sqrt{\frac{4}{4 \cdot 9.8}} = 6.6\,\text{s}.$$

Therefore, we expect the 17 hours per year in which H_s is between 3.75 m and 4.25 m to contain

$$\frac{17 \cdot 3600}{6.6} = 9270 \text{ waves}.$$

The probability of a wave higher than 5 m in a sea state with an H_s of 4 m is calculated by means of (5.48*a*), and proves to be

$$P(5\,\text{m}; H_s = 4\,\text{m}) = \exp\left[-2.31\left(\frac{5}{4}\right)^2\right] = 0.027,$$

so that the number of waves higher than 5 m, which we can expect to occur in the 17 hours in which H_s is between 3.75 m and 4.25 m, is

$$9270 \cdot 0.027 = 250.$$

The conclusion is that

$$R^*(5\,\text{m}) < \frac{365 \cdot 24}{250} = 35\,\text{hours},$$

that is the return period of a wave higher than 5 m is smaller than 35 hours. *Smaller*, clearly because we have counted only the waves in the sea states with H_s between 3.75 m and 4.25 m. If we also count the other waves higher than 5 m, that is to say the waves in the sea states with $H_s < 3.75$ m and in the sea states with $H_s >$ 4.25 m, we get the very nine hours of return period we said above.

The reason why the return period of a wave higher than 5 m seems so small is simply that the definition of $R^*(H)$ does not correspond with our intuitive idea of return period. By return period of a wave higher than 5 m we mean the time that on average elapses between two occurrences of some waves higher than 5 m. While $R^*(5$ m) is the mean time interval between two occurrences of a single wave higher than 5 m; and it is small because even a few hundred waves higher than 5 m may occur together in a rather short time during a storm. Thus we realize that a more effective definition is that of "return period of a storm containing at least one wave higher than the fixed threshold H". We shall call this new return period $R(H)$.

7.7 The return period of a sea storm containing at least one wave higher than a fixed threshold

7.7.1 *Layout of the problem*

Alternatively to the definition given at the end of the foregoing section, $R(H)$ can be also defined as "the return period of a storm whose maximum wave height exceeds the fixed threshold H". The two definitions are equivalent to each other because, if the maximum wave height exceeds the fixed threshold H, then the storm contains at least one wave higher than H; and, conversely, if the storm contains at least one wave higher than H then the maximum wave height exceeds H.

To obtain $R(H)$ it is helpful to refer to a new alternative definition. Specifically it is convenient to pass from *the return period of a storm...* to *the return period of a wave* with the two following features

(i) to be higher than the fixed threshold H,

(ii) to be the highest wave of the storm in which it occurs.

Indeed it is apparent that the number of storms whose maximum wave height exceeds the threshold H is equal to the number of waves having both the properties (i) and (ii).

Then the return period $R(H)$ will be given by

$$R(H) = \frac{\mathscr{T}}{\mathscr{N}(H,\mathscr{T};\max)}, \tag{7.31}$$

where $\mathscr{T}$ is the usual very large time interval and

$$\mathscr{N}(H,\mathscr{T};\max) \equiv \text{number of waves during } \mathscr{T} \text{ which have both the properties (i) and (ii).} \tag{7.32}$$

This number, whatever the H, takes on the same value in the actual sea and in the equivalent sea. Indeed, we saw in sect 6.6 that the equivalent triangular storms and the actual storms have the same probability that the maximum wave height (of the storm) exceeds any fixed threshold H. Therefore, we shall refer again to the equivalent sea, where the storms are triangles like in fig. 7.1.

7.7.2 *Solution for the return period R(H)*

To obtain $\mathscr{N}(H,T;\max)$, we start as usual from a sequence of equalities. The first three equalities are (7.1-2-3) that here we omit to rewrite, hence the sequence goes on as follows

$$[p_A(a)\,\mathrm{d}a\,\mathscr{N}(\mathscr{T})\,p_B(b|a)\,\mathrm{d}b]\,\delta t(h,\mathrm{d}h,a,b)\,\frac{1}{\overline{T}(h)} = \text{number of waves}$$

present in the stages of the storms in which H_s is between h and $h+\mathrm{d}h$, only for the storms whose $H_{s\max}$ is between a and $a+\mathrm{d}a$ and whose duration is between b and $b+\mathrm{d}b$, during $\mathscr{T}$;

$$[p_A(a)\,\mathrm{d}a\,\mathscr{N}(\mathscr{T})\,p_B(b|a)\,\mathrm{d}b]\,\delta t(h,\mathrm{d}h,a,b)\,\frac{1}{\overline{T}(h)}\,p(x;H_s=h)\,\mathrm{d}x =$$

= number of waves with a height between x and $x+\mathrm{d}x$, in the stages of the storms in which H_s is between h and $h+\mathrm{d}h$, only for the storms whose $H_{s\max}$ is between a and $a+\mathrm{d}a$ and whose duration is between b and $b+\mathrm{d}b$, during $\mathscr{T}$;

$$
\begin{gathered}
[p_A(a)\,\mathrm{d}a\,\mathcal{N}(\mathcal{T})\,p_B(b|a)\,\mathrm{d}b]\,\delta t(h,\mathrm{d}h,a,b)\,\frac{1}{\overline{T}(h)}\,p(x;H_s=h)\,\mathrm{d}x\cdot\\
\cdot\{\mathcal{P}(H_{\max}<x;a,b)/[1-P(x;H_s=h)]\} = \text{number of waves which}\\
\text{are the maximum of their storm, with a height between } x \text{ and } x+\mathrm{d}x,\\
\text{in the stages of the storms is which } H_s \text{ is between } h \text{ and } h+\mathrm{d}h,\\
\text{only for the storms whose } H_{s\max} \text{ is between } a \text{ and } a+\mathrm{d}a\\
\text{and whose duration is between } b \text{ and } b+\mathrm{d}b, \text{ during } \mathcal{T}.
\end{gathered}
\tag{7.33}
$$

The term {.} on the l.h.s. of (7.33) is the probability that all other wave heights of the storm (apart from the height between x and $x+\mathrm{d}x$ in the sea state with H_s between h and $h+\mathrm{d}h$) are smaller than x. The proof is left as an exercise. [Hint: keep in mind that $\mathcal{P}(H_{\max}<x;a,b)$ is the probability that all waves of the storm have a height smaller than x, and recall how we got this probability in sect. 6.5.]

$\mathcal{N}(H,\mathcal{T};\max)$ proceeds by integrating the l.h.s. of (7.33) over

$$
\{(x,h,a,b)|x\in(H,\infty),h\in(0,\infty),a\in(h,\infty)\ \text{ and }\ b\in(0,\infty)\}.
$$

Using the formula (7.5) for $\delta t(h,\mathrm{d}h,a,b)$, we obtain

$$
\begin{aligned}
\mathcal{N}(H,\mathcal{T};\max) = \mathcal{N}(\mathcal{T})\int_H^\infty\int_0^\infty\int_h^\infty\int_0^\infty p_A(a)p_B(b|a)\,\frac{b}{a}\,\frac{1}{\overline{T}(h)}\,p(x;H_s=h)\,\cdot\\
\cdot\{\mathcal{P}(H_{\max}<x;a,b)/[1-P(x;H_s=h)]\}\,\mathrm{d}b\,\mathrm{d}a\,\mathrm{d}h\,\mathrm{d}x.
\end{aligned}
$$

Then, using the formula (7.10) for $p_A(a)$ and the equation

$$
\mathcal{P}(H_{\max}<x;a,b) = \exp\left\{\frac{b}{a}\int_0^a\frac{1}{\overline{T}(h')}\ln[1-P(x;H_s=h')]\,\mathrm{d}h'\right\},
$$

we rewrite $\mathcal{N}(H,\mathcal{T};\max)$ in the form

$$
\begin{aligned}
\mathcal{N}(H,\mathcal{T};\max) = \mathcal{T}\int_H^\infty\int_0^\infty\frac{1}{\overline{T}(h)}\,p(x;H_s=h)\,[1-P(x;H_s=h)]^{-1}\,\cdot\\
\cdot\int_h^\infty -\frac{\mathrm{d}p(H_s=a)}{\mathrm{d}a}\,\frac{1}{\overline{b}(a)}\int_0^\infty p_B(b|a)\,b\exp[K(a,x)b]\,\mathrm{d}b\,\mathrm{d}a\,\mathrm{d}h\,\mathrm{d}x,
\end{aligned}
$$

where

$$
K(a,x) \equiv \frac{1}{a}\int_0^a\frac{1}{\overline{T}(h')}\ln[1-P(x;H_s=h')]\,\mathrm{d}h'.
$$

[As for the equation of $\mathcal{P}(H_{\max}<x;a,b)$, we have given it without a comment, since it proceeds immediately from (6.26a).]

From the expression of the number $\mathcal{N}(H, \mathcal{T}; \max)$ and equation (7.31) relating this number to $R(H)$, we get

$$R(H) = \left\{ \int_H^\infty \int_0^\infty \frac{1}{\overline{T}(h)} p(x; H_s = h) \int_h^\infty - \frac{\mathrm{d}p(H_s = a)}{\mathrm{d}a} \frac{1}{\overline{b}(a)} \cdot \right. \tag{7.34}$$
$$\left. \cdot \int_0^\infty p_B(b|a) b \exp[K(a,x) b] \,\mathrm{d}b \,\mathrm{d}a \,\mathrm{d}h \,\mathrm{d}x \right\}^{-1},$$

where the term $[1 - P(x; H_s = h)]^{-1}$ has been neglected because it has little effect on the result, as one can easily verify by numerical evaluation of the integrals.

To proceed, we should know the form of $p_B(b|a)$. However, rather than looking for some formula that fits the data of the equivalent triangular storms it seems more effective to assume two very different forms for $p_B(b|a)$, obtain $R(H)$ for each of these two forms, and see how much the two solutions for $R(H)$ differ from each other. If the difference will be a significant one, of course the data analysis of $p_B(b|a)$ will become necessary, otherwise this analysis will be useless.

As two extreme forms of $p_B(b|a)$ we assume

$$① \quad p_B(b|a) = \delta\left[b - \overline{b}(a)\right],$$

$$② \quad p_B(b|a) = \frac{1}{\overline{b}(a)} \exp\left[-\frac{b}{\overline{b}(a)}\right].$$

The first one (①) is the delta function: all storms of given $H_{s\max} = a$ have the same duration b that coincides with the mean value $\overline{b}(a)$. The second one (②) is the exponential function: the durations of the storms of given $H_{s\max} = a$ have a very large scatter with respect to their mean value $\overline{b}(a)$.

The integral with respect to b in (7.34) proves to be

$$\int_0^\infty p_B(b|a) b \exp[K(a,x) b] \,\mathrm{d}b \begin{cases} = \overline{b}(a) \exp\left[K(a,x)\overline{b}(a)\right] & \text{with form ① of } p_B(b|a), \\ = \overline{b}(a)\left[1 - K(a,x)\overline{b}(a)\right]^{-2} & \text{with form ② of } p_B(b|a). \end{cases}$$

As a consequence we obtain the two following alternative formulae for $R(H)$

$$R(H) = \left\{ \int_H^\infty \int_0^\infty \frac{1}{\overline{T}(h)} p(x; H_s = h) \int_h^\infty - \frac{\mathrm{d}p(H_s = a)}{\mathrm{d}a} \cdot \right.$$
$$① \tag{7.35}$$
$$\left. \cdot \exp\left[\frac{\overline{b}(a)}{a} \int_0^a \frac{1}{\overline{T}(h')} \ln[1 - P(x; H_s = h')] \,\mathrm{d}h'\right] \mathrm{d}a \,\mathrm{d}h \,\mathrm{d}x \right\}^{-1},$$

② $$R(H)=\left\{\int_H^\infty\int_0^\infty \frac{1}{\overline{T}(h)}\,p(x;H_s=h)\int_h^\infty -\frac{\mathrm{d}p(H_s=a)}{\mathrm{d}a}\cdot\right.$$

$$\left.\cdot\left[1-\frac{\overline{b}(a)}{a}\int_0^a\frac{1}{\overline{T}(h')}\ln[1-P(x;H_s=h')]\,\mathrm{d}h'\right]^{-2}\mathrm{d}a\,\mathrm{d}h\,\mathrm{d}x\right\}^{-1},$$

where the integrals can be solved numerically with the criteria suggested in the following sect. 7.7.3.

Table 7.4 compares a few values of $H(R)$ for Ponza ($w=0.874\,\mathrm{m}, u=1.200$) obtained with the two alternative formulae for $R(H)$. We see that the differences due to the use of the two different formulae are very small especially in the range of practical interest for the structure design. This comparison has been done with the same outcome also for other locations of the Mediterranean Sea and of the Oceans. The conclusion is that the *shape* of the distribution of the durations of the storms has only a very small effect on $H(R)$, and thus we can use formula ① or formula ② for $R(H)$ to our liking. We shall use ①, that is to say (7.35) because it is slightly more conservative, indeed it gives a somewhat greater wave height for a given return period.

Table 7.4 *Results obtained from the use of two alternative formulae for R(H)*

FORMULA ① FOR $R(H)$		FORMULA ② FOR $R(H)$	
R [years]	H [m]	R [years]	H [m]
1	8.43	1	8.26
10	10.98	10	10.82
100	13.39	100	13.25
1000	15.72	1000	15.61

7.7.3 *Hints for the calculation of* $R(H)$

The calculation of $R(H)$ by means of (7.35) is much simpler than it may seem. First of all, since the integrand is a regular function, we can use some integration steps greater than we would probably choose: steps Δx, Δh, Δa and $\Delta h'$ of 0.5 m usually prove to be adequate. Moreover all contributions to the integral, apart from a negligible share, come from a small region of the 4-dimensional interval of integration, so that we can replace

$$\int_H^\infty\int_0^\infty\int_h^\infty\int_0^a \ldots\,\mathrm{d}h'\,\mathrm{d}a\,\mathrm{d}h\,\mathrm{d}x,$$

with

$$\int_{H}^{2H}\int_{0.3x}^{0.8x}\int_{h}^{1.5h}\int_{0.6a}^{a} \ldots \mathrm{d}h'\,\mathrm{d}a\,\mathrm{d}h\,\mathrm{d}x\,.$$

Here for simplicity we have treated that of formula (7.35) as a 4-fold integral. Really it is a 3-fold integral with a one-dimensional integral in the argument of the exponential function, but the substance of numerical work is the same.

To fix our ideas, let us think of a location of the Mediterranean Sea, let us take $H = 10$ m, and let us use integration steps Δx, Δh, Δa and $\Delta h'$ of 0.5 m. Then, the first integral is the sum of the number of the highest waves in their respective storms, which have a height x between 10m and 10.5 m, between 10.5 m and 11 m, and so on (recall that the integral multiplied by $\mathscr{T}$ gives the number of waves which are the maximum of their storms and are higher than H, during $\mathscr{T}$). We should go on as far as infinity, but it is intuitive that, in the set of waves higher than 10 m, the amount of those higher than 20 m will be negligible. This is why the upper limit of the first integral passes from ∞ to $2H$ with practically no error.

Let us pass to the second integral, the one with respect to h, and to fix our ideas, let us take $x = 18$ m. This integral is the sum of the number of waves [which are the highest of their storms and have a height of 18 m] which belong to sea states with a significant height between 0 and 0.5 m, between 0.5 m and 1m and so on as far as infinity. Here too it is apparent that the number of these waves occurring in the sea states with a significant wave height between 0 and 0.5 m will be absolutely negligible because the probability of a wave height of 18 m in a sea state with $H_s < 0.5$ m is zero as a matter of fact. Moreover the number of these waves in sea states with a significant wave height between 17.5 m and 18 m is also zero. Indeed, such heavy sea states have a negligible probability of occurring, and furthermore the probability that a wave of height 18 m, in a sea state with a given significant wave height of 18 m, is the highest wave of its storm is practically zero. Hence we realize that the whole contribution to the second integral will come from an interval whose lower limit is greater than zero and whose upper limit is smaller than x. The interval $(0.3x, 0.8x)$ proves to be wide enough.

As to the third integral, to fix our ideas, let us take $x = 18$ m, $h = 11$ m. This integral is the sum of the number of waves [which are the highest of their storms, have a height of 18 m, and belong to a sea state with an H_s of 11 m] which occur in storms whose H_{smax} is between 11 m and 11.5 m, between 11.5 m and 12 m and so on. Thus, fixing an upper limit of integration of $1.5h = 16.5$ m is as if, being informed that the sea storm includes a sea state of $H_s = 11$ m, we said: the probability that H_{smax} has exceeded 16.5 m is negligible. Of course, this suggestion (changing the upper limit of the third integral from ∞ to $1.5h$) is the result of a number of calculations for different areas.

The fourth integral remains, the one in the argument of the exponential function. This exponential function gives the probability $\mathscr{P}(H_{\max} < x)$ for a triangular storm of duration $\bar{b}(a)$ and *maximum* significant wave height a. To fix our ideas let us take $x = 18$ m, $a = 12$ m. Then the integral (multiplied by $\bar{b}(a)/a$) is the sum: [logarithm of the probability that all waves in the stage of the storm where $0 < H_s < 0.5$ m are smaller than 18 m] + [logarithm of the probability that all waves in the stage of the storm where 0.5 m $< H_s <$ 1.0 m are smaller than 18 m] + ... + [logarithm of the probability that all waves in the stage of the storm where 11.5 m $< H_s <$ 12 m are smaller than 18 m]. Clearly, the probability that all waves in the stage of the storm with H_s between 0 and 0.5 m are smaller than 18 m is practically equal to 1, and consequently the logarithm of this probability is zero. Thus we realize that the lower limit of integration can be taken greater than zero without any appreciable consequence on the result. Experience suggests this lower limit can be raised to $0.6a$.

Finally, note that the solution for $R(H)$ is exact provided $p(H_s = h)$ is the p.d.f. of the significant wave height of the equivalent sea. However, we can very well use the actual $p(H_s = h)$, provided H is greater than about $3h_{\text{crit}}$ (cf. sect. 6.1.2). Here, bear in mind that $p(H_s = h)$ of the equivalent sea tends to coincide with the actual $p(H_s = h)$ as h increases (cf. sect. 6.6.3).

7.7.4 *Comparison between $R(H)$ and $R^*(H)$*

The function $R^*(H)$ of fig. 7.10 has been plotted again in fig. 7.11 where it is compared to $R(H)$. The third line gives

$\overline{N}(H) \equiv$ average number of waves higher than a fixed threshold H, in the storms where at least one wave higher than this threshold does occur.

To obtain $\overline{N}(H)$ it suffices to write the two following equalities:

$$\mathcal{T}/R^*(H) = \text{number of waves higher than } H, \text{ in } \mathcal{T},$$

$$\mathcal{T}/R(H) = \text{number of storms containing at least one wave higher than } H, \text{ in } \mathcal{T}.$$

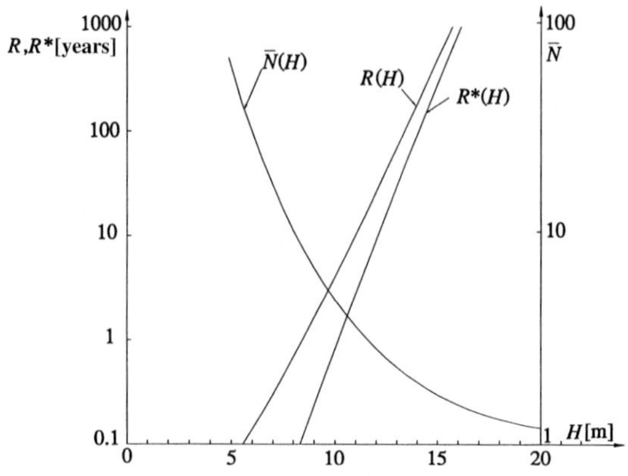

Fig. 7.11 Ponza: return periods $R(H)$ and $R^*(H)$. $\overline{N}(H) = R(H)/R^*(H)$ is the average number of waves exceeding the fixed threshold H in the storms containing at least one of these waves. (Obtained by means of 7.30 and 7.35.)

Indeed $\overline{N}(H)$ is the quotient between these two numbers:

$$\overline{N}(H) = \frac{R(H)}{R^*(H)}. \tag{7.36}$$

At this stage we can definitively understand that strange result of sect. 7.6.2. We have seen that

$$R^*(5\,\text{m}) = 9\,\text{hours};$$

now we can add that

$$R(5\,\text{m}) = 25\,\text{days},$$

and hence

$$\overline{N}(5\,\text{m}) = \frac{25 \cdot 24}{9} = 66.$$

Conclusion: at Ponza, on average, once in 25 days, a storm occurs with wave heights exceeding 5 m, and each of these storms contains, on average, 66 waves higher than 5 m. This means that, on average, we have to wait 25 days to see waves higher than 5 m, but the average time interval between two occurrences of a wave higher than 5 m is only 9 hours, because each time waves higher than 5 m occur, there is an average of 66 of these together.

7.8 Offshore structures: the design wave

7.8.1 *The maximum wave height in the lifetime*

We have to verify that offshore structures are able to withstand the reasonably highest wave in their lifetime. Following the design rules, we have to take the wave height that has a given probability $\mathscr{P}$ of being exceeded in the lifetime L.

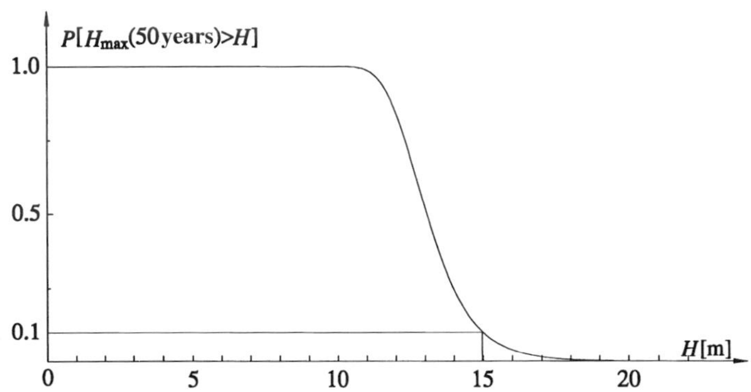

Fig. 7.12 Ponza: probability that the largest wave height in fifty years exceeds any fixed threshold H. (Obtained by means of 7.37.)

The probability that the maximum wave height in the lifetime L exceeds a fixed threshold H is equal to the encounter probability of a storm whose maximum wave height exceeds H:

$$P[H_{\max}(L) > H] = \mathscr{P}[L, R(H)].$$

Indeed, if the maximum wave height in the lifetime L exceeds H, then at least one storm whose maximum wave height exceeds H must occur in this lifetime; and, conversely, if at least one storm whose maximum wave height exceeds H occurs in the lifetime L, then the maximum wave height in this lifetime exceeds H.

$R(H)$ is given by (7.35) and $\mathscr{P}(L, R)$ by (7.29*a*), so that

$$P[H_{\max}(L) > H] = 1 - \exp\left\{-L \int_H^\infty \int_0^\infty \frac{1}{\overline{T}(h)} p(x; H_s = h) \int_h^\infty -\frac{\mathrm{d}p(H_s = a)}{\mathrm{d}a} \cdot \right. \tag{7.37}$$

$$\left. \cdot \exp\left[\frac{\overline{b}(a)}{a} \int_0^a \frac{1}{\overline{T}(h')} \ln[1 - P(x; H_s = h')] \mathrm{d}h'\right] \mathrm{d}a \, \mathrm{d}h \, \mathrm{d}x \right\}.$$

As an example let us consider again the input data

$$L = 50 \, \text{years}, \quad \mathscr{P} = 0.1,$$

for Ponza ($w = 0.874$ m, $u = 1.200$). The $P[H_{\max}(50 \, \text{years}) > H]$ is shown in fig. 7.12. We see this probability is equal to 0.1 (the prescribed value) for $H = 15$ m that, accordingly, is the design wave height, i.e. it is the desired

$$H(L) \equiv \text{maximum wave height in the lifetime of a structure.}$$

Physical meaning: there is a 10% probability that the largest wave height in 50 years at Ponza exceeds 15.0 m.

7.8.2 *The sea state where the highest wave in the lifetime of a structure will occur*

Knowledge of the wave height is not enough for evaluating the wave force. We also need to know the wave period, and hence we must try to foresee the characteristics of the sea state in which the highest wave in the lifetime of the structure will occur. For this goal we reason as follows.

If we retrace the steps that led us to the formula (7.35) for $R(H)$, we realize that

$$\mathscr{T} \frac{1}{\overline{T}(h)} p(x; H_s = h) \int_h^\infty -\frac{\mathrm{d}p(H_s = a)}{\mathrm{d}a} \exp\left[\frac{\overline{b}(a)}{a} \int_0^a \frac{1}{\overline{T}(h')} \cdot \right.$$

$$\left. \cdot \ln[1 - P(x; H_s = h')] \mathrm{d}h'\right] \mathrm{d}a \, \mathrm{d}h \, \mathrm{d}x =$$

$$= \text{number of waves which are the highest of their storms, having a height between } x \text{ and } x + \mathrm{d}x \text{ and belonging to the stages of the storms in which the significant wave height is between } h \text{ and } h + \mathrm{d}h \text{, during } \mathscr{T}, \tag{7.38}$$

and

$$\mathscr{T} \int_0^\infty \frac{1}{\overline{T}(h)} \, p(x; H_s = h) \int_h^\infty - \frac{\mathrm{d}p(H_s = a)}{\mathrm{d}a} \cdot \exp\left[\frac{\overline{b}(a)}{a} \int_0^a \frac{1}{\overline{T}(h')} \ln[1 - P(x; H_s = h')] \, \mathrm{d}h' \right] \mathrm{d}a \, \mathrm{d}h \, \mathrm{d}x =$$

$$= \text{number of waves which are the highest of their storms, having a height between } x \text{ and } x + \mathrm{d}x \text{, during } \mathscr{T}. \tag{7.39}$$

Hence, if we define

$p(H_s = h; H_{\max} = x)\,\mathrm{d}h \equiv$ probability that a wave of given height x, which is the highest wave of its storm, occurs in a sea state with a significant wave height between h and $h + \mathrm{d}h$,

we have

$$p(H_s = h; H_{\max} = x)\,\mathrm{d}h = \text{l.h.s. of (7.38)/l.h.s. of (7.39)},$$

that is

$$p(H_s = h; H_{\max} = x) =$$

$$= \frac{\dfrac{1}{\overline{T}(h)} \, p(x; H_s = h) \displaystyle\int_h^\infty - \frac{\mathrm{d}p(H_s = a)}{\mathrm{d}a} \exp\left[\frac{\overline{b}(a)}{a} \int_0^a \frac{1}{\overline{T}(h')} \ln[1 - P(x; H_s = h')] \, \mathrm{d}h' \right] \mathrm{d}a}{\displaystyle\int_0^\infty \frac{1}{\overline{T}(h)} \, p(x; H_s = h) \int_h^\infty - \frac{\mathrm{d}p(H_s = a)}{\mathrm{d}a} \exp\left[\frac{\overline{b}(a)}{a} \int_0^a \frac{1}{\overline{T}(h')} \ln[1 - P(x; H_s = h')] \, \mathrm{d}h' \right] \mathrm{d}a \, \mathrm{d}h}. \tag{7.40}$$

Let us come back to our worked example. We have a highest wave of 15 m in the lifetime of 50 years. Of course, if this wave is the highest in 50 years, it is also the highest of its storm. As a consequence, the probability density function of the sea state where this wave will occur is given by (7.40) with $x = 15$ m. This probability density function is plotted in fig. 7.13 and proves to be narrow with the peak at $h = 7.2$ m. This value will be denoted by the symbol $H_s(L)$:

$H_s(L) \equiv$ significant wave height of the sea state in which the highest wave in the lifetime of the structure will occur with the highest probability.

In the case under examination, the maximum wave height of 15 m, with the highest probability, will occur in a sea state with an H_s of 7.2 m. Thus the highest wave in the 50 years lifetime will be not only a very high wave for the given location, but also an exceptionally high wave for the very sea state where it will occur.

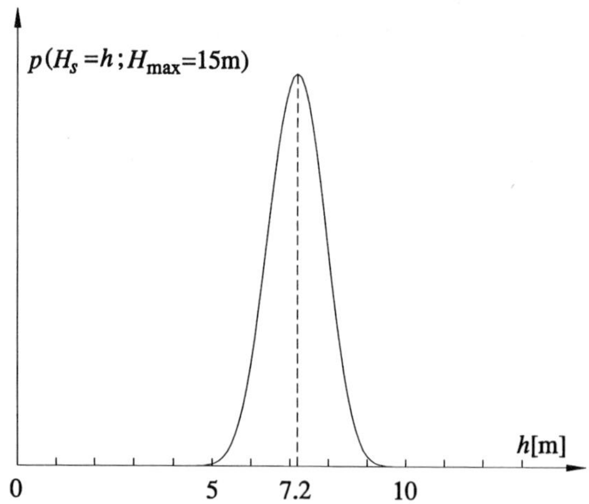

Fig. 7.13 Ponza: p.d.f. of the H_s of the sea state where the maximum wave height in fifty years will occur. (Obtained by means of 7.40 for the maximum wave height of 15 m.)

Very well! Indeed, if the wave is exceptionally high for its own sea state, we can use the quasi-determinism theory, and hence we can foresee that it will have a very precise period. Specifically, we saw in sect. 5.7.2 that a very high wave has a period of 0.92 T_p, if the spectrum is the mean JONSWAP. As to T_p we can use the formula (4.26), or lacking information on Phillips' parameter A, we can use the formula (4.27). Therefore in our case we have

$$T_p = 8.5\,\pi\sqrt{\frac{7.2}{4\cdot 9.8}} = 11.4\,\mathrm{s},$$

and consequently the period of the highest wave, which we will call $T(L)$, is

$$T(L) = 0.92\cdot 11.4 = 10.5\,\mathrm{s}.$$

The conclusion is:

$$\left.\begin{array}{l}\text{Ponza: } w = 0.874\,\mathrm{m},\ \ u = 1.200\\[2ex] \text{design specifications: } L = 50\,\text{years},\ \ \mathscr{P} = 0.1\end{array}\right. \Rightarrow \ \text{design wave}\ \begin{cases}\text{height} = 15.0\,\mathrm{m},\\ \text{period} = 10.5\,\mathrm{s}.\end{cases}$$

Let us note, at the end, that the advantages of being able to use the quasi-determinism theory are not confined to the wave period estimate. Indeed we shall see in chapter 10 that a wave which is exceptionally high with respect to its own sea state, with a very high probability, is the central one of a well defined three-dimensional group at the apex of its development stage. The quasi-determinism theory provides the velocity potential of this group, from which one can draw the flow field and hence the wave forces on structures. Naturally, at a first degree of approximation, the flow field and wave forces can be evaluated even by means of the periodic wave theory.

7.9 Calculations for different wave directions

7.9.1 *Calculation criteria*

There are some offshore structures which put up practically the same resistance to waves whatever the wave direction. Then, there are a few offshore structures and most coastal structures, which put up a different resistance to differently inclined waves. For this second family of structures we have to calculate $h(R;\, \theta_1 < \theta < \theta_2)$ and $H(L; \theta_1 < \theta < \theta_2)$, where the symbols have an apparent meaning: only the waves whose direction is within a given sector are considered.

In the various areas there is usually a main sector for which the probability of exceedance $P(H_s > h; \theta_1 < \theta < \theta_2)$ proves to be nearly equal to $P(H_s > h)$, at least for large values of h. That is to say

$$\text{main sector: } P(H_s > h; \theta_1 < \theta < \theta_2) \cong P(H_s > h). \tag{7.41}$$

It suffices only to read table 6.2 to see that at each location there is a sector whose w_α is very close, sometimes even equal, to w. Recall that $P(H_s > h; \theta_1 < \theta < \theta_2)$ approaches the asymptotic form (6.11) as h grows, and this asymptotic form coincides with $P(H_s > h)$ if $w_\alpha = w$.

As a consequence of (7.41), we have

$$\text{main sector} \begin{cases} h(R; \theta_1 < \theta < \theta_2) \cong h(R), \\ H(L; \theta_1 < \theta < \theta_2) \cong H(L), \end{cases}$$

whereas

$$\text{other sectors} \begin{cases} h(R; \theta_1 < \theta < \theta_2) < h(R), \\ H(L; \theta_1 < \theta < \theta_2) < H(L). \end{cases}$$

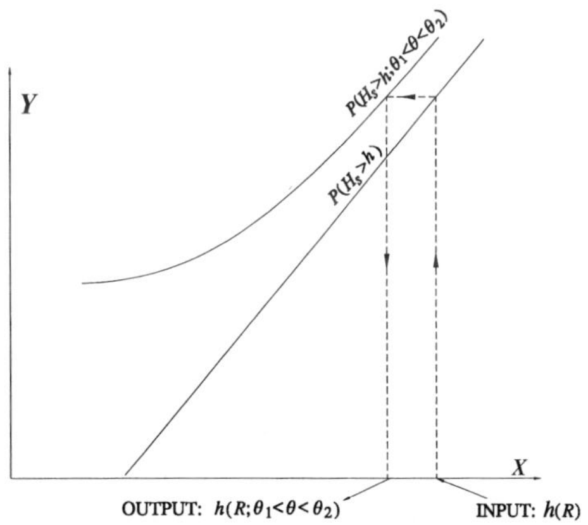

Fig. 7.14 Enter with the significant wave height of given return period, follow the marked path and obtain the significant wave height for the specific direction of wave advance you are interested in.

A practical way to evaluate the wave height for a given sector of the wave direction is the following. We first compute the wave height regardless of the wave direction, that is $h(R)$ and/or $H(L)$ and $H_s(L)$. Then, from $h(R)$ we get $h(R;\theta_1<\theta<\theta_2)$ following the flow of fig. 7.14. Similarly, from $H_s(L)$ we get $H_s(L;\theta_1<\theta<\theta_2)$. Finally, as to $H(L;\theta_1<\theta<\theta_2)$ we assume

$$\frac{H(L;\theta_1<\theta<\theta_2)}{H_s(L;\theta_1<\theta<\theta_2)}=\frac{H(L)}{H_s(L)}. \tag{7.42}$$

In this manner, the duration of the design sea state for a given sector of the wave direction proves to be the same as the duration of the design sea state regardless of the wave direction. The proof of this statement is left as an exercise. [Hint: note that equation (7.18) has a general validity, and take account of the flow of fig. 7.14.]

7.9.2 *Worked example*

For Ponza ($w=0.874$ m, $u=1.200$), with the design specifications: $L=50$ years, $\mathscr{P}=0.1$, which imply $R=475$ years, we have obtained

$$h(R)=7.90 \text{ m}, \ \mathscr{D}=6.6 \text{ hours},$$

$$H(L)=15.0 \text{ m}, \ H_s(L)=7.20 \text{ m}, \ T(L)=10.5 \text{ s}.$$

Here we complete the picture with the wave estimate for the sectors

$$\text{EAST:} \ 90^\circ-11.25^\circ<\theta<90^\circ+11.25^\circ,$$

$$\text{NORTH:} \ -11.25^\circ<\theta<+11.25^\circ.$$

(We mean the waves which advance eastwards and the waves which advance northwards.) In table 6.2 we read

$$\text{EAST:} \ w_\alpha=0.865 \text{ m}, \quad w_\beta=0.754 \text{ m},$$

$$\text{NORTH:} \ w_\alpha=0.639 \text{ m}, \quad w_\beta=0.609 \text{ m}.$$

The east sector is the main one, indeed it has a w_α of 0.865 m which is very close to the w of 0.874 m of the probability $P(H_s>h)$. As already stated, for the main sector we use the same wave heights obtained regardless of the wave direction. In doing so, we overestimate the wave heights of the main sector slightly; in this case, the rise being about 1%.

The calculation must be done for the north sector whose w_α is markedly smaller than w.

The probability of exceedance for this sector, which is

$$P(H_s > h; \text{NORTH}) = \exp\left[-\left(\frac{h}{0.639}\right)^{1.200}\right] - \exp\left[-\left(\frac{h}{0.609}\right)^{1.200}\right] \text{ with } h \text{ in metres,}$$

is shown in fig. 7.15 together with the probability of exceedance for all wave directions:

$$P(H_s > h); = \exp\left[-\left(\frac{h}{0.874}\right)^{1.200}\right] \text{ with } h \text{ in metres.}$$

[As usual the graphs of the probabilities have been given in coordinates (X,Y) defined by (6.3).]

We enter with the value of $h(R)$ obtained regardless of the wave direction, that is $h(R) = 7.90\,\text{m}$. For the relation (6.3) between X and h, we have

$$h(R) = 7.90\,\text{m} \Rightarrow X = 298.3\,.$$

In the probability of exceedance for all wave directions, $X = 298.3$ is associated with $Y = 264.2$ and this value of Y is associated with $X = 263.4$ in the probability of exceedance for the north sector. Hence, using once again the relation (6.3) between X and h, we obtain

$$X = 263.4 \Rightarrow h(R; \text{NORTH}) = 5.57\,\text{m}\,.$$

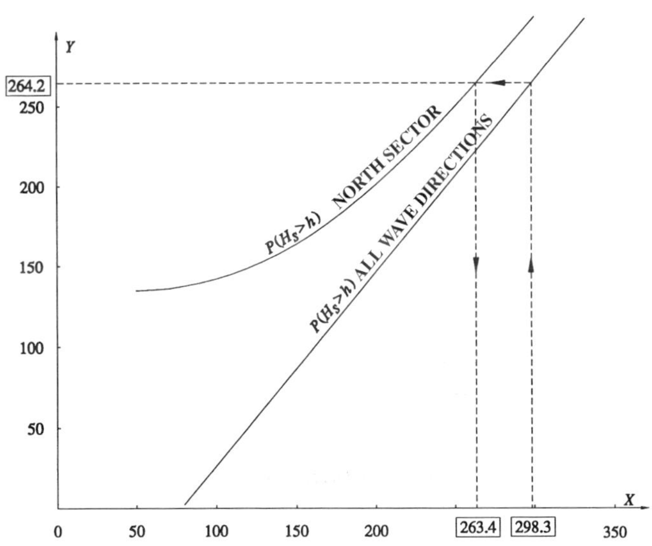

Fig. 7.15 The worked example of sect. 7.9.2.

The same procedure must be followed to get $H_s(L;\text{NORTH})$ from $H_s(L)$, the result being $H_s(L;\text{NORTH}) = 5.03$ m. Then the periods T_p and T_h of a sea state with an H_s of 5.03 m can be evaluated by means of (4.27) and (5.42), and prove to be $T_p = 9.6\,\text{s}$, $T_h = 8.8\,\text{s}$. Finally, the maximum height of a wave which advances northwards in the lifetime of the structure is obtained by means of (7.42) and proves to be 10.5 m. Thus, the picture is

$$h(R;\text{NORTH}) = 5.57\,\text{m}, \qquad \mathscr{D} = 6.6\,\text{hours},$$
$$H(L;\text{NORTH}) = 10.5\,\text{m}, \quad H_s(L;\text{NORTH}) = 5.03\,\text{m}, \quad T(L;\text{NORTH}) = 8.8\,\text{s}.$$

In other words, the design sea state whose dominant wave direction is within the north sector has a significant wave height of 5.57 m and a duration of 6.6 hours. Moreover the design wave for this sector has a height of 10.5 m, a period of 8.8 s and the maximum probability is that it occurs in a sea state with a significant wave height of nearly 5 m. These results are used when dealing with the design of structures whose resistance depends on the wave direction (for example resistance to the waves which advance northwards being different from the resistance to the waves which advance eastwards). Naturally, as was said in the foregoing sections, the design sea state will be used for coastal structures, while the design wave will be used for offshore structures.

7.10 Corollary of risk analysis: a general relation between the confidence interval and the sampling rate

7.10.1 *The terms of the problem*

In sect. 6.2.4 we have shown that the classic confidence interval (based on the assumption of stochastic independence of the outcomes of a random variable) serves only as a first approximation to the actual confidence interval of $P(H_s > h)$. This is because the outcomes of H_s obtained typically with a sampling rate of 1/3 per hour or of 1 per hour are not stochastically independent of one another. In sect. 6.2.4 we have also inferred that the actual confidence interval should be wider and should be located leftward with respect to the classic confidence interval. How much wider, and how much leftward? Now we are able to give some answers to these questions.

The starting point of the reasoning is as follows. Let us imagine we know the $P(H_s > h)$ of a location and we take N measurements of H_s with a sampling interval Δt_{samp} at this location. We aim to determine the probability that the largest measured H_s exceeds a fixed threshold h.

7.10.2 *A general inequality for the confidence interval*

Definitions

OI $\equiv$ observation interval, the length of which is $N \cdot \Delta t_{\text{samp}}$;

$P(H_{s\max} > h;\, N,\, \Delta t_{\text{samp}}) \equiv$ probability that the largest measured H_s in an OI exceeds h.

In addition to these definitions, we shall use the definition of $P(H_s > h)$ given in sect. 6.1.3 and the definition of $\mathscr{P}(L, R)$ given in sect. 7.3.1.

First inequality for the probability of exceedance of the largest measured H_s

The general inequality

$$P(H_{s\max} > h;\, N,\, \Delta t_{\text{samp}}) \le \mathscr{P}\,[N \cdot \Delta t_{\text{samp}},\, R(H_s > h)] \tag{7.43}$$

proceeds from the following reasoning. The probability that the largest measured H_s in an OI exceeds h [l.h.s. of (7.43)] is smaller than (or equal to) the probability that H_s exceeds h during this OI, which in its turn is equal to the probability that at least one sea storm with H_s exceeding h occurs during this OI [r.h.s. of (7.43)].

Inequality (7.43) and inequality (7.27) of the encounter probability yield

$$P(H_{s\max} > h;\, N,\, \Delta t_{\text{samp}}) \le \frac{N \cdot \Delta t_{\text{samp}}}{R(H_s > h)}. \tag{7.44}$$

Second inequality for the probability of exceedance of the largest measured H_s

Let us consider a very large number $\mathscr{N}$ of realizations of the stationary random process $H_s(t)$. Let us fix an OI, and let us consider the ith reading time in this OI. The number of realizations in which H_s on this reading time is greater than the fixed threshold h will be equal to $P(H_s > h)\mathscr{N}$. The same is true of the N reading times belonging to the OI. Therefore, the total number of reading times where H_s exceeds h will be equal to $P(H_s > h)\mathscr{N} \cdot N$.

The number of realizations in which the largest measured H_s in the OI exceeds h is smaller than or equal to the total number of reading times where H_s exceeds h. Hence, the following inequality:

$$P(H_{s\max} > h;\, N,\, \Delta t_{\text{samp}}) \le \frac{P(H_s > h)\mathscr{N} \cdot N}{\mathscr{N}} = P(H_s > h)N$$

which taken together with (7.44) yields

$$P(H_{s\max} > h;\, N,\, \Delta t_{\text{samp}}) \le \text{Min}\left\{\frac{N \cdot \Delta t_{\text{samp}}}{R(H_s > h)},\, P(H_s > h)N\right\}, \tag{7.45}$$

where Min (a, b) stands for "the smaller of a or b".

Consequence: inequality for the limits of the confidence interval

Let h_p denote the threshold that has a given probability $1 - p$ to be exceeded by the largest measured H_s in the OI (we have already used this symbol in sect. 6.2.4 when we dealt with the classic confidence interval). The knowledge of h_p enables us to obtain the confidence interval. As an example the 90% confidence interval has $h_{0.05}$ as lower limit and $h_{0.95}$ as upper limit, as we saw in sect. 6.2.4.

In symbols, h_p is defined by

$$P(H_{s\max} > h_p;\, N,\, \Delta t_{\text{samp}}) \equiv 1 - p\,.$$

From this definition and inequality (7.45), it follows that

$$h_p \le h_p^* \tag{7.46}$$

with h_p^* such that

$$\text{Min}\left[\frac{N \cdot \Delta t_{\text{samp}}}{R(H_s > h_p^*)},\; P(H_s > h_p^*)N\right] = 1 - p\,. \tag{7.47}$$

The l.h.s. of (7.47) is a monotonic decreasing function of h_p^*, so that the h_p^* satisfying (7.47) can be easily evaluated. Hence we have got an upper bound for h_p.

Let us retake the example of sect. 6.2.4. The locality was Mazara del Vallo in the Straits of Sicily. From tables 6.2 and 6.3 we obtain for this locality:

$$w = 1.021\text{ m},\; u = 1.275,\;\; a_{10} = 3.6\text{ m},\;\; b_{10} = 73\text{ hours},$$

$$\bar{b}(a) = 1.12 b_{10} \exp\left(-0.115\,\frac{a}{a_{10}}\right),$$

from which we can evaluate $P(H_s > h)$ by means of (6.2) and $R(H_s > h)$ by means of (7.12). The number N of sampled values of H_s was 14650 and Δt_{samp} was of 3 hours.

From (7.47) we obtain

$$h_{0.05}^* = 5.75\text{ m},\;\; h_{0.95}^* = 7.32\text{ m},\; (\Delta t_{\text{samp}} = 3\text{ hours}).$$

Hence the values

$$h_{0.05} = 5.47\text{ m},\;\; h_{0.95} = 7.43\text{ m},\; \text{(stochastic independence)}, \tag{7.48}$$

which were obtained in sect. 6.2.4 under the classic assumption of *stochastic independence* of the outcomes of H_s, nearly satisfy the general condition (7.46) ($h_{0.05}$ satisfies this condition, and $h_{0.95}$ is not far from satisfying it).

Now, let us reduce Δt_{samp} from 3 hours to 1 hour, under the same number N of sampled values of H_s and for the same locality. From (7.47) we obtain

$$h_{0.05}^* = 5.11\text{ m},\;\; h_{0.95}^* = 6.75\text{ m},\; (\Delta t_{\text{samp}} = 1\text{ hour}),$$

and hence the limits (7.48) of the classic confidence interval do not satisfy the general condition (7.46). This means that, with a Δt_{samp} of 1 hour, we cannot use the classic confidence interval based on the assumption of stochastic independence of the outcomes of H_s. Indeed it is certain that the confidence interval will be located leftward with respect to the classic confidence interval.

To check (7.46-47) reason as follows. If h_p^* satisfies (7.47), then $P(H_{s\max} > h_p^*;\, N,\, \Delta t_{\text{samp}})$ is smaller than (or equal to) $1-p$ because of (7.45). Hence, the h_p making $P(H_{s\max} > h_p; N, \Delta t_{\text{samp}})$ equal to $1-p$ must be smaller than (or equal to) h_p^*.

7.10.3 *Effectiveness of the general inequality for the confidence interval*

It can be proved that the upper limit of the confidence interval is nearly coincident with the h_p^* given by (7.47). On the contrary, the lower limit is strictly smaller than the h_p^* given by (7.47). In other words, (7.46) can be taken as an equality for the upper limit and as an inequality for the lower limit of the confidence interval:

$$\text{upper limit of the confidence interval: } h_p \cong h_p^*,$$

$$\text{lower limit of the confidence interval: } h_p < h_p^*.$$

In this regard, the reader could verify that, for a small $(1-p)$ (upper limit of the confidence interval) h_p^* tends to coincide with h_p both for small Δt_{samp} for which (7.47) is reduced to

$$\frac{N \cdot \Delta t_{\text{samp}}}{R(H_s > h_p^*)} = 1 - p, \tag{7.49}$$

and for large Δt_{samp} for which (7.47) is reduced to

$$P(H_s > h_p^*)N = 1 - p, \tag{7.50}$$

Here below, an outline of the proof.

For small Δt_{samp}, the l.h.s. of (7.43) is nearly equal to the r.h.s. Hence a small value of $1-p$ implies that the encounter probability is also small which in turn implies that the r.h.s. of (7.43) is nearly equal to the r.h.s. of (7.44). Indeed the encounter probability $\mathscr{P}(L,R)$ proves to be nearly coincident with L/R for a small $\mathscr{P}$ (which can be verified using the general form (7.26) of $\mathscr{P}(L,R)$ with a variety of distributions of the interarrival times). The conclusion is that (7.44) becomes an equality, and this equality and (7.49) imply that h_p^* coincides with h_p.

For large Δt_{samp}, the outcomes of H_s become stochastically independent of one another, and hence h_p is given by

$$1 - [1 - P(H_s > h_p)]^N = 1 - p, \tag{7.51}$$

and, for small values of $1-p$, the h_p given by (7.51) approaches the h_p^* given by (7.50). To prove this last statement, multiply and divide by $P(H_s > h_p)$ the exponent N on the l.h.s. of (7.51); bear in mind that $P(H_s > h_p)$ approaches 0 as $(1-p)$ approaches 0; use the limit

$$\lim_{x\to 0}(1-x)^{\frac{1}{x}} = e^{-1},$$

and the equality $e^{-x} \cong 1 - x$ for small $|x|$.

Conclusive note

The solution of sect. 7.3.2 for the encounter probability was introduced into the hydrological practice by Borgman (1963). The reasoning of sect. 7.3.3 leading to the general relation for the encounter probability was developed by the author (1983 and 1986*b*). The solution for $R^*(H)$ (sect. 7.6.1) is due to Jasper (1956). The solutions of sects. 7.1 and 7.7-8 for $R(H_s > h)$, $R(H)$, and corollaries $H(L)$ and $H_s(L)$ were obtained by the author (1986*a-b*). The solution of sect. 7.2.1 for the significant wave height with a given return period and the proof of sect. 7.10 yielding a general relation between the confidence interval and the sampling rate are novelties. Also the algorithm of sect. 7.9 is a new proposal.

In common practice (cf. Tucker, 1989, and Massel, 1996) the following formula is used:

$$R(H_s > h) = \Delta t_{\text{samp}}/P(H_s > h),$$

where Δt_{samp} is the sampling interval of H_s. If $P(H_s > h)$ is given by (6.2) it follows that

$$h(R) = w\left[\ln(R/\Delta t_{\text{samp}})\right]^{1/u}.$$

R is obtained by means of (7.29*b*) from lifetime L and encounter probability $\mathscr{P}$. The design sea state has $h(R)$ as significant wave height, and has a duration equal to Δt_{samp}. The design wave height is the maximum expected wave height in this sea state.

A limit of this approach lies in the subjective choice of Δt_{samp}. Here, note that $P(H_s > h)$ is independent of Δt_{samp} which only affects the confidence intervals, as we have seen in sect. 7.10. Therefore $R(H_s > h)$ proves to be proportional to the subjectively chosen Δt_{samp}.

Another limit of the common approach is that the highest wave is assumed to necessarily occur during the most severe sea state. In order to overcome this second limit, Tucker (1989) essentially suggested to resort to $R^*(H)$ [obtain the return period from L and $\mathscr{P}$, and compute the design wave height by means of (7.30) from this return period]. However, this approach proves to be conservative for the reason we have explained in sect. 7.6.2.

For the same aim (overcoming the second limit) Krogstad (1985) had assumed that the time interval in which H_s is between h and $h + \mathrm{d}h$ during the lifetime was given by the deterministic product $p(H_s = h)\mathrm{d}h\, L$, thus obtaining

$$P[H_{\max}(L) > H] = 1 - \exp\left\{L\int_0^\infty p(H_s = h)\frac{1}{\overline{T}(h)}\ln[1 - P(H; H_s = h)]\mathrm{d}h\right\}.$$

[The reasoning is essentially the same that leads to (6.26*a*) of the probability of exceedance of the maximum wave height in a storm of given history.] This approach becomes exact in the limit as $L \to \infty$, while for some finite L, it proves to be conservative: the more, the smaller L is. The reason is that the time duration in which H_s is between h and $h + \mathrm{d}h$ during the lifetime is a random variable, and $p(H_s = h)\mathrm{d}h L$ is only the mean value of this random variable.

The solutions given in this chapter remove the above mentioned limits, in that they are closed solutions under one assumption: the existence of the equivalent triangular storm with the main property described in sect. 6.6.2. This property, which can be easily checked, has been found to hold for all storm histories we have examined (both from the Mediterranean Sea and from the Oceans).

The concept of equivalent triangular storm yields also the formal relation, of general validity, between $\overline{D}(h)$ and $P(H_s > h)$. [In particular, if $P(H_s > h)$ is given by (6.2), $\overline{D}(h)$ takes on the form (7.19).] Until now, the statistics of the duration given intensity has been dealt with empirically. In particular Graham (1982) and Sobey and Orloff (1999) have published some detailed studies for several areas.

References

Boccotti P., 1983 On wave height probability. *Giornale del Genio Civile*, 165-174 (in Italian);
1986a Wave prediction for the Italian seas. *Proc. 20th Italian Conf. Hydraulics*, 79-91 (in Italian);
1986b On coastal and offshore structure risk analysis. *Excerpta of the Italian Contribution to the Field of Hydraulic Eng.* 1, 19-36.

Borgman L. E., 1963 Risk criteria. *J. Waterways and Harbors. Div. ASCE* 89, 1-35.

Graham C., 1982 The parameterisation and prediction of wave height and wind speed persistence statistics for oil industry operational planning purposes. *Coastal Eng.* 6, 303-329.

Jasper N. H., 1956 Statistical distribution patterns of ocean waves and wave-induced ship stresses and motions, with engineering applications. *Trans. Soc. Nav. Arch. & Mar. Eng.* 64, 375-432.

Krogstad H. E., 1985 Height and period distribution of extreme waves. *Appl. Ocean Res.* 7, 158-165.

Massel S. R., 1996 Ocean surface waves: their physics and prediction. *World Scientific*, 1-491.

Puertos del Estado (Madrid), 1990 Maritime Works Recommendations-ROM 0.2-90.

Sobey R. J. and Orloff L.S., 1999 Intensity-duration-frequency summaries for wave climate. *Coastal Eng.* 36, 37-58.

Tucker M. J., 1989 An improved "Battjes" method for predicting the probability of extreme waves. *Appl. Ocean Res.* 11, 212-218.

Chapter 8

ANALYSIS OF THE SEA STATES IN THE SPACE-TIME

8.1 The concept of homogeneous wave field

Starting from Sverdrup and Munk (1947) and Bretschneider (1952), a few basic ideas were developed about the wave generation by wind. Here we resort to these ideas to introduce the concept of *homogeneous wave field.*

First idea. Let us imagine that a wind starts blowing with a speed u on an initially calm basin, and that the wind speed and direction are reasonably constant. Let us fix a number of cross-sections along an axis parallel to the wind velocity (see fig. 8.1a). The significant wave height as a function of time at these cross-sections will

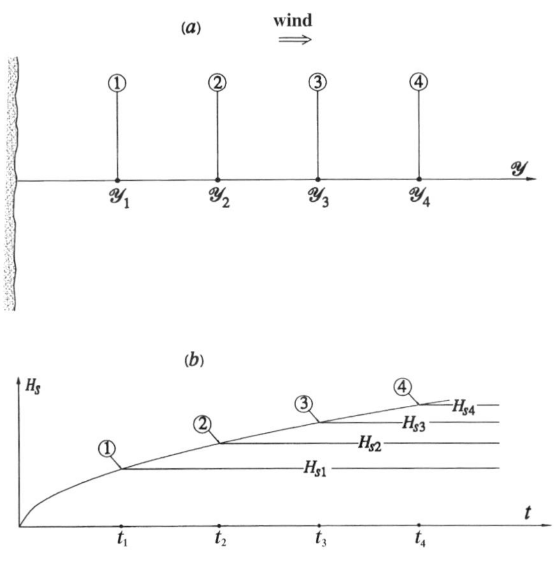

Fig. 8.1 (a) Four fetches. (b) Time histories of H_s at the ends of the four fetches under the action of a steady wind.

be as in fig. 8.1*b*. For a time interval $(0, t_1)$ H_s will grow at the same way at all cross-sections. From an instant t_1, H_s will no longer grow at ①, while it will keep on growing in the same way at all other cross-sections. From an instant t_2, H_s will cease growing also at ②, while it will keep on in the same way at downdrift cross-sections, and so on.

Second idea. The value of H_s of the steady condition, that is H_{s1} at ①, H_{s2} at ②, and so on, depends on the fetch $\mathscr{Y}$ and on the wind speed u through a relation of the kind

$$\frac{gH_s}{u^2} = f\left(\frac{g\mathscr{Y}}{u^2}\right).$$

A similar relation proceeds also from the JONSWAP spectrum, or more exactly, from the set of results of the JONSWAP project (Hasselmann et al., 1973). Indeed the researchers of this project suggested the formulae

$$A = 0.076\left(\frac{g\mathscr{Y}}{u^2}\right)^{-0.22},$$
$$\frac{\omega_p u}{g} = 2\pi\, 3.5\left(\frac{g\mathscr{Y}}{u^2}\right)^{-0.33} \tag{8.1}$$

which, taken with the relation (4.25) between m_0 and ω_p, yield

$$\frac{gH_s}{u^2} = 1.26 \cdot 10^{-3}\left(\frac{g\mathscr{Y}}{u^2}\right)^{0.55}. \tag{8.2}$$

We shall use (8.2) to evaluate how H_s varies along $\mathscr{Y}$. To this end, we define

$$\varDelta \equiv 100\,\frac{\dfrac{\mathrm{d}H_s}{\mathrm{d}\mathscr{Y}}L_{p0}}{H_s}.$$

This gives the per cent variation of H_s in one dominant wavelength along the wind direction. $\varDelta$ can be rewritten in the equivalent form

$$\varDelta \equiv 100\,\frac{\mathrm{d}\left(\dfrac{gH_s}{u^2}\right)}{\mathrm{d}\left(\dfrac{g\mathscr{Y}}{u^2}\right)}\,\frac{gL_{p0}}{u^2}\Big/\frac{gH_s}{u^2},$$

where the formula for the derivative of gH_s/u^2 with respect to $g\mathscr{Y}/u^2$ proceeds from (8.2), and the formula for gL_{p0}/u^2 proceeds from (8.1), with the result that $\varDelta$ is related to $g\mathscr{Y}/u^2$ by the equation

$$\varDelta = 0.715\left(\frac{g\mathscr{Y}}{u^2}\right)^{-0.34}. \tag{8.3}$$

Let us consider a case of practical interest, a fetch of 300 km with a wind of 100 km/hour. We have

$$\frac{g\mathscr{Y}}{u^2} = 3800$$

that together with (8.3) yields

$$\Delta = 0.04 .$$

That is an H_s which varies 0.04% in a space of L_{p0}. Clearly this is the largest variation, the one along the wave direction.

Thus we realize that in the open sea, an area with sides of ten wavelengths can be very well regarded as a homogeneous wave field, that is a field where the variations of H_s from one point to another are negligible.

8.2 The wave field in the open sea

8.2.1 *Basic assumptions and expressions of* η *and* ϕ

According to the theory of the sea states, a steady and homogeneous wave field is thought of as the sum of a very large number N of small periodic waves, with frequencies, directions and phases generally different from one another. Accordingly, the mathematical form to the first order in a Stokes' expansion is

$$\eta(x, y, t) = \sum_{i=1}^{N} a_i \cos(k_i x \sin\theta_i + k_i y \cos\theta_i - \omega_i t + \varepsilon_i), \tag{8.4a}$$

$$\phi(x, y, z, t) = g \sum_{i=1}^{N} a_i \omega_i^{-1} \frac{\cosh[k_i(d+z)]}{\cosh(k_i d)} \sin(k_i x \sin\theta_i + k_i y \cos\theta_i - \omega_i t + \varepsilon_i), \tag{8.4b}$$

where the relation between the wave number k_i and the angular frequency ω_i is

$$k_i \tanh(k_i d) = \frac{\omega_i^2}{g} .$$

The ith term of the sum (8.4a) gives the surface displacement (and the ith term of the sum 8.4b gives the velocity potential) of a periodic wave of amplitude a_i, frequency ω_i and phase ε_i, whose direction of advance makes an angle θ_i with the y-axis. From sect. 1.10 we know that the ith term of the sum (8.4a) and the ith term of the sum (8.4b) satisfy the system of the linear flow equations in the homogeneous form, that is with $f(t) = 0$. It follows that the sum (8.4a) and the sum (8.4b) also satisfy the same system of linear equations in the homogeneous form. Naturally, in order to satisfy the system of linear equations with $f(t)$ generally different from zero, it suffices to add to (8.4b) the term

$$\frac{1}{\rho}\int_0^t f(t')\,\mathrm{d}t'$$

which does not modify the particle velocity nor the pressure fluctuation.

The theory of the sea states, as stated in sect. 4.2, requires that

(i) N tends to infinity;

(ii) ε_i are distributed uniformly on the circle and are stochastically independent of one another;

(iii) ω_i are different from one another;

(iv) a_i are all of the same order.

8.2.2 *η and ϕ are stationary Gaussian processes*

The surface displacement at any fixed point x, y is given by

$$\eta(t) = \sum_{i=1}^{N} a_i \cos(\omega_i t + \tilde{\varepsilon}_i), \tag{8.5}$$

where

$$\tilde{\varepsilon}_i \equiv -(\varepsilon_i + k_i x \sin\theta_i + k_i y \cos\theta_i). \tag{8.6}$$

The $\tilde{\varepsilon}_i$, like the ε_i, are distributed uniformly on the circle and are stochastically independent of one another, so that (8.5) represents a stationary Gaussian process (the proof was given in sect. 5.1).

The velocity potential at any fixed point x, y, z is given by

$$\phi(t) = -\sum_{i=1}^{N} \tilde{a}_i \sin(\omega_i t + \tilde{\varepsilon}_i), \tag{8.7}$$

where

$$\tilde{a}_i \equiv g a_i \omega_i^{-1} \frac{\cosh[k_i(d+z)]}{\cosh(k_i d)}.$$

The fact that the a_i are of the same order implies that also the $\tilde{a}_i$ are of the same order, and hence (8.7) represents again a stationary Gaussian process.

8.2.3 *The pressure head waves beneath the water surface*

The fluctuating pressure head proceeds from the velocity potential (8.4b) through the Bernoulli equation. The result is

$$\eta_{ph}(x,y,z,t) = \sum_{i=1}^{N} a_i \frac{\cosh[k_i(d+z)]}{\cosh(k_i d)} \cos(k_i x \sin\theta_i + k_i y \cos\theta_i - \omega_i t + \varepsilon_i).$$

At any fixed point x, y, z (we mean a point that remains always beneath the water surface), η_{ph} is a random function given by

$$\eta_{ph}(t) = \sum_{i=1}^{N} \tilde{\tilde{a}}_i \cos(\omega_i t + \tilde{\varepsilon}_i), \tag{8.8}$$

where $\tilde{\varepsilon}_i$ has been already defined by (8.6), and

$$\tilde{\tilde{a}}_i \equiv a_i \frac{\cosh[k_i(d+z)]}{\cosh(k_i d)}. \tag{8.9}$$

From the point of view of mathematics, (8.8) represents a new stationary Gaussian process, which can be verified by reasoning as we have done in sect. 8.2.2.

The spectrum of $\eta_{ph}(t)$ is defined by the equation

$$E(\omega; z)\,\delta\omega \equiv \sum_i \frac{1}{2}\tilde{\tilde{a}}_i^2 \quad \text{for } i \text{ such that } \omega < \omega_i < \omega + \delta\omega,$$

which is rewritten in the form

$$E(\omega; z)\,\delta\omega = \frac{\cosh^2[k(d+z)]}{\cosh^2(kd)} \sum_i \frac{1}{2} a_i^2 \text{ for } i \text{ such that } \omega < \omega_i < \omega + \delta\omega, \tag{8.10}$$

because of the relation (8.9). The summation on the r.h.s. of (8.10) is equal to $E(\omega)\,\delta\omega$, and consequently

$$E(\omega; z) = \frac{\cosh^2[k(d+z)]}{\cosh^2(kd)} E(\omega). \tag{8.11}$$

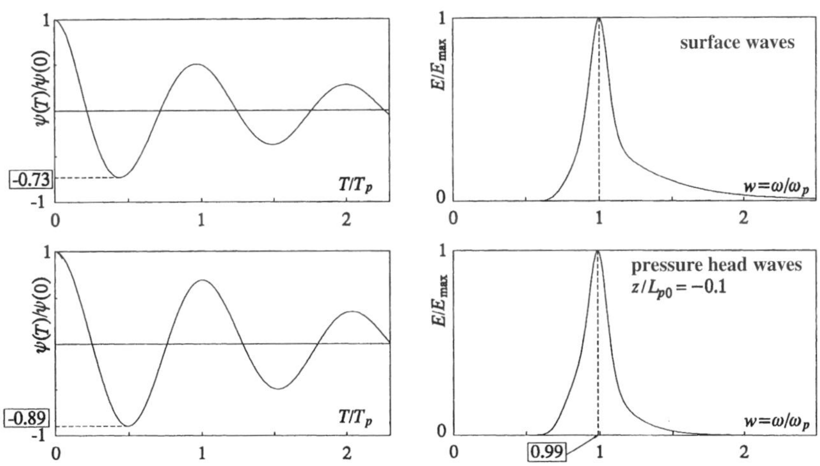

Fig. 8.2 Normalized autocovariance and spectrum of the surface waves and of the pressure head waves at a depth of $L_{p0}/10$ beneath the mean water level, on deep water. (Obtained by means of 8.11.)

Fig. 8.2 shows the spectrum of the surface waves and the spectrum of the pressure head waves at a certain depth beneath the mean water level (equation 8.11). We see that the peak of $E(\omega; z)$ occurs at a frequency slightly smaller than ω_p. Moreover, we see that the spectrum beneath the water surface shrinks, because it sheds its high frequency tail. Both these phenomena are due to the fact that the small wave components undergo an attenuation with depth, which is the larger the greater is frequency.

8.3 The directional spectrum

8.3.1 *Definition and characteristic forms*

The amplitudes, frequencies and directions of the small wave components give rise to a directional spectrum:

$$S(\omega, \theta)\,\delta\omega\,\delta\theta \equiv \sum_i \frac{1}{2} a_i^2 \text{ for } i \text{ such that } \omega < \omega_i < \omega + \delta\omega \text{ and } \theta < \theta_i < \theta + \delta\theta. \quad (8.12)$$

In words: the product $2S(\omega, \theta)\,\delta\omega\,\delta\theta$ represents the sum of the square amplitudes of the small wave components whose frequencies ω_i and angles θ_i fall in the small rectangle $(\omega, \omega + \delta\omega; \theta, \theta + \delta\theta)$ [see fig. 8.3]. The definitions (8.12) of $S(\omega, \theta)$ and (4.2) of $E(\omega)$ yield

$$E(\omega) = \int_0^{2\pi} S(\omega, \theta)\,\mathrm{d}\theta. \quad (8.13)$$

The directional spectrum is generally given in the form

$$S(\omega, \theta) = E(\omega)\,D(\theta; \omega), \quad (8.14)$$

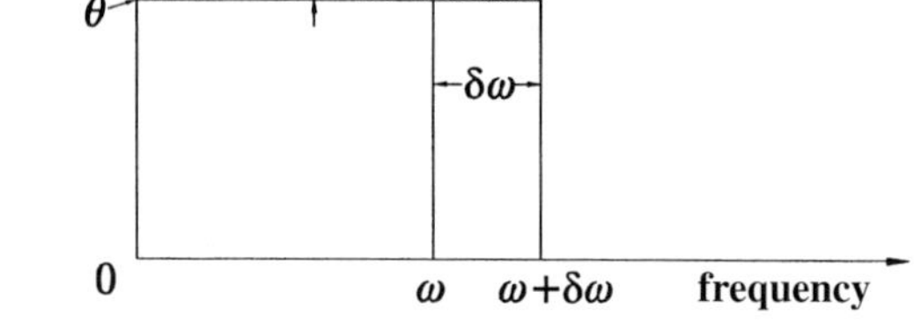

Fig. 8.3 Definition of directional spectrum: the product $S(\omega, \theta)\,\delta\omega\,\delta\theta$ is equal to the partial sum $\sum_i \frac{1}{2} a_i^2$ for i such that ω_i and θ_i belong to the small rectangle.

where $D(\theta;\omega)$ is the directional spreading function. Clearly, the relation (8.13) between $E(\omega)$ and $S(\omega,\theta)$ implies

$$\int_0^{2\pi} D(\theta;\omega)\,\mathrm{d}\theta = 1. \tag{8.15}$$

The directional spreading function of the wind waves on deep water has a typical shape that usually is fitted by the cosine-power function

$$D(\theta;\omega) = K(n)\cos^{2n}\left[\frac{1}{2}(\theta-\overline{\theta})\right] \text{ with } n = n(\omega), \tag{8.16}$$

where $K(n)$ is the normalizing factor

$$K(n) \equiv \left[\int_0^{2\pi} \cos^{2n}\frac{1}{2}\,\theta\,\mathrm{d}\theta\right]^{-1}, \tag{8.17}$$

being necessary to comply with (8.15). $\overline{\theta}$ is the angle the dominant wave direction makes with the y-axis. As to the parameter n, it generally depends on frequency ω and this is why we have written $D(\theta;\omega)$. Typically, n takes on its largest value at the peak frequency ω_p, and consequently the minimum directional spread is associated with this frequency. Mitsuyasu et al. (1975) suggest

$$n \begin{cases} = n_p(\omega/\omega_p)^5 & \text{if } \omega \le \omega_p, \\ = n_p(\omega_p/\omega)^{2.5} & \text{if } \omega > \omega_p, \end{cases} \tag{8.18}$$

$$n_p = 7.5\cdot 10^{-3}\left(\frac{g\mathscr{Y}}{u^2}\right)^{0.825}.$$

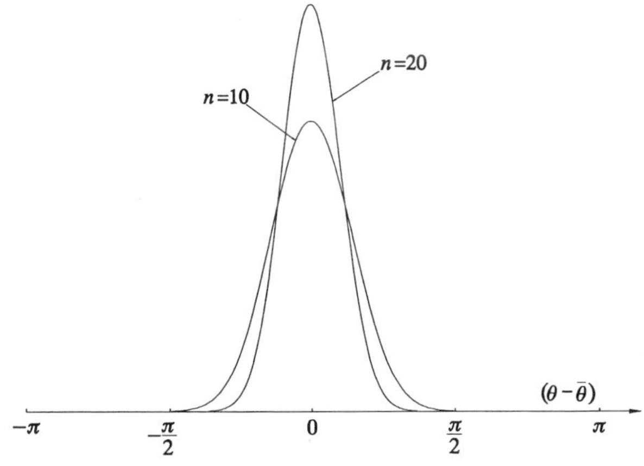

Fig. 8.4 The cosine power ($\cos^{2n}$) directional spreading function.

The graph of function (8.16) is a "bell-shaped" curve where the sharper the peak, the larger is n[see fig. 8.4]. One more remark: n is a real number, as we can infer from (8.18), and the cosine of $\frac{1}{2}(\theta - \bar{\theta})$ may be negative, so that $\cos^{2n}$ is understood to be the nth power of the square cosine. Of course a more appropriate form is

$$\left|\cos\left[\frac{1}{2}(\theta - \bar{\theta})\right]\right|^{2n}.$$

8.3.2 *The nondimensional directional spectrum*

It is convenient to resort to the nondimensional directional spectrum

$$\mathscr{S}(w, \theta) \equiv S(w\omega_p, \theta)/(Ag^2\omega_p^{-5}). \tag{8.19}$$

With the JONSWAP spectrum and the directional spreading function of Mitsuyasu et al. (1975), we have

$$\mathscr{S}(w, \theta) \equiv \mathscr{E}(w) K(n) \cos^{2n}\left[\frac{1}{2}(\theta - \bar{\theta})\right], \tag{8.20}$$

where $\mathscr{E}(w)$ is the nondimensional frequency spectrum defined by (4.24), and

$$n\begin{cases} = n_p w^5 & \text{if} \quad w \leq 1, \\ = n_p w^{-2.5} & \text{if} \quad w > 1. \end{cases} \tag{8.21}$$

Thus the nondimensional directional spectrum $\mathscr{S}(w, \theta)$ depends on three parameters: χ_1 and χ_2 of the JONSWAP spectrum and n_p of the directional spreading function. For the illustrations henceforth we shall take the following characteristic values

$$\chi_1 = 3, \quad \chi_2 = 0.08, \quad n_p = 20.$$

8.3.3 *A typical double integral concerning the directional spectrum*

Dealing with the mathematical analysis of a sea state we have to evaluate

$$\text{typical integral} = \int_0^\infty \int_0^{2\pi} f(kX, kY, kz, \omega T, \theta)\, S(\omega, \theta)\, \mathrm{d}\theta\, \mathrm{d}\omega, \tag{8.22}$$

where f is some function depending on the special problem, and X, Y, z, T are space-time variables. Replacement of ω with $w = \omega/\omega_p$ and use of the nondimensional wave number

$$\mathscr{k}(w) \equiv \frac{k}{2\pi/L_{p0}} \tag{8.23}$$

prove to be very helpful for this job.

From the definition (8.23) and the linear dispersion rule (1.23), we obtain

$$\mathscr{k}(w)\tanh\left[2\pi\mathscr{k}(w)\frac{d}{L_{p0}}\right] = w^2, \tag{8.24}$$

so that $\mathscr{k}(w)$ is equal to the x for which the function

$$f(x) \equiv x\tanh\left(2\pi\frac{d}{L_{p0}}x\right)$$

is equal to w^2. This function is increasing, so that the abscissa associated with a given value of the ordinate is easily obtained numerically.

Using (8.19) and (8.23) we rewrite (8.22) in the form

$$\text{typical integral} = Ag^2\omega_p^{-4}\int_0^\infty\int_0^{2\pi} f\left[2\pi\mathscr{k}(w)\frac{X}{L_{p0}}, 2\pi\mathscr{k}(w)\frac{Y}{L_{p0}}, 2\pi\mathscr{k}(w)\frac{z}{L_{p0}},\right.$$
$$\left., 2\pi w\frac{T}{T_p}, \theta\right]\mathscr{S}(w,\theta)\,\mathrm{d}\theta\,\mathrm{d}w.$$

8.4 Shoaling and refraction of the wind-generated waves

8.4.1 *Supplementary notes on the refraction of a periodic wave*

Area ⓪ of fig. 8.5 is on deep water, area ① is on shallow water. The contour lines are parallel to the x-axis. The bottom slope is small, such that the water in area ① is assumed to have a nearly constant depth d.

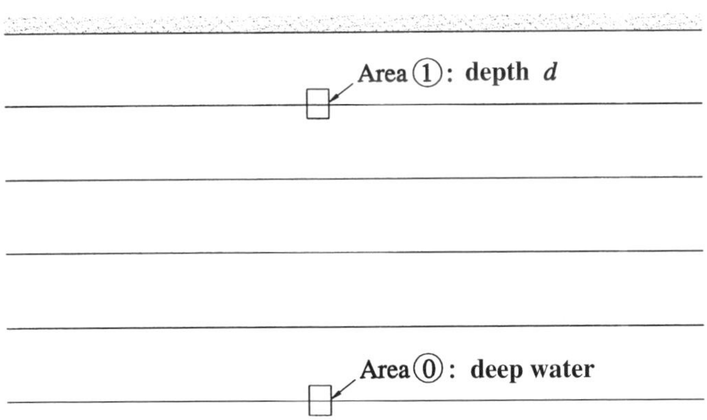

Fig. 8.5 In area ⓪ the waves are given by (8.25) and in area ① by (8.26). The relation between the wave heights in the two areas is known [see chap. 2]. The problem of the relative phase of two fixed points, one of area ⓪ and one of area ①, is unsolved at this time.

If in area ⓪ we have

$$\eta(x,y,t) = a_0 \cos\left(\frac{\omega^2}{g} x \sin\theta_0 + \frac{\omega^2}{g} y \cos\theta_0 - \omega t + \varepsilon_0\right), \tag{8.25}$$

then in area ① we have

$$\eta(x,y,t) = a \cos(kx \sin\theta + ky \cos\theta - \omega t + \varepsilon), \tag{8.26}$$

where k satisfies the linear dispersion rule (1.23), and

$$a = a_0 \sqrt{\frac{\sinh(2kd)}{\tanh(kd)\,[\sinh(2kd) + 2kd]}} \sqrt[4]{\frac{1 - \sin^2\theta/\tanh^2(kd)}{1 - \sin^2\theta}}, \tag{8.27a}$$

$$\sin\theta = \sin\theta_0 \tanh(kd), \tag{8.27b}$$

$$\cos\theta \cos\theta_0 > 0. \tag{8.27c}$$

Equation (8.27a) is simply an alternative form of (2.45). Inequality (8.27c) is the mathematical way to say: if the wave advances landward on deep water, then it advances landward also on shallow water. Equation (8.27b) and inequality (8.27c) imply that θ and θ_0 belong to the same quadrant, and they also imply that

$$|\sin\theta| < \tanh(kd). \tag{8.28}$$

From (8.27b) we get

$$\theta_0 = \arcsin\left[\frac{\sin\theta}{\tanh(kd)}\right], \tag{8.29}$$

and hence:

$$\mathrm{d}\theta_0 = \frac{|\cos\theta|/\tanh(kd)}{\sqrt{1 - \sin^2\theta/\tanh^2(kd)}}\, \mathrm{d}\theta. \tag{8.30}$$

This last relation shows how does the angle vary on deep water because of a variation $\mathrm{d}\theta$ of the angle on water depth d. This relation will be used shortly.

Let us now complete the analysis of (8.25-26), noting that ε is linked to ε_0 in a not easy way. We have

$$\varepsilon = \varepsilon_0 + \Delta\varepsilon,$$

where $\Delta\varepsilon$ depends on angle θ_0, on frequency ω and on how the water depth changes from area ⓪ to area ①. For getting $\Delta\varepsilon$ we should be able to solve the problem: what is the phase angle between two fixed points P_0 and P_1, one in area ⓪ and the other in area ①? This problem is very difficult because the known expression of the propagation speed c, being exact on a constant depth, leads to an error for a sloping bottom; and this error should have some nonnegligible effects on $\Delta\varepsilon$, even in the limit for the bottom slope approaching zero (given that, in this limit, the distance between P_0 and P_1 tends to infinity). But luckily, the solution to the $\Delta\varepsilon$ problem is useless for our goals. Indeed, it suffices to know that ε is equal to ε_0 plus some terms depending on angle θ_0, frequency ω and function $d(y)$.

8.4.2 *The step from periodic waves to wind-generated waves*

A sea state is the sum of a very large number N of small wave components with frequencies, directions and phases generally different from one another. Therefore the sea state is given by

$$\eta(x,y,t) = \sum_{i=1}^{N} a_{0i} \cos\left(\frac{\omega_i^2}{g} x \sin\theta_{0i} + \frac{\omega_i^2}{g} y \cos\theta_{0i} - \omega_i t + \varepsilon_{0i}\right) \text{in area } ⓪,$$

$$\eta(x,y,t) = \sum_{i=1}^{N} a_i \cos(k_i x \sin\theta_i + k_i y \cos\theta_i - \omega_i t + \varepsilon_i) \text{ in area } ①,$$

where the amplitudes a_i are related to the amplitudes a_{0i} by (8.27*a*), and the angles θ_i are related to the angles θ_{0i} by (8.27*b-c*).

If the small wave components in area ⓪ satisfy the four assumptions (i)-(iv) of sect. 8.2.1, then also the small wave components in area ① do satisfy these assumptions. Specifically, since ε_i is equal to ε_{0i} plus a term independent of ε_{0i}, the ε_i are uniformly distributed over the circle and are stochastically independent of one another like the ε_{0i}. Therefore $\eta(t)$ at any fixed point x, y represents a stationary Gaussian process.

8.4.3 *The relation between the directional spectrum on a given water depth and the directional spectrum on deep water*

From (8.27*a*), (8.28), (8.29), and (8.30), the following relation proceeds between the directional spectrum $S(\omega,\theta)$ on a water depth d and the deep water spectrum $S_0(\omega,\theta)$:

$$S(\omega,\theta)\begin{cases} = \dfrac{\sinh(2kd)}{\tanh^2(kd)[\sinh(2kd)+2kd]} S_0\left(\omega, \arcsin\left[\dfrac{\sin\theta}{\tanh(kd)}\right]\right) & \text{if } |\sin\theta| < \tanh(kd), \\ = 0 \text{ otherwise.} \end{cases} \tag{8.31}$$

In using this relation, bear in mind that angles

$$\theta \quad \text{and} \quad \arcsin\left[\frac{\sin\theta}{\tanh(kd)}\right]$$

belong to the same quadrant.

If S_0 is the JONSWAP-Mitsuyasu spectrum, the directional spectrum S on water depth d becomes

$$S(\omega,\theta)\begin{cases} = \dfrac{\sinh(2kd)}{\tanh^2(kd)[\sinh(2kd)+2kd]}\,Ag^2\omega^{-5}\exp\left[-\dfrac{5}{4}\left(\dfrac{\omega_p}{\omega}\right)^4\right]\cdot \\ \cdot\exp\left\{\ln\chi_1\exp\left[-\dfrac{(\omega-\omega_p)^2}{2\chi_2^2\omega_p^2}\right]\right\}K(n)\cdot \\ \cdot\cos^{2n}\left\{\dfrac{1}{2}\arcsin\left[\dfrac{\sin\theta}{\tanh kd}\right]-\dfrac{1}{2}\bar{\theta}_0\right\}\ \text{if}\ |\sin\theta|<\tanh(kd), \\ = 0\ \text{otherwise}, \end{cases} \tag{8.32}$$

where $\bar{\theta}_0$ denotes the dominant direction on deep water. [Strictly speaking, ω_p should be replaced by ω_{p0} that is the peak frequency on deep water. However the difference between ω_p (on the given water depth d) and ω_{p0}, usually, proves to be very small so that it is convenient to reason as if the peak frequency does not change from deep to shallow water.]

The nondimensional directional spectrum on water depth d proceeds from (8.32) and the definition (8.19):

$$\mathscr{S}(w,\theta)\begin{cases} = \dfrac{\sinh\left[4\pi\ell(w)\dfrac{d}{L_{p0}}\right]}{\tanh^2\left[2\pi\ell(w)\dfrac{d}{L_{p0}}\right]\left\{\sinh\left[4\pi\ell(w)\dfrac{d}{L_{p0}}\right]+4\pi\ell(w)\dfrac{d}{L_{p0}}\right\}}\mathscr{E}(w)K(n)\cdot \\ \cdot\cos^{2n}\left\{\dfrac{1}{2}\arcsin\left[\dfrac{\sin\theta}{\tanh\left(2\pi\ell(w)\dfrac{d}{L_{p0}}\right)}\right]-\dfrac{1}{2}\bar{\theta}_0\right\}\ \text{if}\ |\sin\theta|<\tanh\left[2\pi\ell(w)\dfrac{d}{L_{p0}}\right], \\ = 0\quad\text{otherwise}, \end{cases} \tag{8.33}$$

where $\mathscr{E}(w)$ is given by (4.24), and $\ell(w)$ by (8.24).

We should use the form (8.20) of $\mathscr{S}(w,\theta)$ on deep water, and the form (8.33) on shallow water (provided that the contour lines are straight). In fact, in what follows we shall use (8.20) on both deep water and on shallow water. This is for appreciating the effects of the bottom depth, under the same directional spectrum. However, for the illustration of the shoaling-refraction effects we shall resort to (8.33).

To check (8.31), use the two following equations:

$$S(\omega,\theta)\,\delta\omega\,\delta\theta\equiv\left\{\sum_i\frac{1}{2}a_i^2\quad\text{for } i \text{ such that } \omega<\omega_i<\omega+\delta\omega\quad\text{and}\quad\theta<\theta_i<\theta+\delta\theta\right\}=$$

$$\begin{cases} = \left\{ \dfrac{\sinh(2kd)}{\tanh(kd)[\sinh(2kd)+2kd]} \sqrt{\dfrac{1-\sin^2\theta/\tanh^2(kd)}{1-\sin^2\theta}} \sum_i \dfrac{1}{2} a_{0i}^2 \quad \text{for } i \text{ such that } \omega < \omega_i < \omega + \delta\omega \text{ and} \right. \\ \left. \arcsin\left[\dfrac{\sin\theta}{\tanh(kd)}\right] < \theta_{0i} < \arcsin\left[\dfrac{\sin\theta}{\tanh(kd)}\right] + \dfrac{|\cos\theta|/\tanh(kd)}{\sqrt{1-\sin^2\theta/\tanh^2(kd)}}\delta\theta \quad \text{if } |\sin\theta| < \tanh(kd) \right\}, \\ = 0 \quad \text{otherwise}, \end{cases} \tag{8.34}$$

$$S_0(\omega,\theta)\,\delta\omega\,\delta\theta \equiv \sum_i \frac{1}{2} a_{0i}^2 \quad \text{for } i \text{ such that } \omega < \omega_i < \omega + \delta\omega \quad \text{and} \quad \theta < \theta_{0i} < \theta + \delta\theta .$$

You obtain the rightmost equality (8.34), using the relation (8.27*a*) to replace a_i with a_{0i}. Then use equation (8.29) which relates the angle on deep water to the angle on water depth d. Finally, use (8.30) which gives the variation of the deep water angle for a given variation of the angle on water depth d.

To achieve (8.31) you have to express the summation $\sum_i \frac{1}{2} a_{0i}^2$ on the r.h.s. of (8.34) in terms of S_0, and clear the fractions.

8.4.4 *Shoaling-refraction effects on the wind waves*

The nondimensional directional spectrum (8.33) can be used to get the shoaling-refraction curves for the wind waves. Indeed the quotient between the variance on some given water depth d and the variance on deep water is

$$\frac{\sigma^2}{\sigma_0^2} = \frac{\displaystyle\int_0^\infty \int_0^{2\pi} \mathscr{S}(w,\theta)\,\mathrm{d}\theta\,\mathrm{d}w \qquad \text{for the given } \frac{d}{L_{p0}}}{\displaystyle\int_0^\infty \int_0^{2\pi} \mathscr{S}(w,\theta)\,\mathrm{d}\theta\,\mathrm{d}w \qquad \text{as } \frac{d}{L_{p0}} \to \infty} . \tag{8.35}$$

Fig. 8.6 shows the shoaling-refraction curve, that is to say the square root of (8.35) as a function of d/L_{p0}, for two different angles of the dominant wave direction on deep water.

We can get also the frequency spectrum on the given water depth:

$$E(w\omega_p) = A g^2 \omega_p^{-5} \int_0^{2\pi} \mathscr{S}(w,\theta)\,\mathrm{d}\theta .$$

As an example, fig. 8.7 shows the spectrum on water depth $d = L_{p0}/10$, for two different angles of the dominant wave direction on deep water.

From fig. 8.6 we see that the shoaling-refraction curve of the wind waves is very close to the corresponding curve of the periodic waves. Then, from fig. 8.7 we see that shoaling-refraction has only a small effect on the spectrum *shape*. The consequence is that the frequency spectrum on shallow water can be taken as equal as the deep water spectrum multiplied by the constant factor σ^2/σ_0^2 (this being constant with respect to frequency ω).

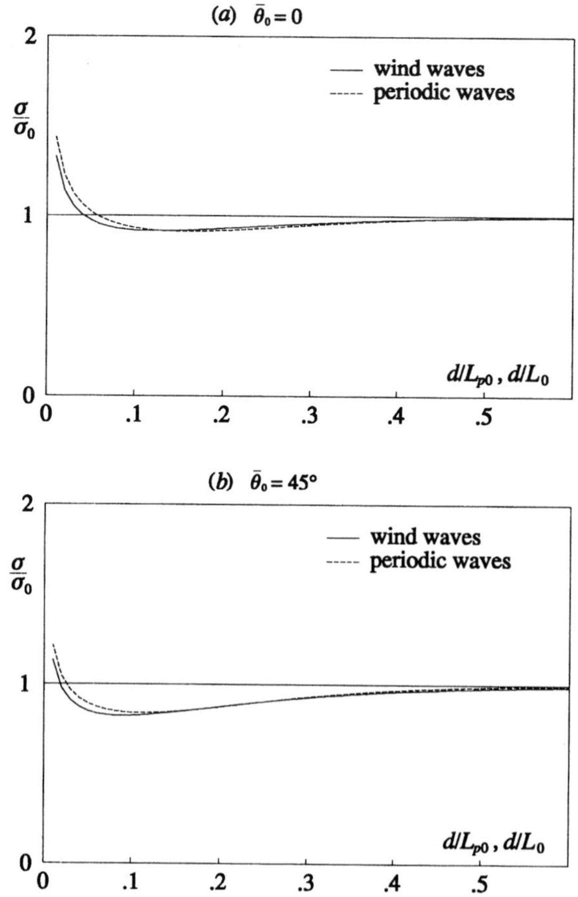

Fig. 8.6 Shoaling-refraction of the periodic waves and of the wind waves.

Perhaps this will seem a disappointing picture, in the sense that, after rather hard work to get the directional spectrum on shallow water, we find the shoaling-refraction effects to be nearly obvious (fig. 8.6) or nearly negligible (fig. 8.7). Really, shoaling-refraction of the wind waves gets interesting essentially when one deals with the length of the wave crest. This face of shoaling-refraction will be illustrated in chap. 10, using the quasi-determinism theory and the nondimensional directional spectrum (8.33).

Note that those of fig. 8.7 are normalized spectra, indeed each spectrum is related to its maximum value. Accordingly, the figure enables us to conclude the shallow water spectrum to be nearly equal to the deep water spectrum apart from a constant factor. That is it enables us to conclude, as we have done, that the *shape* of the shallow water spectrum is nearly equal to the shape of the deep water spectrum.

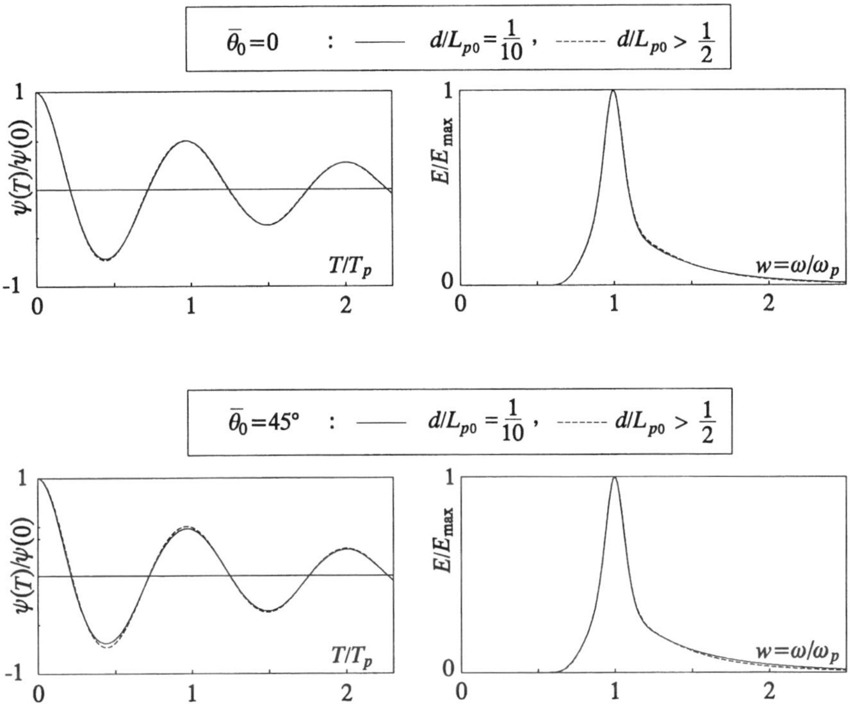

Fig. 8.7 Normalized autocovariance and spectrum on shallow water and on deep water. Shallow water: continuous line. Deep water: dashed line. We see that the differences are very small.

8.5 Reflection of the wind-generated waves

8.5.1 *A big difference with respect to the periodic waves*

The surface displacement $\eta(x, y, t)$ and the velocity potential $\phi(x, y, z, t)$ of a periodic wave before a vertical reflecting wall are given by (1.57*a*-*b*). Hence η and ϕ of a sea state, which is the sum of a very large number N of small periodic waves, are given by

$$\eta(x, y, t) = 2\sum_{i=1}^{N} a_i \cos(k_i x \sin\theta_i - \omega_i t + \varepsilon_i)\cos(k_i y \cos\theta_i), \tag{8.36a}$$

$$\phi(x, y, z, t) = 2g\sum_{i=1}^{N} a_i \omega_i^{-1} \frac{\cosh[k_i(d+z)]}{\cosh(k_i d)} \sin(k_i x \sin\theta_i - \omega_i t + \varepsilon_i)\cos(k_i y \cos\theta_i). \tag{8.36b}$$

In other words: if we put a vertical reflecting wall in the plane $y = 0$, the random wave field (8.4*a*-*b*) takes on the form (8.36*a*-*b*).

Reasoning as usual, we can easily verify that both $\eta(t)$ and $\phi(t)$ at any fixed point are stationary Gaussian processes. But an important novelty arises: these random processes are no longer homogeneous in space. Indeed it is apparent that the characteristics of $\eta(t)$ (as well as of $\phi(t)$) depend on the distance y from the wall. In particular the very variance of $\eta(t)$ depends on y, so that we shall denote it by the symbol $\sigma^2(y)$.

From (8.36*a*) we have

$$\sigma^2(y) = 4\sum_{i=1}^{N}\sum_{j=1}^{N} a_i a_j \cos(k_i y \cos\theta_i)\cos(k_j y\cos\theta_j)\cdot$$
$$\cdot < \cos(k_i x \sin\theta_i - \omega_i t + \varepsilon_i)\cos(k_j x\sin\theta_j - \omega_j t + \varepsilon_j) >$$

that reduces itself to

$$\sigma^2(y) = 4\sum_{i=1}^{N}\frac{1}{2}a_i^2\cos^2(k_i y\cos\theta_i), \tag{8.37}$$

since $\omega_i \neq \omega_j$ if $i \neq j$. The contribution to the summation (8.37) from the small components with $\omega < \omega_i < \omega + \delta\omega$ and $\theta < \theta_i < \theta + \delta\theta$ is equal to

$$4S(\omega,\theta)\,\delta\omega\,\delta\theta\cos^2(ky\cos\theta),$$

and hence

$$\sigma^2(y) = 4\int_0^\infty\int_0^{2\pi} S(\omega,\theta)\cos^2(ky\cos\theta)\,\mathrm{d}\theta\,\mathrm{d}\omega.$$

Finally, replacing the dimensional variable ω with the nondimensional w, we arrive at

$$\frac{\sigma^2(y)}{\sigma^2(0)} = \frac{\displaystyle\int_0^\infty\int_0^{2\pi}\mathscr{S}(w,\theta)\cos^2\left[2\pi\mathscr{k}(w)\frac{y}{L_{p0}}\cos\theta\right]\mathrm{d}\theta\,\mathrm{d}w}{\displaystyle\int_0^\infty\int_0^{2\pi}\mathscr{S}(w,\theta)\,\mathrm{d}\theta\,\mathrm{d}w}. \tag{8.38}$$

Fig. 8.8 shows the function (8.38) for the basic condition of deep water and waves attacking the wall orthogonally (we mean that the dominant direction of the incident waves is orthogonal to the wall). For a comparison, the figure also shows

$$\frac{\sigma^2(y)}{\sigma^2(0)} = \cos^2\left(2\pi\frac{y}{L}\right)$$

of the periodic waves (for the same condition of wave attack orthogonal to the wall).

This time the difference between wind waves and periodic waves is really amazing! The variance of the periodic waves fluctuates between maxima equal to the maximum at the wall and minima equal to zero, and the fluctuations go on as far as an infinite distance from the wall. The variance of the wind generated waves is a damped oscillatory function of the distance from the wall; and, at only one or two

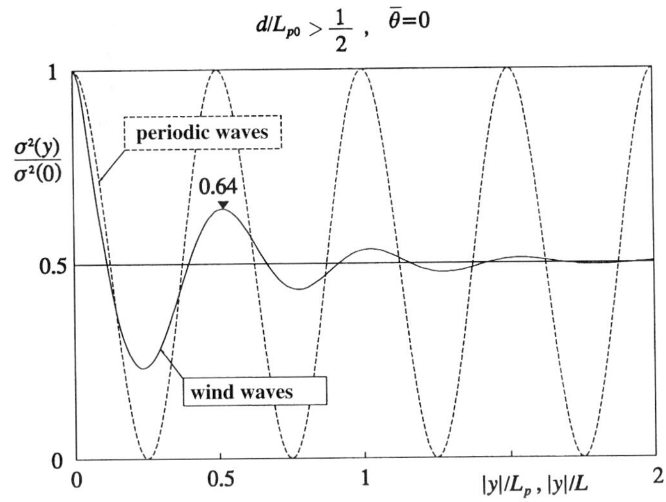

Fig. 8.8 Graph of $\sigma^2(y)/\sigma^2(0)$ [σ^2= variance of the surface displacement, y = distance from the wall]. Continuous line: wind waves. Dashed line: periodic waves. It is the basic case of deep water and orthogonal attack.

wavelengths from the wall, it approaches a constant value being equal to half the value at the wall.

The difference is so big that the person who found it could hardly believe it. Indeed one of the basic notions of the wave theory is that nodes and antinodes follow one another as far as an infinite distance from the wall. Now it emerged that the wind waves have a pseudo-node at a quarter wavelength from the wall, a pseudo-antinode at a half-wavelength from the wall, and then nodes and antinodes disappear all together.

The first reaction was that the formula (8.38) or the calculations are wrong! But it didn't take long to realize that it was not so [the reader is invited to verify it himself, checking slowly the few steps from (8.36*a*) to (8.38), and evaluating numerically the two integrals, with the formulae (8.20) for $\mathscr{S}(w,\theta)$ and (8.24) for $\mathscr{k}(w)$]. Then the question arises, how is it possible that the sea waves differ so greatly from the periodic waves where reflection is concerned?

The answer to this question came from the quasi-determinism theory. Hence, we shall give it in chap. 10. Here we conclude with two further remarks.

Firstly, under the same spectrum, the pseudo-nodes and antinodes get more apparent on shallow water (cf. fig. 8.9*a* with fig. 8.8). In particular, we see that the first local maximum of $\sigma^2(y)/\sigma^2(0)$, the one at a half-wavelength from the wall, on shallow water ($d/L_{p0} = 0.1$) is equal to 0.71, while on deep water it is equal to 0.64.

Secondly, under the same water depth, the pseudo-nodes and antinodes get even less apparent if the dominant direction of the incident waves is not wall-orthogonal (cf. fig. 8.9*b* with fig. 8.8). In particular, we see that the first local maximum of

$\sigma^2(y)/\sigma^2(0)$ after the wall, in the case of the oblique reflection ($\bar{\theta} = 45°$) is equal to 0.53, whereas in the case of the orthogonal reflection it is equal to 0.64.

We shall see in chap. 10 that even these second order differences (that is the variations of the differences between wind waves and periodic waves due to the water depth or the angle of wave attack) find a clear explanation in the light of the quasi-determinism theory.

8.5.2 *Diffraction coefficients*

The diffraction coefficient C_d of the waves interacting with some solid obstacle is defined as the ratio between the root mean square surface displacement at a given

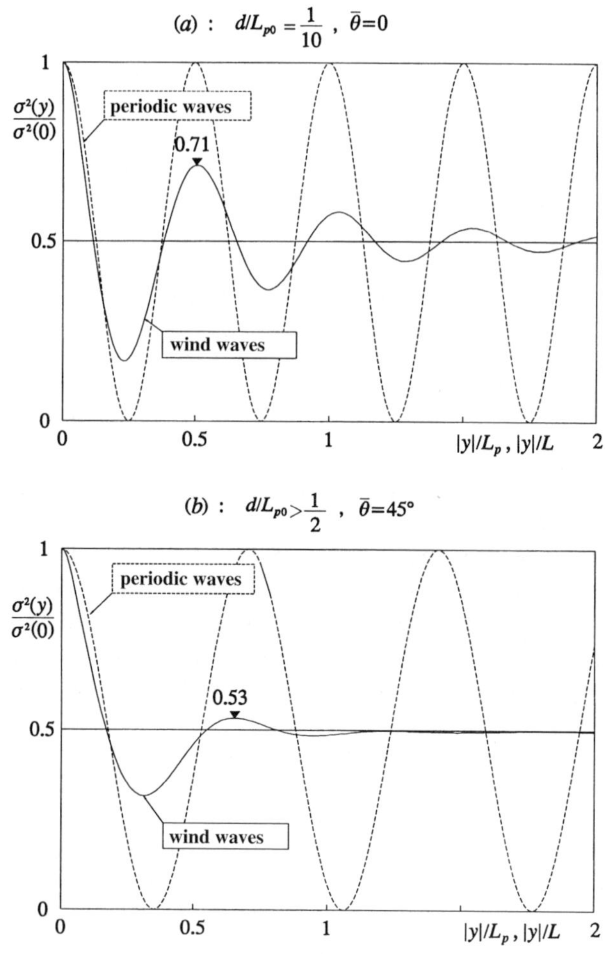

Fig. 8.9 Once again $\sigma^2(y)/\sigma^2(0)$ for: (*a*) shallow water and orthogonal attack; (*b*) deep water and inclined attack.

point and the root mean square surface displacement of the incident waves (for the periodic waves, C_d coincides with the ratio between the wave height at the given point and the height of the incident waves, as we saw in chap. 1). In the case of the infinitely long wall we have

$$C_d(y) \equiv \frac{\sigma(y)}{\sigma} = \frac{\sigma(y)}{\sigma(0)} \frac{\sigma(0)}{\sigma},$$

where σ is the root mean square surface displacement of the wave field (8.4a) that would be there without the wall. The quotient $\sigma(0)/\sigma$ is equal to 2 and hence we have the relation

$$C_d(y) = 2\frac{\sigma(y)}{\sigma(0)}$$

that enables us to deduce C_d from the function $\sigma^2(y)/\sigma^2(0)$ of figs. 8.8 and 8.9a-b. At the wall C_d is equal to 2, and starting from about one wavelength from the wall, where $\sigma^2(y)/\sigma^2(0)$ takes on the constant value $\frac{1}{2}$, it becomes equal $2\sqrt{\frac{1}{2}} = \sqrt{2}$.

8.6 Diffraction of the wind-generated waves

If we put a semi-infinite vertical wall in the plane $y = 0$, with the origin at $x = 0$, the wave field (8.4a-b) becomes

$$\eta(r,\beta,t) = \sum_{i=1}^{N} a_i [F(r,\beta;\omega_i,\theta_i)\cos(\omega_i t + \varepsilon_i) + G(r,\beta;\omega_i,\theta_i)\sin(\omega_i t + \varepsilon_i)], \tag{8.39a}$$

$$\phi(r,\beta,z,t) = g\sum_{i=1}^{N} a_i\,\omega_i^{-1}\frac{\cosh[k_i(d+z)]}{\cosh(k_i d)}[G(r,\beta;\omega_i,\theta_i)\cos(\omega_i t + \varepsilon_i) +$$

$$-F(r,\beta;\omega_i,\theta_i)\sin(\omega_i t + \varepsilon_i)], \tag{8.39b}$$

where r and β are the polar coordinates [see the definition sketch of fig. 1.17]. The ith terms of these two summations represent the flow field due to an incident wave of amplitude a_i, frequency ω_i, phase ε_i, whose direction makes an angle θ_i with the y-axis. The functions $F(r,\beta;\omega,\theta)$ and $G(r,\beta;\omega,\theta)$ were defined in sect. 1.12.1.

Here too, it can be easily verified that both $\eta(t)$ and $\phi(t)$, at any fixed point, represent some stationary Gaussian processes. Of course, they are non-homogeneous random processes, and the variance $\sigma^2(r,\beta)$ of the surface displacement here can vary very much from one point to another, and especially from a point before the wall to a point behind the wall.

From (8.39a) we get

$$\sigma^2(r,\beta) = \sum_{i=1}^{N}\frac{1}{2}a_i^2[F^2(r,\beta;\omega_i,\theta_i) + G^2(r,\beta;\omega_i,\theta_i)],$$

which can be expressed in terms of the directional spectrum $S(\omega,\theta)$ of the incident waves:

$$\sigma^2(r,\beta) = \int_0^\infty \int_0^{2\pi} S(\omega,\theta)[F^2(r,\beta;\omega,\theta) + G^2(r,\beta;\omega,\theta)]\,d\theta\,d\omega.$$

Hence, the diffraction coefficient, that is equal to the square root of the quotient $\sigma^2(r,\beta)/\sigma^2$ [σ^2 being the variance of the incident waves], proves to be

$$C_d(r,\beta) = \sqrt{\frac{\int_0^\infty \int_0^{2\pi} \mathscr{S}(w,\theta)[F^2(r,\beta;w,\theta) + G^2(r,\beta;w,\theta)]\mathrm{d}\theta\,\mathrm{d}w}{\int_0^\infty \int_0^{2\pi} \mathscr{S}(w,\theta)\,\mathrm{d}\theta\,\mathrm{d}w}}, \tag{8.40}$$

where functions F and G have been expressed in terms of w rather than of ω because of the relation $w = \omega/\omega_p$.

Figs. 8.10 and 8.11 show the diffraction coefficient as a function of the nondimensional coordinates x/L_p and y/L_p, respectively, for the orthogonal reflection ($\bar{\theta} = 0$) and for a case of oblique reflection [$\bar{\theta} = 45°$, $\mathscr{S}(w,\theta)$ given by (8.20) for $|\theta - \bar{\theta}| < 45°$, otherwise $\mathscr{S}(w,\theta) = 0$]. The two figures retain their validity for every depth (under the same spectrum).

Figs. 8.10 and 8.11 should be compared respectively with fig. 1.19 and fig. 1.20. The comparison shows that

(i) in the more sheltered area of the geometric shadow (that is, close to the wall and far from the wall's tip), the C_d of the wind waves is practically coincident with the C_d of the periodic waves;

(ii) within the dark part of the geometric shadow [see figs. 8.10 and 8.11], the C_d of the wind waves is at least 50% greater than the C_d of the periodic waves;

(iii) along the wave-beaten wall, the C_d of the wind waves takes on the constant value 2 at a short distance from the breakwater's tip, while the C_d of the periodic waves shows an infinite sequence of local maxima greater than 2 and local minima smaller than 2.

It is apparent that the really important matter is (ii), that is the very big difference between the C_d of the wind waves and the C_d of the periodic waves in the dark area. We shall see in chap. 10 that also this new difference between the wind waves and the periodic waves gets a clear explanation in the light of the theory of quasi-determinism.

Note that it suffices to specify r/L_{p0}, β, w, θ, and d/L_{p0} to obtain functions F and G. In order to realize this point, please go back to the equations (1.60*a*-*f*) defining these functions, and verify that

$$u_1 \equiv \sqrt{8\mathscr{k}(w)r/L_{p0}}\sin\left[\frac{1}{2}\left(\beta+\theta-\frac{\pi}{2}\right)\right], \quad u_2 \equiv -\sqrt{8\mathscr{k}(w)r/L_{p0}}\sin\left[\frac{1}{2}\left(\beta-\theta+\frac{\pi}{2}\right)\right], \tag{8.41a}$$

$$q_1 \equiv 2\pi\mathscr{k}(w)\frac{r}{L_{p0}}\sin(\beta+\theta), \quad q_2 \equiv -2\pi\mathscr{k}(w)\frac{r}{L_{p0}}\sin(\beta-\theta). \tag{8.41b}$$

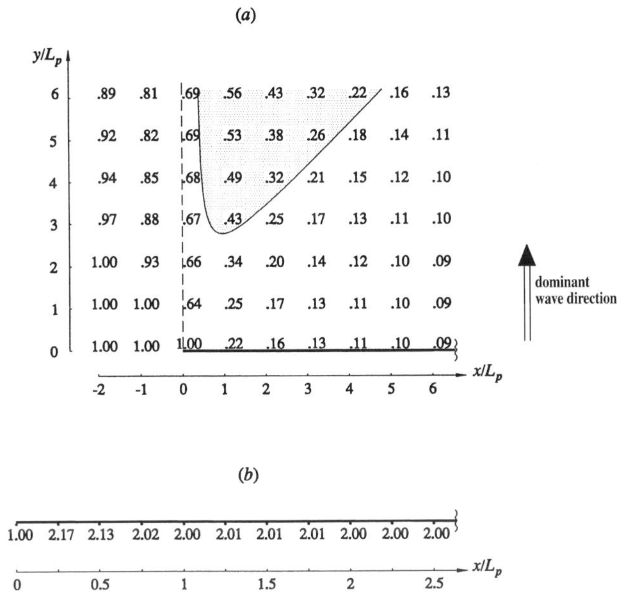

Fig. 8.10 Diffraction coefficient C_d of the wind waves for the orthogonal attack. Comparing this figure with fig. 1.19, you will note that, in the dark area, the C_d of the wind waves is at least 50% greater than the C_d of the periodic waves.

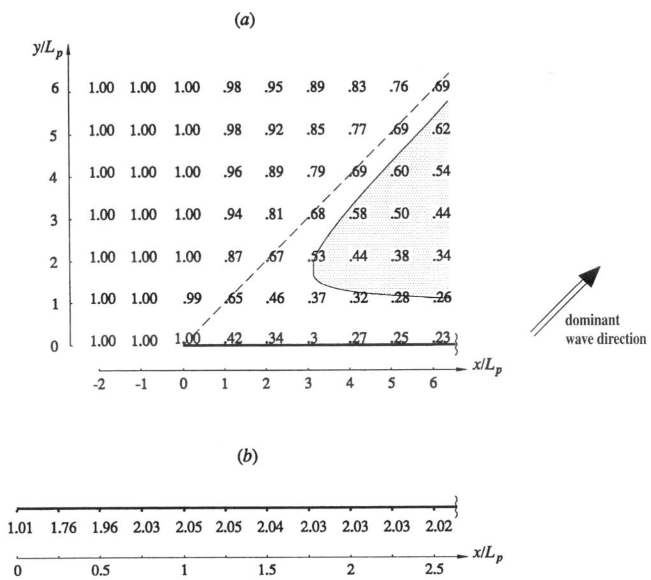

Fig. 8.11 Diffraction coefficient C_d of the wind waves for an inclined attack. Here too, in the dark area, the C_d of the wind waves is at least 50% greater than the C_d of the periodic waves (compare with fig. 1.20).

8.7 Long-crested random waves: the link between periodic waves and wind-generated waves

8.7.1 *Two-dimensional and three-dimensional random waves*

The wind-generated waves are random and short-crested. Random, because each wave is generally different from the following wave. Short-crested, because they are three-dimensional. The differences in size and shape are due to the bandwidth; the short-crestedness is due to the directional spread. With regard to this, the reader will have noted that for all problems concerning the wave variability (e.g. wave height distribution and period-height regression) the directional spread was not used. Indeed, as far as chap. 7, it was not necessary to introduce the directional spectrum.

Now, in the two foregoing sections we have met two great differences between the wind waves and the periodic waves. We have seen that in the wind waves the nodes and antinodes disappear starting from about one wavelength from the breakwater; and we have seen that in the sheltered area there is a wide region where the C_d of the wind waves is at least 50% larger than the C_d of the periodic waves. Thus we wonder, do these great differences depend on the bandwidth or on the directional spread? Of course, if we succeed in answering this question, we shall be able also to predict the different effects of different sea states. For example, if the bandwidth grows under the same directional spread do nodes and antinodes get even less apparent? And do the wave heights in the sheltered region increase further?

The question, whether the great differences between wind waves and periodic waves depend on the bandwidth or on the directional spread, is also useful to find our way in the world of the wave tanks. Indeed the laboratory tanks, for a given frequency spectrum, can generate waves with no directional spread (long-crested waves or 2-D random waves), and waves with a directional spread (short-crested waves, or 3-D random waves). It is apparent that the first ones are cheaper than the second ones, because the 2-D random waves are generated by a single wave paddle, while the 3-D random waves require a more complex array of small independent paddles. The reader can probably see where we are driving at. On the one hand it is useless resorting to the expensive 3-D waves to study some phenomenon that depends only on the frequency spectrum; on the other hand it is misleading to resort to the 2-D waves to study a phenomenon that is affected by the directional spread. For example, if we wanted to study how the height of the pressure head wave reduces itself from the water surface to the seabed, it would certainly be sufficient to resort to the 2-D waves. Indeed, we have seen in sect. 8.2 this attenuation depends only on the frequency spectrum.

8.7.2 *Special analytical forms for the two-dimensional random waves*

The mathematical treatment of the 2-D random waves can be done using the formulae for the 3-D random waves with a very small directional spread, that is with a very large value of the exponent n of the directional spreading function $D(\theta;\omega)$. Of course, we can also obtain some special solutions for the 2-D waves, which are valuable because of a greater simplicity. For example, the 2-D random waves before a vertical breakwater are expressed by

$$\eta(x,y,t) = 2\sum_{i=1}^{N} a_i \cos(k_i x \sin\overline{\theta} - \omega_i t + \varepsilon_i)\cos(k_i y \cos\overline{\theta}), \tag{8.42a}$$

$$\phi(x,y,z,t) = 2g\sum_{i=1}^{N} a_i \omega_i^{-1} \frac{\cosh[k_i(d+z)]}{\cosh(k_i d)} \sin(k_i x \sin\overline{\theta} - \omega_i t + \varepsilon_i)\cos(k_i y \cos\overline{\theta}). \tag{8.42b}$$

These are the formulae (8.36*a-b*) with:

$$\theta_i = \overline{\theta} \quad \forall i.$$

The variance of the surface displacement (8.42*a*) is

$$\sigma^2(y) = 4\sum_{i=1}^{N} \frac{1}{2} a_i^2 \cos^2(k_i y \cos\overline{\theta}),$$

from which it follows

$$\frac{\sigma^2(y)}{\sigma^2(0)} = \frac{\displaystyle\int_0^\infty \mathscr{E}(w)\cos^2\left[2\pi \mathscr{k}(w)\frac{y}{L_{p0}}\cos\overline{\theta}\right]\mathrm{d}w}{\displaystyle\int_0^\infty \mathscr{E}(w)\,\mathrm{d}w}.$$

Similarly, the 2-D random waves interacting with a semi-infinite vertical breakwater are given by the formulae (8.39*a-b*) with $\theta_i = \overline{\theta}$, and consequently the variance at a fixed point r, β is

$$\sigma^2(r,\beta) = \sum_{i=1}^{N} \frac{1}{2} a_i^2 \left[F^2(r,\beta;\omega_i,\overline{\theta}) + G^2(r,\beta;\omega_i,\overline{\theta})\right],$$

from which it follows

$$C_d(r,\beta) = \sqrt{\frac{\displaystyle\int_0^\infty \mathscr{E}(w)\left[F^2(r,\beta;w,\overline{\theta}) + G^2(r,\beta;w,\overline{\theta})\right]\mathrm{d}w}{\displaystyle\int_0^\infty \mathscr{E}(w)\,\mathrm{d}w}}.$$

8.7.3 *Are the two-dimensional random waves closer to the wind waves or to the periodic waves?*

Fig. 8.12 shows the quotient $\sigma^2(y)/\sigma^2(0)$ before an infinitely long vertical breakwater for

(*a*) the periodic waves;

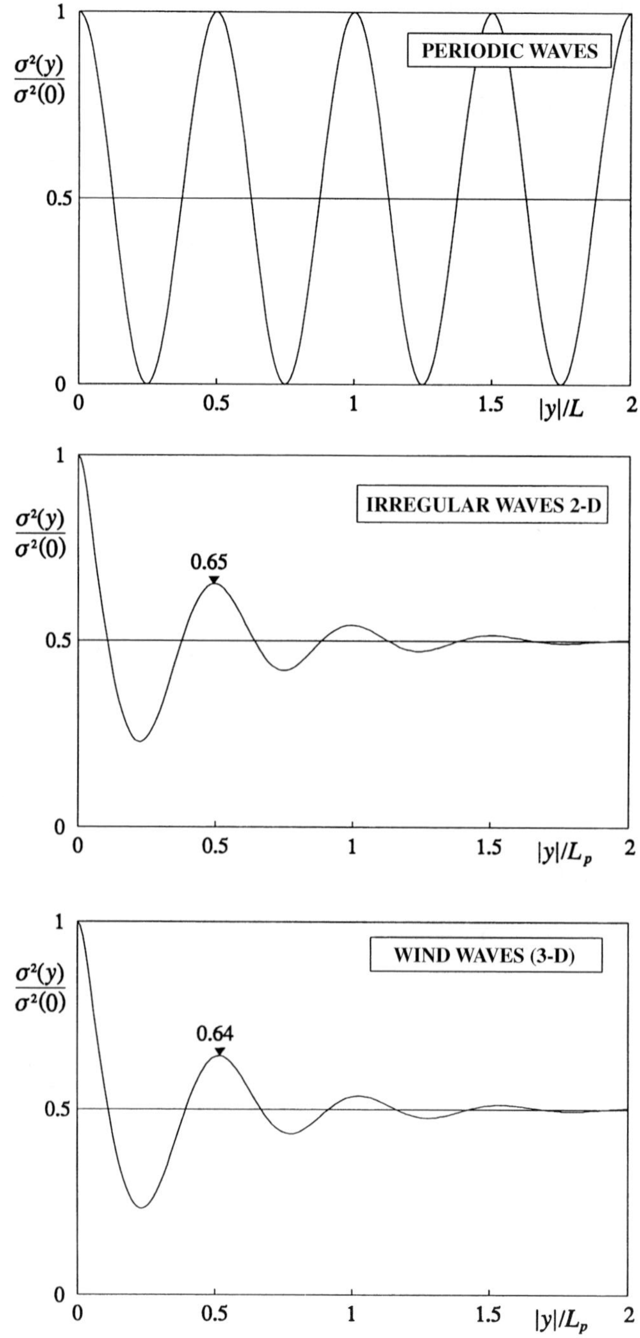

Fig. 8.12 Comparison between periodic waves, long crested irregular waves and wind waves, with regard to reflection. We see the long crested waves behave like the wind waves.

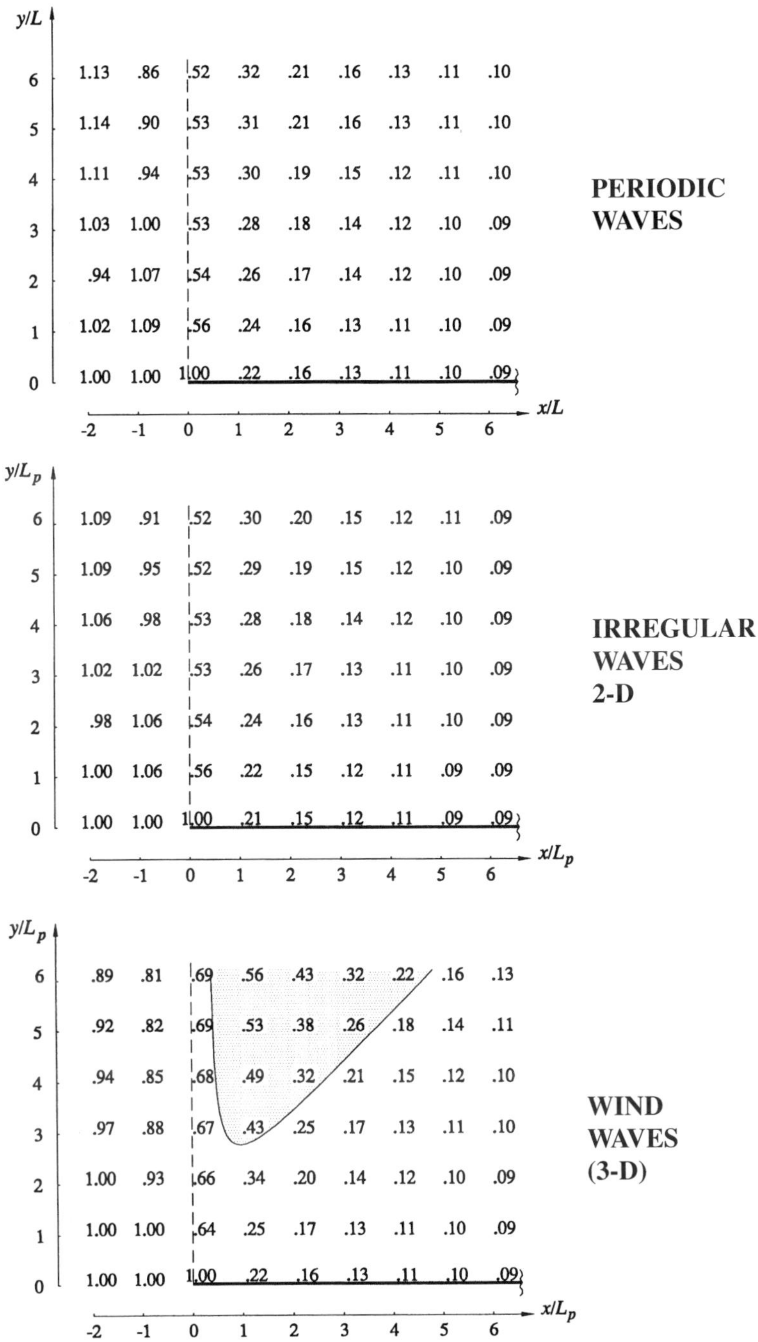

Fig. 8.13 Comparison between periodic waves, long crested irregular waves and wind waves, with regard to diffraction. Here the long crested waves behave like the periodic waves.

(*b*) the 2-D random waves;
(*c*) the wind waves (3-D random waves).

We see the function $\sigma^2(y)/\sigma^2(0)$ to be practically the same for the 2-D random waves and for the wind waves. Hence, we can conclude that the great difference between wind waves and periodic waves, as far as reflection is concerned, is due to the bandwidth. Even in the absence of directional spread, the aforesaid great difference would remain unchanged. [Really, if the dominant direction of the incident waves is not wall-orthogonal, the directional spread also somewhat affects $\sigma^2(y)/\sigma^2(0)$].

Fig. 8.13 extends the comparison between periodic waves, 2-D random waves and wind waves into diffraction. Specifically, the figure shows the diffraction coefficient C_d in the sheltered area behind a semi-infinite vertical breakwater. Here, as we can see, the matter changes. The difference between wind waves and 2-D random waves is big, while there is a little difference between the 2-D random waves and the periodic waves. Indeed we see that the C_d of the 2-D random waves are nearly coincident with the C_d of the periodic waves.

Conclusion: concerning reflection, the 2-D random waves are very close to the wind waves, while for diffraction, the 2-D random waves are very close to the periodic waves. Referring to the organic evolution we think of the periodic waves as the ancestors, the wind waves as the descendants, and the 2-D waves as the link between ancestors and descendants.

8.8 Direct proportion between the maximum expected wave height and the diffraction coefficient

As we have seen, the waves on the time domain in each of the fields we have considered represent some stationary Gaussian processes. As a consequence, the probability of exceedance of the wave height at any fixed point is given by (5.45), and the maximum expected wave height in a given interval Δt is given by (5.57-59).

Naturally, if the wave field is non-homogeneous because the wind-generated waves interact with some solid body, the significant wave height generally varies from one point to another. That is we have

$$H_s(x,y) = hC_d(x,y),$$

with h being here the significant height of the incident waves. Therefore the formula (5.57-59) for $\overline{H_{\max}}$ becomes

$$\overline{H_{\max}}(x,y) = C_d(x,y)\,h\int_0^\infty 1-\left\{1-\exp\left[-\frac{4}{1+\psi^*(x,y)}\,u^2\right]\right\}^{\frac{\Delta t}{\overline{T}(x,y)}}\mathrm{d}u, \qquad (8.43)$$

where we write $\psi^*(x,y)$ and $\overline{T}(x,y)$ because the narrow-bandedness parameter and the mean period generally vary from one point to another.

From (8.43) we obtain the ratio between the maximum expected wave height at any fixed point x, y and the maximum expected wave height if the solid obstacle (breakwater or something else) was not there. This ratio is

$$\frac{\overline{H_{\max}}(x,y)}{\overline{H_{\max}}} = \frac{\displaystyle\int_0^\infty 1-\left\{1-\exp\left[-\frac{4}{1+\psi^*(x,y)}u^2\right]\right\}^{\frac{\Delta t}{\overline{T}(x,y)}} du}{\displaystyle\int_0^\infty 1-\left\{1-\exp\left[-\frac{4}{1+\psi^*}u^2\right]\right\}^{\frac{\Delta t}{\overline{T}}} du} C_d(x,y), \quad (8.44)$$

where, of course, ψ^* and $\overline{T}$ without the specification (x, y) refer to the incident waves. Typically, the quotient between the two integrals on the r.h.s. (8.44) proves to be very close to 1, so that we can assume

$$\frac{\overline{H_{\max}}(x,y)}{\overline{H_{\max}}} \cong C_d(x,y). \quad (8.45)$$

For the step from (8.44) to (8.45) we reason on the autocovariance only, given that also $\overline{T}$ can be expressed in terms of the autocovariance. Indeed, equations (4.6), (4.12) and (5.24) taken together yield

$$\overline{T} = 2\pi\sqrt{\frac{\psi(0)}{|\ddot{\psi}(0)|}}. \quad (8.46)$$

Let us consider the wind waves interacting with a semi-infinite vertical breakwater, which are extremely non-homogeneous (think of the very big differences in the wave size between the wave-beaten side and the lee side). The autocovariance, at a fixed point of polar coordinates r, β, is given by

$$\psi(T) = \sum_{i=1}^{N} \frac{1}{2} a_i^2 (F_i^2 + G_i^2) \cos(\omega_i T), \quad (8.47)$$

where

$$F_i \equiv F(r,\beta;\omega_i,\theta_i), \qquad G_i \equiv G(r,\beta;\omega_i,\theta_i).$$

In terms of the directional spectrum of the incident waves, (8.47) becomes

$$\psi(T) = \int_0^\infty \int_0^{2\pi} S(\omega,\theta)\,[F^2(r,\beta;\omega,\theta) + G^2(r,\beta;\omega,\theta)]\cos(\omega T)\,d\theta\,d\omega, \quad (8.48)$$

and hence we have

$$\frac{\psi(T)}{\psi(0)} = \frac{\displaystyle\int_0^\infty \int_0^{2\pi} \mathscr{S}(w,\theta)[F^2(r,\beta;w,\theta) + G^2(r,\beta;w,\theta)]\cos(2\pi w T/T_p)\,d\theta\,dw}{\displaystyle\int_0^\infty \int_0^{2\pi} \mathscr{S}(w,\theta)\,[F^2(r,\beta;w,\theta) + G^2(r,\beta;w,\theta)\,d\theta\,dw} \quad (8.49)$$

which enables us to obtain the ψ^* at the given point r, β. Then, the formulae (8.46) for $\overline{T}$ and (8.48) for $\psi(T)$ lead to

$$\overline{T} = T_p \sqrt{\frac{\int_0^\infty \int_0^{2\pi} \mathscr{S}(w,\theta)[F^2(r,\beta;w,\theta) + G^2(r,\beta;w,\theta)]\,\mathrm{d}\theta\,\mathrm{d}w}{\int_0^\infty \int_0^{2\pi} \mathscr{S}(w,\theta)\,w^2[F^2(r,\beta;w,\theta) + G^2(r,\beta;w,\theta)]\,\mathrm{d}\theta\,\mathrm{d}w}} \tag{8.50}$$

which gives the mean wave period at point r, β.

At this stage, we can evaluate the quotient between the two integrals on the r.h.s. of (8.44), and verify this quotient to range between 0.98 and 1.02 in the area covered by figs. 8.10 and 8.11. Hence, the validity of the simplified equation (8.45).

8.9 Space-time covariances

We shall see in chap. 9 that the quasi-determinism theory is expressed in terms of the space-time covariances

$$\Psi(X,Y,T;x_o,y_o) \equiv <\eta(x_o,y_o,t)\,\eta(x_o+X,y_o+Y,t+T)>\,, \tag{8.51a}$$

$$\Phi(X,Y,z,T;x_o,y_o) \equiv <\eta(x_o,y_o,t)\,\phi(x_o+X,y_o+Y,z,t+T)>\,, \tag{8.51b}$$

where x_o, y_o is any fixed point. To use this theory it is first necessary to obtain the expressions of Ψ and Φ in terms of the spectrum $S(\omega,\theta)$. Of course, the expression of Ψ (or Φ) varies according to whether the wave field is (8.4*a*-*b*) or (8.36*a*-*b*) or (8.39*a*-*b*) or something else. The logical path to obtain these expressions is the same we have repeatedly traced in this chapter [e.g. to obtain $\sigma^2(y)$ in sect. 8.5.1 where the various steps were illustrated in detail]. For the special problem of Ψ,

(i) substitute in (8.51*a*) η by its expression, so getting a temporal mean of a summation on i and j;

(ii) execute the temporal mean with the usual procedure: the mean value of the product of two cosine (or sine) functions with a different frequency is zero; the mean value of a square cosine (or sine) is $\frac{1}{2}$;

(iii) doing so, you obtain an equation like this:

$$\Psi(X,Y,T;x_o,y_o) = \sum_{i=1}^{N} \frac{1}{2}\,a_i^2 f(X,Y,T;x_o,y_o,\omega_i,\theta_i)\,, \tag{8.52}$$

where f denotes a function that will be different from one wave field to another;

(iv) consider the partial sum of the terms whose frequency ω_i is within a small interval $\omega, \omega + \delta\omega$ and angle θ_i is within a small interval $\theta, \theta + \delta\theta$; this can be written in the form

$$f(X,Y,T;x_o,y_o,\omega,\theta)\sum_i \frac{1}{2}a_i^2 \quad \text{for } i \text{ such that } \omega<\omega_i<\omega+\delta\omega \quad \text{and} \quad \theta<\theta_i<\theta+\delta\theta=$$
$$=f(X,Y,T;x_o,y_o,\omega,\theta)\,S(\omega,\theta)\,\delta\omega\,\delta\theta; \tag{8.53}$$

(v) integrating the r.h.s. of (8.53) with respect to ω over $(0,\infty)$ and with respect to θ over $(0,2\pi)$, you arrive at

$$\Psi(X,Y,T;x_o,y_o)=\int_0^\infty\int_0^{2\pi} S(\omega,\theta)f(X,Y,T;x_o,y_o,\omega,\theta)\,\mathrm{d}\theta\,\mathrm{d}\omega;$$

(vi) replace the dimensional variable ω with the nondimensional w, and express k in terms of $\mathscr{k}(w)$ through the definition (8.23).

A sequence of the same kind leads to the expression of Φ.

A few expressions of Ψ and Φ are given in the following frames. Such expressions can be used with the formula (8.20) or (8.33) for $\mathscr{S}(w,\theta)$, and the equation (8.24) of $\mathscr{k}(w)$.

WAVE FIELD (8.4*a*-*b*)

$$\Psi(X,Y,T;x_o,y_o)=Ag^2\omega_p^{-4}\int_0^\infty\int_0^{2\pi}\mathscr{S}(w,\theta)\cos\left[2\pi\mathscr{k}(w)\frac{X}{L_{p0}}\sin\theta+\right.$$
$$\left.+2\pi\mathscr{k}(w)\frac{Y}{L_{p0}}\cos\theta-2\pi w\frac{T}{T_p}\right]\mathrm{d}\theta\,\mathrm{d}w, \tag{8.54a}$$

$$\Phi(X,Y,z,T;x_o,y_o)=Ag^3\omega_p^{-5}\int_0^\infty\int_0^{2\pi}\mathscr{S}(w,\theta)w^{-1}\left\{\cosh\left[2\pi\mathscr{k}(w)\left(\frac{d}{L_{p0}}+\frac{z}{L_{p0}}\right)\right]\right/$$
$$\left/\cosh\left[2\pi\mathscr{k}(w)\frac{d}{L_{p0}}\right]\right\}\sin\left[2\pi\mathscr{k}(w)\frac{X}{L_{p0}}\sin\theta+2\pi\mathscr{k}(w)\frac{Y}{L_{p0}}\cos\theta-2\pi w\frac{T}{T_p}\right]\mathrm{d}\theta\,\mathrm{d}w. \tag{8.54b}$$

WAVE FIELD (8.36*a-b*)

$$\Psi(X,Y,T;x_o,y_o) = 4Ag^2\omega_p^{-4}\int_0^\infty\int_0^{2\pi}\mathscr{S}(w,\theta)\cos\left[2\pi k(w)\frac{X}{L_{p0}}\sin\theta + -2\pi w\frac{T}{T_p}\right]\cos\left[2\pi k(w)\frac{y_o}{L_{p0}}\cos\theta\right]\cos\left[2\pi k(w)\left(\frac{y_o}{L_{p0}}+\frac{Y}{L_{p0}}\right)\cos\theta\right]d\theta\,dw, \tag{8.55a}$$

$$\Phi(X,Y,z,T;x_o,y_o) = 4Ag^3\omega_p^{-5}\int_0^\infty\int_0^{2\pi}\mathscr{S}(w,\theta)w^{-1}\left\{\cosh\left[2\pi k(w)\left(\frac{d}{L_{p0}}+\frac{z}{L_{p0}}\right)\right]\Big/\cosh\left[2\pi k(w)\frac{d}{L_{p0}}\right]\right\}\sin\left[2\pi k(w)\frac{X}{L_{p0}}\sin\theta - 2\pi w\frac{T}{T_p}\right]\cos\left[2\pi k(w)\frac{y_o}{L_{p0}}\cos\theta\right]\cdot \cdot\cos\left[2\pi k(w)\left(\frac{y_o}{L_{p0}}+\frac{Y}{L_{p0}}\right)\cos\theta\right]d\theta\,dw. \tag{8.55b}$$

WAVE FIELD (8.39*a-b*)

$$\Psi(X,Y,T;x_o,y_o) = Ag^2\omega_p^{-4}\int_0^\infty\int_0^{2\pi}\mathscr{S}(w,\theta)\Big\{[F_o(w,\theta)F_1(w,\theta)+G_o(w,\theta)G_1(w,\theta)]\cdot \cdot\cos\left(2\pi w\frac{T}{T_p}\right) + [F_o(w,\theta)G_1(w,\theta)-G_o(w,\theta)F_1(w,\theta)]\sin\left(2\pi w\frac{T}{T_p}\right)\Big\}d\theta\,dw, \tag{8.56a}$$

$$\Phi(X,Y,z,T;x_o,y_o) = Ag^3\omega_p^{-5}\int_0^\infty\int_0^{2\pi}\mathscr{S}(w,\theta)w^{-1}\left\{\cosh\left[2\pi k(w)\left(\frac{d}{L_{p0}}+\frac{z}{L_{p0}}\right)\right]\Big/\cosh\left(2\pi k(w)\frac{d}{L_{p0}}\right)\right\}\Big\{-[F_o(w,\theta)F_1(w,\theta)+G_o(w,\theta)G_1(w,\theta)]\sin\left(2\pi w\frac{T}{T_p}\right) + +[F_o(w,\theta)G_1(w,\theta)-G_o(w,\theta)F_1(w,\theta)]\cos\left(2\pi w\frac{T}{T_p}\right)\Big\}d\theta\,dw, \tag{8.56b}$$

where

$$x_o, y_o \Leftrightarrow r_o, \beta_o,$$

$$x_o + X, y_o + Y \Leftrightarrow r_1, \beta_1,$$

$$F_o(w,\theta) \equiv F(r_o,\beta_o;w,\theta), \quad G_o(w,\theta) \equiv G(r_o,\beta_o;w,\theta),$$

$$F_1(w,\theta) \equiv F(r_1,\beta_1;w,\theta), \quad G_1(w,\theta) \equiv G(r_1,\beta_1;w,\theta),$$

F and G being defined by (1.60*a-d*) and (8.41*a-b*).

Note that Ψ and Φ of the wave field (8.4*a-b*) in the open sea are independent of x_o and y_o, which is correct given that the wave field is homogeneous. Then, the Ψ and Φ of the wave field (8.36*a-b*) are independent of x_o, which is again correct, it being the wave field before an infinitely long breakwater parallel to the x-axis.

Conclusive note

As said at the end of chap. 4, the theory of the sea states in space-time was due to the contributions of several authors, among whom: Longuet-Higgins (1957, 1963), Phillips (1967), and Borgman (1969). Goda et al. (1978) probably were the first to evaluate the diffraction coefficient of the wind waves. The author (1988) pointed out the great difference between the wind waves and the periodic waves in what concerns reflection (figs. 8.8 and 8.9). This great difference was then verified through a small scale field experiment (1993).

References

Boccotti P., 1988 Refraction, reflection and diffraction of irregular gravity waves. *Excerpta of the Italian Contribution to the Field of Hydraulic Eng.* 3, 47-88.

Boccotti P., Barbaro G., Fiamma V. et al., 1993 An experiment at sea on the reflection of the wind waves. *Ocean Eng.* 20, 493-507.

Borgman L. E., 1969 Ocean wave simulation for engineering design. *J. Waterways and Harbors Div. ASCE*, 557-583.

Bretschneider C. L., 1952 Revised wave forecasting relationships. *Proc. 2nd Conf. Coastal Eng. ASCE.*

Goda Y., Takayama T. and Suzuki Y., 1978 Diffraction diagrams for directional random waves. *Proc. 16th Conf. Coastal Eng. ASCE*, 628-650.

Hasselmann K., Barnett T. P., Bouws E. et al., 1973 Measurements of wind wave growth and swell decay during the Joint North Sea Wave Project (JONSWAP). *Deut. Hydrogr. Zeit. A8*, 1-95.

Longuet-Higgins M. S., 1957 On the transformation of the continuous spectrum by refraction. *Proc. Cambridge Phil. Soc.* 53, 226-229;
1963 The effects of non-linearities on statistical distributions in the theory of sea waves. *J. Fluid Mech.* 17, 459-480.

Mitsuyasu H., Tasai F., Suhara T. et al., 1975 Observation of directional spectrum of ocean waves using a clover-leaf buoy. *J. Phys. Oceanogr.* 5, 750-760.

Phillips O. M., 1967 The theory of wind-generated waves. *Advances in Hydroscience* 4, 119-149.

Sverdrup H. U. and Munk W. H., 1947 Wind sea and swell; theory of relations for forecasting. *U.S. Navy Hydrogr. Office, Publ.* 601.

Chapter 9

THE THEORY OF QUASI-DETERMINISM

9.1 A sufficient condition for occurrence of a wave of given height very large

9.1.1 *Condition (9.1)*

Let us consider the surface displacement $\eta(t)$ at any fixed point $\mathbf{x}_o = (x_o, y_o)$ in a random wave field, it being homogeneous like (8.4*a*-*b*) or non-homogeneous like (8.36*a*-*b*) or (8.39*a*-*b*). From the theory of the sea states, we know that $\eta(t)$ represents a stationary random Gaussian process.

Let us analyse the p.d.f. of the surface displacement at any fixed time, given the condition

$$\boxed{\eta(t_o) = \frac{1}{2}H, \quad \eta(t_o + T^*) = -\frac{1}{2}H}, \tag{9.1}$$

where t_o is an arbitrary time instant, H an arbitrary height and T^* is the abscissa of the absolute minimum of the autocovariance function, which is assumed to be also the first local minimum of this function on the positive domain.

9.1.2 *The probability of $\eta(t_o + T)$ given condition (9.1)*

The p.d.f. of the surface displacement at time $t_o + T$ (T being any fixed time lag), given condition (9.1), is

$$\begin{aligned} &p\left[\eta(t_o + T) = u \,\middle|\, \eta(t_o) = \frac{1}{2}H, \eta(t_o + T^*) = -\frac{1}{2}H\right] = \\ &= \frac{p\left[\eta(t_o) = \frac{1}{2}H, \eta(t_o + T^*) = -\frac{1}{2}H, \eta(t_o + T) = u\right]}{p\left[\eta(t_o) = \frac{1}{2}H, \eta(t_o + T^*) = -\frac{1}{2}H\right]} \end{aligned} \tag{9.2}$$

(u being the independent variable). The general form of the probability density functions like the ones on the r.h.s. of (9.2) was given in sect. 5.1. Specifically, we have

$$p\left[\eta(t_o)=\frac{1}{2}H,\,\eta(t_o+T^*)=-\frac{1}{2}H,\eta(t_o+T)=u\right]= \tag{9.3a}$$

$$=\frac{1}{(2\pi)^{3/2}\sqrt{A}}\exp\left\{-\frac{1}{2A}\left[A_{33}u^2+2(A_{13}-A_{23})\frac{1}{2}Hu+(A_{11}+A_{22}-2A_{12})\frac{1}{4}H^2\right]\right\},$$

$$p\left[\eta(t_o)=\frac{1}{2}H,\eta(t_o+T^*)=-\frac{1}{2}H\right]=\frac{1}{2\pi\sqrt{\tilde{A}}}\exp\left[-\frac{1}{2\tilde{A}}(\tilde{A}_{11}+\tilde{A}_{22}-2\tilde{A}_{12})\frac{1}{4}H^2\right], \tag{9.3b}$$

where A_{ij} and A are, respectively, the i, j cofactor and the determinant of the covariance matrix of $\eta(t_o),\eta(t_o+T^*),\eta(t_o+T)$:

$$\begin{matrix}A_{ij}\equiv i,\,j\text{ cofactor}\\ A\equiv\text{determinant of}\end{matrix}\begin{pmatrix}\psi(0) & \psi(T^*) & \psi(T)\\ \psi(T^*) & \psi(0) & \psi(T-T^*)\\ \psi(T) & \psi(T-T^*) & \psi(0)\end{pmatrix},$$

and $\tilde{A}_{ij}$ and $\tilde{A}$ are, respectively, the i, j cofactor and the determinant of the covariance matrix of $\eta(t_o),\eta(t_o+T^*)$:

$$\begin{matrix}\tilde{A}_{ij}\equiv i,\,j\text{ cofactor}\\ \tilde{A}\equiv\text{determinant of}\end{matrix}\begin{pmatrix}\psi(0) & \psi(T^*)\\ \psi(T^*) & \psi(0)\end{pmatrix}.$$

With (9.3a) and (9.3b), equation (9.2) of the conditional p.d.f. becomes

$$p\left[\eta(t_o+T)=u\,\middle|\,\eta(t_o)=\frac{1}{2}H,\,\eta(t_o+T^*)=-\frac{1}{2}H\right]=\sqrt{\frac{A_{33}}{2\pi A}}\exp\left\{-\frac{1}{2A}\left[A_{33}u^2+\right.\right. \tag{9.4}$$

$$\left.\left.+2(A_{13}-A_{23})\frac{1}{2}Hu+(A_{11}+A_{22}-2A_{12})\frac{1}{4}H^2-\frac{A}{A_{33}}(\tilde{A}_{11}+\tilde{A}_{22}-2\tilde{A}_{12})\frac{1}{4}H^2\right]\right\},$$

where we have used the equality

$$\tilde{A}=A_{33}.$$

The argument $\arg(u)$ of the exponential function on the r.h.s. of (9.4), that is to say

$$\begin{aligned}\arg(u)\equiv-\frac{1}{2A}&\left[A_{33}u^2+2(A_{13}-A_{23})\frac{1}{2}Hu+\right.\\ &\left.+(A_{11}+A_{22}-2A_{12})\frac{1}{4}H^2-\frac{A}{A_{33}}(\tilde{A}_{11}+\tilde{A}_{22}-2\tilde{A}_{12})\frac{1}{4}H^2\right],\end{aligned} \tag{9.5}$$

is rewritten in the form

$$\arg(u) = -\frac{A_{33}}{2A}(u - u_m)^2 + \arg(u_m), \tag{9.6}$$

where u_m is the abscissa of the maximum of this function:

$$u_m = \frac{A_{23} - A_{13}}{A_{33}}\frac{H}{2}. \tag{9.7}$$

(The determinant and the i, i cofactors of a covariance matrix are positive, and this is why we have concluded that the function arg(u) has a maximum.)

The maximum arg(u_m) proves to be zero. Hence, the conditional p.d.f. is reduced to the simple form

$$p\left[\eta(t_o + T) = u \,\middle|\, \eta(t_o) = \frac{1}{2}H, \eta(t_o + T^*) = -\frac{1}{2}H\right] = \sqrt{\frac{A_{33}}{2\pi A}}\exp\left[-\frac{A_{33}}{2A}(u - u_m)^2\right]. \tag{9.8}$$

Verify that

$$\arg(u_m) = 0. \tag{9.9}$$

You have two alternative ways to do this. First way: from (9.5) and (9.7) obtain

$$\arg(u_m) = -\frac{1}{2AA_{33}}\left[-(A_{13} - A_{23})^2 + (A_{11} + A_{22} - 2A_{12})A_{33} - (\tilde{A}_{11} + \tilde{A}_{22} - 2\tilde{A}_{12})A\right]\frac{1}{4}H^2.$$

Then, replacing A, A_{ij} and $\tilde{A}_{ij}$ with their own expressions, you will find that the sum within square parentheses on the r.h.s. of this equation is zero.

Second way (more elegant): rewrite (9.4) in the form

$$p\left[\eta(t_o + T) = u \,\middle|\, \eta(t_o) = \frac{1}{2}H, \eta(t_o + T^*) = -\frac{1}{2}H\right] = \sqrt{\frac{A_{33}}{2\pi A}}\exp\left[-\frac{A_{33}}{2A}(u - u_m)^2 + \arg(u_m)\right]$$

which implies

$$\int_{-\infty}^{+\infty} p\left[\eta(t_o + T) = u \,\middle|\, \eta(t_o) = \frac{1}{2}H, \eta(t_o + T^*) = -\frac{1}{2}H\right]\mathrm{d}u =$$

$$= \sqrt{\frac{A_{33}}{2\pi A}}\int_{-\infty}^{+\infty}\exp\left[-\frac{A_{33}}{2A}(u - u_m)^2 + \arg(u_m)\right]\mathrm{d}u = \exp[\arg(u_m)]$$

which in its turn implies (9.9).

9.1.3 *The mean value and the standard deviation of $\eta(t_o + T)$ given condition (9.1)*

Equation (9.8) implies that the mean value of $\eta(t_o + T)$ (given condition 9.1) is u_m:

$$\overline{\eta}(t_o + T) = u_m.$$

The formula for u_m is (9.7) which is rewritten as

$$\overline{\eta}(t_o + T) = F(T)\,\frac{H}{2}\,, \tag{9.10}$$

on defining

$$F(T) \equiv \frac{\psi(T) - \psi(T - T^*)}{\psi(0) - \psi(T^*)}\,.$$

Equation (9.8) implies also that the standard deviation of $\eta(t_o + T)$, given condition (9.1), is

$$\sigma(t_o + T) = \sqrt{\frac{A}{A_{33}}} =$$

$$= \sqrt{\psi(0)\left[1 - \frac{\psi^2(T) + \psi^2(T - T^*) - 2\psi(T)\,\psi(T - T^*)\,[\psi(T^*)/\psi(0)]}{\psi^2(0) - \psi^2(T^*)}\right]}$$

from which it follows

$$\sigma(t_o + T) < \sigma = \sqrt{\psi(0)}\,, \tag{9.11}$$

since

$$\psi^2(0) - \psi^2(T^*) > 0 \quad \text{and} \quad |\psi(T^*)/\psi(0)| < 1\,.$$

Equality (9.10) and inequality (9.11) imply

$$\frac{\sigma(t_o + T)}{\overline{\eta}(t_o + T)} \to 0 \quad \text{as} \quad \frac{H}{\sigma} \to \infty\,,$$

provided $F(T)$ is nonzero. In words: given condition (9.1), the quotient between the standard deviation and the mean value of $\eta(t_o + T)$ approaches zero as $H/\sigma \to \infty$. Since T is arbitrary, it follows that, given condition (9.1), *the random function $\eta(t_o + T)$ is asymptotically equal to the deterministic function $\overline{\eta}(t_o + T)$, as $H/\sigma \to \infty$.* Of course, this conclusion is valid for finite values of T, given that

$$F(T) \to 0 \quad \text{and} \quad \sigma(t_o + T) \to \sigma \quad \text{as} \quad |T| \to \infty\,.$$

Note, σ is the standard deviation of the random process $\eta(t)$, while $\sigma(t_o + T)$ is a conditional standard deviation. Specifically, $\sigma(t_o + T)$ is the standard deviation of η at time $t_o + T$, given the condition (9.1).

9.1.4 *Proof that (9.1) is a sufficient condition*

So far we have not used the property of T^*, so that T^* can be any fixed time lag. Now it is time for the peculiarity of T^* to come into the picture.

Since $\psi(0)$ and $\psi(T^*) = \psi(-T^*)$ are, respectively, the absolute maximum and the absolute minimum of $\psi(T)$, $\overline{\eta}$ has its absolute maximum (equal to $\frac{1}{2}H$) at $T = 0$

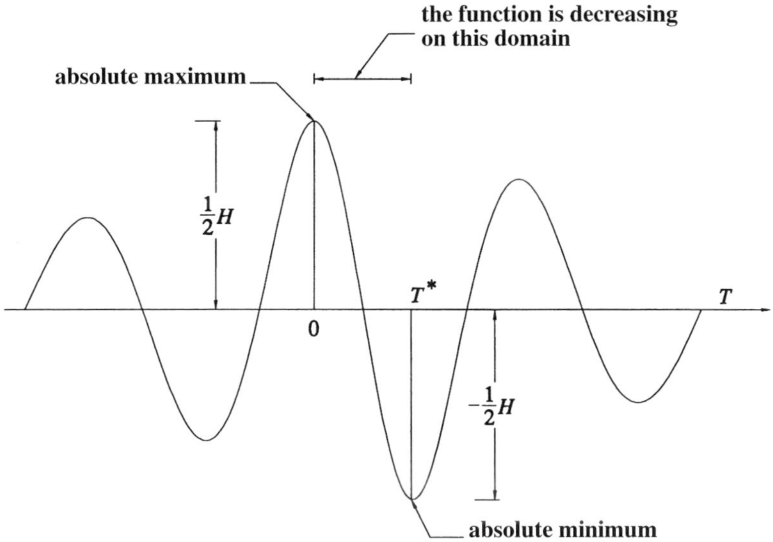

Fig. 9.1 General characteristics of $\overline{\eta}(t_o + T)$ [function (9.10)].

and its absolute minimum (equal to $-\frac{1}{2}H$) at $T = T^*$, and hence at $T = 0$ there is a wave crest, and at $T = T^*$ there is a wave trough.

Since T^* is the abscissa of the first local minimum of $\psi(T)$ after $T = 0$, the derivative $\dot{\psi}(T)$ is negative on the interval $(0, T^*)$:

$$\dot{\psi}(T) < 0 \quad \text{on}\,(0, T^*), \tag{9.12}$$

which implies

$$\dot{\psi}(T) > 0 \quad \text{on}\,(-T^*, 0),$$

(since ψ is an even function), or alternatively

$$\dot{\psi}(T - T^*) > 0 \quad \text{on}\,(0, T^*). \tag{9.13}$$

The inequalities (9.12), (9.13) imply that $\overline{\eta}$ is strictly decreasing for T in $[0, T^*]$. Therefore, the wave crest at $T = 0$ and the wave trough at $T = T^*$ belong to the same wave, and this wave has a crest-to-trough height equal to H [see fig. 9.1]. Hence, the conclusion

as $H/\sigma \to \infty$, condition (9.1)
becomes sufficient for the occurrence of a wave of given height H.

9.2 A necessary condition for occurrence of a wave of given height very large

9.2.1 *A general necessary condition*

A general necessary condition for occurrence of a wave of given height H is that the surface displacement is ξH, with ξ in (0,1), at a time t_o and is $(\xi - 1)H$ at a later time $t_o + \tau$ (t_o being the instant of the wave crest and $t_o + \tau$ the instant of the wave trough). The mathematical form of this general necessary condition is

$$\eta(t_o) = \xi H, \quad \eta(t_o + \tau) = (\xi - 1)H \qquad \text{with } 0 < \xi < 1.$$

For focusing the general necessary condition, look at fig. 5.5 [sect. 5.7.3] which shows two waves with a fixed height H and different values of ξ and τ.

9.2.2 *The probability* $\mathscr{P}(H, \tau, \xi)$

Let us consider the probability that the surface displacement at an instant t_o falls between

$$\xi H \text{ and } \xi H + \mathrm{d}\eta_1$$

and, at a later instant $t_o + \tau$, falls between

$$(\xi - 1)H \text{ and } (\xi - 1)H + \mathrm{d}\eta_2$$

t_o, H, τ, ξ being arbitrarily fixed, and $\mathrm{d}\eta_1$, $\mathrm{d}\eta_2$ being two fixed small intervals. The probability under examination is given by

$$\mathscr{P}(H, \tau, \xi) = \frac{1}{2\pi\sqrt{\tilde{\tilde{A}}}} \exp\left\{-\frac{1}{2\tilde{\tilde{A}}}\left[\tilde{\tilde{A}}_{11}\,\xi^2 H^2 + \tilde{\tilde{A}}_{22}(\xi - 1)^2 H^2 + 2\tilde{\tilde{A}}_{12}\,\xi(\xi - 1)H^2\right]\right\} \mathrm{d}\eta_1\,\mathrm{d}\eta_2, \tag{9.14}$$

where $\tilde{\tilde{A}}_{ij}$ and $\tilde{\tilde{A}}$ are, respectively, the i, j cofactor and the determinant of the covariance matrix of $\eta(t_o)$ and $\eta(t_o + \tau)$:

$$\begin{aligned} \tilde{\tilde{A}}_{ij} &\equiv i, j \text{ cofactor} \\ \tilde{\tilde{A}} &\equiv \text{determinant of} \end{aligned} \quad \begin{pmatrix} \psi(0) & \psi(\tau) \\ \psi(\tau) & \psi(0) \end{pmatrix}.$$

Equation (9.14) is rewritten in the form

$$\mathscr{P}(H, \tau, \xi) = \frac{1}{2\pi\sqrt{\psi^2(0) - \psi^2(\tau)}} \exp\left[-\frac{1}{2} f(\tau, \xi)\left(\frac{H}{\sigma}\right)^2\right] \mathrm{d}\eta_1\,\mathrm{d}\eta_2, \tag{9.15}$$

with

$$f(\tau, \xi) \equiv \frac{1}{2}\,\frac{\psi(0)}{\psi(0) - \psi(\tau)} + 2\,\frac{\psi(0)}{\psi(0) + \psi(\tau)}\left(\xi - \frac{1}{2}\right)^2. \tag{9.16}$$

Check (9.16). First, compare the exponential function on the r.h.s. of (9.14) with the exponential function on the r.h.s. of (9.15), so obtaining

$$f(\tau,\xi)=\frac{\psi(0)}{\tilde{\tilde{A}}}\left[\tilde{\tilde{A}}_{11}\xi^2+\tilde{\tilde{A}}_{22}(\xi-1)^2+2\tilde{\tilde{A}}_{12}\xi(\xi-1)\right].$$

Then put $\tilde{\tilde{A}}$ and $\tilde{\tilde{A}}_{ij}$ in terms of $\psi(0)$ and $\psi(\tau)$, obtaining

$$f(\tau,\xi)=\frac{\psi(0)}{\psi^2(0)-\psi^2(\tau)}\left[2(\psi(0)-\psi(\tau))(\xi^2-\xi)+\psi(0)\right].$$

Hence, using the equality

$$\xi^2-\xi=\left(\xi-\frac{1}{2}\right)^2-\frac{1}{4},$$

you will easily arrive at (9.16).

9.2.3 *Analysis of the function $f(\tau,\xi)$*

The first term on the r.h.s. of (9.16), that is to say

$$\frac{1}{2}\,\frac{\psi(0)}{\psi(0)-\psi(\tau)},$$

is independent of ξ, and its absolute minimum on the domain $\tau>0$ occurs at $\tau=T^*$ [given that T^* is the abscissa of the absolute minimum of $\psi(\tau)$]. The second term on the r.h.s. of (9.16), that is to say

$$2\,\frac{\psi(0)}{\psi(0)+\psi(\tau)}\left(\xi-\frac{1}{2}\right)^2,$$

is zero if $\xi=\frac{1}{2}$, and is greater than zero if $\xi\neq\frac{1}{2}$. Therefore, the absolute minimum of the function $f(\tau,\xi)$ for τ in $(0,\infty)$ and ξ in $(-\infty,\infty)$ occurs at $\tau=T^*$ and $\xi=\frac{1}{2}$.

9.2.4 *Condition (9.1) is necessary*

The fact that the absolute minimum of $f(\tau,\xi)$ occurs at $\tau=T^*$, $\xi=\frac{1}{2}$ implies

$$\frac{\mathscr{P}(H,\tau,\xi)}{\mathscr{P}\left(H,T^*,\frac{1}{2}\right)}\to 0\quad\text{as}\quad\frac{H}{\sigma}\to\infty,$$

for every fixed pair τ,ξ with $\tau\neq T^*$ and/or $\xi\neq\frac{1}{2}$. In words: as $H/\sigma\to\infty$, the probability that the surface displacement is ξH at an instant t_o and is $(\xi-1)H$ at an instant $t_o+\tau$, with $\tau\neq T^*$ and/or $\xi\neq\frac{1}{2}$, is negligible with respect to the probability that the surface displacement is $\frac{1}{2}H$ at t_o and is $-\frac{1}{2}H$ at $t_o + T^*$. Hence, the idea that a wave with a given height H has necessarily $\tau = T^*$ and $\xi = \frac{1}{2}$, as $H/\sigma \to \infty$. That is

as $H/\sigma \to \infty$, condition (9.1) becomes necessary for the occurrence of a wave of given height H.

Really, we have got a clue that condition (9.1) is necessary for a wave of given height H very large (and this is why we have written *hence the idea that* rather than *hence it follows that*). The formal proof calls for some harder work and yields as a corollary the closed solution for the wave height distribution. Such a proof will be dealt with later. Now we go on with the main stream of our reasoning.

9.3 The water surface on space-time, if a wave of given height very large occurs at a fixed point

9.3.1 *The conditional p.d.f. of the surface displacement at point $\boldsymbol{x}_o + \boldsymbol{X}$ and time $t_o + T$ given condition (9.1)*

If a wave with a very large height H occurs at a fixed point $\mathbf{x}_o (= x_o, y_o)$, how will the water surface all around this point be?

To say "a wave with a height H occurs at $\mathbf{x}_o$", as $H/\sigma \to \infty$, is equivalent to say "the surface displacement at $\mathbf{x}_o$ is equal to $\frac{1}{2}H$ at an instant t_o and is equal to $-\frac{1}{2}H$ at instant $t_o + T^*$". This is the gist of sects. 9.1 and 9.2. Hence, to answer the opening question of this section we have to examine the p.d.f. of the surface displacement at a point $\mathbf{x}_o + \mathbf{X}$ at an instant $t_o + T$ (both the point and instant being arbitrarily fixed) given the condition

$$\eta(\mathbf{x}_o, t_o) = \frac{1}{2}H, \quad \eta(\mathbf{x}_o, t_o + T^*) = -\frac{1}{2}H \tag{9.17}$$

which is the same as (9.1), only that we write $\eta(\mathbf{x}_o, t_o)$ in place of $\eta(t_o)$ because now we are dealing with the waves on space-time.

Through the same reasoning we used to arrive at (9.8), we arrive also at

$$p\left[\eta(\mathbf{x}_o + \mathbf{X}, t_o + T) = u \,\middle|\, \eta(\mathbf{x}_o, t_o) = \frac{1}{2}H, \eta(\mathbf{x}_o, t_o + T^*) = -\frac{1}{2}H\right] =$$
$$= \sqrt{\frac{B_{33}}{2\pi B}} \exp\left[-\frac{B_{33}}{2B}(u - u_m)^2\right], \tag{9.18}$$

where

$$u_m = \frac{B_{23} - B_{13}}{B_{33}} \frac{H}{2}, \tag{9.19}$$

and where B_{ij} and B are the i, j cofactor and determinant of the covariance matrix of $\eta(\mathbf{x}_o, t_o)$, $\eta(\mathbf{x}_o, t_o + T^*)$, $\eta(\mathbf{x}_o + \mathbf{X}, t_o + T)$:

$$\begin{array}{l} B_{ij} \equiv i,\, j \text{ cofactor} \\ B \equiv \text{determinant of} \end{array} \begin{pmatrix} \sigma^2 & \Psi(\mathbf{0}, T^*) & \Psi(\mathbf{X}, T) \\ \Psi(\mathbf{0}, T^*) & \sigma^2 & \Psi(\mathbf{X}, T - T^*) \\ \Psi(\mathbf{X}, T) & \Psi(\mathbf{X}, T - T^*) & \sigma_{\mathbf{X}}^2 \end{pmatrix}.$$

In this matrix σ^2 and $\sigma_{\mathbf{X}}^2$ are the variances of the surface displacement, respectively, at $\mathbf{x}_o$ and at $\mathbf{x}_o + \mathbf{X}$, and Ψ is the space-time covariance defined by (8.51*a*), which here has been written in the compact form

$$\Psi(\mathbf{X}, T) \equiv \Psi(X, Y, T; x_o, y_o). \tag{9.20}$$

9.3.2 *The mean value and the standard deviation of the surface displacement at point $\mathbf{x}_o + \mathbf{X}$ and time $t_o + T$ given condition (9.1)*

The formula (9.18) for the conditional p.d.f implies that the mean value of $\eta(\mathbf{x}_o + \mathbf{X}, t_o + T)$ given condition (9.17) is

$$\overline{\eta}(\mathbf{x}_o + \mathbf{X}, t_o + T) = \frac{\Psi(\mathbf{X}, T) - \Psi(\mathbf{X}, T - T^*)}{\Psi(\mathbf{0}, 0) - \Psi(\mathbf{0}, T^*)} \frac{H}{2}. \tag{9.21}$$

Then (9.18) also implies that the standard deviation of $\eta(\mathbf{x}_o + \mathbf{X}, t_o + T)$, given condition (9.17), is

$$\sigma(\mathbf{x}_o + \mathbf{X}, t_o + T) = \sqrt{\frac{B}{B_{33}}} =$$

$$= \sqrt{\sigma_{\mathbf{X}}^2 - \sigma^2 \frac{\Psi^2(\mathbf{X}, T) + \Psi^2(\mathbf{X}, T - T^*) - 2\Psi(\mathbf{X}, T)\Psi(\mathbf{X}, T - T^*)[\Psi(\mathbf{0}, T^*)/\Psi(\mathbf{0}, 0)]}{\Psi^2(\mathbf{0}, 0) - \Psi^2(\mathbf{0}, T^*)}}$$

which yields

$$\sigma(\mathbf{x}_o + \mathbf{X}, t_o + T) < \sigma_{\mathbf{X}}, \tag{9.22}$$

since

$$\Psi^2(\mathbf{0}, 0) - \Psi^2(\mathbf{0}, T^*) > 0 \quad \text{and} \quad |\Psi(\mathbf{0}, T^*)/\Psi(\mathbf{0}, 0)| < 1.$$

Check (9.21). Equation (9.18) shows that the mean value is u_m. Hence, use the definition (9.19) of u_m, and obtain the expressions of B_{13}, B_{23}, and B_{33} from the covariance matrix.

Note the differences between the three standard deviations in this section:

(i) σ is the standard deviation of the random process $\eta(t)$ at point $\mathbf{x}_o$,
(ii) $\sigma_{\mathbf{X}}$ is the standard deviation of the random process $\eta(t)$ at point $\mathbf{x}_o + \mathbf{X}$,
(iii) $\sigma(\mathbf{x}_o + \mathbf{X}, t_o + T)$ is the standard deviation of η at point $\mathbf{x}_o + \mathbf{X}$ at time $t_o + T$, given condition (9.17).

9.3.3 *An alternative form for the mean value $\overline{\eta}(\mathbf{x}_o+\mathbf{X},t_o+T)$ given condition (9.1)*

The mean value (9.21) is rewritten in the form

$$\overline{\eta}(\mathbf{x}_o+\mathbf{X},t_o+T)=F(\mathbf{X},T)\,\sigma_{\mathbf{X}}\,\frac{H}{\sigma}\,, \tag{9.23}$$

where

$$F(\mathbf{X},T)\equiv\frac{1}{2}\,\frac{\widetilde{\Psi}(\mathbf{X},T)-\widetilde{\Psi}(\mathbf{X},T-T^*)}{1-\widetilde{\Psi}(\mathbf{0},T^*)}\,,$$

and where $\widetilde{\Psi}(\mathbf{X},T)$ is the normalized space-time covariance:

$$\widetilde{\Psi}(\mathbf{X},T)\equiv\Psi(\mathbf{X},T)/(\sigma\,\sigma_{\mathbf{X}})\,.$$

$\widetilde{\Psi}(\mathbf{X},T)$ and, consequently, $F(\mathbf{X},T)$ do not depend on σ nor on $\sigma_{\mathbf{X}}$; they depend only upon the degree of correlation of the random waves at $\mathbf{x}_o$ and at $\mathbf{x}_o+\mathbf{X}$ after a time lag T. Specifically, $\widetilde{\Psi}(\mathbf{X},T)$ takes on values within -1 and 1 for every σ and $\sigma_{\mathbf{X}}$. It is 1 if η/σ at $\mathbf{x}_o$ is always equal to $\eta/\sigma_{\mathbf{X}}$ at $\mathbf{x}_o+\mathbf{X}$ after a time lag T; it is -1 if η/σ at $\mathbf{x}_o$ is always equal to $-\eta/\sigma_{\mathbf{X}}$ at $\mathbf{x}_o+\mathbf{X}$ after a time lag T. Finally $\widetilde{\Psi}(\mathbf{X},T)$ approaches zero if η/σ at $\mathbf{x}_o$ at time t is uncorrelated with $\eta/\sigma_{\mathbf{X}}$ at $\mathbf{x}_o+\mathbf{X}$ at time $t+T$.

What is the aim of the alternative form (9.23)? Let us consider the case of an extremely non-homogeneous wave field like that of fig. 9.2, where σ is much smaller than $\sigma_{\mathbf{X}}$. Well, (9.22) and (9.23) reveal at once that the free surface displacement at point $\mathbf{x}_o+\mathbf{X}$ at time t_o+T is very close to the deterministic value $\overline{\eta}(\mathbf{x}_o+\mathbf{X},t_o+T)$,

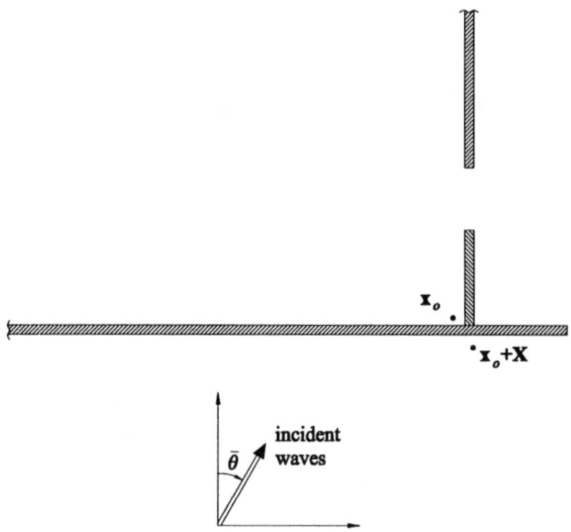

Fig. 9.2 If an exceptionally high wave occurs at $\mathbf{x}_o$, the surface displacement at $\mathbf{x}_o+\mathbf{X}$ approaches a well precise deterministic form. This, even if the waves at $\mathbf{x}_o$ are much smaller than at $\mathbf{x}_o+\mathbf{X}$.

if a wave with a height H exceptionally large with respect to σ occurs at $\mathbf{x}_o$. This, even if H is very small with respect to the wave heights at $\mathbf{x}_o + \mathbf{X}$.

9.3.4 *The main statement of the theory*

In conclusion, formula (9.23) for the mean value and inequality (9.22) for the standard deviation imply that

$$\frac{\sigma(\mathbf{x}_o + \mathbf{X}, t_o + T)}{\overline{\eta}(\mathbf{x}_o + \mathbf{X}, t_o + T)} \to 0 \quad \text{as} \quad \frac{H}{\sigma} \to \infty,$$

provided $F(\mathbf{X}, T)$ is nonzero. Since $\mathbf{X}$ and T are arbitrary it follows that, given condition (9.17), the random function $\eta(\mathbf{x}_o + \mathbf{X}, t_o + T)$ (function of $\mathbf{X}$ and T) is asymptotically equal to the deterministic function $\overline{\eta}(\mathbf{x}_o + \mathbf{X}, t_o + T)$, as $H/\sigma \to \infty$. Of course, this is true within a finite area and for a finite span of time, given that $\widetilde{\Psi}(\mathbf{X}, T)$, and consequently $F(\mathbf{X}, T)$, approaches zero as $|\mathbf{X}|$ and/or $|T|$ tend to infinity.

The conclusion is: *if a wave with a given height H occurs at a fixed point $\mathbf{x}_o$ and H is very large with respect to the mean wave height at this point, we can expect the water surface near $\mathbf{x}_o$ to be very close to the deterministic form* (9.21-23). (9.21 and 9.23 being two alternative forms of the same deterministic function.)

We shall see in chap. 10 that, if $\mathbf{x}_o$ is in the open sea, (9.21-23) represent a three-dimensional wave group whose centre passes at $\mathbf{x}_o$; if $\mathbf{x}_o$ is before a breakwater, (9.21-23) represent the collision of two wave groups, one approaching the breakwater and one going back seaward after reflection. In short, (9.21-23) show what happens if a very high wave occurs at the fixed point $\mathbf{x}_o$.

In what follows we shall use (9.21) since it seems more suitable for the practical applications.

9.4 The velocity potential if a wave of given height very large occurs at a fixed point

Associated with the deterministic configuration (9.21) is a distribution of velocity potential in the water, which to the lowest order in a Stokes' expansion is given by

$$\bar{\phi}(\mathbf{x}_o + \mathbf{X}, z, t_o + T) = \left\{ \frac{\Phi(\mathbf{X}, z, T) - \Phi(\mathbf{X}, z, T - T^*)}{\Psi(\mathbf{0}, 0) - \Psi(\mathbf{0}, T^*)} \right\} \frac{H}{2}, \tag{9.24}$$

where $\mathbf{X}$, z and T are the independent variables and Φ is the covariance of the surface displacement and velocity potential of the random wave field (sea state), which is defined by (8.51b) and here has been rewritten in the compact form

$$\Phi(\mathbf{X}, z, T) \equiv \Phi(X, Y, z, T; x_o, y_o). \tag{9.25}$$

The surface displacement (9.21) and the velocity potential (9.24) satisfy the linear flow equations. In particular we shall prove that $\overline{\eta}$ and $\overline{\phi}$ satisfy the first linear flow equation under the hypothesis that η and ϕ satisfy this equation. That is to say we shall prove that

$$\overline{\eta} = -\frac{1}{g}\left(\frac{\partial \overline{\phi}}{\partial T}\right)_{z=0}, \tag{9.26}$$

provided that

$$\eta = -\frac{1}{g}\left(\frac{\partial \phi}{\partial t}\right)_{z=0}. \tag{9.27}$$

With the formulae (9.21) and (9.24) for $\overline{\eta}$ and $\overline{\phi}$ and the definitions of the covariances Ψ and Φ, the equality (9.26) (to be proved) takes on the form

$$\begin{aligned} &\frac{H}{2}\,\frac{<\eta(\mathbf{x}_o,t)\,\eta(\mathbf{x}_o+\mathbf{X},t+T)> - <\eta(\mathbf{x}_o,t)\,\eta(\mathbf{x}_o+\mathbf{X},t+T-T^*)>}{<\eta^2(\mathbf{x}_o,t)> - <\eta(\mathbf{x}_o,t)\,\eta(\mathbf{x}_o,t+T^*)>} = \\ &= -\frac{1}{g}\frac{\partial}{\partial T}\left\{\frac{H}{2}\,\frac{<\eta(\mathbf{x}_o,t)\,\phi(\mathbf{x}_o+\mathbf{X},z,t+T)> - <\eta(\mathbf{x}_o,t)\,\phi(\mathbf{x}_o+\mathbf{X},z,t+T-T^*)>}{<\eta^2(\mathbf{x}_o,t)> - <\eta(\mathbf{x}_o,t)\,\eta(\mathbf{x}_o,t+T^*)>}\right\}_{z=0}, \end{aligned} \tag{9.28}$$

where the common term

$$<\eta^2(\mathbf{x}_o,t)> - <\eta(\mathbf{x}_o,t)\,\eta(\mathbf{x}_o,t+T^*)>$$

can be cancelled since it does not depend on T. With the definition of temporal mean, the equality (9.28) becomes

$$\begin{aligned} &\lim_{\mathcal{T}\to\infty}\frac{1}{\mathcal{T}}\int_0^{\mathcal{T}}\eta(\mathbf{x}_o,t)\,[\eta(\mathbf{x}_o+\mathbf{X},t+T)-\eta(\mathbf{x}_o+\mathbf{X},t+T-T^*)]\,\mathrm{d}t = \\ &= \lim_{\mathcal{T}\to\infty}\frac{1}{\mathcal{T}}\int_0^{\mathcal{T}}\eta(\mathbf{x}_o,t)\left(-\frac{1}{g}\right)\frac{\partial}{\partial T}[\phi(\mathbf{x}_o+\mathbf{X},z,t+T)-\phi(\mathbf{x}_o+\mathbf{X},z,t+T-T^*)]_{z=0}\,\mathrm{d}t, \end{aligned}$$

and is satisfied since

$$\eta(\mathbf{x}_o+\mathbf{X},t+T) = -\frac{1}{g}\frac{\partial}{\partial T}\phi(\mathbf{x}_o+\mathbf{X},z,t+T)_{z=0}, \tag{9.29a}$$

$$\eta(\mathbf{x}_o+\mathbf{X},t+T-T^*) = -\frac{1}{g}\frac{\partial}{\partial T}\phi(\mathbf{x}_o+\mathbf{X},z,t+T-T^*)_{z=0}, \tag{9.29b}$$

as a consequence of (9.27). [(9.29*a-b*) simply say that η and ϕ satisfy the first linear flow equation at point $\mathbf{x}_o+\mathbf{X}$ at time $t+T$ and at time $t+T-T^*$.]

As an exercise, the reader could verify that the deterministic functions $\overline{\eta}$ and $\overline{\phi}$ fulfill also the other linear flow equations. It will suffice to use the fact that the random functions η and ϕ satisfy those linear equations.

9.5 Theory's generality and consistency with Stokes' theory

9.5.1 *The theory holds for an arbitrary solid boundary*

Following the theory of the sea states, we assumed that the surface displacement at any fixed point represented a stationary Gaussian random process of time. However, we did not call for the process to be homogeneous in space. Indeed, we assumed that the variance $\sigma_{\mathbf{X}}^2$ of the surface displacement at $\mathbf{x}_o + \mathbf{X}$ was generally different from the variance σ^2 of the surface displacement at $\mathbf{x}_o$, which is the case if some solid body causes wave reflection or diffraction.

Thus, the theory holds whichever the configuration of the solid boundary (provided, of course, the flow can be regarded as ideal). As a confirmation of this important property, let us prove that the deterministic velocity potential $\overline{\phi}$ automatically fulfils any solid boundary condition, provided the random velocity potential ϕ fulfils this boundary condition.

Let a point $\mathbf{x}', z'$ belong to a solid surface and $\mathbf{n}$ be the unit vector orthogonal to the solid surface at this point. Knowing that the velocity potential ϕ of the random waves fulfils the solid boundary condition at $\mathbf{x}', z'$ that is knowing that

$$\mathbf{n} \cdot \nabla\phi = 0 \quad \text{for } \mathbf{x} = \mathbf{x}' \quad \text{and} \quad z = z', \tag{9.30}$$

we aim to prove that the deterministic velocity potential $\overline{\phi}$ also fulfils this solid boundary condition. In other words we aim to prove that

$$\mathbf{n} \cdot \nabla\overline{\phi} = 0 \quad \text{for } \mathbf{X} = \mathbf{x}' - \mathbf{x}_o \quad \text{and} \quad z = z'. \tag{9.31}$$

With the formula (9.24) for $\overline{\phi}$, the equality (9.31) to be proved becomes

$$\mathbf{n} \cdot \nabla \left\{ \frac{< \eta(\mathbf{x}_o, t)\, \phi(\mathbf{x}_o + \mathbf{X}, z, t + T) > - < \eta(\mathbf{x}_o, t)\, \phi(\mathbf{x}_o + \mathbf{X}, z, t + T - T^*) >}{< \eta^2(\mathbf{x}_o, t) > - < \eta(\mathbf{x}_o, t)\, \eta(\mathbf{x}_o, t + T^*) >} \frac{H}{2} \right\} = 0$$

$$\text{for } \mathbf{X} = \mathbf{x}' - \mathbf{x}_o \text{ and } z = z'.$$

Given that $\mathbf{X}$ and z are the only space variables, this equality is rewritten in the form

$$< \eta(\mathbf{x}_o, t)\, \mathbf{n} \cdot \nabla\phi(\mathbf{x}_o + \mathbf{X}, z, t + T) > - < \eta(\mathbf{x}_o, t)\, \mathbf{n} \cdot \nabla\phi(\mathbf{x}_o + \mathbf{X}, z, t + T - T^*) > = 0$$

$$\text{for } \mathbf{X} = \mathbf{x}' - \mathbf{x}_o \text{ and } z = z',$$

and is proved since

$$\mathbf{n} \cdot \nabla\phi(\mathbf{x}_o + \mathbf{X}, z, t) = 0 \qquad \text{for } \mathbf{X} = \mathbf{x}' - \mathbf{x}_o \text{ and } z = z',$$

for every t, as a straightforward consequence of (9.30).

9.5.2 *The theory holds for nearly arbitrary spectra*

The theory is general not only because the solid boundary is arbitrary, but also because the spectrum is nearly arbitrary. Indeed the only restriction is that the autocovariance must have an absolute minimum which is also the first local minimum after $T = 0$. We already saw in chap. 4 that this condition is satisfied by the wind waves, and might not be satisfied in special cases of wind waves superimposed on swells of somewhat greater height and very different period.

9.5.3 *Consistency with Stokes' theory*

At first glance one could catch a discrepancy between the assumptions of the quasi-determinism theory and those of Stokes' theory. Indeed the quasi-determinism theory calls for the wave height H to be very large, and, on the contrary, Stokes' theory calls for the wave height H to be very small. But it takes short time to realize that there is no discrepancy.

In Stokes' theory the wave height H must be very small with respect to *wavelength L and depth d*, while in the quasi-determinism theory H must be very large with respect to the *standard deviation σ of the surface displacement in the sea state.* Hence, the two assumptions are not inconsistent with each other, given that H can well be very small with respect to d and L, even if it is very large with respect to σ.

9.6 Formal proof of the necessary condition. Part I: symbols and assumptions

The formal proof of the conclusive statement of sect. 9.2: "condition (9.1) is necessary..." remains to be given. It was not given in sect. 9.2 to make easier the overall view of the theory.

We adopt the compact symbols

$$\eta_t \text{ and } \psi_T \text{ in place of } \eta(t) \text{ and } \psi(T),$$

and

$$\alpha \equiv H/\sigma.$$

Then, aiming to make the algorithms easier, we assume without loss of generality that both the zeroth and the second moment of the spectrum are equal to 1:

$$m_0 = \sigma^2 = 1, \quad m_2 = 1, \tag{9.32}$$

which implies

$$\psi_0 = 1, \quad \ddot{\psi}_0 = -1, \tag{9.33a}$$

$$|\psi_T| < 1 \text{ and } |\ddot{\psi}_T| < 1 \quad \text{if } T \neq 0. \tag{9.33b}$$

The moment m_4 will appear in some intermediate steps of the proof, and this could suggest the idea that our analysis does not cover spectra like JONSWAP which has not m_4 (the moments of the JONSWAP spectrum exist as far as m_3); but the matter is not so. Let us see why.

Let us fix ω_{sup} such that

$$\frac{\int_0^{\omega_{\text{sup}}} \omega^2 E(\omega)\,\mathrm{d}\omega}{\int_0^{\infty} \omega^2 E(\omega)\,\mathrm{d}\omega} \cong 1, \tag{9.34}$$

and let us define the spectrum

$$\widetilde{E}(\omega)\begin{cases} \equiv E(\omega) & \text{if} \quad \omega \le \omega_{\text{sup}}, \\ \equiv 0 & \text{if} \quad \omega > \omega_{\text{sup}}, \end{cases}$$

which obviously has the moments of any order. What we have done is to cut off a part of the high frequency tail, of negligible energy content. Indeed (9.34) implies

$$\int_{\omega_{\text{sup}}}^{\infty} E(\omega)\,\mathrm{d}\omega \ll \int_0^{\infty} E(\omega)\,\mathrm{d}\omega.$$

Passing from a random process with $E(\omega)$ (=JONSWAP spectrum) to the random process with $\widetilde{E}(\omega)$ (=JONSWAP spectrum with a high frequency cutoff) is like passing from the function

$$\eta(t) = a\cos(\omega t) + \frac{a}{n^{1.5}}\cos(n\omega t) \quad \text{as} \quad n \to \infty, \tag{9.35}$$

to the function

$$\eta(t) = a\cos(\omega t).$$

Indeed, the high frequency term

$$\frac{a}{n^{1.5}}\cos(n\omega t)$$

gives a negligible contribution to m_0 and m_2, while it makes infinitely large m_4:

$$m_0 = \frac{1}{2}a^2 + \frac{1}{2}\frac{a^2}{n^3},$$

$$m_2 = \frac{1}{2}a^2\omega^2 + \frac{1}{2}\frac{a^2}{n}\omega^2,$$

$$m_4 = \frac{1}{2}a^2\omega^4 + \frac{1}{2}n\,a^2\omega^4.$$

From (9.35) we easily realize that the high frequency term does not alter the crest elevation, nor the trough depth, nor the time interval between the crest and trough, nor the wave period. It simply ruffles the wave surface with a lot of very small ripples. Therefore the high frequency term can be neglected when the goal of the analysis is concerned with the wave height, or the crest elevation, or the trough depth, or the time lag between crest and trough, or the wave period.

This is why in the proof we assume the high frequency cutoff and the existence of m_4. Then, given that

$$m_4 = \left(\frac{\mathrm{d}^4\psi}{\mathrm{d}T^4}\right)_{T=0},$$

$$\left|\frac{\mathrm{d}^4\psi}{\mathrm{d}T^4}\right| < \left(\frac{\mathrm{d}^4\psi}{\mathrm{d}T^4}\right)_{T=0} \quad \text{for } T \neq 0,$$

the existence of m_4 implies also the existence of the fourth derivative of $\psi(T)$.

9.7 Formal proof of the necessary condition. Part II: core of the proof

9.7.1 *The definition and the general form of $EX(\alpha, \tau, \xi)$*

Let us consider the expected number per unit time

$$EX(\alpha, \tau, \xi)\,\mathrm{d}\alpha\,\mathrm{d}\tau\,\mathrm{d}\xi$$

of local maxima of the surface displacement $\eta(t)$ (at the fixed point $\mathbf{x}_o$) whose elevation is between

$$\xi\alpha \text{ and } (\xi + \mathrm{d}\xi)\alpha,$$

and which are followed by a local minimum with an elevation between

$$(\xi - 1)\alpha \quad \text{and} \quad (\xi - 1)\alpha - \mathrm{d}\alpha,$$

after a time lag between

$$\tau \text{ and } \tau + \mathrm{d}\tau$$

(α, τ, ξ being arbitrarily fixed). Calling $\mathscr{P}$ the probability that a fixed small interval $\mathrm{d}t$ contains such a local maximum of $\eta(t)$ (i.e. a local maximum of the given elevation being followed after the fixed time lag by a local minimum of the given elevation), we have

$$EX(\alpha, \tau, \xi)\,\mathrm{d}\alpha\,\mathrm{d}\tau\,\mathrm{d}\xi = \frac{\mathscr{P}\mathscr{T}/\mathrm{d}t}{\mathscr{T}}, \tag{9.36}$$

where $\mathscr{P}\mathscr{T}/\mathrm{d}t$ is the number of the local maxima with the given characteristics which occur in a very large time $\mathscr{T}$.

Equation (9.36) shows that our expected number per unit time represents a Rice's problem (cf. sect. 5.6). The solution of this problem leads to

$$EX(\alpha, \tau, \xi) = \alpha \int_{-\infty}^{0} \int_{0}^{\infty} |u|\, w p\left[\eta_0 = \xi\alpha, \dot{\eta}_0 = 0, \ddot{\eta}_0 = u, \eta_\tau = (\xi - 1)\alpha, \dot{\eta}_\tau = 0, \ddot{\eta}_\tau = w\right] \mathrm{d}w\,\mathrm{d}u.$$

This is rewritten in the form

$$EX(\alpha,\tau,\xi) = \alpha\, p\left[\eta_0 = \xi\alpha, \dot{\eta}_0 = 0, \dot{\eta}_\tau = 0, \eta_\tau = (\xi-1)\alpha\right] \int_{-\infty}^{0}\int_{0}^{\infty} |u|\, w \cdot \\ \cdot\, p\left[\ddot{\eta}_0 = u, \ddot{\eta}_\tau = w \mid \eta_0 = \xi\alpha, \dot{\eta}_0 = 0, \dot{\eta}_\tau = 0, \eta_\tau = (\xi-1)\alpha\right] \mathrm{d}w\, \mathrm{d}u, \tag{9.37}$$

and the joint p.d.f. before the integral is expressed in the form

$$p\left[\eta_0 = \xi\alpha, \dot{\eta}_0 = 0, \dot{\eta}_\tau = 0, \eta_\tau = (\xi-1)\alpha\right] = \frac{1}{(2\pi)^2\sqrt{M}} \exp\left[-\frac{1}{4}\widehat{f}(\tau,\xi)\alpha^2\right], \tag{9.38}$$

where

$$\widehat{f}(\tau,\xi) \equiv \frac{4(M_{11}+M_{14})(\xi^2-\xi)+2M_{11}}{M}, \tag{9.39}$$

and where M_{ij} and M are respectively the i, j cofactor and determinant of the covariance matrix of $\eta_0, \dot{\eta}_0, \dot{\eta}_\tau, \eta_\tau$:

$$\begin{array}{l} M_{ij} \equiv i, j \text{ cofactor} \\ M \equiv \text{determinant of} \end{array} \quad \begin{pmatrix} 1 & 0 & \dot{\psi}_\tau & \psi_\tau \\ 0 & 1 & -\ddot{\psi}_\tau & -\dot{\psi}_\tau \\ \dot{\psi}_\tau & -\ddot{\psi}_\tau & 1 & 0 \\ \psi_\tau & -\dot{\psi}_\tau & 0 & 1 \end{pmatrix}.$$

Check the solution to Rice's problem. To this end, you have to obtain the probability $\mathscr{P}$ on the r.h.s. of (9.36). This probability is equal to the integral over

$$\{(u, w) | u \in (-\infty, 0) \text{ and } w \in (0, \infty)\}$$

of the probability that

(i) η_0 falls between $\xi\alpha$ and $(\xi + \mathrm{d}\xi)\alpha$,
(ii) $\ddot{\eta}_0$ falls between u and $u + \mathrm{d}u$ $(u < 0)$,
(iii) $\dot{\eta}_0$ falls between $-|u|\mathrm{d}t/2$ and $|u|\mathrm{d}t/2$,
(iv) η_τ falls between $(\xi-1)\alpha$ and $(\xi-1)\alpha - \mathrm{d}\alpha$,
(v) $\ddot{\eta}_\tau$ falls between w and $w + \mathrm{d}w$ $(w > 0)$,
(vi) $\dot{\eta}_\tau$ falls between $-w\mathrm{d}\tau/2$ and $w\mathrm{d}\tau/2$.

The logic is the same as we developed in sects. 5.3 and 5.4.

Check also (9.38-39) of the joint p.d.f. [hint: use the general form (5.10) of these probability density functions, and note that $M_{44} = M_{11}$]. Finally, check the covariance matrix, reasoning as we did in sect. 5.2.2.

9.7.2 *The exact form of the conditional p.d.f. on the r.h.s. of (9.37) as* $\alpha \to \infty$

Let us consider preliminarily the p.d.f. of the second derivative $\ddot{\eta}_0$, given the condition

$$\eta_0 = \xi\alpha, \dot{\eta}_0 = 0, \dot{\eta}_\tau = 0, \eta_\tau = (\xi-1)\alpha. \tag{9.40}$$

It can be given the compact form

$$p[\ddot{\eta}_0 = u \,|\, \eta_0 = \xi\alpha, \dot{\eta}_0 = 0, \dot{\eta}_\tau = 0, \eta_\tau = (\xi - 1)\alpha] = \\ = \sqrt{\frac{C_{55}}{2\pi C}} \exp\left\{-\frac{C_{55}}{2C}\left[u - \frac{-C_{15}\xi - C_{45}(\xi - 1)}{C_{55}}\alpha\right]^2\right\}, \tag{9.41}$$

which proceeds from essentially the same steps leading to (9.8). Here it will be understood that C_{ij} and C are the i, j cofactor and the determinant of the covariance matrix of η_0, $\dot{\eta}_0$, $\dot{\eta}_\tau$, η_τ, $\ddot{\eta}_0$:

$$\begin{matrix} C_{ij} \equiv i, j \text{ cofactor} \\ C \equiv \text{determinant of} \end{matrix} \quad \begin{pmatrix} 1 & 0 & \dot{\psi}_\tau & \psi_\tau & -1 \\ 0 & 1 & -\ddot{\psi}_\tau & -\dot{\psi}_\tau & 0 \\ \dot{\psi}_\tau & -\ddot{\psi}_\tau & 1 & 0 & \dddot{\psi}_\tau \\ \psi_\tau & -\dot{\psi}_\tau & 0 & 1 & \ddot{\psi}_\tau \\ -1 & 0 & \dddot{\psi}_\tau & \ddot{\psi}_\tau & m_4 \end{pmatrix}.$$

Equation (9.41) implies that the mean value of $\ddot{\eta}_0$ given the condition (9.40) is

$$\overline{\ddot{\eta}_0} = K_1(\tau, \xi)\alpha, \tag{9.42}$$

where

$$K_1(\tau, \xi) \equiv \frac{-C_{15}\xi - C_{45}(\xi - 1)}{C_{55}}, \tag{9.43}$$

and that the standard deviation is

$$\text{standard deviation of } \ddot{\eta}_0 = \sqrt{\frac{C}{C_{55}}}. \tag{9.44}$$

Equations (9.42) and (9.44) imply that

$$\left.\begin{matrix} \overline{\ddot{\eta}_0/\alpha} = K_1(\tau, \xi), \\ (\text{standard deviation of } \ddot{\eta}_0/\alpha) \to 0 \end{matrix}\right\} \text{ as } \alpha \to \infty,$$

so that we write

$$p[(\ddot{\eta}_0/\alpha) = u \,|\, \eta_0 = \xi\alpha, \dot{\eta}_0 = 0, \dot{\eta}_\tau = 0, \eta_\tau = (\xi - 1)\alpha] = \delta[u - K_1(\tau, \xi)] \text{ as } \alpha \to \infty, \tag{9.45}$$

where, as usual, $\delta(u - u_m)$ represents the delta function at $u = u_m$.

With the same sequence of steps leading to (9.45) we also arrive at

$$p[(\ddot{\eta}_\tau/\alpha) = w \,|\, \eta_0 = \xi\alpha, \dot{\eta}_0 = 0, \dot{\eta}_\tau = 0, \eta_\tau = (\xi - 1)\alpha] = \delta[w - K_2(\tau, \xi)] \text{ as } \alpha \to \infty, \tag{9.46}$$

where

$$K_2(\tau,\xi) \equiv \frac{-D_{15}\,\xi - D_{45}\,(\xi-1)}{D_{55}}, \tag{9.47}$$

and where D_{ij} is the i, j cofactor of the covariance matrix of η_0, $\dot{\eta}_0$, $\dot{\eta}_\tau$, η_τ, $\ddot{\eta}_\tau$:

$$D_{ij} \equiv \ i,j \text{ cofactor of} \begin{pmatrix} 1 & 0 & \dot{\psi}_\tau & \psi_\tau & \ddot{\psi}_\tau \\ 0 & 1 & -\ddot{\psi}_\tau & -\dot{\psi}_\tau & -\dddot{\psi}_\tau \\ \dot{\psi}_\tau & -\ddot{\psi}_\tau & 1 & 0 & 0 \\ \psi_\tau & -\dot{\psi}_\tau & 0 & 1 & -1 \\ \ddot{\psi}_\tau & -\dddot{\psi}_\tau & 0 & -1 & m_4 \end{pmatrix}.$$

Equations (9.45) and (9.46) yield

$$p\,[(\ddot{\eta}_0/\alpha) = u,\, (\ddot{\eta}_\tau/\alpha) = w \,|\, \eta_0 = \xi\alpha,\, \dot{\eta}_0 = 0,\, \dot{\eta}_\tau = 0,\, \eta_\tau = (\xi-1)\,\alpha] =$$
$$= \delta\,[u - K_1(\tau,\xi),\, w - K_2(\tau,\xi)] \text{ as } \alpha \to \infty,$$

where $\delta\,(u - u_m, w - w_m)$ is the delta function of two variables.

9.7.3 *The evaluation of the double integral on the r.h.s. of (9.37), as* $\alpha \to \infty$

The last equation enables us to evaluate the double integral on the r.h.s. of (9.37). The result is

$$\text{as } \alpha \to \infty, \int_{-\infty}^{0}\int_{0}^{\infty} |u|\, w\, p\,[\ddot{\eta}_0 = u, \ddot{\eta}_\tau = w \,|\, \eta_0 = \xi\alpha, \dot{\eta}_0 = 0, \dot{\eta}_\tau = 0, \eta_\tau = (\xi-1)\,\alpha]\, \mathrm{d}w\, \mathrm{d}u =$$
$$= \begin{cases} = |K_1(\tau,\xi)|\, K_2(\tau,\xi)\,\alpha^2 & \text{if} \quad K_1(\tau,\xi) < 0 \quad \text{and} \quad K_2(\tau,\xi) > 0, \\ = 0 & \text{otherwise}, \end{cases} \tag{9.48}$$

which is rewritten in the alternative form

$$\text{as } \alpha \to \infty, \int_{-\infty}^{0}\int_{0}^{\infty} |u|\, w\, p\,[\ddot{\eta}_0 = u, \ddot{\eta}_\tau = w \,|\, \eta_0 = \xi\alpha, \dot{\eta}_0 = 0, \dot{\eta}_\tau = 0, \eta_\tau = (\xi-1)\,\alpha]\, \mathrm{d}w\, \mathrm{d}u =$$
$$= \begin{cases} = |K_1(\tau,\xi) K_2(\tau,\xi)|\,\alpha^2 & \text{if } \tau = T^* \text{ and } \xi = \dfrac{1}{2}, \\ \le |K_1(\tau,\xi) K_2(\tau,\xi)|\,\alpha^2 & \text{if } \tau \ne T^* \text{ and/or } \xi \ne \dfrac{1}{2}, \end{cases} \tag{9.49}$$

given that

$$K_1\left(T^*, \frac{1}{2}\right) = -\frac{1}{2}\,\frac{1+\ddot{\psi}_{T^*}}{1-\psi_{T^*}} < 0, \quad K_2\left(T^*, \frac{1}{2}\right) = \frac{1}{2}\,\frac{1+\ddot{\psi}_{T^*}}{1-\psi_{T^*}} > 0. \tag{9.50}$$

Check (9.48). To this end, divide the plane u-w into the four quadrants (1st quadrant: $u>0$, $w>0$; 2nd quadrant: $u>0$, $w<0$; 3rd quadrant: $u<0$, $w<0$; 4th quadrant: $u<0$, $w>0$). The double integral covers the fourth quadrant. The conditional p.d.f. in this integral gathers at point $u=K_1(\tau,\xi)\alpha$, $w=K_2(\tau,\xi)\alpha$. Therefore, if this point belongs to the fourth quadrant, the integral is equal to

$$|K_1(\tau,\xi)|\alpha K_2(\tau,\xi)\alpha,$$

or else it is zero.

Then check (9.50). It suffices to use the definitions (9.43) and (9.47) and obtain the expressions of the cofactors C_{ij} and D_{ij} for $\tau=T^*$.

9.7.4 *The final form of* $EX(\alpha,\tau,\xi)$

Equation (9.49) and formula (9.38) for the joint p.d.f. enable us to rewrite (9.37) in the form

$$\text{as } \alpha\to\infty,\ EX(\alpha,\tau,\xi)\begin{cases} = |K_1(\tau,\xi)K_2(\tau,\xi)|\dfrac{\alpha^3}{(2\pi)^2\sqrt{M(\tau)}}\exp\left[-\dfrac{1}{4}\widehat{f}(\tau,\xi)\alpha^2\right] \\ \qquad\text{if } \tau=T^* \text{ and } \xi=\dfrac{1}{2}, & (9.51a) \\ \le |K_1(\tau,\xi)K_2(\tau,\xi)|\dfrac{\alpha^3}{(2\pi)^2\sqrt{M(\tau)}}\exp\left[-\dfrac{1}{4}\widehat{f}(\tau,\xi)\alpha^2\right] \\ \qquad\text{if } \tau\neq T^* \text{ and/or } \xi\neq\dfrac{1}{2}, & (9.51b)\end{cases}$$

where we write $M(\tau)$, in place of M, to outline that the determinant M of the covariance matrix depends on τ.

9.8 Formal proof of the necessary condition. Part III: the central inequality

9.8.1 *Theorem*

The function $\widehat{f}(\tau,\xi)$ $\{\tau\in(0,\infty),\ \xi\in(-\infty,+\infty)\}$ given by (9.39) satisfies the inequality

$$\widehat{f}(\tau,\xi)>\widehat{f}\left(T^*,\frac{1}{2}\right) \quad\text{if } \tau\neq T^* \text{ and/or } \xi\neq\frac{1}{2}. \tag{9.52}$$

Proof

$\widehat{f}(\tau,\xi)$ thought of as a function of ξ, for any fixed τ, has its absolute minimum at $\xi=\dfrac{1}{2}$:

$$\widehat{f}(\tau,\xi)>\widehat{f}\left(\tau,\frac{1}{2}\right) \quad\text{if } \xi\neq\frac{1}{2}. \tag{9.53}$$

This absolute minimum is

$$\widehat{f}\left(\tau, \frac{1}{2}\right) = \frac{M_{11} - M_{14}}{M} \tag{9.54}$$

and is a function of τ.

The numerator and the denominator on the r.h.s. of (9.54) are given by

$$\begin{aligned} M_{11} - M_{14} &= 1 - \dot{\psi}_\tau^2 - \ddot{\psi}_\tau^2 + \psi_\tau + \dot{\psi}_\tau^2 \ddot{\psi}_\tau - \psi_\tau \ddot{\psi}_\tau^2, \\ M &= 1 - \ddot{\psi}_\tau^2 - 2\dot{\psi}_\tau^2 + \dot{\psi}_\tau^4 - 2\psi_\tau \dot{\psi}_\tau^2 \ddot{\psi}_\tau - \psi_\tau^2 + \psi_\tau^2 \ddot{\psi}_\tau^2, \end{aligned} \tag{9.55}$$

and are related to each other by

$$M = (M_{11} - M_{14})(1 - \psi_\tau) - \dot{\psi}_\tau^2 \left[(1 + \psi_\tau)(1 + \ddot{\psi}_\tau) - \dot{\psi}_\tau^2\right]. \tag{9.56}$$

In the Appendix (sect. 9.8.3) we prove the inequality

$$\left[(1 + \psi_\tau)(1 + \ddot{\psi}_\tau) - \dot{\psi}_\tau^2\right] > 0, \tag{9.57}$$

which together with the equation (9.56) yields

$$M \leq (M_{11} - M_{14})(1 - \psi_\tau),$$

and this inequality and equation (9.54) taken together give

$$\widehat{f}\left(\tau, \frac{1}{2}\right) \geq \frac{1}{1 - \psi_\tau}.$$

Finally, this last inequality implies

$$\widehat{f}\left(\tau, \frac{1}{2}\right) > \widehat{f}\left(T^*, \frac{1}{2}\right) \quad \text{if } \tau \neq T^*, \tag{9.58}$$

given that

$$\psi_\tau > \psi_{T^*} \quad \text{if } \tau \neq T^*,$$

$$\widehat{f}\left(T^*, \frac{1}{2}\right) = \frac{1}{1 - \psi_{T^*}}. \tag{9.59}$$

The two inequalities (9.53) and (9.58) prove the theorem.

Note: the p.d.f. (9.38) must approach zero as $|\xi| \to \infty$, which implies that $\widehat{f}(\tau, \xi)$ must approach $+\infty$ as $|\xi| \to \infty$, or equivalently that $M_{11} + M_{14}$ must be greater than zero for every $\tau (\tau \neq 0)$. And that is why we have assumed the one at $\xi = \dfrac{1}{2}$ to be the minimum rather than the maximum of $\widehat{f}(\tau, \xi)$ thought of as function of ξ, for any fixed τ.

9.8.2 *Corollary*

The equality (9.51*a*) and the inequalities (9.51*b*) and (9.52) imply that

$$\frac{EX(\alpha,\tau,\xi)}{EX\left(\alpha,T^*,\frac{1}{2}\right)} \to 0 \quad \text{as} \quad \alpha \to \infty, \tag{9.60}$$

for every fixed pair τ, ξ such that $\tau \neq T^*$ and/or $\xi \neq \frac{1}{2}$.

In words: $EX(\alpha,\tau,\xi)$ looks like a delta function at $\tau = T^*$, $\xi = \frac{1}{2}$, as $\alpha \to \infty$.

9.8.3 *Appendix: proof of the inequality (9.57)*

Originally, it was guessed that the inequality (9.57) should hold good for any $\tau \neq 0$, whatever the spectrum. Then, the formal proof followed after a few attempts. First it was found that the function on the l.h.s. of (9.57) is related to the cofactors M_{22} and M_{23} by

$$\left[(1+\psi_\tau)(1+\ddot{\psi}_\tau) - \dot{\psi}_\tau^2\right] = \frac{M_{22}+M_{23}}{1-\psi_\tau}.$$

Then, the sum $M_{22}+M_{23}$ was shown to be always greater than zero, which yields (9.57). The reasoning made to show that $M_{22}+M_{23}$ is always greater than zero was as follows.

The joint p.d.f.

$$p(\eta_0 = 0, \dot{\eta}_0 = w, \dot{\eta}_\tau = u\,w, \eta_\tau = 0) = \frac{1}{(2\pi)^2\sqrt{M}} \exp\left[-\frac{1}{2M}\left(M_{33}u^2 + 2M_{23}u + M_{22}\right)w^2\right],$$

where u and w are two independent variables, must approach zero, for every u, as $|w| \to \infty$. This implies that the function of u

$$M_{33}u^2 + 2M_{23}u + M_{22}$$

must be greater than zero, for every u. This function has a minimum (bearing in mind that the cofactors *i*, *i* are greater than zero). This minimum is

$$\left(M_{22}M_{33} - M_{23}^2\right)/M_{33},$$

and hence the inequality

$$M_{22}M_{33} - M_{23}^2 > 0 \tag{9.61}$$

follows. Since

$$M_{33} = M_{22},$$

the inequality (9.61) is rewritten in the form

$$(M_{22}+M_{23})(M_{22}-M_{23})>0$$

which implies

$$M_{22}+M_{23}>0, \qquad M_{22}-M_{23}>0, \tag{9.62}$$

since M_{22} is greater than zero. The inequality (9.62) completes the proof.

9.9 Formal proof of the necessary condition. Part IV: conclusion

9.9.1 *Definitions of $EX_{s.w.}(\alpha,\tau,\xi)$ and $EX(\alpha)$*

Let us define

$$EX_{s.w.}(\alpha,\tau,\xi)\,\mathrm{d}\alpha\,\mathrm{d}\tau\,\mathrm{d}\xi,$$

the expected number per unit time of local maxima of the surface displacement $\eta(t)$ (at the fixed point $\mathbf{x}_o$), maxima whose elevation falls between $\xi\alpha$ and $(\xi+\mathrm{d}\xi)\alpha$ and are followed by local minima of elevation between $(\xi-1)\alpha$ and $(\xi-1)\alpha-\mathrm{d}\alpha$ after a time lag between τ and $\tau+\mathrm{d}\tau$ (α,τ and ξ being arbitrarily fixed), with the *local maximum and the local minimum which must be, respectively, the crest and trough of the same wave*. The underlined condition will be called *the same wave condition* (symbol *s.w.*).

Let us define also $EX(\alpha)\mathrm{d}\alpha$, the expected number per unit time of waves whose height falls within a fixed small interval $\alpha,\alpha+\mathrm{d}\alpha$.

The following relation proceeds straightforwardly from the two definitions:

$$EX(\alpha)=\int_0^\infty\int_0^1 EX_{s.w.}(\alpha,\tau,\xi)\,\mathrm{d}\xi\,\mathrm{d}\tau. \tag{9.63}$$

9.9.2 *The relationship between $EX_{s.w.}(\alpha,\tau,\xi)$ and $EX(\alpha,\tau,\xi)$*

From the definitions of EX and $EX_{s.w.}$ it follows that

$$EX_{s.w.}(\alpha,\tau,\xi)\le EX(\alpha,\tau,\xi),$$

$$EX_{s.w.}(\alpha,\tau,\xi)/EX(\alpha,\tau,\xi)\to 0 \quad \text{as} \quad \tau\to\infty.$$

(We have the second of these relations, because the probability that the *same wave condition* is fulfilled for large values of the time lag τ is negligible.)

One more relation between $EX_{s.w.}$ and EX proceeds from sect. 9.1. Indeed, a corollary of sect. 9.1 is that, given a local maximum of elevation $\frac{1}{2}\alpha$, and given a local minimum of elevation $-\frac{1}{2}\alpha$ after a time lag T^*, these are the crest and trough of the same wave, as $\alpha \to \infty$. Therefore

$$EX_{s.w.}\left(\alpha, T^*, \frac{1}{2}\right) = EX\left(\alpha, T^*, \frac{1}{2}\right) \quad \text{as } \alpha \to \infty.$$

Finally the set of these relations between $EX_{s.w.}$ and EX, and the limit (9.60) imply

$$\frac{EX_{s.w.}(\alpha, \tau, \xi)}{EX_{s.w.}\left(\alpha, T^*, \frac{1}{2}\right)} \to 0 \quad \text{as } \alpha \to \infty,$$

for every fixed pair τ, ξ such that $\tau \neq T^*$ and/or $\xi \neq \frac{1}{2}$.

9.9.3 *Condition (9.1) is necessary*

The consequence is that the whole contribution to the integral on the r.h.s. of (9.63), apart from a negligible share, proceeds from a neighborhood of point $\tau = T^*$, $\xi \neq \frac{1}{2}$. In this neighborhood, $EX_{s.w.}$ coincides with EX which is given by (9.51*a*):

$$EX_{s.w.}\left(\alpha, T^* + \delta\tau, \frac{1}{2} + \delta\xi\right) = \left|K_1\left(T^*, \frac{1}{2}\right)K_2\left(T^*, \frac{1}{2}\right)\right| \frac{\alpha^3}{(2\pi)^2\sqrt{M(T^*)}} \cdot$$
$$\cdot \exp\left[-\frac{1}{4}\hat{f}\left(T^* + \delta\tau, \frac{1}{2} + \delta\xi\right)\alpha^2\right] \qquad \text{as } \alpha \to \infty.$$

This is rewritten in the form

$$EX_{s.w.}\left(\alpha, T^* + \delta\tau, \frac{1}{2} + \delta\xi\right) = \left|K_1\left(T^*, \frac{1}{2}\right)K_2\left(T^*, \frac{1}{2}\right)\right| \frac{\alpha^3}{(2\pi)^2\sqrt{M(T^*)}} \cdot$$
$$\cdot \exp\left[-\frac{1}{4}\hat{f}\left(T^*, \frac{1}{2}\right)\alpha^2\right]\exp\left[-\frac{1}{8}(K_\tau^*\delta\tau^2 + K_\xi^*\delta\xi^2)\alpha^2\right] \quad \text{as } \alpha \to \infty. \quad (9.64)$$

where the formulae for K_1, K_2, M, $\hat{f}\left(T^*, \frac{1}{2}\right)$, K_τ^*, and K_ξ^* will be gathered in the next section.

Since K_τ^* and K_ξ^* are nonzero,

$$EX_{s.w.}\left(\alpha, T^* + \delta\tau, \frac{1}{2} + \delta\xi\right)$$

has the same order as

$$EX_{s.w.}\left(\alpha, T^*, \frac{1}{2}\right),$$

only if $\delta\tau$ and $\delta\xi$ have an order smaller than or equal to α^{-1}. Therefore the neighborhood of point $\tau = T^*, \xi = \dfrac{1}{2}$ making the whole contribution (apart from a negligible share) to the integral on the r.h.s. of (9.63) has a radius of order α^{-1}.

The conclusion is that a wave of height $\alpha \to \infty$ has necessarily the two following characteristics

(i) time lag between crest and trough equal to T^*, apart from a random difference $\delta\tau$ of order α^{-1};

(ii) quotient between the crest elevation and the crest-to-trough height equal to $\dfrac{1}{2}$, apart from a random difference $\delta\xi$ of order α^{-1}.

This proves that condition (9.1) is necessary for occurrence of a wave of given height H if the quotient $\alpha = H/\sigma$ tends to infinity. Also the meaning of *necessary condition* becomes more clear: as $H/\sigma \to \infty$, the wave of given height H fulfils the condition apart from some random difference of a smaller order.

9.9.4 *Details of equation (9.64)*

The terms on the r.h.s. of (9.64) have the following expressions:

$$\left|K_1\left(T^*, \frac{1}{2}\right)\right| = \left|K_2\left(T^*, \frac{1}{2}\right)\right| = \frac{1}{2}\frac{1+\ddot{\psi}_{T^*}}{1-\psi_{T^*}}, \tag{9.65a}$$

$$M(T^*) = (1-\psi_{T^*}^2)(1-\ddot{\psi}_{T^*}^2), \tag{9.65b}$$

$$\widehat{f}\left(T^*, \frac{1}{2}\right) = \frac{1}{1-\psi_{T^*}}, \tag{9.65c}$$

$$K_\tau^* \equiv \left(\frac{\partial^2 \widehat{f}}{\partial \tau^2}\right)_{\tau=T^*,\, \xi=\frac{1}{2}} = \frac{\ddot{\psi}_{T^*}(1+\ddot{\psi}_{T^*})}{(1-\psi_{T^*})^2(1-\ddot{\psi}_{T^*})}, \tag{9.65d}$$

$$K_\xi^* \equiv \left(\frac{\partial^2 \widehat{f}}{\partial \xi^2}\right)_{\tau=T^*,\, \xi=\frac{1}{2}} = \frac{8}{1+\psi_{T^*}}. \tag{9.65e}$$

Note also that

$$\left(\frac{\partial^2 \widehat{f}}{\partial \tau\, \partial \xi}\right)_{\tau=T^*,\, \xi=\frac{1}{2}} = 0$$

which has been used for the second exponential function on the r.h.s. of (9.64).

The formulae for $K_1\left(T^*, \dfrac{1}{2}\right)$, $K_2\left(T^*, \dfrac{1}{2}\right)$ and $\widehat{f}\left(T^*, \dfrac{1}{2}\right)$ have been simply rewritten for the convenience of the reader, indeed they were already given [cf. (9.50) and (9.59)]. The formula for $M(T^*)$ proceeds from (9.55) with $\tau = T^*$. Finally, the formulae for K_τ^* and K_ξ^* are obtained by differentiating twice $\widehat{f}(\tau, \xi)$.

The formula for the second derivative of $\widehat{f}(\tau, \xi)$ with respect to ξ is easily obtained from the definition (9.39) of this function. For obtaining the second derivative with respect to τ at $\tau = T^*$, $\xi = \frac{1}{2}$, it is convenient working on (9.54) of $\widehat{f}\left(\tau, \frac{1}{2}\right)$. The result is

$$\left(\frac{\partial^2 \widehat{f}}{\partial \tau^2}\right)_{\xi=1/2} = \frac{1}{M^3}\left[(\ddot{M}_{11} - \ddot{M}_{14})M^2 - (M_{11} - M_{14})M\ddot{M} \underline{-2(\dot{M}_{11} - \dot{M}_{14})M\dot{M} + 2(M_{11} - M_{14})\dot{M}^2}\right].$$

The derivatives $\dot{M}$, $\dot{M}_{11} - \dot{M}_{14}$, $\ddot{M}$ and $\ddot{M}_{11} - \ddot{M}_{14}$ are obtained from the formulae for M and $M_{11} - M_{14}$ [see sect. 9.8.1]. Once these derivatives will be obtained, put $\tau = T^*$, and since $\dot{\psi}_\tau = 0$ at $\tau = T^*$ you will be able to cancel all the terms containing $\dot{\psi}_\tau$. Then you will realize that a further big simplification is possible. Specifically, all terms from the underlined part of the formula for $\partial^2\widehat{f}/\partial\tau^2$ will cancel out. Moreover, all terms containing the third derivative and/or the fourth derivative of ψ_τ will cancel out, and with such a large number of cancelled terms, the compact form (9.65*d*) will be achieved.

9.10 Corollary: the closed solution for the wave height distribution

9.10.1 *Solution for the expected number $EX(\alpha)$*

$EX(\alpha)$ is deduced by integration of (9.64) over the neighborhood (with a radius of order α^{-1}) of point $\tau = T^*, \xi = \frac{1}{2}$. With the new variables

$$u \equiv \alpha\,\delta\tau, \quad w \equiv \alpha\,\delta\xi,$$

we have

$$EX(\alpha) = \left|K_1\left(T^*, \frac{1}{2}\right)K_2\left(T^*, \frac{1}{2}\right)\right| \frac{\alpha}{(2\pi)^2\sqrt{M(T^*)}} \exp\left[-\frac{1}{4}\widehat{f}\left(T^*, \frac{1}{2}\right)\alpha^2\right] \cdot$$

$$\cdot \int_{-\infty}^{+\infty}\int_{-\infty}^{+\infty} \exp\left[-\frac{1}{8}(K_\tau^* u^2 + K_\xi^* w^2)\right] \mathrm{d}w\,\mathrm{d}u \quad \text{as } \alpha \to \infty.$$

Then, using twice the fact that the integral over $(-\infty, +\infty)$ of $\exp(-x^2)$ is equal to $\sqrt{\pi}$, we readily evaluate the double integral. The result is

$$EX(\alpha) = \frac{2\left|K_1\left(T^*, \frac{1}{2}\right)K_2\left(T^*, \frac{1}{2}\right)\right|}{\pi\sqrt{M(T^*)K_\tau^* K_\xi^*}} \alpha \exp\left[-\frac{1}{4}\widehat{f}\left(T^*, \frac{1}{2}\right)\alpha^2\right], \text{ as } \alpha \to \infty, \quad (9.66)$$

where the constants are given by (9.65*a*-*e*).

9.10.2 *Solution for the p.d.f.* $p(\alpha)$

The probability $p(\alpha)\mathrm{d}\alpha$ that a wave height falls within a fixed small interval $(\alpha, \alpha+\mathrm{d}\alpha)$ is

$$p(\alpha)\mathrm{d}\alpha = EX(\alpha)\mathrm{d}\alpha / EX_+, \tag{9.67}$$

where EX_+ is the expected number per unit time of zero up-crossings (0_+) of the surface displacement $\eta(t)$. This expected number is the reciprocal of the mean wave period:

$$EX_+ = 1/\overline{T} = 1/2\pi,$$

where the rightmost equality proceeds from (5.24) of $\overline{T}$ and from the fact that we have taken m_0 and m_2 as units.

With the formula (9.66) for $EX(\alpha)$ and the value of EX_+, equation (9.67) yields

$$p(\alpha) = \frac{1+\ddot{\psi}_{T^*}}{\sqrt{2\ddot{\psi}_{T^*}(1-\psi_{T^*})}} \frac{\alpha}{2(1-\psi_{T^*})} \exp\left[-\frac{\alpha^2}{4(1-\psi_{T^*})}\right] \text{ as } \alpha \to \infty. \tag{9.68}$$

9.10.3 *Solution for the probability* $P(\alpha)$

The probability of exceedance $P(\alpha)$ is obtained by integration over (α, ∞) of the probability density function. The result is

$$P(\alpha) = \frac{1+\ddot{\psi}_{T^*}}{\sqrt{2\ddot{\psi}_{T^*}(1-\psi_{T^*})}} \exp\left[-\frac{\alpha^2}{4(1-\psi_{T^*})}\right] \text{ as } \alpha \to \infty. \tag{9.69}$$

Since we have taken $m_0 = 1$, $m_2 = 1$, α is the quotient between the wave height and $\sqrt{m_0}$, and ψ_{T^*}, $\ddot{\psi}_{T^*}$ represent the quotients

$$\psi_{T^*} = \psi(T^*)/m_0 = -\psi^*,$$

$$\ddot{\psi}_{T^*} = \ddot{\psi}(T^*)/m_2.$$

The formula (9.69) was given in advance in sect. 5.7.3. There, it was written in a few alternative ways, and the constant before the exponential function was generically denoted by K. Now we see that

$$K \equiv \frac{1+\ddot{\psi}_{T^*}}{\sqrt{2\ddot{\psi}_{T^*}(1-\psi_{T^*})}}.$$

In sect. 5.7.3 we also said that K can be taken to be 1 for the practical applications. On the contrary, the exact value of K must be applied for a careful test of the formula against the data from numerical simulations.

9.10.4 *Tests against data from numerical simulations of Gaussian random processes*

Forristall (1984) published a few data of

$$\mathcal{Q}(P) \equiv \alpha(P)_{\text{given spectrum}} \Big/ \alpha(P)_{\text{infinitely narrow spectrum}},$$

where $\alpha(P)$ is the inverse function of $P(\alpha)$. These data were obtained from a careful numerical simulation of Gaussian random processes. Now we have got the exact form of $\mathcal{Q}(P)$ as $\alpha \to \infty$ or $P \to 0$:

$$\mathcal{Q}(P) = \left[\frac{(1-\psi_{T^*})\ln(K/P)}{2\ln(1/P)}\right]^{\frac{1}{2}} \quad \text{as } P \to 0 \tag{9.70}$$

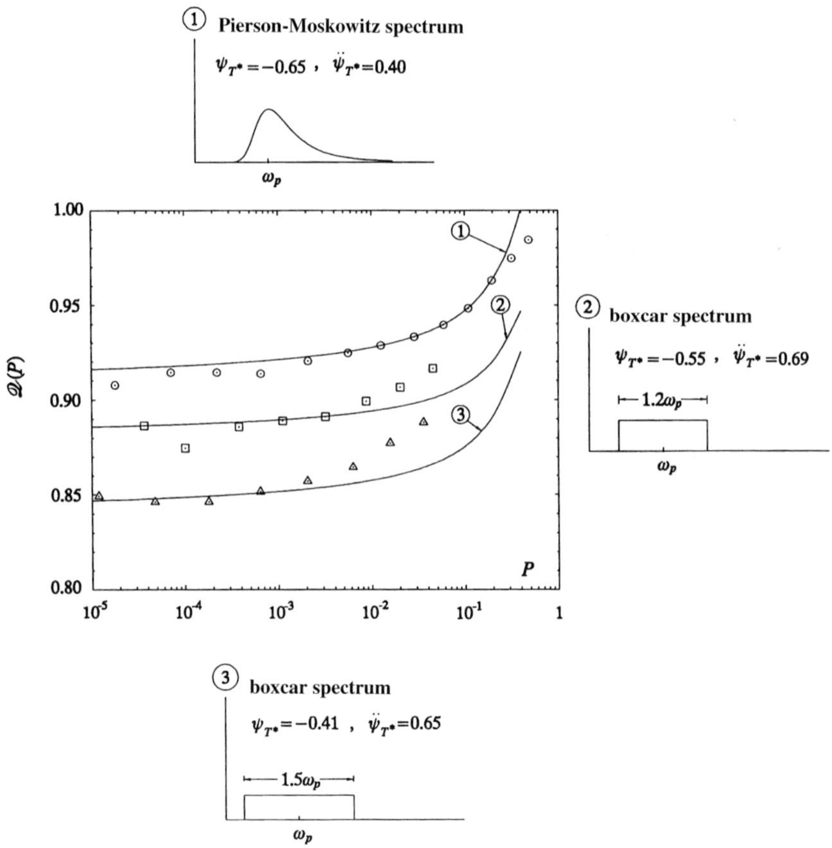

Fig. 9.3 Check of the closed solution for the probability of the wave height. The data points were obtained by Forristall (1984) through numerical simulations of random Gaussian processes. Abscissa: the probability of exceedance; ordinate: the quotient between the wave height with the given spectrum and the wave height with the infinitely narrow spectrum.

(which straightforwardly proceeds from 9.69), and we can check this formula against Forristall's data.

The data of $\mathcal{Q}(P)$ (see fig. 9.3) were obtained for three distinct shapes of the spectrum. The lines represent the function $\mathcal{Q}(P)$ obtained by means of (9.70) for these spectra. Basing ourselves on the theory, we expect that the convergence of the data points to the theoretical lines occurs as $P \to 0$, that is on the left side of the figure; and this is just what happens. But the figure tells us something more. It shows that the convergence of the data points on the asymptotic form (*asymptotic* in that it is exact as $P \to 0$) is quicker, the narrower the spectrum is: for the very broad rectangular spectrum ($\psi^* = 0.41$), the convergence occurs for P between 10^{-4} and 10^{-3}; for the somewhat narrower rectangular spectrum ($\psi^* = 0.55$), the convergence occurs for P between 10^{-3} and 10^{-2}; for the characteristic spectrum of the wind waves ($\psi^* = 0.65$) the convergence occurs even for $P = 0.20$. In other words, in a Gaussian random process with a characteristic spectrum of the sea waves, equation (9.69) proves to be fully effective for $P < 0.20$ which corresponds to $\alpha > 3.5$.

Hence, the first important evidence that the theory of quasi-determinism, which is exact as $H/\sigma \to \infty$, is effective to predict the properties of the realistically high waves.

Conclusive note

This is included at the end of chap. 10.

References

These are incorporated in those at the end of chap. 10.

Chapter 10

USES AND CONSEQUENCES OF THE QUASI-DETERMINISM THEORY

10.1 The first way to employ the theory

10.1.1 *The technique of calculation with theoretical spectra*

The two basic equations are (9.21) and (9.24). In chap. 9 they were given with the compact symbols (9.20) and (9.25) used to ease the mathematics. With the full symbols, the aforesaid equations become

$$\overline{\eta}(x_o+X,y_o+Y,t_o+T)=\frac{\Psi(X,Y,T;x_o,y_o)-\Psi(X,Y,T-T^*;x_o,y_o)}{\Psi(0,0,0;x_o,y_o)-\Psi(0,0,T^*;x_o,y_o)}\frac{H}{2}, \qquad (10.1a)$$

$$\overline{\phi}(x_o+X,y_o+Y,z,t_o+T)=\frac{\Phi(X,Y,z,T;x_o,y_o)-\Phi(X,Y,z,T-T^*;x_o,y_o)}{\Psi(0,0,0;x_o,y_o)-\Psi(0,0,T^*;x_o,y_o)}\frac{H}{2}. \qquad (10.1b)$$

The formulae for the covariances Ψ and Φ were obtained in chap. 8 for three wave fields: (8.4*a-b*) (open sea), (8.36*a-b*) (interaction with an infinitely long breakwater), (8.39*a-b*) (interaction with a semi-infinite breakwater).

The first operation in applying (10.1*a-b*) is to evaluate T^*, that is the abscissa of the absolute minimum (which is assumed to be also the first local minimum) of the autocovariance $\psi(T)$ at point x_o, y_o. For the waves in the open sea, $\psi(T)$ is the same for every point of the wave field and can be obtained from the frequency spectrum; and indeed in chap. 4 we have already seen that T^* is equal to 0.44 T_p for the mean JONSWAP spectrum [cf. fig. 4.17].

At all events, $\psi(T)$ can be easily obtained from the formula for the space-time covariance with $X=Y=0$. As an example, for the waves in the open sea, the space-time covariance is given by (8.54*a*) from which we obtain.

$$\frac{\psi(T)}{\psi(0)}=\frac{\Psi(0,0,T;x_o,y_o)}{\Psi(0,0,0;x_o,y_o)}=\frac{\displaystyle\int_0^\infty\int_0^{2\pi}\mathscr{S}(w,\theta)\cos\left(2\pi w\frac{T}{T_p}\right)\mathrm{d}\theta\,\mathrm{d}w}{\displaystyle\int_0^\infty\int_0^{2\pi}\mathscr{S}(w,\theta)\,\mathrm{d}\theta\,\mathrm{d}w}.$$

The integrals of the rightmost equality can be evaluated numerically, which yields $\psi(T)/\psi(0)$ as a function of T/T_p, and hence we obtain T^*/T_p.

The second operation is to obtain the functions $\overline{\eta}$ and $\overline{\phi}$ for the special case of interest. In particular for the waves in the open sea Ψ and Φ are given by (8.54*a-b*), and hence (10.1*a-b*) take on the form

$$\overline{\eta}(x_o + X, y_o + Y, t_o + T) = \frac{H}{2} \int_0^\infty \int_0^{2\pi} \mathscr{S}(w, \theta) \{ \cos[ARG1(w, \theta)] + \\ - \cos[ARG2(w, \theta)] \} \, \mathrm{d}\theta \, \mathrm{d}w / D0, \tag{10.2a}$$

$$\overline{\phi}(x_o + X, y_o + Y, z, t_o + T) = g \frac{H}{2} \omega_p^{-1} \int_0^\infty \int_0^{2\pi} \mathscr{S}(w, \theta) w^{-1} AT1(w) \{ \sin[ARG1(w, \theta)] + \\ - \sin[ARG2(w, \theta)] \} \, \mathrm{d}\theta \, \mathrm{d}w / D0. \tag{10.2b}$$

These equations have been written in a suitable form for PC, thanks to the following definitions:

$$ARG1(w, \theta) \equiv 2\pi \left[k(w) \frac{X}{L_{p0}} \sin\theta + k(w) \frac{Y}{L_{p0}} \cos\theta - w \frac{T}{T_p} \right],$$

$$ARG2(w, \theta) \equiv 2\pi \left[k(w) \frac{X}{L_{p0}} \sin\theta + k(w) \frac{Y}{L_{p0}} \cos\theta - w \frac{(T - T^*)}{T_p} \right],$$

$$D0 \equiv \int_0^\infty \int_0^{2\pi} \mathscr{S}(w, \theta) \left[1 - \cos\left(2\pi w \frac{T^*}{T_p} \right) \right] \mathrm{d}\theta \, \mathrm{d}w,$$

$$AT1(w) \equiv \cosh\left[2\pi k(w) \left(\frac{d}{L_{p0}} + \frac{z}{L_{p0}} \right) \right] \Big/ \cosh\left(2\pi k(w) \frac{d}{L_{p0}} \right),$$

$$AT2(w) \equiv \sinh\left[2\pi k(w) \left(\frac{d}{L_{p0}} + \frac{z}{L_{p0}} \right) \right] \Big/ \cosh\left(2\pi k(w) \frac{d}{L_{p0}} \right).$$

Of course, from the formula for $\overline{\phi}$ we obtain particle velocities and accelerations and pressure fluctuations. In particular, from (10.2*b*) we have

$$\overline{v}_y(x_o + X, y_o + Y, z, t_o + T) = \frac{\partial \overline{\phi}}{\partial Y} = g \frac{H}{2} \omega_p^{-1} \frac{2\pi}{L_{p0}} \int_0^\infty \int_0^{2\pi} \mathscr{S}(w, \theta) w^{-1} AT1(w) k(w) \cdot \\ \cdot \cos\theta \{ \cos[ARG1(w, \theta)] - \cos[ARG2(w, \theta)] \} \, \mathrm{d}\theta \, \mathrm{d}w / D0, \tag{10.3a}$$

$$\overline{v}_x(x_o + X, y_o + Y, z, t_o + T) = \frac{\partial \overline{\phi}}{\partial X} = g \frac{H}{2} \omega_p^{-1} \frac{2\pi}{L_{p0}} \int_0^\infty \int_0^{2\pi} \mathscr{S}(w, \theta) w^{-1} AT1(w) k(w) \cdot \\ \cdot \sin\theta \{ \cos[ARG1(w, \theta)] - \cos[ARG2(w, \theta)] \} \, \mathrm{d}\theta \, \mathrm{d}w / D0, \tag{10.3b}$$

$$\overline{v}_z(x_o+X, y_o+Y, z, t_o+T) = \frac{\partial \overline{\phi}}{\partial z} = g\,\frac{H}{2}\,\omega_p^{-1}\,\frac{2\pi}{L_{p0}} \int_0^\infty \int_0^{2\pi} \mathscr{S}(w,\theta)\, w^{-1} AT2(w)\, k(w) \cdot$$
$$\cdot \{\sin[ARG1(w,\theta)] - \sin[ARG2(w,\theta)]\}\, d\theta\, dw/D0, \tag{10.3c}$$

$$\overline{a}_y(x_o+X, y_o+Y, z, t_o+T) = \frac{\partial^2 \overline{\phi}}{\partial Y \partial T} = g\,\frac{H}{2}\,\frac{2\pi}{L_{p0}} \int_0^\infty \int_0^{2\pi} \mathscr{S}(w,\theta)\, AT1(w)\, k(w) \cdot$$
$$\cdot \cos\theta \{\sin[ARG1(w,\theta)] - \sin[ARG2(w,\theta)]\}\, d\theta\, dw/D0, \tag{10.3d}$$

$$\overline{a}_x(x_o+X, y_o+Y, z, t_o+T) = \frac{\partial^2 \overline{\phi}}{\partial X\, \partial T} = g\,\frac{H}{2}\,\frac{2\pi}{L_{p0}} \int_0^\infty \int_0^{2\pi} \mathscr{S}(w,\theta)\, AT1(w)\, k(w) \cdot$$
$$\cdot \sin\theta \{\sin[ARG1(w,\theta)] - \sin[ARG2(w,\theta)]\}\, d\theta\, dw/D0, \tag{10.3e}$$

$$\overline{a}_z(x_o+X, y_o+Y, z, t_o+T) = \frac{\partial^2 \overline{\phi}}{\partial z\, \partial T} = -g\,\frac{H}{2}\,\frac{2\pi}{L_{p0}} \int_0^\infty \int_0^{2\pi} \mathscr{S}(w,\theta)\, AT2(w)\, k(w) \cdot$$
$$\cdot \{\cos[ARG1(w,\theta)] - \cos[ARG2(w,\theta)]\}\, d\theta\, dw/D0, \tag{10.3f}$$

$$\overline{\Delta p}(x_o+X, y_o+Y, z, t_o+T) = -\rho \frac{\partial \overline{\phi}}{\partial T} = \rho g \frac{H}{2} \int_0^\infty \int_0^{2\pi} \mathscr{S}(w,\theta)\, AT1(w) \cdot$$
$$\cdot \{\cos[ARG1(w,\theta)] - \cos[ARG2(w,\theta)]\}\, d\theta\, dw/D0. \tag{10.3g}$$

We see that $\overline{\eta}$, $\overline{v}_x$, $\overline{v}_y$, $\overline{v}_z$, $\overline{a}_x$, $\overline{a}_y$, $\overline{a}_z$, and $\overline{\Delta p}$ are given by double integrals with respect to w and θ. These integrals can be numerically evaluated even with some relatively large steps $\Delta w, \Delta\theta$. Generally, $\Delta w = 0.02$ and $\Delta\theta = 0.07$ prove to be suitable; even if $\Delta w = 0.05$ is large enough at least over the domain $|X| < 2L_p$, $|Y| < 2L_p$, $|T| < 2T_p$. As to the limits of integration, we suggest taking $w_{\text{inf}} = 0.5$ and $w_{\text{sup}} = 2.5$, which means that the interval of integration is reduced to $0.5\omega_p$, $2.5\omega_p$ (this interval being wide enough for the characteristic spectra of the wind waves). Finally, if we apply the directional spreading function (8.16) the interval of integration with respect to θ usually can be reduced from the whole circle to

$$\theta_{\text{inf}} = \overline{\theta} - 1\,\text{rad}, \quad \theta_{\text{sup}} = \overline{\theta} + 1\,\text{rad}.$$

10.1.2 *A few suggestions for calculations*

Let us consider the set of w and θ values used for the numerical integration. It is convenient to associate an integer index with each of these values: e.g. $w_{\text{inf}} \Rightarrow i = 1$, $w_{\text{inf}} + \Delta w \Rightarrow i = 2$ ..., and $\theta_{\text{inf}} \Rightarrow j = 1$, $\theta_{\text{inf}} + \Delta\theta \Rightarrow j = 2$, ... Hence, the nondimensional spectrum $\mathscr{S}(w,\theta)$ and the nondimensional wave number $\mathscr{k}(w)$ will be stored in memory, respectively, as a matrix and as a vector. As to $\mathscr{S}(w,\theta)$, we shall use (8.20) or (8.33). As to $\mathscr{k}(w)$ it can be obtained by means of (8.24) with d/L_{p0} as input.

Typically, the calculation has to be performed for many pairs X, Y and several values of T. This is why storing in memory $\mathscr{S}(w,\theta)$ and $\mathscr{k}(w)$ is useful: the calculation of $\mathscr{S}(w,\theta)$ and $\mathscr{k}(w)$ will be done only once rather than a number of times equal to the product of the number of pairs X, Y by the number of T values. Similarly, one easily recognizes that it is highly convenient to store in memory a vector for $\sin\theta$ and a vector for $\cos\theta$.

Of course, the more complex algorithms are those of the diffracted waves. Indeed (10.1a) with (8.56a) of Ψ (interaction with a semi-infinite breakwater) becomes of the type

$$\overline{\eta}(x_o+X,y_o+Y,t_o+T)=$$
$$=\frac{H}{2}\,\frac{\displaystyle\int_0^\infty\int_0^{2\pi} f_1(w,\theta;X,Y;x_o,y_o)\cos\left(2\pi\, w\frac{T}{T_p}\right)+f_2(w,\theta;X,Y;x_o,y_o)\sin\left(2\pi\, w\frac{T}{T_p}\right)\mathrm{d}\theta\,\mathrm{d}w}{\displaystyle\int_0^\infty\int_0^{2\pi} f_3(w,\theta;x_o,y_o)\,\mathrm{d}\theta\,\mathrm{d}w}, \tag{10.4}$$

where functions f_1, f_2 and f_3 have some rather long expressions which include Fresnel integrals. The machine time for the denominator is rather short because there is only one double integral to be evaluated. On the contrary, the numerator calls for the evaluation of a number of double integrals equal to the product of the number of points X, Y and the number of times T.

Here, a great saving of machine time is achieved on defining:

$$I_1(w;X,Y;x_o,y_o)\equiv\int_0^{2\pi} f_1(w,\theta;X,Y;x_o,y_o)\,\mathrm{d}\theta, \tag{10.5a}$$

$$I_2(w;X,Y;x_o,y_o)\equiv\int_0^{2\pi} f_2(w,\theta;X,Y;x_o,y_o)\,\mathrm{d}\theta, \tag{10.5b}$$

and rewriting (10.4) in the form

$$\overline{\eta}(x_o+X,y_o+Y,t_o+T)=$$
$$=\frac{H}{2}\,\frac{\displaystyle\int_0^\infty I_1(w;X,Y;x_o,y_o)\cos\left(2\pi\, w\frac{T}{T_p}\right)+I_2(w;X,Y;x_o,y_o)\sin\left(2\pi\, w\frac{T}{T_p}\right)\mathrm{d}w}{\displaystyle\int_0^\infty\int_0^{2\pi} f_3(w,\theta;x_o,y_o)\,\mathrm{d}\theta\,\mathrm{d}w}. \tag{10.6}$$

If, for each point X, Y, we shall store in memory the vectors $I_1(w)$ and $I_2(w)$, we shall avoid repeating the calculation of the same pair f_1, f_2. As an example, let us imagine we have to compute $\overline{\eta}$ at 10 instants T, at 6000 points X, Y, with 100 values of w and 30 values of θ (which is a realistic condition). With (10.4) we should execute

$$10 \cdot 6000 \cdot 100 \cdot 30 = 180 \text{ million}$$

calculations of f_1 and f_2. While, with the sequence (10.5*a-b*) + (10.6) we should execute

$$6000 \cdot 100 \cdot 30 = 18 \text{ million}$$

calculations of f_1 and f_2. Since practically the whole machine time is needed for the calculations of the functions f_1 and f_2, passing from 180 million to 18 million pairs f_1, f_2 to be calculated means reducing the machine time to one tenth.

Of course, it is useful also to store in memory the Fresnel integrals. Coming back to the previous example, we have to calculate 18 million pairs f_1, f_2, each of them calling for 4 Fresnel integrals, which makes

$$18 \text{ million} \cdot 4 = 72 \text{ million}$$

Fresnel integrals to be evaluated. Thus it is convenient to store in memory $S_{FR}(x)$ (the Fresnel integral with integrand sine) and $C_{FR}(x)$ (the Fresnel integral with integrand cosine) with some fixed increment. As an example if the increment is 0.01 and we have to evaluate $S_{FR}(4.932)$ we shall simply find this value by interpolation, using the two known values of $S_{FR}(4.93)$ and $S_{FR}(4.94)$. Following these suggestions, a calculation like that of the example (10 instants T, 6000 points X, Y) for the more complex case (reflection + diffraction) takes nearly five minutes of Pentium PC time.

10.1.3 *Uses of the theory*

The theory of quasi determinism can be used in place of the periodic wave theory. It is valuable because it predicts what happens on the space-time just when the highest waves do occur. As an example x_o, y_o could be the location of an offshore structure and H the maximum expected wave height in the lifetime of this structure. Then the theory is able to predict the configuration of the water surface and the particle velocities and accelerations and the pressure fluctuations when this maximum wave height occurs at the structure.

But the theory can be also used for some interactive processing. Let us see an example. In chap. 8 we have found an amazing difference between the wind waves and the periodic waves in what concerns reflection. How is it possible that nodes and antinodes practically disappear in the wind waves? Here we shall use the theory starting from the question "what happens if a wave with an exceptionally large height H should occur at a point x_o, y_o at the breakwater?". The answer will bring us into the core of the phenomenon. If we shall not realize at once what we are interested to (specifically why nodes and antinodes disappear) we shall choose a new point x_o, y_o and get the new answer (hence the interactive processing). Generally, the answers of the theory are simple and enlightening, as we shall verify in the next sections.

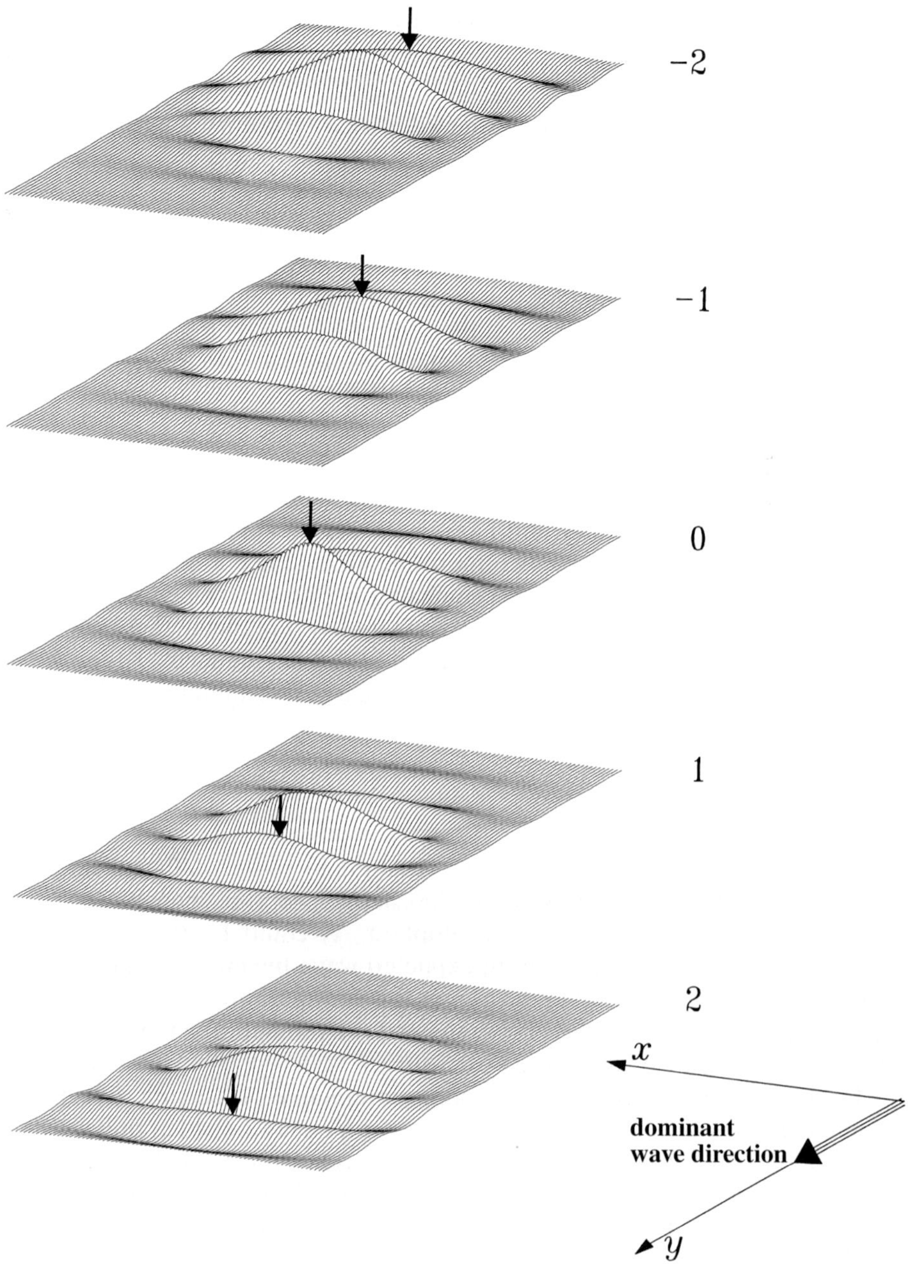

Fig. 10.1 We know that a wave of exceptionally large height H occurs at time t_o at point x_o, y_o at the centre of the framed area. The pictures show what we can expect to occur before and after t_o. The nondimensional time lag T/T_p is given on the right side of each picture. The wave of given height H is that at the central position of the group in the third picture ($T = 0$). The arrow points to this wave during its evolution. [Obtained by means of (10.2*a*).]

These are the more conventional ways to use the theory, starting from some theoretical spectrum. Then there is a new way we shall deal with in sects. 10.8-9-10.

10.2 A three dimensional wave group

What happens if a wave of exceptionally large height H should occur at a given point x_o, y_o in the open sea? The answer is given by fig. 10.1. The framed area is the same in each picture, and point x_o, y_o is at the centre of this area. The input data are: deep water, nondimensional spectrum (8.20) with $\chi_1 = 3$ and $n_p = 20$.

We see a three dimensional wave group passing at x_o, y_o. Thus, the theory implicitly reveals the existence of a well defined physical unit that is the 3-D wave group. The theory reveals also two basic features of this group. First: the individual waves have a propagation speed greater than the group's speed, so that each of them runs along the envelope from the tail where it is born to the head where it dies (the arrow points to an individual wave during its evolution). Second: the wave group has a development stage which is followed by a decay stage; in the development stage the three-dimensional envelope shrinks, so that the height of the central wave grows up to a maximum.

As we see, the answer of the theory is simple and clear: "If you record an exceptionally high wave at a point at sea, you can expect that most probably it is the centre of a well defined group at the apex of its development".

The wave groups are similar to the human families. Let us think of the house of Medici of Florence. Throughout the fourteenth and fifteenth century this family had a development stage up to a maximum at the age of Lorenzo the Magnificent. Then, during the two following centuries the family had a progressive decay. In the four centuries from 1300 AD to 1700 AD a lot of individuals of this family were born, grew up and died. The same is true of the waves: the group is like the family, the individual waves are like the members of the family. The wave group at the apex of the development stage is like the house of Medici at the age of Lorenzo the Magnificent, and the wave at the centre of the group at the apex of its development is like Lorenzo the Magnificent in the years of his full maturity.

The propagation speed of the wave group is nearly equal to the group celerity c_G [cf. sect. 2.9] associated with the period T_p, and the propagation speed of each individual wave is nearly equal to the wave celerity c associated with T_p. Therefore, on deep water, the propagation speed of each individual wave is nearly twice the propagation speed of the wave group. The fact that each individual wave moves along the envelope leads to a few amazing transformations: the wave that goes to occupy the central position of the group at the apex of the development stage nearly doubles its height in a run of only one wavelength! Moreover, as an individual wave approaches the central position of the group, its wave period and its wavelength reduces itself. As it leaves the central position, a compensating stretch occurs.

An essential picture of wave replacement at the central position of the group is given in fig. 10.2, where the waves are sketched as vertical segments (it is the conventional representation we already gave in fig. 2.17). The three configurations

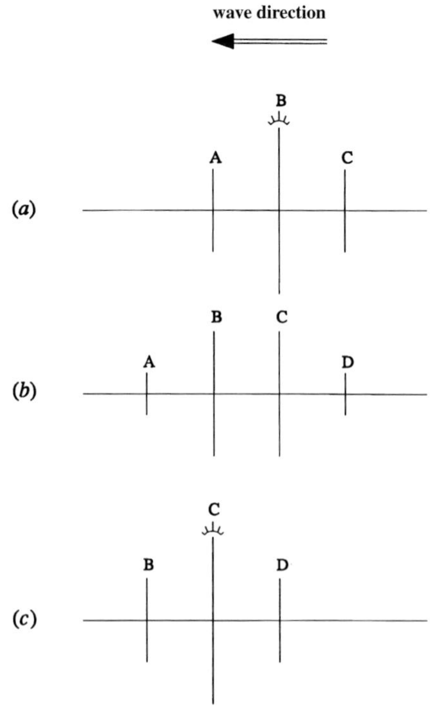

Fig. 10.2 Replacement of a wave at the envelope centre on deep water. The waves are represented by vertical segments like in fig. 2.17.

(*a*), (*b*) and (*c*) have been taken one period apart from each other. They show wave B being replaced by wave C at the group's centre.

At sea, the replacement at the central position of a group can become well evident. This occurs if a high wave like B in the first picture of fig. 10.2*a* is spilling near the crest. Then this wave leaves the central position (second picture) and it sheds its whitecap. Then the next wave (C) takes the central position and in its turn it starts spilling near the crest. In these cases the whitecap is like a crown passing from one wave to the next one.

10.3 The waves are higher on the time domain than on the space domain!

Fig. 10.3*a* shows the wave group on the time domain at point x_o, y_o. Fig. 10.3*b* shows the wave group on the space domain at time $T = T^*/2$, when the envelope centre passes at x_o, y_o. Henceforth we shall call fig. 10.3*a* "*record*" since it is the ideal record of the wave group of fig. 10.1 taken by an ideal gauge at point x_o, y_o; and we shall call fig. 10.3*b* "*snapshot*" since it catches the group at a fixed time instant. Finally, we shall refer to the point x_o, y_o simply as y_o given that we deal only with the waves along the y-axis (propagation axis).

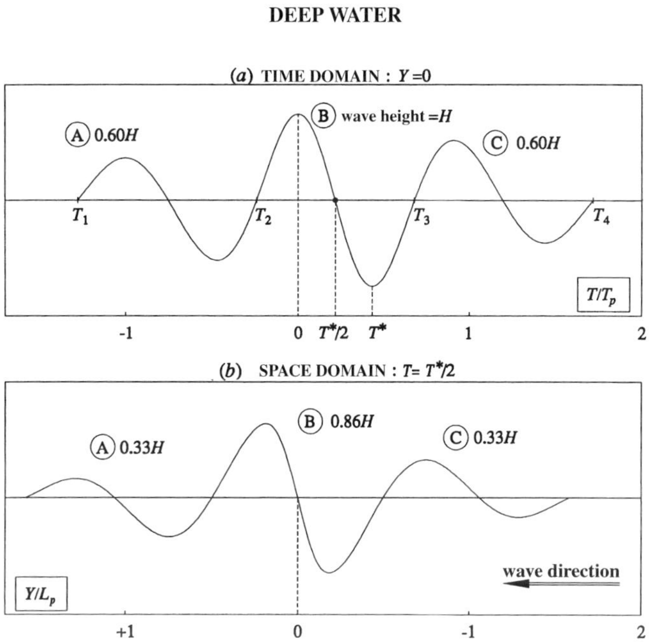

Fig. 10.3 (*a*) Ideal record of the group of fig. 10.1 taken at point x_o, y_o. (*b*) Ideal snapshot of the group taken when the envelope centre passes at x_o, y_o. The heights of the waves on the space domain (panel *b*) are smaller than the heights of the waves on the time domain (panel *a*). This is a general property due to the fact that each single wave moves along the envelope.

We note at once that the wave heights are smaller in the snapshot than in the record. This is an unexpected feature, and indeed hitherto a cornerstone of the wave theory was: the wave height is the same on the time domain and on the space domain. The new finding can be given a precise explanation as shown below.

Wave A is covered by the record for a time interval starting at T_1 and ending at T_2, while it is caught by the snapshot at a later time ($T^*/2$). This is why A is smaller in the snapshot: both the snapshot and the record show wave A in its decay stage (the wave has left the central position of the group) but the snapshot catches A after it is recorded.

Wave C is covered by the record for a time interval starting at T_3 and ending at T_4, while it is caught by the snapshot at a previous time ($T^*/2$). This is why C is smaller in the snapshot than in the record: both the snapshot and the record show wave C in its development stage (the wave has not yet reached the central position of the group) but the snapshot catches C before it is recorded.

Fig. 10.4 should help us to better understand this phenomenon. Indeed it shows a few snapshots of wave A from the time when this wave passes at y_o to the time of fig. 10.3*b*. Moreover, the figure shows a few snapshots of wave C from the time of fig. 10.3*b* to the time when it passes at y_o.

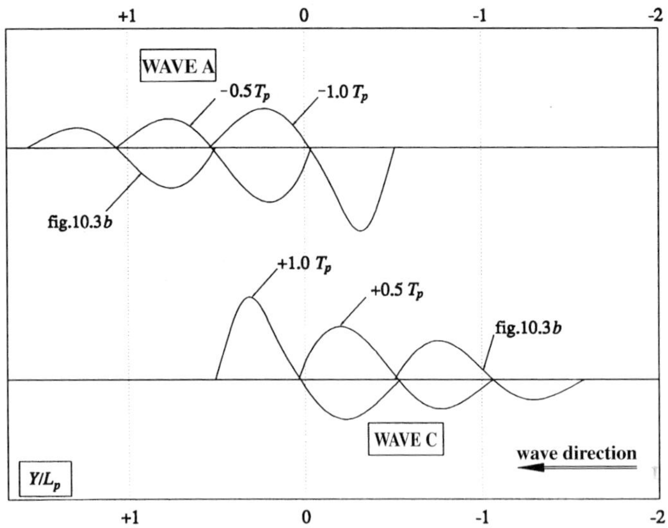

Fig. 10.4 Two "snapshots" of wave *A before* the time instant of fig. 10.3*b*, and two "snapshots" of wave C *after* this time instant.

The same difference between space domain and time domain, even if in a more attenuated form, affects wave B at the group centre. Let us observe the snapshot: the wave crest is in front of the envelope centre. Hence shortly before, while passing at y_o, the wave crest was somewhat higher than in the snapshot. Then we see in the snapshot that the trough of wave B is behind the envelope centre. Hence, some time later, when it passes at y_o it will be somewhat deeper than in the snapshot.

This phenomenon is general: if we photograph a wave group passing at any point and we compare the snapshot to the record taken at this point, we shall find that the waves are smaller in the snapshot than in the record. This is simply because the waves in the group's head are decreasing and the waves in the group's tail are increasing, so that the record catches the decreasing waves before the snapshot and the increasing waves after the snapshot. The result being that both the decreasing waves and the increasing waves are higher in the record than in the snapshot.

10.4 Effects of water depth and of spectrum shape on the wave group

10.4.1 *Effects of the water depth*

We have seen that the crest-to-trough heights on the space domain are smaller than on the time domain, as a consequence of the fact that the individual waves move along the envelope. Thus we expect that the greater the difference between the space domain and the time domain (as to the wave height) will be, the larger the relative propagation speed of the individual waves is with respect to the envelope.

Since this relative speed is $\frac{1}{2}c$ on deep water and approaches zero on shallow water (cf. sect. 2.9.3), we expect that the difference between time domain and space domain will be greater on deep water and will tend to disappear on very shallow water.

Fig. 10.5 shows the wave group on the time domain (GT) and the wave group on the space domain (GS), on shallow water ($d/L_{p0} = 0.1$); for the rest, the input data are the same as in fig. 10.3, and in particular the spectrum is the same. In line with our prediction, we see that the difference between GS and GT is smaller on shallow water (fig. 10.5) than on deep water (fig. 10.3).

Since

(i) the difference between GS and GT is smaller on shallow water than on deep water,

(ii) under the same spectrum, GT does not change from deep water to shallow water,

it follows that GS changes from deep water to shallow water. Specifically, the wave heights are larger in the GS of shallow water than in the GS of deep water,

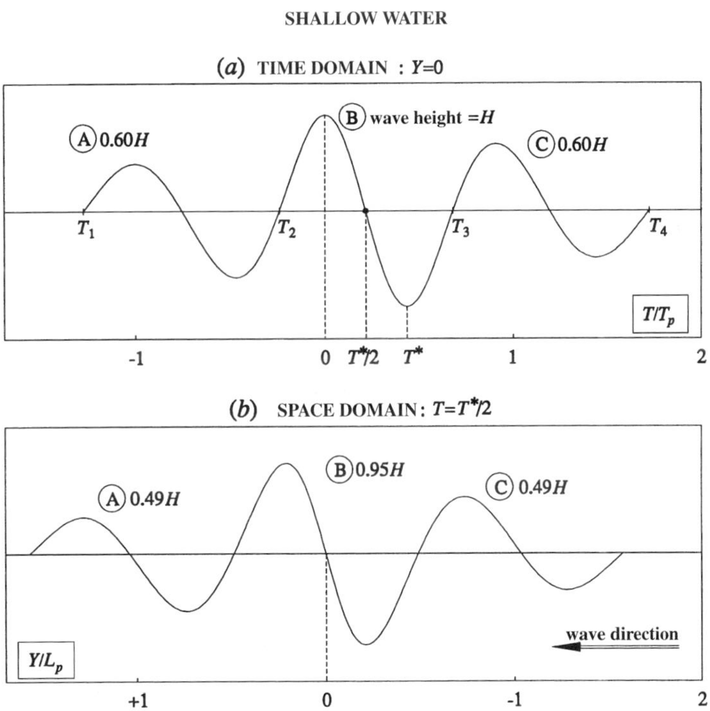

Fig. 10.5 The same as in fig. 10.3 with the only difference that the waves here are on shallow water ($d/L_{p0} = 0.1$).

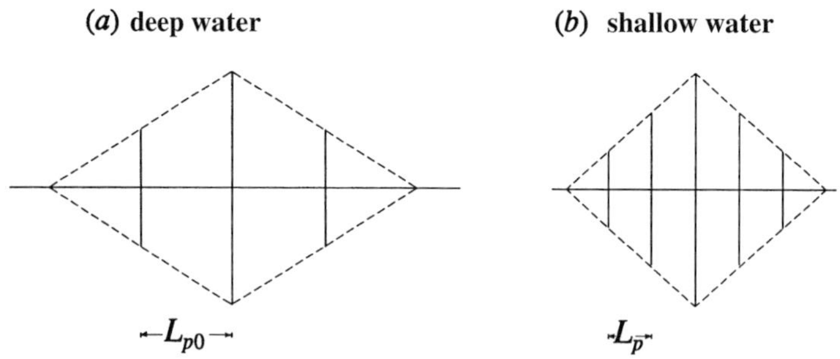

Fig. 10.6 The wave group on the space domain: (*a*) on deep water, (*b*) on shallow water.

and this difference is greater for the waves of the group's tail and of the group's head. The effect being that the GS of shallow water consists of a greater number of waves (or, more precisely, of a greater number of waves high enough to be noticed). The scheme of fig. 10.6 should help to realize this item.

The fact that the group on shallow water consists of more waves than on deep water is clearly evident in the three dimensional pictures (fig. 10.7). A comparison between fig. 10.7 (shallow water) and fig. 10.1 (deep water) reveals that the length of the wave crest and the wavelength are both smaller on shallow water. This, naturally, under the same spectrum.

10.4.2 *Effects of the spectrum shape*

The basic properties of the wave group are independent of the spectrum shape; in particular, whichever the spectrum, the wave group has a development stage in which the three dimensional envelope shrinks gradually, and a decay stage in which the envelope stretches. The spectrum shape affects the conformation of the envelope. Specifically, if the frequency spectrum gets narrower, the envelope becomes longer; if the directional spread decreases, the wave crest widens. An idea of the effect of the directional spread on the length of the wave crest is given by fig. 10.8. (To appreciate this effect better, we have resorted to an extremely narrow spreading direction function, with $n_p = 50$.)

At this stage it should be clear that the directional spread of the spectrum is responsible for the short-crestedness. We have already realized this item in sect. 8.7 and now we can appraise it from the three dimensional pictures. It remains only to wonder which is the relation between the directional spread of the spectrum and the directions of propagation of the waves.

There is a general belief that the directions of propagation of the waves in a sea state are very different from one another. According to this, we should expect that an exceptionally high wave at a fixed point x_o, y_o is due to a collision of wave groups with different directions; similar to that shown in fig. 10.9*a*. But, as we have seen, the matter is not so, at least for the wind waves. Indeed the quasi-determinism

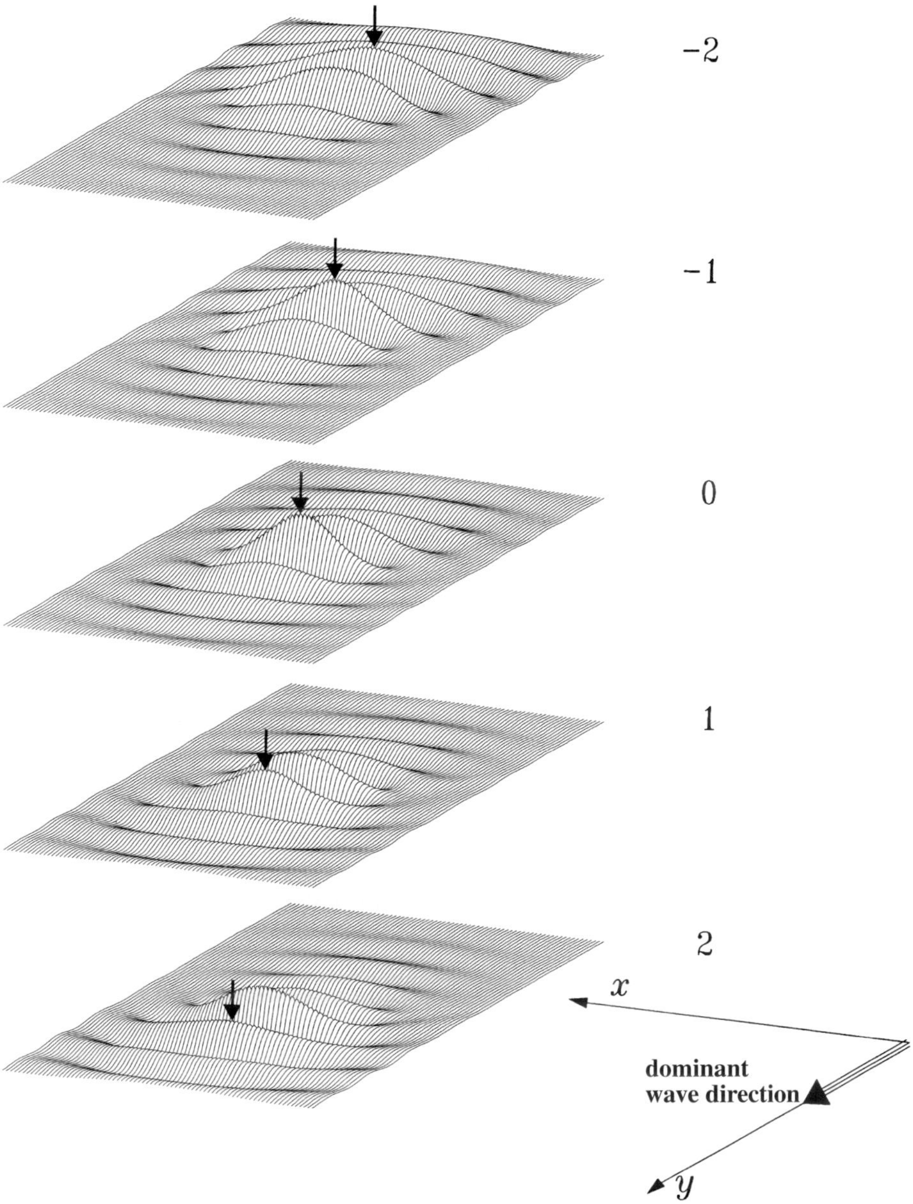

Fig. 10.7 The same as in fig. 10.1 with the only difference being that the waves here are on shallow water ($d/L_{p0} = 0.1$). We see that the wavelength is smaller than on deep water, the wave crest is shorter and the wave group consists of a greater number of individual waves.

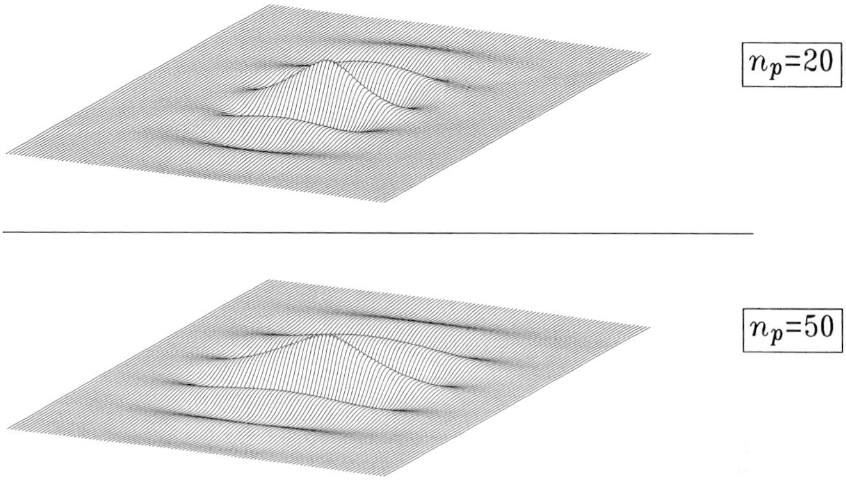

Fig. 10.8 With a smaller directional spread, the wave crest gets longer. Here is what happens on passing from $n_p = 20$ to $n_p = 50$.

theory shows that an exceptionally high wave, most probably, is due to a single wave group travelling in the dominant direction of the spectrum. Indeed, both in fig. 10.1 and in fig. 10.7, we assumed $\bar{\theta} = 0$ and found that the wave group travels along the y-axis.

10.5 Shoaling and refraction of the wave group

Because of refraction the fan of wave directions shrinks. Indeed, whichever the direction on deep water, the direction of propagation close to the shoreline tends to become orthogonal to this line [cf. fig. 10.10].

We saw in sect. 10.4 that a smaller water depth under the same spectrum leads to some shorter wave crests; and we saw also that a smaller directional spread of the spectrum, under the same water depth, leads to some longer wave crests. Now, from deep to shallow water the waves experience both these phenomena, because both the water depth and the directional spread of the spectrum get smaller. Owing to the reduction of the water depth the wave crest becomes shorter. Conversely, owing to the reduction of the directional spread, the wave crest widens. Generally, these two opposite trends are nearly equivalent to each other, so that the length of the wave crest remains nearly unchanged from deep to shallow water.

Fig. 10.11 shows three pictures of wave groups at the apex of their development stage (time $T = 0$) on different water depths. Here, the shoaling-refraction effects have been taken into account, and the formula (8.33) for $\mathscr{S}(w, \theta)$ has been used. We can clearly see that the length of the wave crest is practically constant from deep to shallow water.

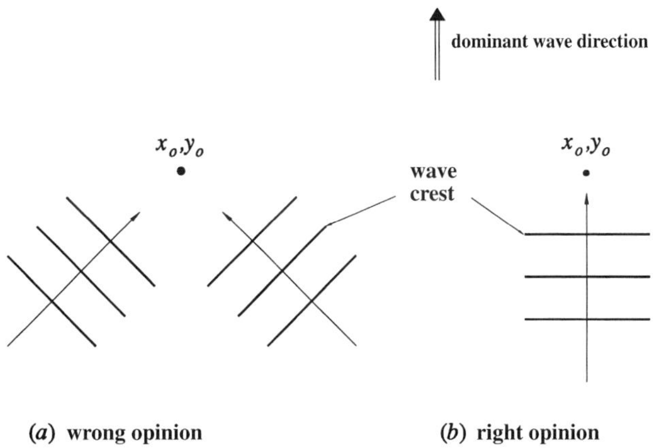

Fig. 10.9 (*a*) If the directions of the wave groups of a sea state were markedly different from one another, an exceptionally high wave would probably occur because of the collision of two groups. (*b*) The quasi-determinism theory removes this idea, showing that an exceptionally high wave most probably occurs because of the transit of a single wave group which travels in the dominant direction of the spectrum.

In the case of fig. 10.11 the dominant direction on deep water was orthogonal to the contour lines, and this is why the direction of advance of the wave groups on the various depths was nearly orthogonal to the contour lines. If the dominant direction on deep water is not orthogonal to the contour lines, then the direction of the wave groups changes gradually from deep to shallow water (see an example in fig. 10.12). These changes in direction are well predicted by equation (2.43) with T_p as the wave period.

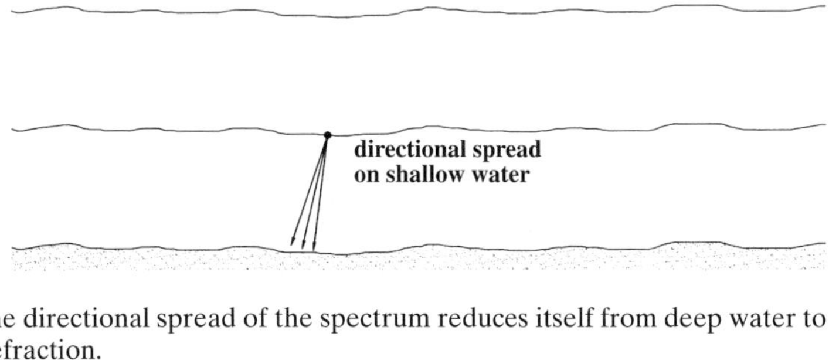

Fig. 10.10 The directional spread of the spectrum reduces itself from deep water to shallow water because of refraction.

10.6 Explanation of the first big difference between sea waves and periodic waves

10.6.1 *Formation of an exceptionally high wave in the water sheet before a breakwater*

The first big difference between sea waves and periodic waves concerns reflection. You will remember that, starting from about one wavelength from the wall, the variance $< \eta^2(t) >$ of the sea waves takes on a constant value equal to half the value at the wall [cf. fig. 8.8]. This implies that the diffraction coefficient C_d takes on the constant value $\sqrt{2}$ rather than fluctuating between 0 and 2 like in the periodic waves. As a further consequence, the maximum expected wave height in a time interval Δt, at any point, starting from about one wavelength from the wall, is $1/\sqrt{2}$ times smaller than at the wall (bearing in mind from sect. 8.8 that the maximum expected wave height at any fixed point is proportional to the C_d of this point). Whereas, in the periodic waves, the wave height at the antinodes (even very far from the wall) is the same as at the wall.

Why such a big a difference between sea waves and periodic waves? The quasi-determinism theory enables us to answer this question.

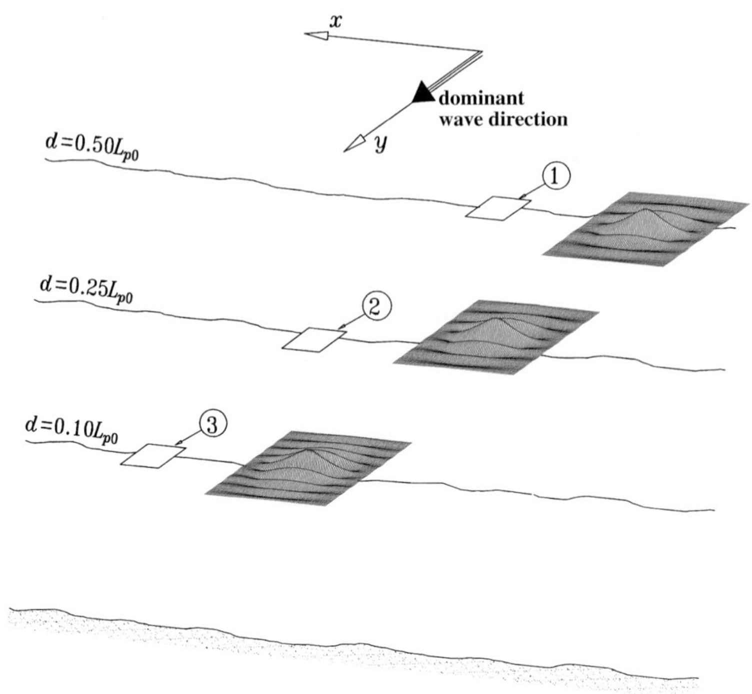

Fig. 10.11 Three "snapshots" of groups at the apex of their development on decreasing depths. (Obtained by means of 10.2*a* and 8.34.)

Let us see first what happens when the maximum expected wave height occurs at a fixed point x_o, y_o of the breakwater; then let us see what happens when the maximum expected wave height occurs at a fixed point x_o, y_o at some distance, say $3L_p$, from the breakwater. Clearly, we shall use the formula (10.1*a*) with the (8.55*a*) of Ψ.

Fig. 10.13 answers the question "what happens if a wave of given exceptionally large height occurs at a point x_o, y_o at the breakwater?". This point is at the centre of the framed part of the breakwater, and the input data are the basic ones [nondimensional spectrum (8.20) with $\chi_1 = 3$ and $n_p = 20$, $\bar{\theta} = 0$, deep water]. We see a wave group that approaches the breakwater, hits it, is reflected, and goes back seaward. It will have been noted that, while the wave group is approaching the wall its envelope gradually shrinks; and, on the contrary, while the wave group goes back seaward, the envelope stretches. This means that the wave group is at the apex of its development stage when it arrives at the breakwater. In short, the answer is: if you record an exceptionally high wave at the wall, most probably, it is the central wave of a group hitting the breakwater, at the apex of its development stage.

Let us pass to the second question: what happens if a wave of given exceptionally large height occurs at a point x_o, y_o, $3L_p$ before the breakwater? The answer is given

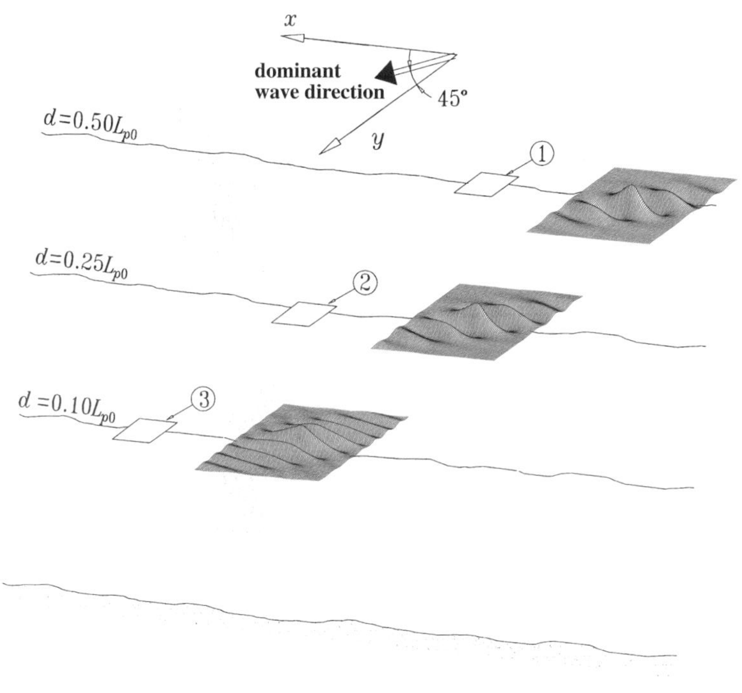

Fig. 10.12 A new series of three "snapshots" of wave groups at the apex of their development. Here the dominant direction on deep water is no longer orthogonal to the coastline.

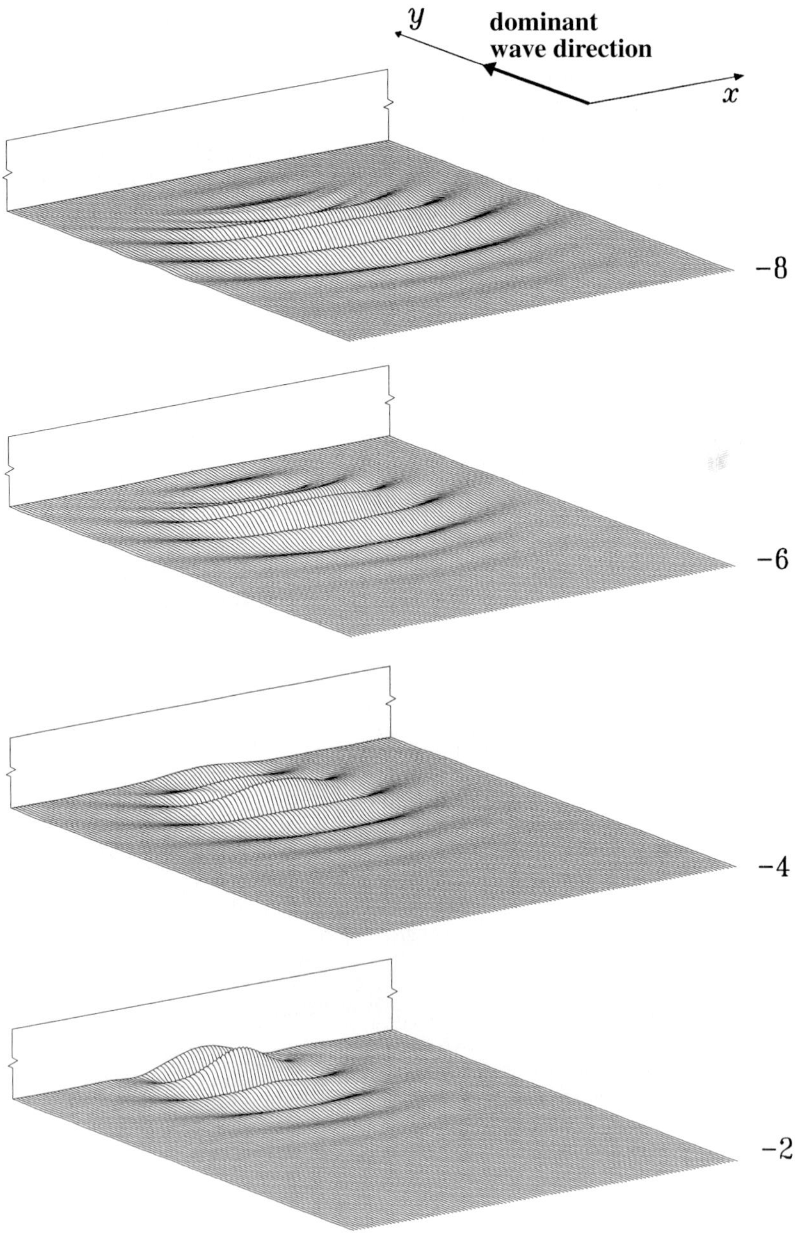

Fig. 10.13 Formation of an exceptionally high wave at a point of an upright breakwater. A wave group at the apex of its development hits the breakwater, is reflected, and goes back seaward. (Obtained by means of 10.1*a* and 8.55*a*.) *Continued opposite page.*

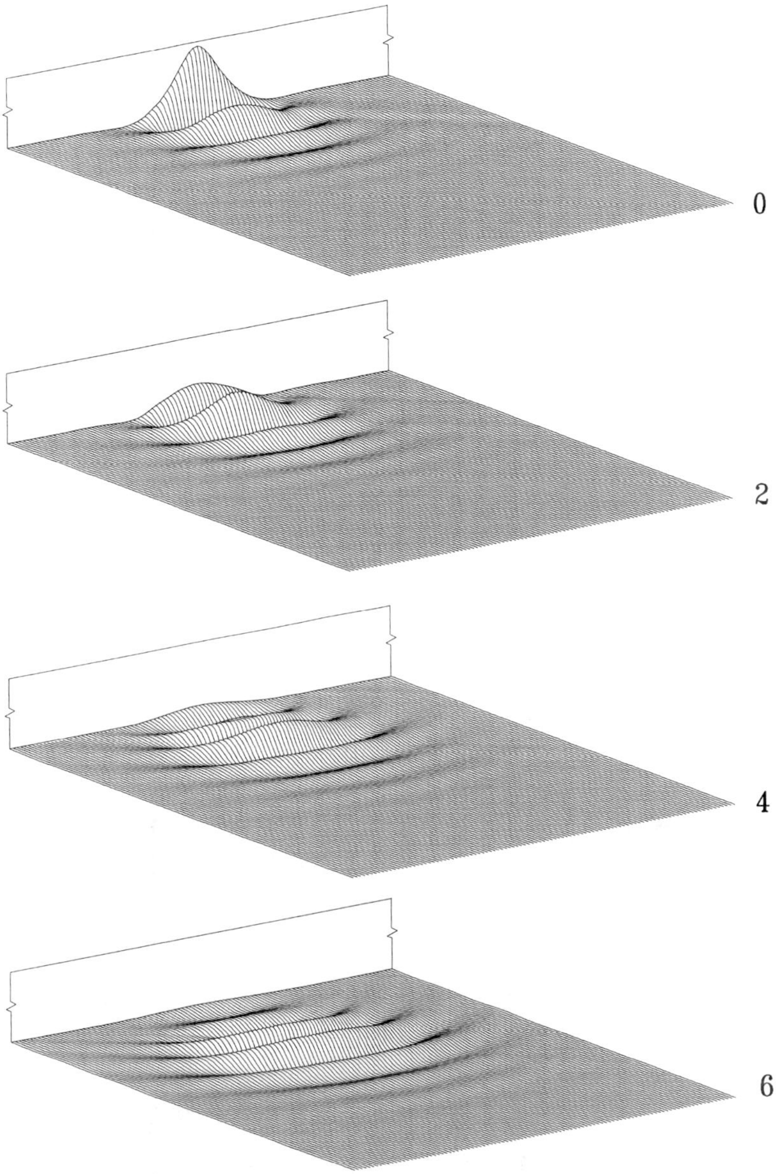

Fig. 10.13 *Continued.*

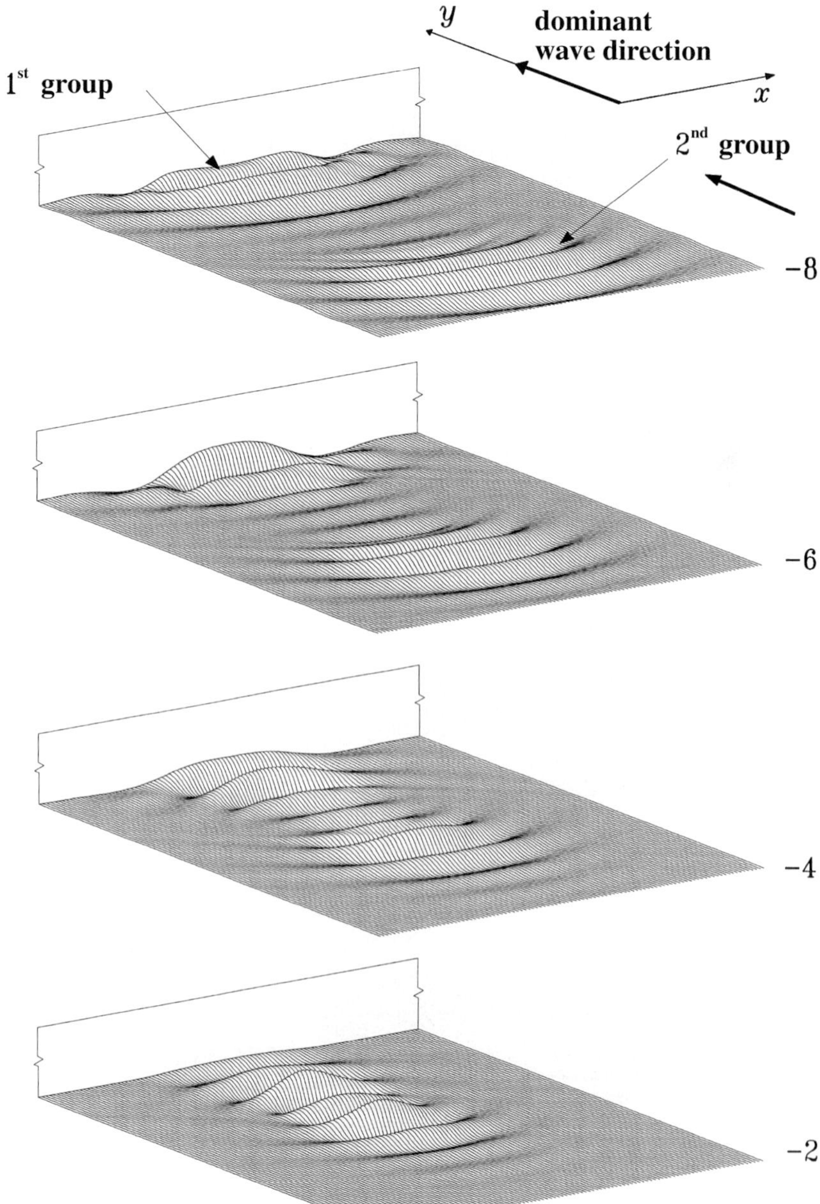

Fig. 10.14 Formation of an exceptionally high wave at a point x_o, y_o three wavelengths before a breakwater. Here we see two wave groups, the first one is reflected and goes to collide with the second wave group approaching the breakwater. The exceptionally high wave is produced by the overlapping of the central waves of the two groups. (Obtained by means of 10.1a and 8.55a.) *Continued opposite page.*

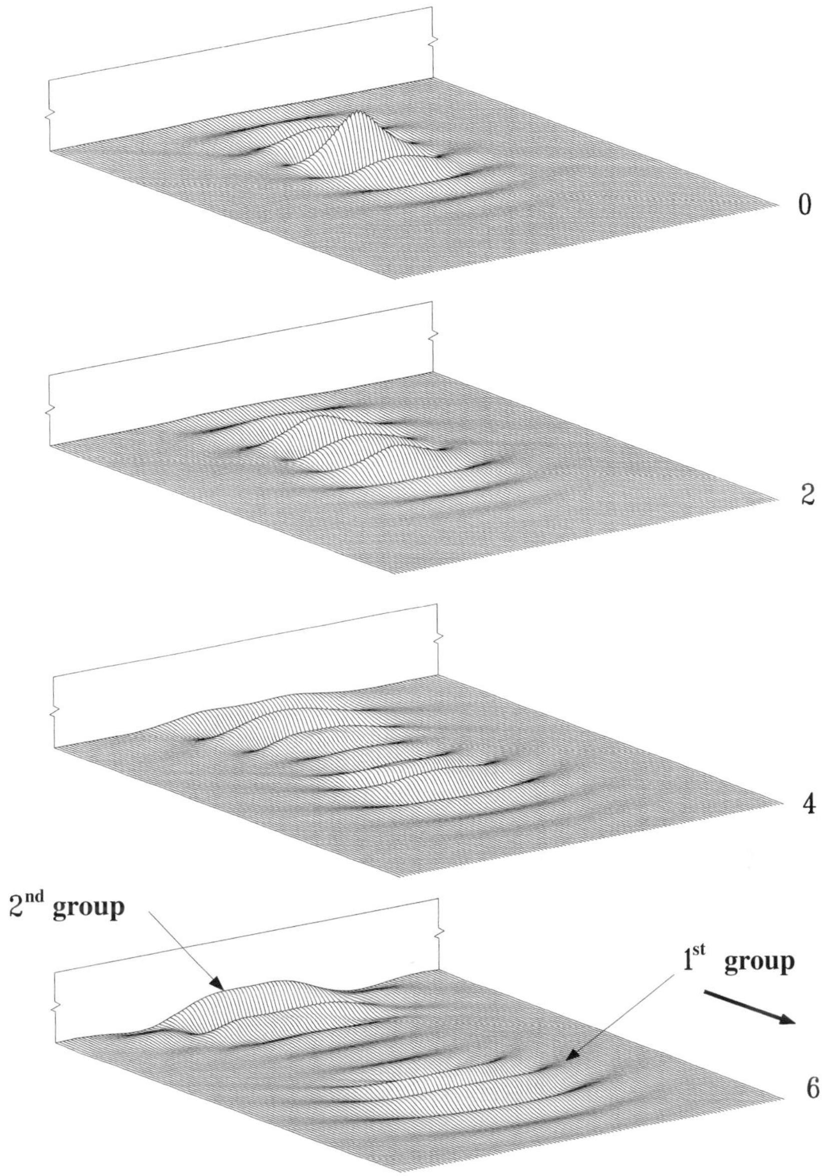

Fig. 10.14 *Continued.*

in fig. 10.14 which shows a big novelty: two groups! The first wave group hits the breakwater at time $T = -6T_p$, is reflected, goes back seaward and collides with the second wave group approaching the breakwater. The central waves of the two groups overlap fully at the fixed point x_o, y_o, yielding the given exceptionally large wave height. This wave height is very large because both wave groups reach the apex of their development when they come into collision (the two envelopes being at the extreme contraction). Hence, the answer is: if you record an exceptionally high wave at a point far from the breakwater, most probably it is the outcome of two wave groups at the apex of their development, which strike each other in full at this point.

These two answers of the quasi-determinism theory remove the curtain from the reflection phenomenon.

10.6.2 *Why the maximum expected wave height at any point far from the breakwater is smaller than at the breakwater*

Occurrence of an exceptionally high wave at a point of the breakwater requires only that a group at the apex of its development strikes this point. The very high wave will be the central one of the group, whose height doubles because of reflection.

Occurrence of an exceptionally high wave at a point x_o, y_o far from the breakwater calls for an event that has a much smaller probability to occur: two wave groups at the apex of their development must strike each other in full just at this point. The very high wave will be produced by the overlapping of the central waves of the two groups, and its height will be the sum of the heights of these two waves.

This is why the maximum expected wave height at a point far from the breakwater is smaller than the maximum expected wave height at a point on the breakwater.

10.6.3 *Why nodes and antinodes disappear*

We have seen how a very high wave occurs at a point ($3L_p$ before the wall) which, according to the classic theory of the periodic waves, should be an antinode. What should we find if the fixed point was the location of a node, say $3.25L_p$ before the wall? The answer is straightforward: just the same process; that is, two wave groups at the apex of their development come into collision at this point. There is only one minor difference: the time lag between the two groups. Indeed, if the two groups have to strike each other in full $3L_p$ before the wall, it is necessary that the time lag is $12T_p$ [cf. fig. 10.15]. While, if the collision has to occur $3.25L_p$ before the breakwater, the time lag must be of $13T_p$.

Of course, the probability of two wave groups at $12T_p$ from each other is equal to the probability of two wave groups at $13T_p$ from each other. Consequently, the probability of a given wave height very large at $3L_p$ from the wall is equal to the probability of the same wave height at $3.25L_p$ from the wall. Or in other words:

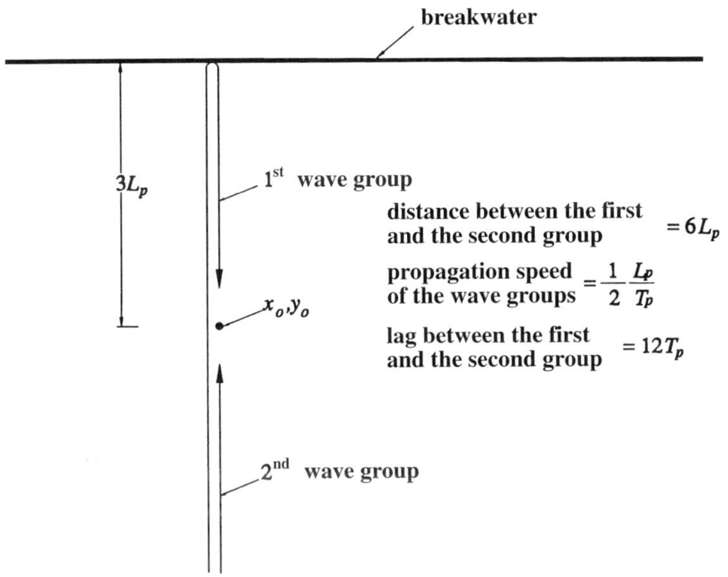

Fig. 10.15 The trajectories of the two wave groups from fig. 10.14.

the maximum expected wave height at $3L_p$ from the wall (where an antinode was expected) is equal to the maximum expected wave height at $3.25L_p$ from the wall (where a node was expected).

Nodes and antinodes would survive also in the wind waves, were the wave groups longer than they actually are. If the spectrum was infinitely narrow, the wave group would consist of a sequence of many waves of nearly the same height. In this case, at the point $y_o = -3L_p$, the waves of the same group would overlap in phase

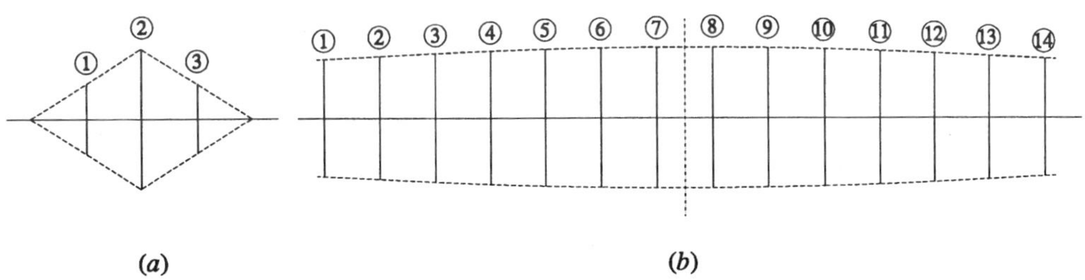

Fig. 10.16 (a) A typical sea wave group. (b) How the wave groups would be if the spectra were much narrower than they are. The waves of group (a) overlap one another up to one wavelength from the breakwater. The waves of group (b) overlap one another up to many wavelengths from the breakwater.

coherence: wave crest ① returning seaward + wave crest ⑦ approaching the wall, wave crest ② + wave crest ⑧ and so on [see fig. 10.16*b*]. Hence, the wave height at this point would be $2H$ as at the wall. Conversely, at point $y_o = -3.25L_p$, the waves of the same group would overlap in opposition: crest of wave returning seaward + trough of wave approaching the wall, and vice-versa. Hence, the wave height at this point would be zero: node.

Thus, we realize that the disappearance of nodes and antinodes is due to the small length of the wave groups. And, given that the length of a wave group depends only on the frequency spectrum, we realize also why nodes and antinodes disappear in the long-crested waves as well.

10.6.4 *Why a few pseudo-nodes and pseudo-antinodes remain near the breakwater*

A half-wavelength before the wall, the crest of the group's central wave approaching the wall overlaps in phase coherence the crest of the group's head wave returning seaward; and the crest of the group's central wave returning seaward overlaps in phase coherence the crest of the group's rear wave approaching the wall. Of course, the largest wave height at this point (a half-wavelength from the wall) will be smaller than the largest wave height at the wall. Indeed, the largest wave height at the wall will be twice the height of the group's central wave, while the largest wave height at a distance of a half-wavelength from the wall will be the sum of the heights of the group's central wave and of the preceding (or following) wave.

In conclusion, the maximum wave height at the distance of $\frac{1}{2}L_p$ from the wall will be rather large because two waves of the same group necessarily overlap one another in phase coherence at this point. But this maximum wave height will also be necessarily smaller than the maximum wave height at the wall. Hence, the origin of the pseudo-antinode.

10.6.5 *Why the pseudo-antinode becomes more evident on shallow water*

Take a snapshot of the wave group on deep water, e.g. that of fig. 10.3*b*. Measure the elevations of the central wave crest and of the preceding wave crest. Then add up these two elevations. Doing so, you will obtain the elevation of the wave crest at the first pseudo-antinode if the wave group hits a breakwater. Indeed, the crest elevation a half-wavelength before the wall is equal to the sum of the elevations of two consecutive crests taken at the *same time*.

Now do likewise with fig. 10.5*b* that shows the wave group, under the same spectrum, on shallow water. You will obtain a greater crest elevation because the wave heights are greater in fig. 10.5*b* than in fig. 10.3*b* (the reason for this, having been explained in sect. 10.4). Hence, we realize why, under the same spectrum, the maximum expected wave height at a half-wavelength from the wall is greater on shallow water than on deep water.

10.6.6 *Why the pseudo-antinode becomes less evident in the oblique reflection*

Fig. 10.17 shows what happens if a wave of given exceptionally large height occurs at a point x_o, y_o on the breakwater, in a case of inclined wave attack: the dominant direction of the spectrum of the incident waves makes a 25° angle with the wall-orthogonal. We see that the wave group travels in the dominant direction of the spectrum, hits the wall (at the apex of its development stage), is subjected to specular reflection and returns seaward.

In this manner the centre of the wave crest approaching the breakwater does not collide with the centre of the wave crest returning seaward [see the scheme of fig. 10.18]. Hence, we realize why the maximum expected wave height at the first pseudo-antinode is smaller than in the orthogonal reflection, as we found on comparing fig. 8.9*b* with fig. 8.8.

10.7 Explanation of the second big difference between sea waves and periodic waves

The second big difference between sea waves and periodic waves concerns diffraction. The C_d (and consequently the maximum expected wave height) is surprisingly high in the dark area [see figs. 8.10 and 8.11]. On the other hand the C_d in the more sheltered area is practically the same as the C_d of the periodic waves. What is the reason for this?

To answer this question let us see what happens when the maximum expected wave height in a very large time interval occurs at a point x_o, y_o of the more sheltered area; and what happens when the maximum expected wave height occurs at a point x_o, y_o of the less sheltered area (dark area).

More sheltered area

As to the point of the more sheltered area, we fix

$$x_o = 1.5L_p, \quad y_o = 0_+,$$

where 0_+ means in contact with the lee wall. Fig. 10.19 shows what happens when the maximum expected wave height in a very large time interval occurs at this point: a wave group strikes the breakwater's tip in full; thus one half of the wave crest hits the wall and one half enters the sheltered area and produces the exceptionally large wave height at the fixed point.

The configuration of the water surface has been obtained by means of (10.1*a*) with (8.56*a*) of Ψ. Since the C_d of the fixed point is 0.2, while the C_d of the wave beaten wall is 2, the height H of fig. 10.19 has been taken ten times smaller than the height H of fig. 10.13. (Always, bear in mind that the maximum expected wave height at a point is proportional to the C_d of this point.)

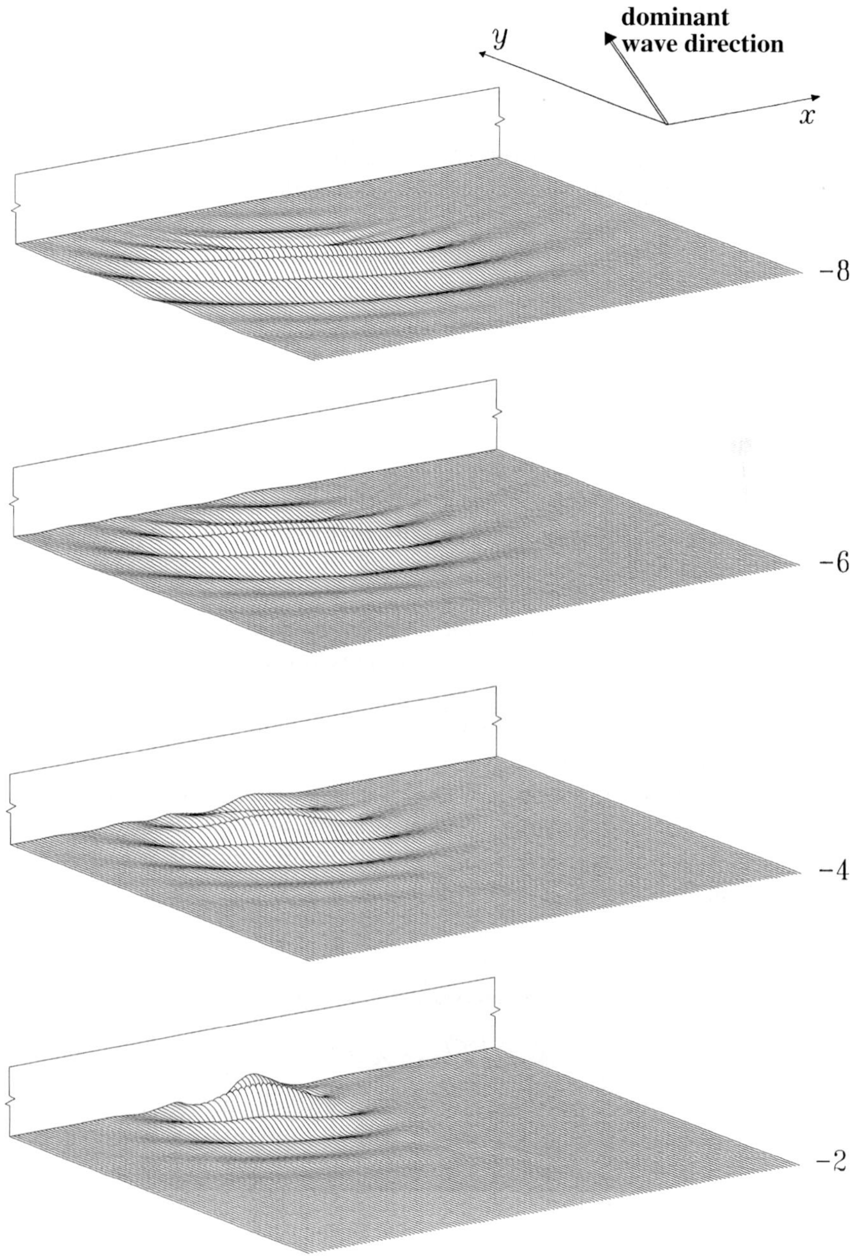

Fig. 10.17 Formation of an exceptionally high wave at a point x_o, y_o of a breakwater, if the dominant direction is not wall-orthogonal. A wave group at the apex of its development hits the breakwater and is subjected to specular reflection. (Obtained by means of 10.1*a* and 8.55*a*.) *Continued opposite page.*

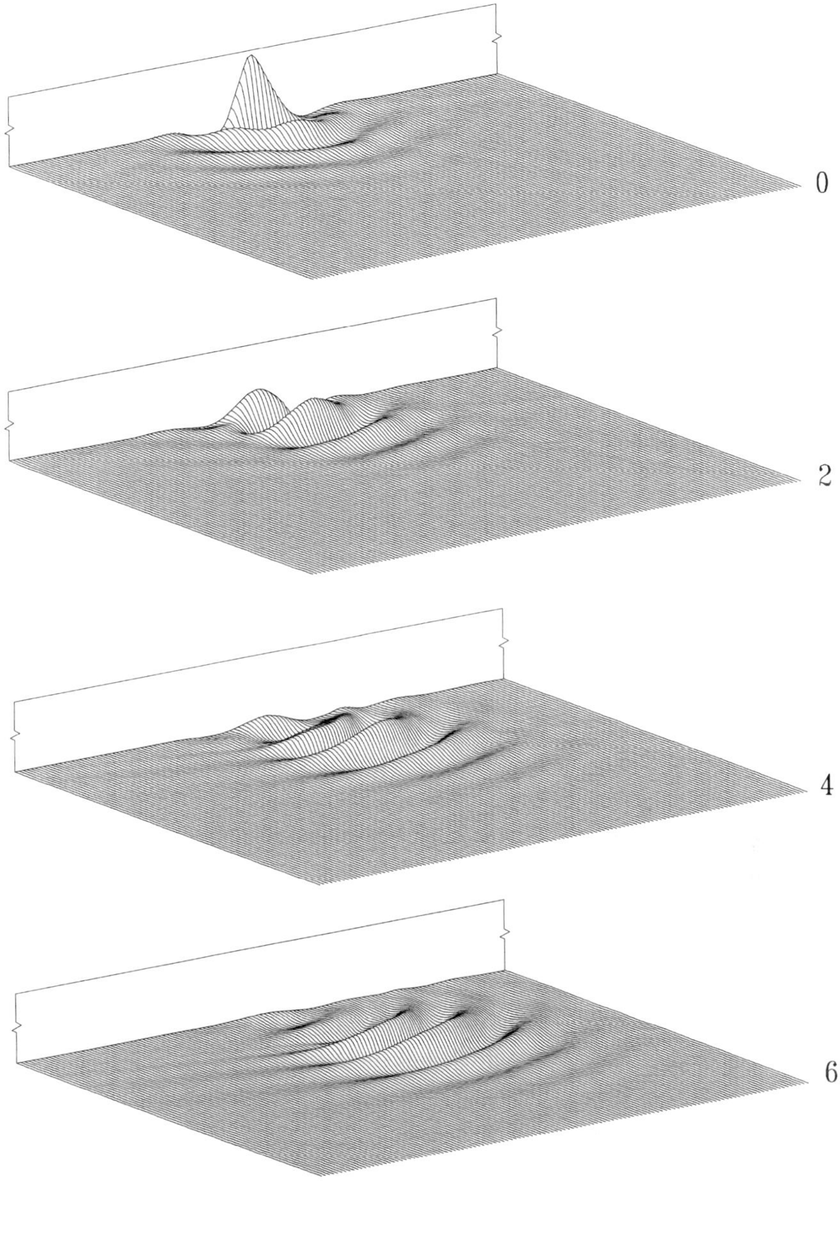

Fig. 10.17 *Continued.*

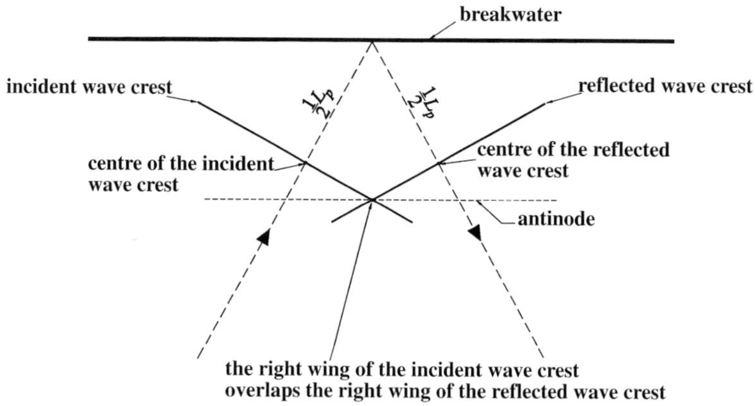

Fig. 10.18 If the waves are inclined with respect to the wall-orthogonal, the incident and reflected waves of a group do not strike each other in full.

Less sheltered area

As to the point of the less sheltered area, we fix

$$x_o = L_p, \quad y_o = 4L_p.$$

Note that the C_d of the wind waves at this point is 0.49 while the C_d of the periodic waves is 0.30. Fig. 10.20 answers the question "what happens when the maximum expected wave height occurs at this point?" Here we see a great novelty: for the first time the quasi-determinism theory shows a wave group that does not travel in the dominant direction of the spectrum; the dominant direction is wall-orthogonal, while the group attacks from the left side.

To explain this novelty, here is a short story. Two researchers, Paul and John analyse a wave record taken at the fixed point x_o, y_o of the less sheltered area, and find an exceptionally high wave in this record. Then, John reasons as follows: "The dominant direction of the spectrum is wall-orthogonal, which means: the highest wave groups, most probably, approach the barrier orthogonally. Hence, our exceptionally high wave, most probably, has been produced by a wave group which stroke the breakwater's tip in full." He takes the pencil and sketches the trajectory of this wave group (see fig. 10.21*a*).

Paul objects to this reasoning: "Suppose it was a wave group, so to say, anomalous, that is a group whose direction did not coincide with the dominant direction of the spectrum. With this trajectory – (see fig. 10.21*b*) – the wave group would have nearly struck our point in full; indeed, it would have only grazed the tip of the breakwater with its right wing." Paul concludes: "It's true, the percentage of groups like mine which do not travel in the dominant direction is very small; but it's also true that even with a small deviation from the dominant direction, the wave height considerably grows at our point x_o, y_o. Weighing the pros and cons, I am inclined to believe that the group's trajectory was as in my sketch."

It is not easy to decide who's right, but the quasi-determinism theory takes a definite position: Paul is right! Had the recorded wave height H been infinitely great with respect to the r.m.s. surface displacement at the fixed point x_o, y_o, the probability that the matter is as in Paul's opinion would even approach 1.

This is, of course, if the point x_o, y_o belongs to the less sheltered area. Indeed, John's interpretation holds good if the exceptionally high wave is recorded at a point x_o, y_o of the more sheltered area (the case of fig. 10.19). And, in fact, even a very inclined wave group would be far from striking a point in the more sheltered area in full.

The conclusion is evident: the heights of the wind waves in the less sheltered area (the dark area) are unexpectedly great, because a few wave groups are able to nearly strike this area in full. We realize also that the long crested waves, having no directional spread, have some much smaller heights in the dark area.

10.8 The second way to employ the theory

10.8.1 *Technique of calculation with the time series data of the surface displacement*

Let us fix two points ① and ② at sea. The distance between these two points should be within about ten wavelengths (this condition guarantees that in the open sea the two points belong to a homogeneous wave field, cf. sect. 8.1). We allow the wave field to be generally non-homogeneous because of reflection, diffraction and/or refraction.

Let us take a record of a sea state, simultaneously, at point ① and ②. Then we wonder "what happens at point ② if a wave of given very large height H should occur at point ①?". We can use the time series data to answer this question. Let us see how.

Equation (10.1*a*) is rewritten in the form

$$\overline{\eta}(x_o+X, y_o+Y, t_o+T) = \\ = \frac{<\eta(x_o, y_o, t)\eta(x_o+X, y_o+Y, t+T)> - <\eta(x_o, y_o, t)\eta(x_o+X, y_o+Y, t+T-T^*)>}{<\eta^2(x_o, y_o, t)> - <\eta(x_o, y_o, t)\eta(x_o, y_o, t+T^*)>}\frac{H}{2}. \tag{10.7}$$

Therefore, if we take point ① as x_o,y_o and point ② as x_o+X, y_o+Y [see fig. 10.22], we have

$$\overline{\eta}_2(t_o+T) = \frac{<\eta_1(t)\eta_2(t+T)> - <\eta_1(t)\eta_2(t+T-T^*)>}{<\eta_1^2(t)> - <\eta_1(t)\eta_1(t+T^*)>}\frac{H}{2}, \tag{10.8}$$

where

$$\eta_1(t) \equiv \eta(t) \text{ recorded at ①},$$

$$\eta_2(t) \equiv \eta(t) \text{ recorded at ②}.$$

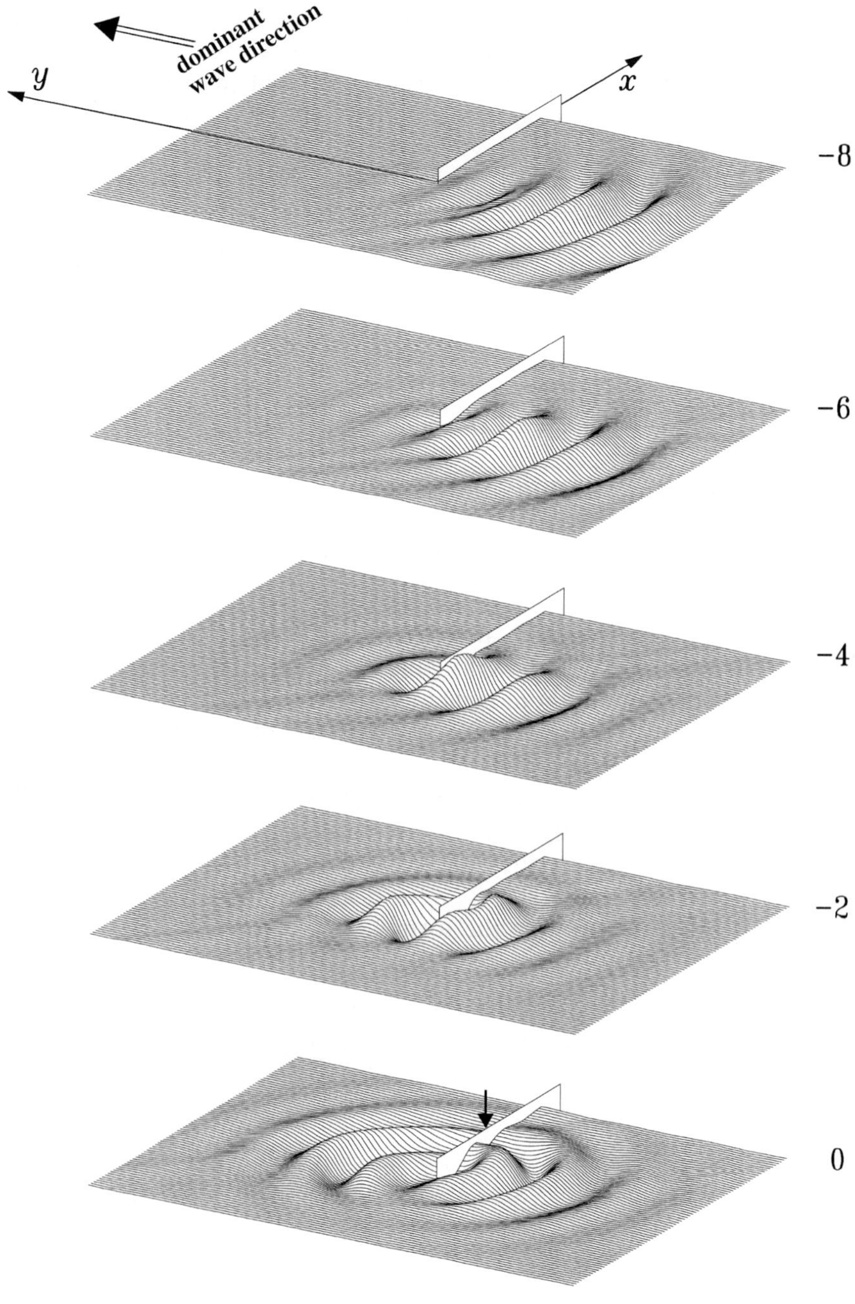

Fig. 10.19 Formation of an exceptionally high wave at a point x_o, y_o on the lee wall, one and a half wavelengths from the tip. The exceptionally high wave (we mean that it is exceptionally high for the point where it occurs) is pointed by an arrow. It is due to a wave group striking the breakwater's tip in full. (Obtained by means of 10.1*a* and 8.56*a*.)

$\overline{\eta}_2(t_o+T)$ is the expected surface displacement at point ②, if a wave of given exceptionally large height H occurs at ① (t_o being the instant when the crest of this wave passes at point ①). T^* in (10.8) is the abscissa of the absolute minimum of the autocovariance $\psi(T)$ at point ①:

$$\psi(T) = < \eta_1(t)\eta_1(t+T) > .$$

Of course, the calculation can be done for an arbitrarily large number of points, and the expected surface displacement at point j if a wave of given very large height H occurs at point i is given by

$$\overline{\eta}_j(t_o+T) = \frac{< \eta_i(t)\eta_j(t+T) > - < \eta_i(t)\eta_j(t+T-T^*) >}{< \eta_i^2(t) > - < \eta_i(t)\eta_i(t+T^*) >}\,\frac{H}{2},$$

where T^* is the abscissa of the absolute minimum of the autocovariance at point i:

$$\psi(T) = < \eta_i(t)\eta_i(t+T) > .$$

Let us take sampling interval Δt_{samp} of 0.1s, wave record of 300s, two wave gauges like in fig. 10.22. Then, as an example, let us consider the average $< \eta_1(t)\eta_2(t+T) >$ for $T = 0.5$s [this average being necessary for obtaining $\overline{\eta}_2$ by means of (10.8)]. It is given by

$$\frac{1}{2996}\left[\eta_1(0)\eta_2(0.5\text{s}) + \eta_1(0.1\text{s})\eta_2(0.6\text{s}) + \ldots + \eta_1(299.5\text{s})\eta_2(300\,\text{s})\right]$$

(the sampled data are 3001 from each of the two gauges, and the products $\eta_1 \cdot \eta_2$ available for this average are 2996).

Note, it is not necessary to specify the location of ② relative to ①, in that X and Y do not appear in the formula for $\overline{\eta}_2(t_o+T)$.

10.8.2 *Technique of calculation with time series data of the surface displacement and of the fluctuating pressure head*

Equation (10.1*b*) is rewritten in the form

$$\overline{\phi}(x_o+X, y_o+Y, z, t_o+T) =$$
$$= \frac{< \eta(x_o,y_o,t)\phi(x_o+X,y_o+Y,z,t+T) > - < \eta(x_o,y_o,t)\phi(x_o+X,y_o+Y,z,t+T-T^*) >}{< \eta^2(x_o,y_o,t) > - < \eta(x_o,y_o,t)\eta(x_o,y_o,t+T^*) >}\,\frac{H}{2}.$$

This gives the expected velocity potential if a wave of given very large height H occurs at point x_o, y_o. Hence, we can obtain the expected fluctuating pressure head by means of

$$\overline{\eta_{ph}}(x_o+X, y_o+Y, z, t_o+T) = -\frac{1}{g}\frac{\partial}{\partial T}\overline{\phi}(x_o+X, y_o+Y, z, t_o+T) = \tag{10.9}$$
$$= \frac{<\eta(x_o,y_o,t)\eta_{ph}(x_o+X,y_o+Y,z,t+T)> - <\eta(x_o,y_o,t)\eta_{ph}(x_o+X,y_o+Y,z,t+T-T^*)>}{<\eta^2(x_o,y_o,t)> - <\eta(x_o,y_o,t)\eta(x_o,y_o,t+T^*)>}\,\frac{H}{2},$$

where η_{ph} is the fluctuating pressure head of the random sea waves.

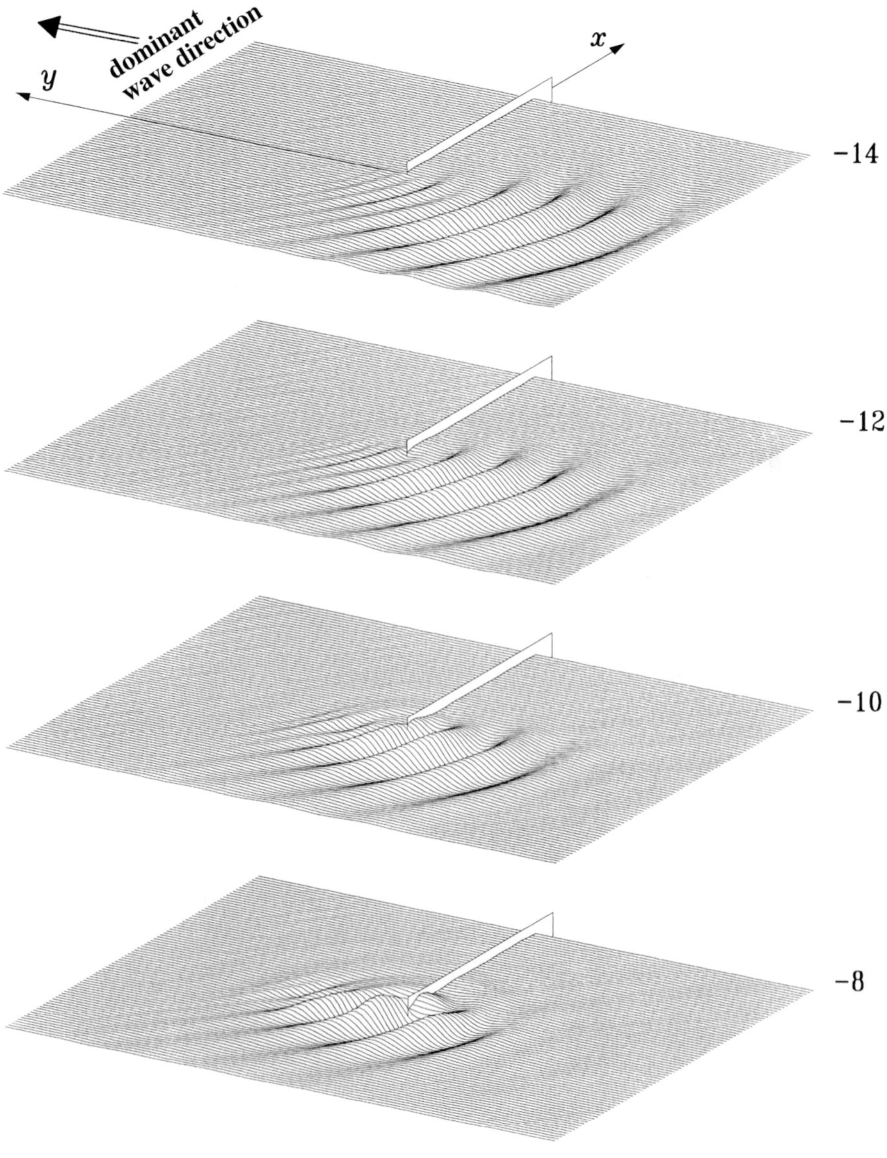

Fig. 10.20 Formation of an exceptionally high wave at a point x_o, y_o of the area where the diffraction coefficients proved to be surprisingly great (dark area of fig. 8.10). Here is a novelty: the wave group travels in a direction slightly different from the dominant wave direction (the dominant wave direction is wall-orthogonal while the group attacks from the left). *Continued opposite page.*

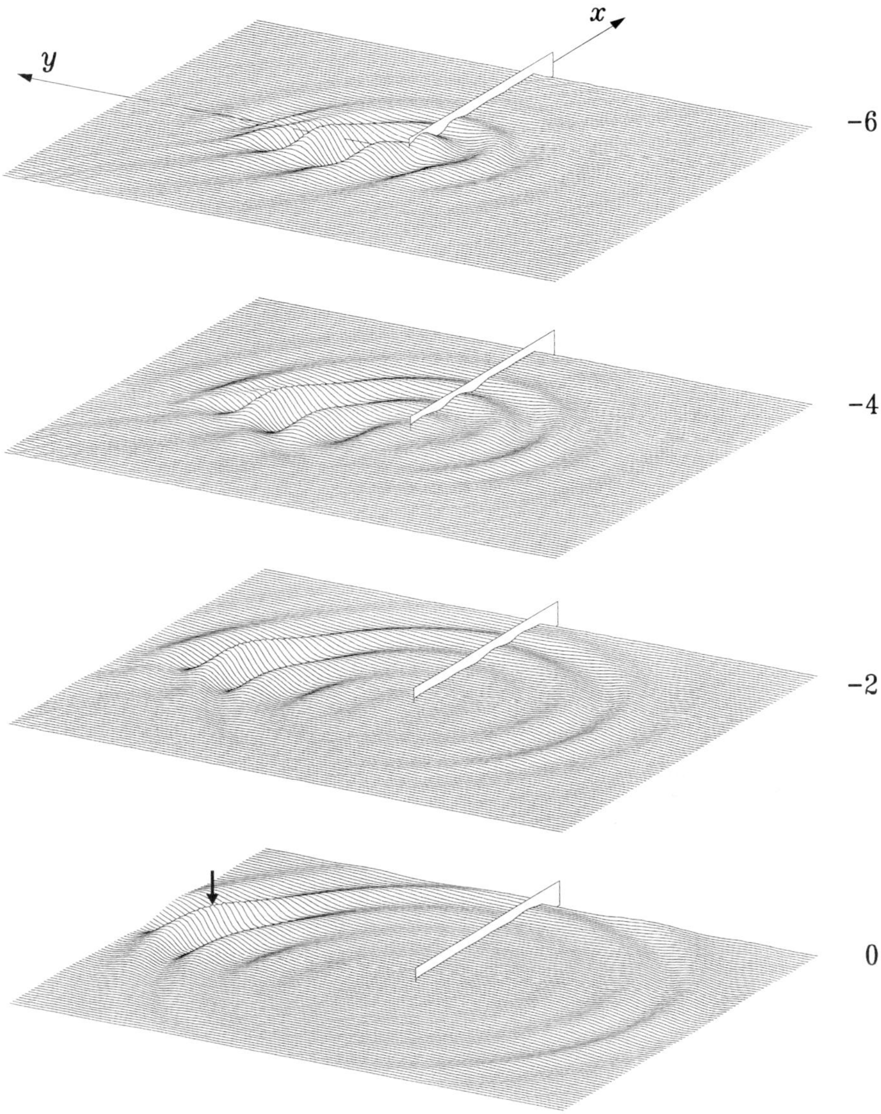

Fig. 10.20 *Continued.*

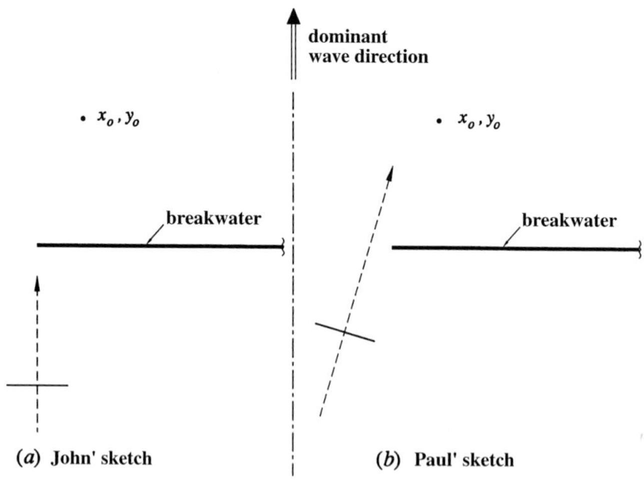

Fig. 10.21 Paul and John notice an exceptionally high wave in a record taken at point x_o, y_o behind the breakwater. They give two different explanations of how this wave occurred. Who's right?

Let us imagine we record the surface displacement at a point ① and the fluctuating pressure head at a point ② beneath the water surface. Then, equation (10.9) enables us to obtain the expected fluctuating pressure head at ② if a wave of given very large height H occurs at ①. Indeed, taking ① as x_o, y_o and ② as $x_o + X, y_o + Y, z$ we have

$$\overline{\eta_{ph2}}(t_o + T) = \frac{< \eta_1(t)\eta_{ph2}(t+T) > - < \eta_1(t)\eta_{ph2}(t+T-T^*) >}{< \eta_1^2(t) > - < \eta_1(t)\eta_1(t+T^*) >} \frac{H}{2},$$

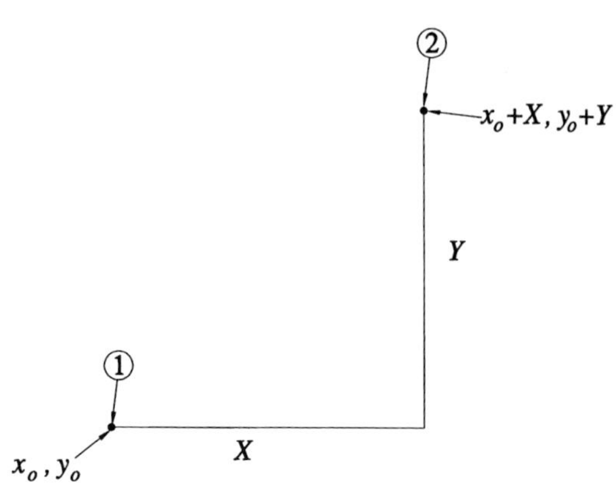

Fig. 10.22 Graphic aid for the step from (10.7) to (10.8).

where $\eta_1(t)$ is the time series data of the surface displacement at ①, and $\eta_{ph2}(t)$ is the time series data of the fluctuating pressure head at ②, and T^* is the abscissa of the absolute minimum of the autocovariance of $\eta_1(t)$.

Note, equation (10.9) requires that the point $x_o + X, y_o + Y, z$ is permanently beneath the water surface. Indeed, to obtain this equation, we have assumed that the Bernoulli equation holds at this point at every time instant.

10.8.3 *Technique of calculation with the time series data of the fluctuating pressure head*

If a pressure head wave with a given exceptionally large height H occurs at a fixed point x_o, y_o, z_o, at a time t_o, we can expect that the pressure head waves and the surface waves near x_o, y_o, z_o are given, respectively, by

$$\begin{aligned}
&\overline{\eta_{ph}}(x_o + X, y_o + Y, z_o + Z, t_o + T) = \\
&= \frac{H}{2}\Big[< \eta_{ph}(x_o, y_o, z_o, t)\,\eta_{ph}(x_o + X, y_o + Y, z_o + Z, t + T) > - < \eta_{ph}(x_o, y_o, z_o, t)\,\cdot \\
&\cdot\,\eta_{ph}(x_o + X, y_o + Y, z_o + Z, t + T - T^*) >\Big] \Big/ \Big[< \eta_{ph}^2(x_o, y_o, z_o, t) > + \\
&- < \eta_{ph}(x_o, y_o, z_o, t)\,\eta_{ph}(x_o, y_o, z_o, t + T^*) >\Big],
\end{aligned} \tag{10.10a}$$

$$\begin{aligned}
&\overline{\eta}(x_o + X, y_o + Y, t_o + T) = \\
&= \frac{H}{2}\Big[< \eta_{ph}(x_o, y_o, z_o, t)\,\eta(x_o + X, y_o + Y, t + T) > - < \eta_{ph}(x_o, y_o, z_o, t)\,\cdot \\
&\cdot\,\eta(x_o + X, y_o + Y, t + T - T^*) >\Big] \Big/ \Big[< \eta_{ph}^2(x_o, y_o, z_o, t) > + \\
&- < \eta_{ph}(x_o, y_o, z_o, t)\,\eta_{ph}(x_o, y_o, z_o, t + T^*) >\Big],
\end{aligned} \tag{10.10b}$$

where T^* is the abscissa of the absolute minimum of the autocovariance $\psi(T)$ of the pressure head waves at x_o, y_o, z_o:

$$\psi(T) \equiv < \eta_{ph}(x_o, y_o, z_o, t)\,\eta_{ph}(x_o, y_o, z_o, t + T) > .$$

The use of the theory with only the time series data of the fluctuating pressure head is very convenient. Indeed the pressure transducers have a very high precision and low cost, and work well also in sea water.

As to the employment of (10.10*a*), let us suppose to record the fluctuating pressure head at two fixed points ① and ② beneath the water surface. Then the expected fluctuating pressure head at ②, if a pressure head wave of given exceptionally large height H occurs at ①, is given by

$$\overline{\eta_{ph2}}(t_o + T) = \frac{< \eta_{ph1}(t)\,\eta_{ph2}(t + T) > - < \eta_{ph1}(t)\,\eta_{ph2}(t + T - T^*) >}{< \eta_{ph1}^2(t) > - < \eta_{ph1}(t)\,\eta_{ph1}(t + T^*) >}\,\frac{H}{2},$$

where $\eta_{ph1}(t)$ and $\eta_{ph2}(t)$ are the time series data of the fluctuating pressure head obtained at the two fixed points, and T^* is the abscissa of the absolute minimum of the autocovariance of the pressure head waves at ①:

$$\psi(T) \equiv < \eta_{ph1}(t)\eta_{ph1}(t+T) >.$$

Check equations (10.10*a-b*). First, consider the pressure head waves at the fixed point x_o, y_o, z_o, and prove that the condition

$$\eta_{ph}(t_o) = \frac{H}{2}, \quad \eta_{ph}(t_o + T^*) = -\frac{H}{2}$$

becomes necessary and sufficient for the occurrence of a pressure head wave of given height H, as $H/\sigma \to \infty$ (here σ being the r.m.s. fluctuating pressure head at the fixed point). To this end, bear in mind that $\eta_{ph}(t)$ at any fixed point represents a stationary Gaussian random process, and retrace all the steps of sects. 9.1 and 9.2.

Then, to achieve (10.10*a*) retrace step by step also sect. 9.3. The only difference being that now you have to consider the conditional p.d.f. of the fluctuating pressure head at point $x_o + X, y_o + Y, z_o + Z$, at time $t_o + T$, given the condition

$$\eta_{ph}(x_o, y_o, z_o, t_o) = \frac{H}{2}, \quad \eta_{ph}(x_o, y_o, z_o, t_o + T^*) = -\frac{H}{2}. \tag{10.11}$$

Use the same procedure also to achieve (10.10*b*). This time, consider the conditional p.d.f. of the surface displacement at point $x_o + X, y_o + Y$, at time $t_o + T$, given the condition (10.11). All these conditional p.d.f.s. take on the general form (9.18), and each time you have only to specify the proper entries of the covariance matrix.

10.8.4 *The use of the theory with time series data is very easy even with a very complex solid boundary!*

In the writer's opinion the most remarkable property of the quasi-determinism theory is that it holds whatever the configuration of the solid boundary (of course, provided the flow is frictionless). This property is fully appreciated when working with the time series data of η and/or η_{ph}. It does not matter at all how complex the solid boundary is, and, more important, it does not matter at all whether the problem of the interaction between waves and this solid boundary has been solved. We have simply to obtain a few time-series data of η and/or η_{ph} and use equations (10.7), or (10.9) or (10.10*a-b*), with these time series, to predict what happens if a wave of given exceptionally large height occurs at a given point.

The answers, as we shall see in the next two sections, are surprisingly clear and enable us to understand the essential features of the wave motion under examination.

10.9 The "genetic code" of the sea waves

Fig. 10.23 shows the formation of a wave of a given exceptionally large height at the centre of the framed area in the open sea. The phenomenon is the same as we

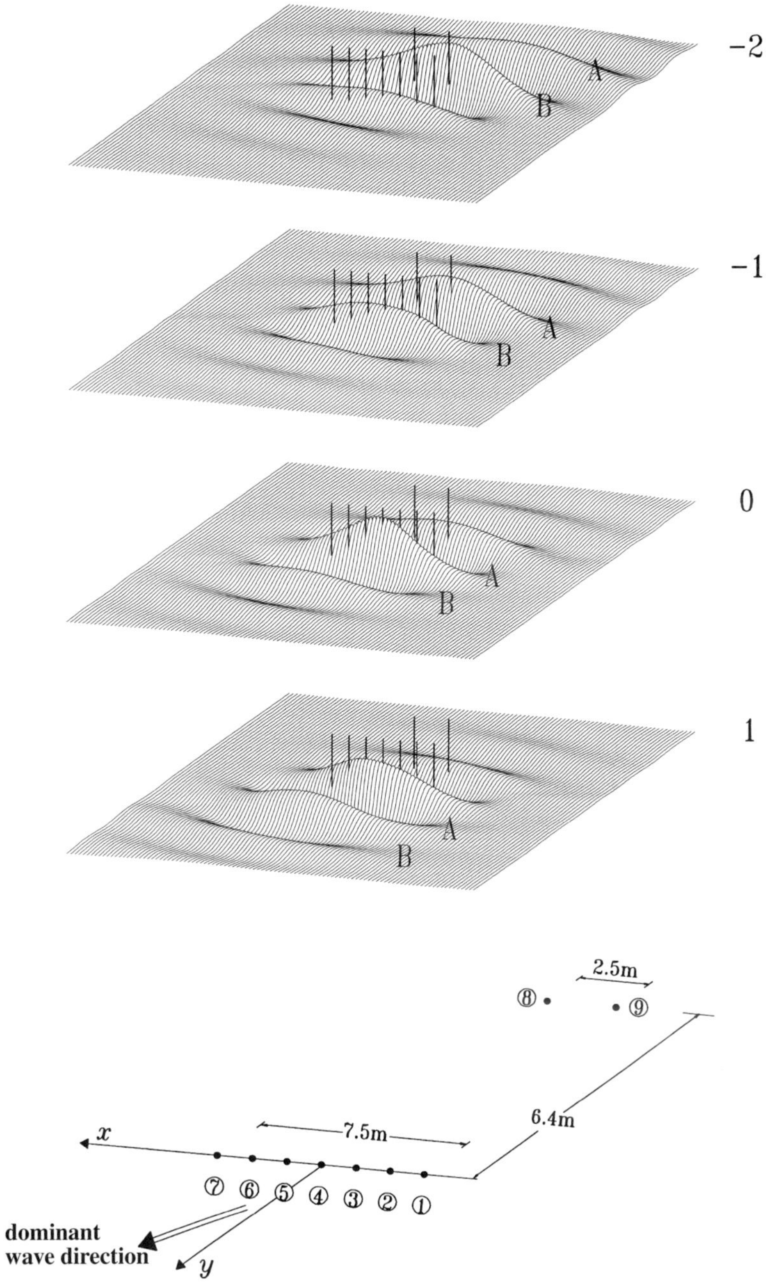

Fig. 10.23 Question: what happens if an exceptionally high wave should be recorded at pile ④? Answer: the three dimensional picture.

have already seen in fig. 10.1, with the only difference being that the wave attack is from the right side and there is an array of nine piles. These nine piles of small sections were actually put at sea for the experiment RC 1990, and their plan is shown in the lowest part of fig. 10.23.

Let us observe the first picture of this figure: wave B occupies the central position of the group and is at piles ⑧-⑨. After a wave period T_p, the envelope centre falls between wave B and wave A: wave B has become smaller because it has left the central position of the group, and wave A has become higher because it is going to occupy the central position of the group. One more period T_p (third picture) and wave A passes the array ①-⑦ and occupies the central position of the group.

Replacement of wave A with wave B at the envelope centre is evident also in fig. 10.24 which shows the records made by an ideal array of gauges at the piles. We see that wave B occupies the central position of the group at points ⑧ and ⑨, while at points ①-⑦ wave A has replaced B. The centre of the wave crest passes at

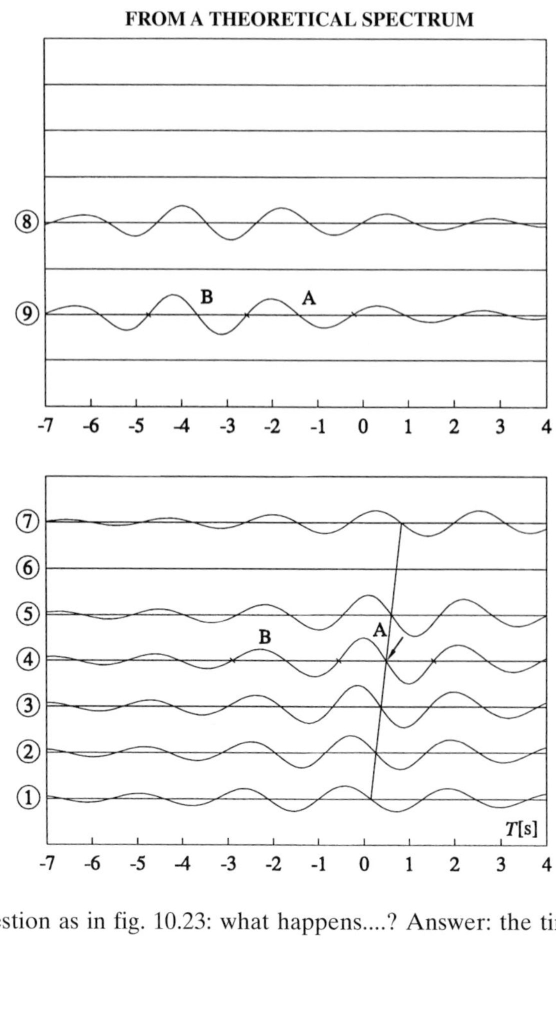

Fig. 10.24 Same question as in fig. 10.23: what happens....? Answer: the time series at the piles.

point ⑨ and at point ④, and from ⑨ to ④, wave B becomes smaller and longer, and, conversely, wave A becomes higher and shorter.

Fig. 10.25, like fig. 10.24, shows what happens at points ①-⑨ if a wave of given exceptionally large height should occur at point ④. Fig. 10.24 was obtained by means of (10.2*a*) with a theoretical spectrum, while fig. 10.25 was obtained by means of (10.7) with the time series data of a sea state. The likeness of the two figures is really amazing!

What we see in fig. 10.25 is something like the *genetic code* of the sea state. It contains the information on what happens if an exceptionally high wave should occur at a given point (specifically, at point ④). The genetic code can be extracted from the time series data of a gauge array with the technique described in sect. 10.8.1, and it is not necessary that these time series include some very high

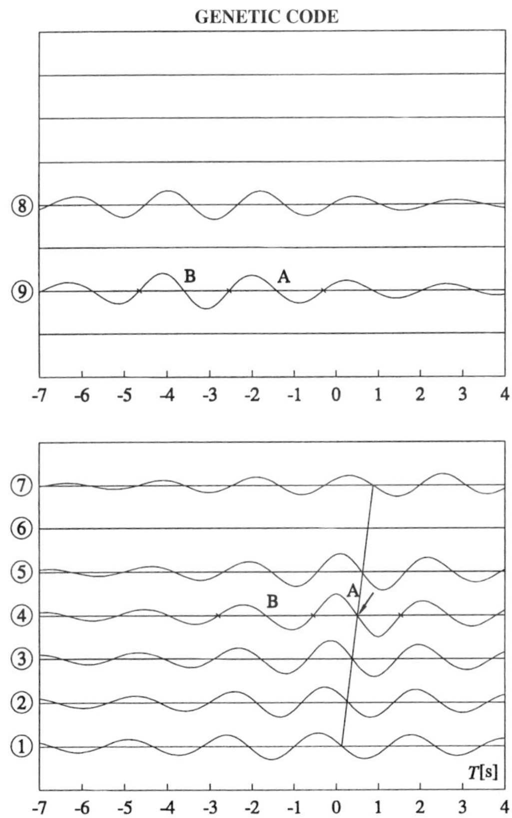

Fig. 10.25 Same question as in fig. 10.23: what happens....? Answer: once again the time series at the piles. The previous figure (fig. 10.24) was obtained by means of (10.2*a*) with the theoretical spectrum (8.20), while this figure was obtained by means of (10.7) from the time series data of the experiment RC 1990!! Specifically, the time series data of a sea state (540s, 250 individual waves) were used to obtain the averages on the r.h.s. of equation (10.7).

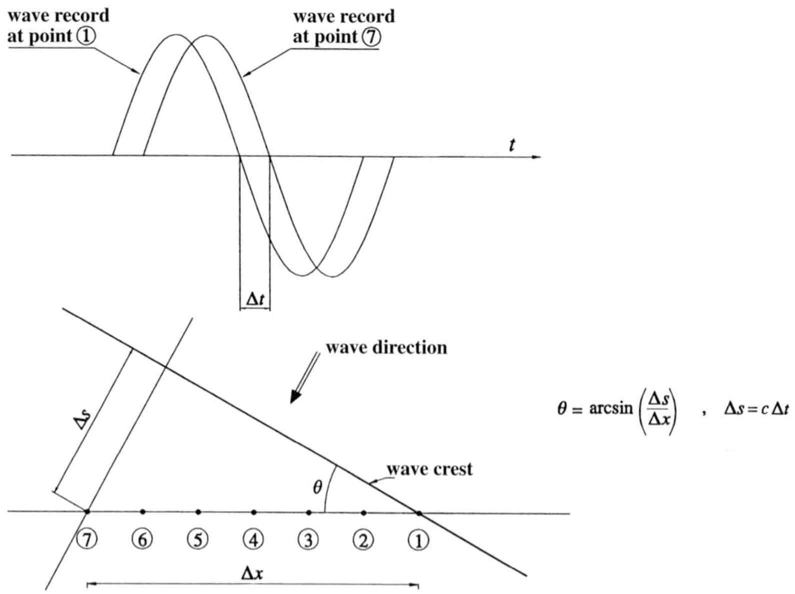

Fig. 10.26 The dominant direction is estimated with precision from the relative phase of the points ①-⑦. Indeed the group's central crest (in the genetic code) is perfectly straight.

wave. In this, the genetic code of the sea state is like the genetic code of the human beings. Indeed, for obtaining the genetic code of a human being, even a few cells are enough; and, similarly, for obtaining the genetic code of a sea state, even a few scores of consecutive waves are enough.

It will have been noted that the genetic code (fig. 10.25) fully confirms replacement of wave A with wave B at the central position of the group, and the related big transformations: wave B becoming smaller and longer, and wave A growing and shrinking. This confirmation comes out in the whole set of records of the experiment RC 1990.

Of course the input data for the theoretical group of figs. 10.23 and 10.24 are those of the sea state from which we extracted the genetic code of fig. 10.25: $d = 3\,\mathrm{m}$, $T_p = 2.20\,\mathrm{s}$, $\bar{\theta} = 20°$. As to angle $\bar{\theta}$, it can be evaluated very carefully from the relative phases of wave A at the points ①-⑦ [cf. the scheme of fig. 10.26].

10.10 The determinism arises from within the random waves

10.10.1 *Comparison between records of exceptionally high waves and the genetic code*

In the following pages a few records of very high waves are shown together with the genetic codes of their sea states. These are the highest waves of four small scale field experiments in the natural laboratory of Reggio Calabria. In each of these experiments a gauge array was used to record the surface displacement and/or the fluctuating pressure head at some fixed points.

By *the highest wave of an experiment* we mean the wave with the highest quotient H/σ in the whole set of zero up-crossing waves and zero down-crossing waves obtained by the gauge array. According to the zero up-crossing definition, which was introduced in sect. 4.1 and hitherto followed, an individual wave is the se-

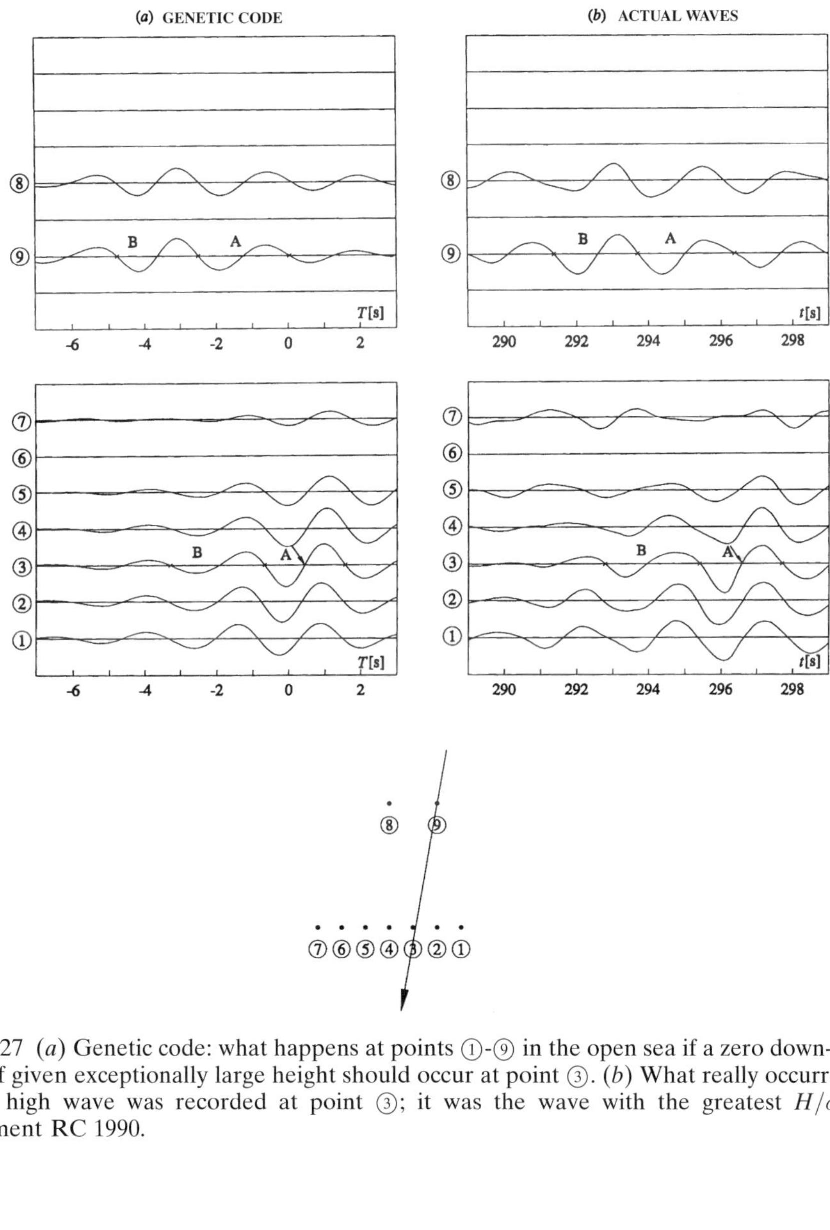

Fig. 10.27 (*a*) Genetic code: what happens at points ①-⑨ in the open sea if a zero down-crossing wave of given exceptionally large height should occur at point ③. (*b*) What really occurred when a very high wave was recorded at point ③; it was the wave with the greatest H/σ of the experiment RC 1990.

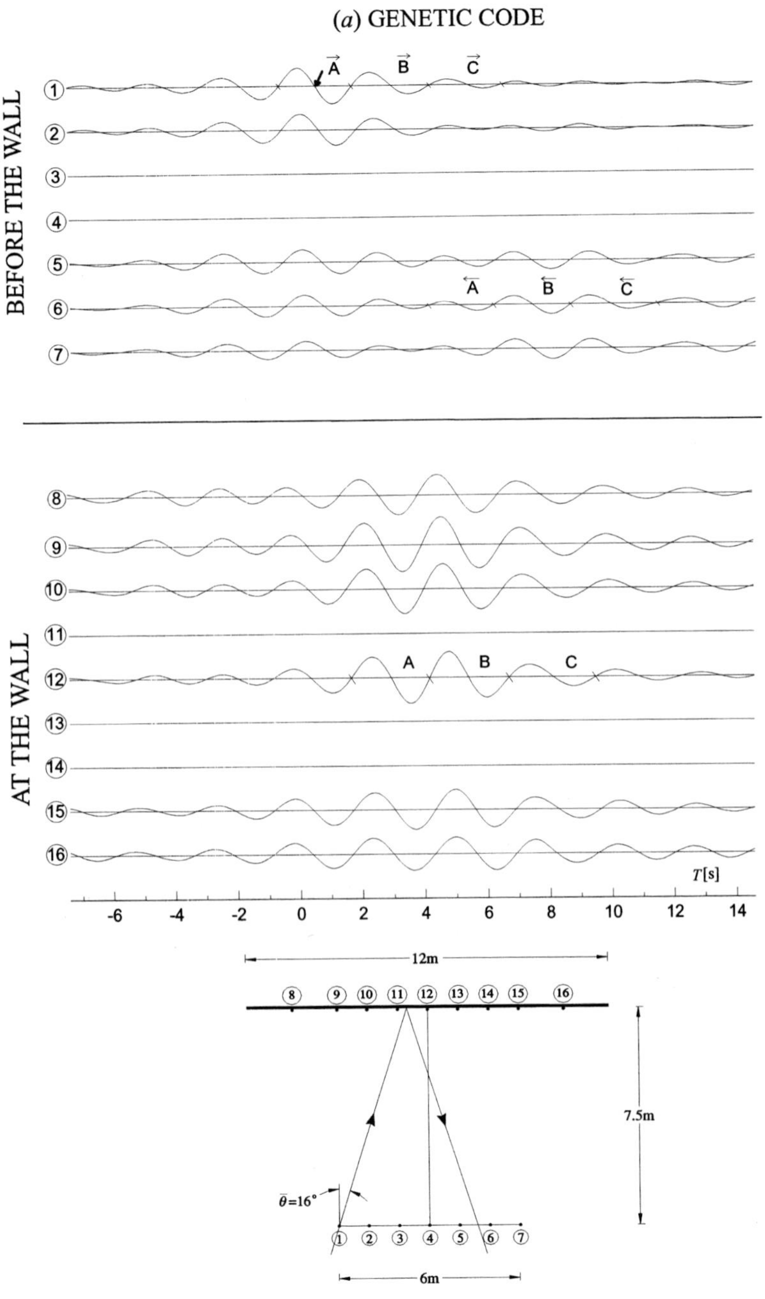

Fig. 10.28 (*a*) Genetic code: what happens at the various points before the breakwater if a wave of given exceptionally large height should occur at point ① (we mean that the wave is exceptionally high with respect to the mean wave height at this point). *See panel (b) of this figure on the next page.*

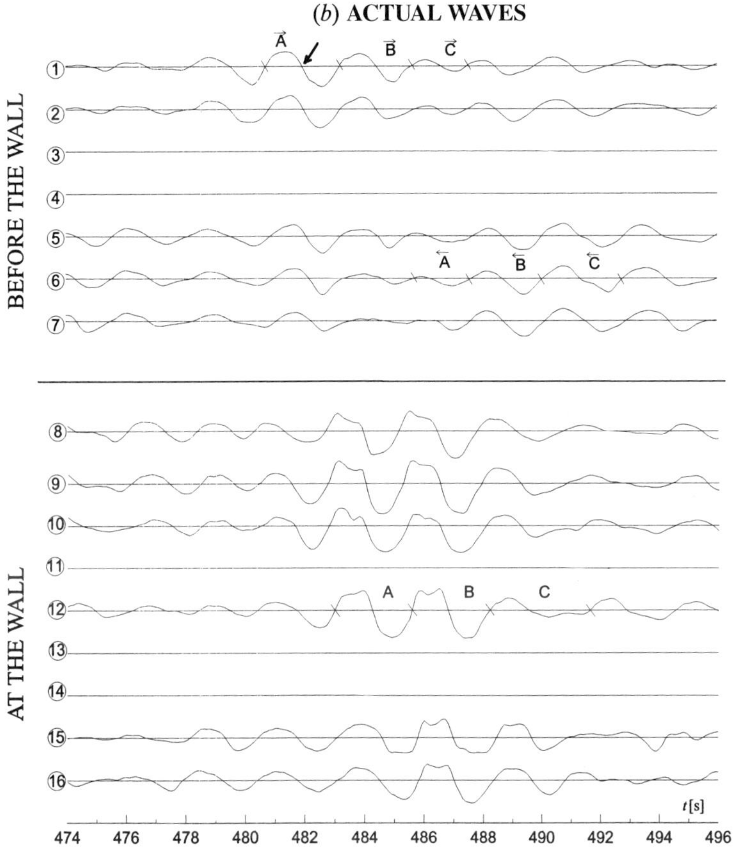

Fig. 10.28 (b) What really occurred when a very high wave was recorded at point ①; it was the wave with the largest H/σ of the experiment RC 1991. *See panel (a) of this figure on the previous page.*

quence of a crest and the following trough; according to the zero down-crossing definition an individual wave consists of a trough and the following crest. Of course, the quasi-determinism theory holds also with the zero down-crossing definition; only that the r.h.s. of (10.1a-b) must be multiplied by -1.

Experiment RC 1990

The highest wave was recorded at point ③. It was a zero down-crossing wave with $H = 9.6\sigma$. Fig. 10.27b shows the record made by the gauge array, and the arrow points to the highest wave. Then fig. 10.27a shows what happens if a wave of given very large height H should occur at point ③. It is the genetic code of the sea state where the exceptionally high wave was recorded.

Let us examine this code. The relative phases at the traverse of points ①-⑦ reveal that $\bar{\theta}$ was about 10°. Hence, if the centre of the wave crest passed at point ③ it had to pass very close to point ⑨. The replacement of wave A with wave B at the central position of the group is well evident; note the characteristic big transformations from point ⑨ to ③ along the path of the crest centre.

The actual waves (record) are more irregular, but the essential features are the same as in the genetic code.

There were nine piles, each with an ultrasonic probe above the water surface and a pressure transducer beneath the water surface. The sampling device was able to receive 16 signals simultaneously, and hence two gauges had to be disconnected (the disconnected gauges were those of pile 6). The figure shows only the pressure head waves, the surface waves can be seen in the paper of Boccotti et al. (1993).

Experiment RC 1991

A small upright breakwater of 12 m was built and an array of pressure transducers was assembled before this breakwater [see the sketch in the lowest part of fig. 10.28*a*]. These gauges were 0.60 m below the mean water level and were permanently beneath the water surface. The gauges far from the wall were supported by horizontal beams of small section.

The highest wave was a zero up-crossing with $H = 9.6\sigma$. It was recorded at point ①, in the course of record 51. Fig. 10.28*a* shows what happens at the various points if a wave of given exceptionally large height H should occur at point ① (of course, we mean *exceptionally large for this point*). Fig. 10.28*b* shows what really happened. The likeness is evident.

A few comments for a deeper understanding. The time histories of the genetic code are consistent with one another in depicting the following scene: a wave group approaching the wall from the left strikes point ① in full, then the centre of the wave crest hits the wall (nearly in the centre), and returns seaward passing close to point ⑥. [The angle between the direction of wave advance and the wall-orthogonal can be estimated with precision from the relative phases at the traverse of gauges ①-⑦, and it proves to be of 16°.] Note that wave A at the wall has left the central position of the group and wave B is going to replace it. As a consequence, the wall is beaten by a sequence of two big waves of nearly the same height.

The double peak of the high wave crest at the wall is very characteristic of the pressure head waves and is associated with the occurrence of a high vertical water jet. These water jets damaged the wire connections, and this is why a few time series are lacking.

In the case under examination, the genetic code does not foresee that the exceptionally high wave is due to the collision of two groups, although the point x_o, y_o is far from the wall. The reason is that, if the centres of two groups collided at location ①, the centre of the first group would hit just the left tip of the breakwater and hence only a part of the group's energy would be reflected. The genetic code foresees that the exceptionally high wave is due to the collision of two groups only if we take x_o, y_o to the right of point ①, for example if we take ④ as x_o, y_o. Indeed if the centre of the reflected group has to pass at this point, it must hit the wall at a certain distance from the tips.

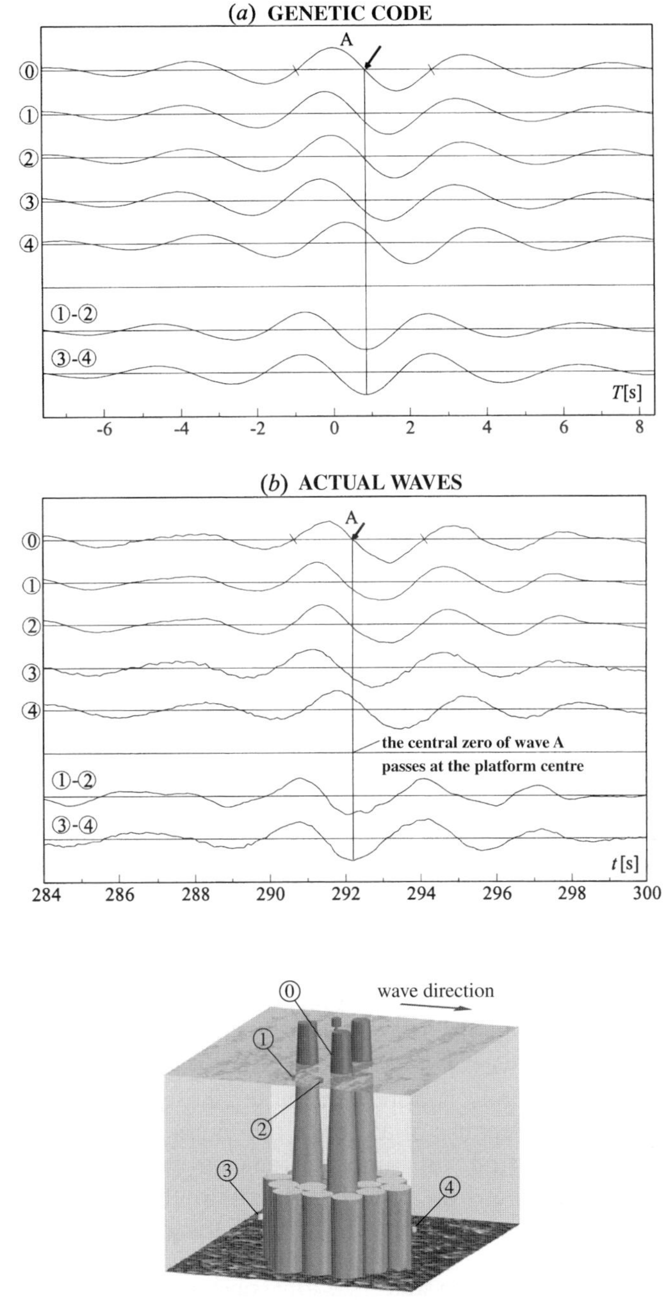

Fig. 10.29 (*a*) Genetic code: what happens if a wave of given exceptionally large height should pass through the platform. (*b*) What really occurred when a very high wave was recorded amid the columns; it was the wave with the largest H/σ of the experiment RC 1992.

Experiment RC 1992

The field laboratory of this experiment is shown in fig. 11.1. The structure is the 1:50 scale model of a gravity offshore platform.

The highest wave which passed through the platform was a zero up-crossing with a height somewhat greater than 8σ. Fig. 10.29*a* [obtained by means of (10.7) and (10.9) from the time series data] shows what happens if a wave of given exceptionally large height should pass at point ⓪ amid the columns. Specifically, it shows the expected surface displacement at point ⓪ and the expected pressure fluctuations at four points of the structure. The differences

$$\Delta p_{1-2} = p_1 - p_2,$$

$$\Delta p_{3-4} = p_3 - p_4,$$

give an idea of the fluctuating horizontal force. All time histories here are given in a normalized form. This means that a surface wave of, say, 8 times the r.m.s. surface displacement has the same height (in the figure) as a pressure head wave of 8 times the r.m.s. fluctuating pressure head.

Here too, the records taken in the highest wave are close to the expected time histories of the genetic code.

Experiment RC 1993

The field laboratory of this experiment is shown in fig. 11.6. The structure is the 1:30 scale model of a submerged tunnel.

The highest wave was a zero down-crossing of the force-process $F_y(t)$. The height was $H = 10\sigma_{F_y}$ [H here has the dimensions of a force and σ_{F_y} is the standard deviation of the force-process $F_y(t)$]. Fig. 10.30*a* shows the expected time histories of the surface displacement above the tunnel, the pressure fluctuations at a few points, and the forces $F_y(t)$ and $F_z(t)$, when a wave of the process $F_y(t)$ with a given exceptionally large height occurs.

According to the genetic code the exceptionally large trough-to-crest displacement of the horizontal force should occur because a wave group passes over the tunnel when a wave (B) is going to replace the preceding one (A) at the central position.

As usual, panel (*b*) of the figure shows what really occurred when the exceptionally high wave was recorded; and once again the likeness with the prediction of the genetic code is apparent. Note, the crest of the second highest surface wave is deformed because of a flaw of the ultrasonic probe (this probe does not work well with spilling or almost breaking waves). Nevertheless, the configuration of the wave group is well evident in the record of the pressure transducer ① beneath the water surface at the same location of the ultrasonic probe.

According to the theory of the sea states, the force on a solid body represents a stationary Gaussian random process of time (provided the flow is frictionless). Therefore the quasi-determinism theory can be applied also to the force on the tunnel. In particular, if a wave of given exceptionally large trough-to-crest height of the process $F_y(t)$ (horizontal force) occurs at an instant t_o (we mean that t_o is the instant of the positive peak), the expected time

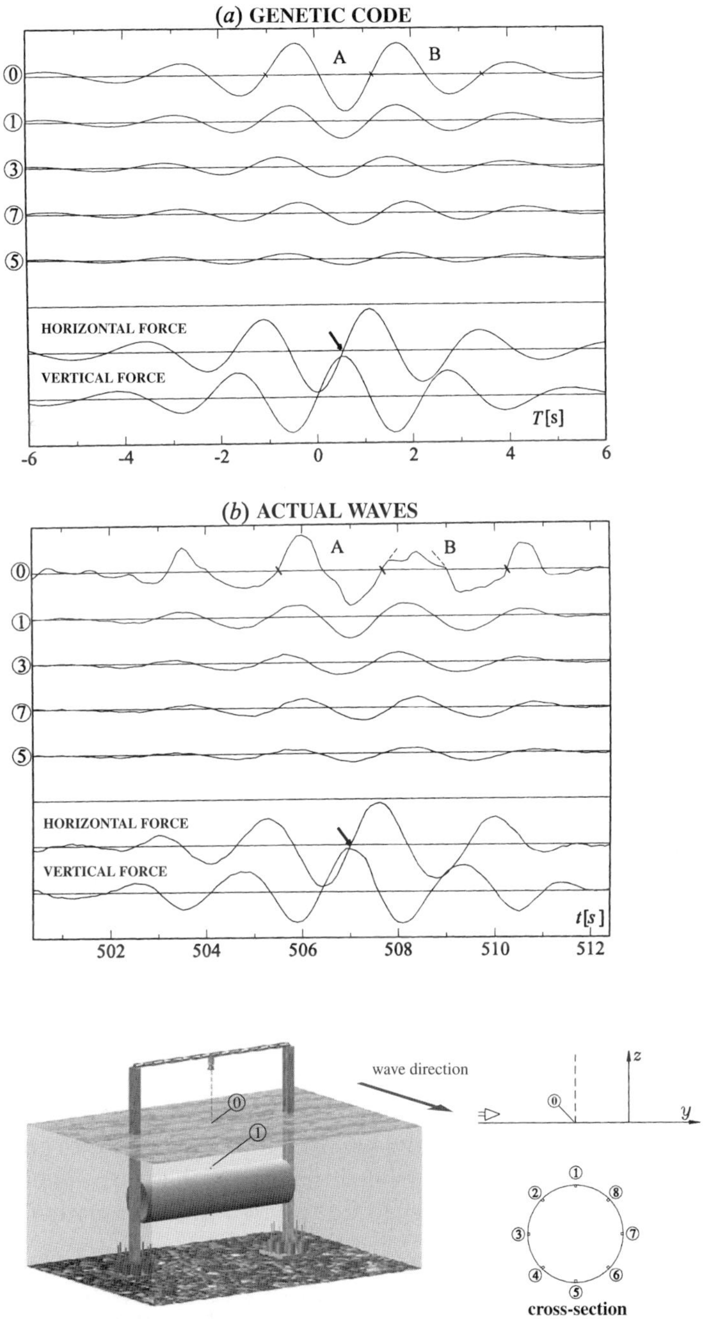

Fig. 10.30 (*a*) Genetic code: what happens if an exceptionally high wave of $F_y(t)$ [horizontal force on the cylinder] should occur. (*b*) What really occurred when a very high wave of $F_y(t)$ was recorded.

histories of the surface displacement at any point x, y, the pressure fluctuation at any point x, y, z, and the horizontal and vertical forces are given, respectively, by

$$\overline{\eta}(x, y, t_o + T) = \frac{< F_y(t)\eta(x, y, t+T) > - < F_y(t)\eta(x, y, t+T-T^*) >}{< F_y^2(t) > - < F_y(t)\, F_y(t+T^*) >} \frac{H}{2},$$

$$\overline{\Delta p}(x, y, z, t_o + T) = \frac{< F_y(t)\Delta p(x, y, z, t+T) > - < F_y(t)\Delta p(x, y, z, t+T-T^*) >}{< F_y^2(t) > - < F_y(t)F_y(t+T^*) >} \frac{H}{2},$$

$$\overline{F_y}(t_o + T) = \frac{< F_y(t)F_y(t+T) > - < F_y(t)F_y(t+T-T^*) >}{< F_y^2(t) > - < F_y(t)F_y(t+T^*) >} \frac{H}{2},$$

$$\overline{F_z}(t_o + T) = \frac{< F_y(t)F_z(t+T) > - < F_y(t)F_z(t+T-T^*) >}{< F_y^2(t) > - < F_y(t)F_y(t+T^*) >} \frac{H}{2},$$

where T^* is the abscissa of the absolute minimum of the autocovariance of $F_y(t)$.

10.10.2 *The genetic code is linear*

Of course, the very high waves, besides a residual random nature, also exhibit some non-linearity effects. A clear example is the double peak of the pressure head waves of fig. 10.28*b*, which is associated with the occurrence of a vertical water jet at the breakwater.

The genetic code is unable to foresee these non-linearity effects. Indeed, it proves to be strictly linear even if it is obtained from the time series data. In other words, the genetic code is able to foresee the essential features of sea waves: the existence of the three-dimensional groups, the characteristic mechanics of these groups in the open sea, and the way of interaction of these groups with any solid body. But, it is not able to foresee the local deformations of the individual waves due to non-linearity.

10.10.3 *Conclusion*

Comparing fig. 10.27*b* with 10.27*a*, 10.28*b* with 10.28*a*, 10.29*b* with 10.29*a*, and 10.30*b* with 10.30*a*, we see the chaotic (wind-generated) waves approach a deterministic archetype. This archetype is contained in the very chaos of the sea storm, and it can be extracted and seen. The highest waves tend to this archetype but, necessarily, retain something of the chaos where they are born. The quasi-determinism theory says that the deterministic archetype would stand out from the chaos only with an infinitely high energy concentration ($H/\sigma \to \infty$). The deterministic archetype is like Michelangelo's prisoner searching the extreme energy to free himself from the marble's grasp.

Conclusive note of chapters 9 and 10

The quasi-determinism theory was introduced by the author in a series of papers from 1981 to 1989. The theory for the time domain, including the formula for the wave height probability, was given in the early papers of 1981, 1982. The experimental verification off the beach of Reggio Calabria was made by the author and his students.

Fig. 10.31 Michelangelo Buonarroti: Prisoner named Atlantis; Florence, Academy's gallery (see mention in sect. 10.10.3). Photo by Alinari, Florence.

References

Boccotti P., 1981 On the highest waves in a stationary Gaussian process. *Atti Acc. Ligure di Scienze e Lettere* 38, 271-302;

1982 On the highest sea waves. *Monograph Inst. of Hydraulics, University of Genoa*, 1-161;

1982 On ocean waves with high crests. *Meccanica* 17, 16-19;

1983 Some new results on statistical properties of wind waves. *Appl. Ocean Res.* 5, 134;

1984 Sea waves and quasi-determinism of rare events in random processes. *Atti Acc. Naz. Lincei Rendiconti* 76, 119-127;

1988 Refraction, reflection and diffraction of irregular gravity waves. *Excerpta of the Italian Contribution to the Field of Hydraulic Eng.* 3, 47-89;

1989 On mechanics of irregular gravity waves. *Atti Acc. Naz. Lincei Memorie* 19, 110-170;

1989 Quasi-determinism of sea wave groups. *Meccanica* 24, 3-14;

1997 A general theory of three-dimensional wave groups. *Ocean Eng.* 24, 265-300;

Boccotti P., Barbaro G. and Mannino L., 1993 A field experiment on the mechanics of irregular gravity waves. *J. Fluid Mech.* 252, 173-186.

Forristall G. Z., 1984 The distribution of measured and simulated heights as a function of spectra shape. *J. Geophys. Res.* 89, 10547-10552.

Chapter 11
ANALYSIS OF THE WAVE FORCES ON OFFSHORE STRUCTURES

11.1 Wave forces on gravity offshore platforms

For the experiment RC 1992, besides the array of transducers at the platform, there was also a second array of transducers with the same configuration. The second array was assembled at a certain distance from the platform model, where the waves were not disturbed. The transducers of the second array were supported by piles of small section and high stiffness [see fig. 11.1].

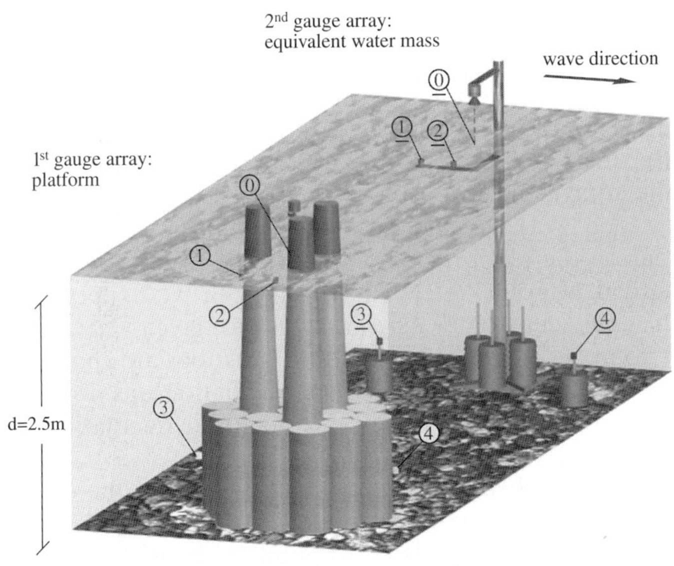

Fig. 11.1 The field laboratory for the experiment RC 1992. Foreground: the 1:50 scale model (with some minor variations) of the support structure of a gravity offshore platform. The prototype was described by Dawson (1983), and the 1:120 scale model was tested in a wave flume by Garrison et al. (1974).

Thanks to the two arrays of gauges it was possible to compare the force on the platform with the force on an equivalent water mass (i.e. one with the same volume and shape of the solid body). Specifically, it was possible to evaluate the horizontal diffraction coefficient

$$C_{do} \equiv \frac{\sqrt{< F_o^2(t) >}}{\sqrt{< \underline{F}_o^2(t) >}}, \tag{11.1}$$

where $F_o(t)$ is the horizontal force on the solid body and $\underline{F}_o(t)$ is the horizontal force on the equivalent water mass, which is called the Froude-Krylov force. The C_{do} of the columns was close to 2, while the C_{do} of the base consisting of nineteen cylinders ranged between 1.2 and 1.6 according to the wave characteristics. The waves were the 1:50 scale model (same Froude number) of ocean waves with an H_s ranging from 10 m to 20 m.

The analysis of a few hundred records showed that the amplitudes of the pressure fluctuations, at corresponding points of the platform and of the equivalent water mass, were nearly equivalent to each other. In particular the standard deviation

$$\sqrt{< \Delta p_3^2(t) >}$$

of the pressure fluctuation at point ③ was nearly coincident with the standard deviation

$$\sqrt{< \Delta p_{\underline{3}}^2(t) >}$$

of the pressure fluctuation at point $\underline{③}$. Thus the question came: how is it possible that the wave force on the solid body is markedly larger than the wave force on the equivalent water mass if the amplitude of the pressure fluctuations on the solid body and on the equivalent water mass are nearly equal to each other?

The "genetic code" was investigated. Two questions were put. First: what can we expect will occur on the platform's boundary if a wave of given very large height should be recorded at point ⓪? Second: what can we expect will occur on the boundary of the equivalent water mass, if a wave of given very large height should be recorded at point $\underline{⓪}$? The answers to these questions are shown, respectively, in figs. 11.2*a*-11.3*a* and in figs. 11.2*b*-11.3*b*. The calculation was done with (10.9) from the time series data of the surface displacement and the fluctuating pressure head of one of the records.

Note that the pressure head wave takes 0.134 s to cover the distance from point ① to point ② at the platform's column, while it takes only 0.067 s (half the time!) to cover the same distance from point $\underline{①}$ to point $\underline{②}$ in the undisturbed wave field. Then, the pressure head wave takes 0.60 s to cover the distance from point ③ to point ④ at the platform's base, while it takes 0.30 s (again half the time) to cover the same distance from point $\underline{③}$ to point $\underline{④}$ in the undisturbed wave field.

The result is that between the first half and the second half of the solid body

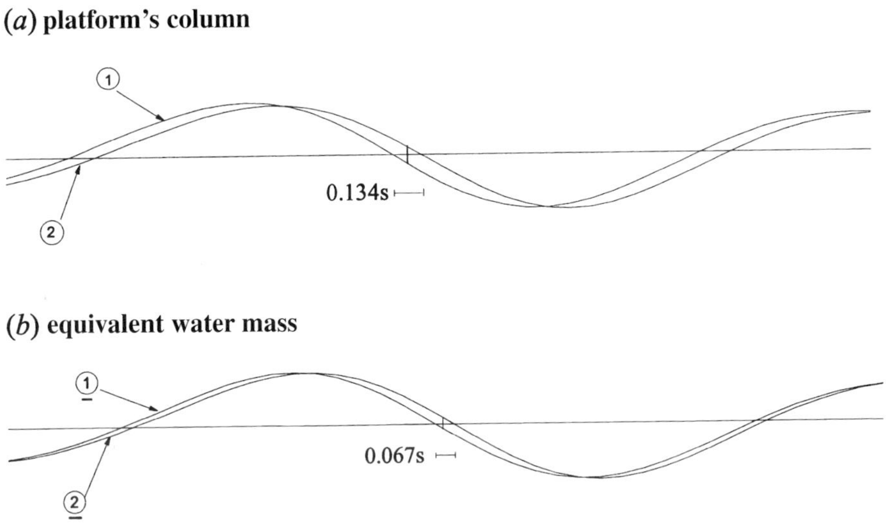

Fig. 11.2 (*a*) Pressure head waves at points ① and ② of the columns, if a wave of given very large height passes at point ⓪ amid the columns. (*b*) Pressure head waves at points ① and ② of the equivalent water volume, if a wave of given very large height passes at point ⓪.

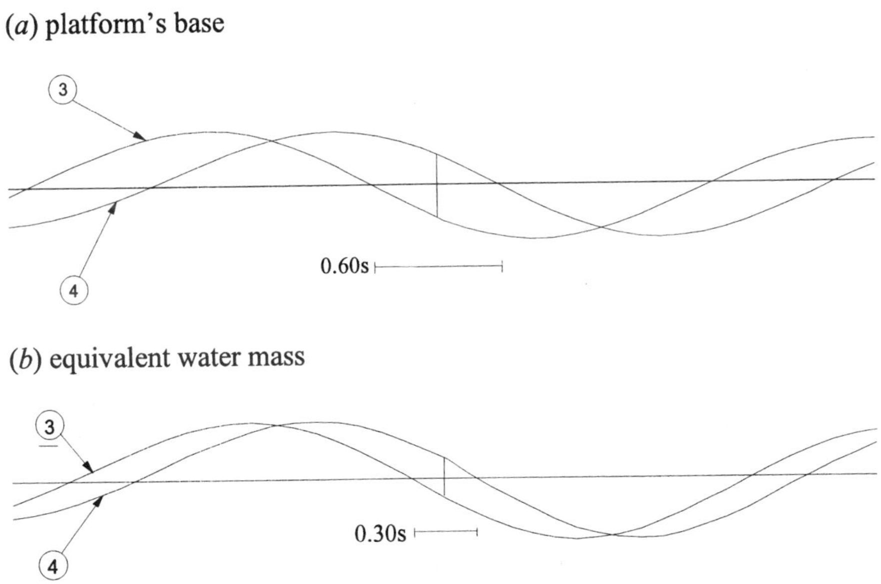

Fig. 11.3 (*a*) Pressure head waves at points ③ and ④ of the platform's base if a wave of given very large height passes at point ⓪ amid the columns. (*b*) Pressure head waves at points ③ and ④ of the equivalent water volume, if a wave of given very large height passes at point ⓪.

there is a phase angle which is twice as large as the phase angle between the first half and the second half of the equivalent water mass (the first half being the wave beaten one, and the second half being the sheltered one). The larger phase angle implies a larger pressure difference between the first half and the second half of the solid body.

Note that if we redouble the phase angle between two points (where the pressure fluctuations have the same amplitude and frequency) we increase the pressure difference between these two points only if the original phase angle is smaller than 120°. This condition is usually satisfied by pairs of points like ③ and ④ at the platform's base, at least in the design sea states which have some very large wavelengths. Even more so, the aforesaid condition (the original phase angle being smaller than 120°) is satisfied by pairs of points like ① and ② which are only one column's diameter apart.

To summarize, the horizontal forces on the base and on the columns of the platform are larger than the horizontal forces on the equivalent water masses, essentially because the propagation speed of the pressure head waves drops at the base and columns. The drop in the propagation speed is the same at the column and at the base. However, the percentage rise of the horizontal force due to this drop is greater for the column, since it has a smaller diameter. In sect. 11.4.3 we shall give the relation between the force enhancement and the drop in the propagation speed.

11.2 Local perturbation of the flow field at an offshore structure

11.2.1 *Why does the propagation speed drop?*

Let us examine the variations of the flow field from a water vertical cylinder to a solid cylinder of the same diameter. To this end, let us assume ideal flow and small wave amplitude, which means neglecting the shear stress and the flux of momentum with respect to the local inertia.

Let us consider the water vertical cylinder. The pressure fluctuation at point A of the cylinder's boundary [see fig. 11.4] has some zero up-crossings at the instants t_{A1} and t_{A3}. At these instants, the control volume before point A has a positive local inertia

$$I_y \equiv \rho \int_W \frac{\partial v_y}{\partial t}\, \mathrm{d}W.$$

If the water cylinder is replaced with a solid cylinder of the same diameter, the absolute value of I_y reduces itself ($\partial v_y/\partial t$ becomes zero at point A of the solid boundary); while the pressure at point $\overline{A}$ [the left end of the control volume] does not change, provided the length Δy of the control volume is such that the local perturbation gets negligible at point $\overline{A}$. Therefore, the pressure at point A at instants t_{A1} and t_{A3} grows if the water cylinder is replaced with the solid cylinder.

Through the same reasoning we conclude the pressure at point A, at time t_{A2} of the zero down-crossing, is reduced if the ideal water cylinder is replaced with the solid cylinder (this time the local inertia I_y is negative and its absolute value gets smaller because of the presence of the solid body).

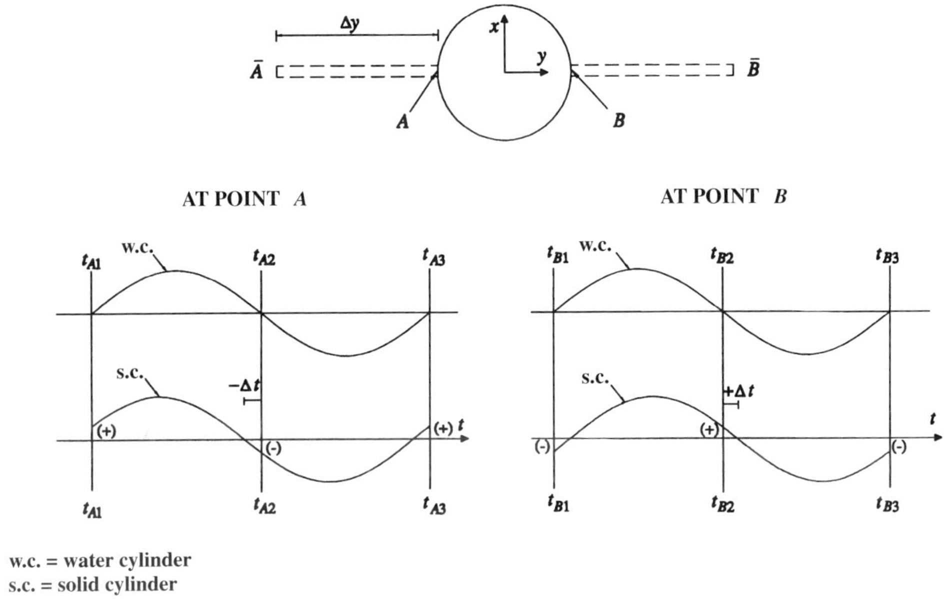

Fig. 11.4 The time taken by the pressure head wave to cover the distance from A to B at the vertical water cylinder is $t_{B2} - t_{A2}$. If A and B belong to the solid cylinder, the time gets longer: $t_{B2} - t_{A2} + 2\Delta t$. This can be proved by applying the linear momentum equation to the small control volumes A-$\bar{A}$ and B-$\bar{B}$.

Because of the rise at times t_{A1} and t_{A3} and the fall at time t_{A2}, the pressure fluctuation at point A is subjected to a negative time shift $(-\Delta t)$ if the ideal water cylinder is replaced with the solid cylinder. Similarly, through the analysis of the control volume before point B, we can conclude that replacing the ideal water cylinder with the solid cylinder we would find a positive time shift $(+\Delta t)$ of the pressure fluctuation at point B.

The conclusion, as we see from fig. 11.4, is that

(i) in the case of the ideal water cylinder the pressure head wave takes the time $t_{B2} - t_{A2}$ to pass from point A to point B;

(ii) in the case of the solid cylinder, the pressure head wave takes the time $(t_{B2} - t_{A2}) + 2\Delta t$ to cover the same distance from point A to point B.

11.2.2 *Is this a good physical interpretation?*

The reasoning we have done for a vertical cylinder is true also for a horizontal cylinder. It suffices to think of the circle of fig. 11.4 as the cross section of a horizontal cylinder. Hence, if the reasoning is correct, we should expect to find the propagation speed of the pressure head waves also dropping at a submerged horizontal cylinder.

In addition, if we extend the analysis done for the two horizontal control volumes A-$\overline{A}$ and B-$\overline{B}$ to the vertical control volumes C-$\overline{C}$ and D-$\overline{D}$ (see fig. 11.5), we arrive at a new interesting prediction on what should occur on a horizontal cylinder. Let us see it here below.

When a wave crest passes over the horizontal cylinder, the local inertia I_z of the control volume above C is negative. If the horizontal water cylinder is replaced with a solid cylinder of the same diameter, the absolute value of I_z decreases, while the pressure at the upper limit $\overline{C}$ of the control volume does not change, provided the height Δz is large enough to make the local perturbation negligible at point $\overline{C}$. Therefore, the pressure at point C, at time t_1 when the wave crest is over this point, must grow if the water cylinder is replaced with the solid cylinder. Then the analysis of the control volume below point D shows that the pressure is reduced at this point of the solid cylinder at time t_1.

Finally, applying the linear momentum equation to the two control volumes at time t_2 of the wave trough, we can conclude that replacement of the ideal water cylinder with the solid cylinder causes pressure decrease at C and pressure increase at D. Hence, from the ideal water cylinder to the solid cylinder, the amplitude of the pressure fluctuation grows at point C of the upper half-cylinder and decreases at point D of the lower half-cylinder.

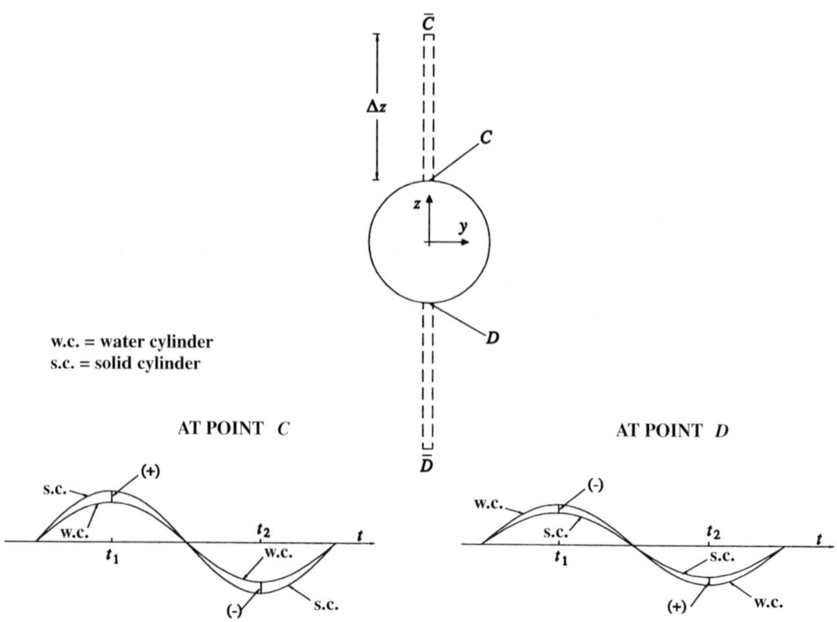

Fig. 11.5 Passing from an ideal water cylinder to a solid cylinder (here we are dealing with a horizontal cylinder), the amplitude of the pressure head waves grows at the upper half-cylinder and decreases at the lower-half. This is proved by applying the linear momentum equation to the small control volumes C-$\overline{C}$ and D-$\overline{D}$. The method of reasoning is the same that we followed to explain the drop in the propagation speed.

These predictions were made after having observed the characteristic drop in the propagation speed at the base and at the columns of the offshore platform. Thus, a new experiment (RC 1993) was designed to verify

(i) whether or not the propagation speed of the pressure head waves also drops at the horizontal cylinder;

(ii) whether or not the amplitude of the pressure fluctuation grows at the upper half-cylinder and decreases at the lower half-cylinder.

11.3 Wave forces on submerged tunnels

11.3.1 *Horizontal forces*

For analysing the wave forces on a horizontal cylinder, the model of a submerged tunnel was assembled. The scale of the model was 1:30 since the wave heights in the natural laboratory of Reggio Calabria are roughly 1:30 of the wave heights of heavy Mediterranean storms. Two models were studied: first, the model of a big tunnel

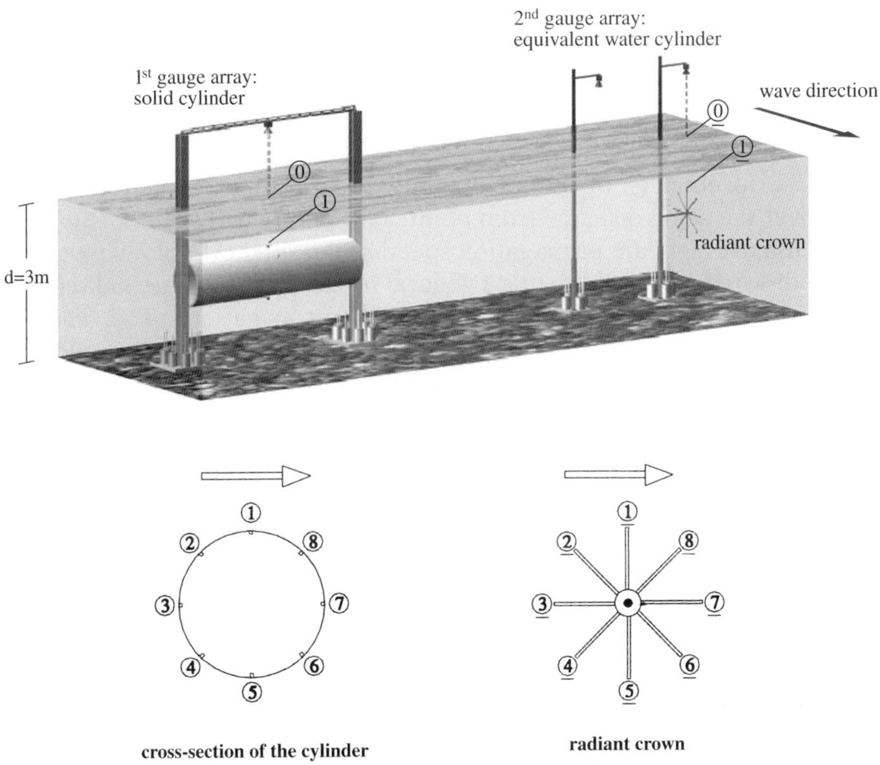

Fig. 11.6 The field laboratory for the experiment RC 1993. (The ultrasonic probe to the left of probe ⓪ was needed only for a careful estimate of the wave direction.)

containing a railway and highway (prototype diameter: 27 m), then the model of a relatively small tunnel containing only the railway (prototype diameter: 13.5 m). The model of the larger tunnel is shown in fig. 11.6.

Once again there were two arrays of gauges with the same configuration. The first one was at the model of the tunnel and the second one at a certain distance from it, where the waves were not disturbed. The aim, of course, was to compare the force on the cylinder with the Froude-Krylov force, that is to say, with the force on an ideal water cylinder equivalent to the solid cylinder (which we shall call simply the *ideal water cylinder*).

Fig. 11.7*a* shows the expected fluctuations of the pressure head on the boundary of the solid cylinder if a wave of given very large height passes at point ⓪ above this solid cylinder. Similarly, fig. 11.7*b* shows the expected fluctuations of the pressure head on the boundary of the ideal cylinder, if a wave of given very large height passes at point ⓪ above this cylinder. The calculations were made by means of (10.9) from the time series data of the surface displacement and the fluctuating pressure head of one of the records.

We can see the full confirmation of the drop in the propagation speed of the pressure head waves. Specifically, we see that the pressure head wave takes 0.43 s to pass from point ③ to point ⑦ of the solid cylinder; while it takes 0.22 s (half the time!) to cover the same distance from point ③ to point ⑦ in the undisturbed wave field. Then we see that the time taken to pass from point ② to point ⑧ of the solid cylinder is 1.7 times greater than the time taken to cover the same distance from point ② to point ⑧ of the undisturbed wave field. Finally, the time taken to pass from point ④ to point ⑥ of the solid cylinder is 2.3 times greater than the time taken to cover the same distance from point ④ to point ⑥ of the undisturbed wave field. Thus, on average, the propagation speed at the solid cylinder is twice as small as in the undisturbed wave field. It is a general property: 380 out of 380 records with the big tunnel showed a drop of about 50% in the propagation speed. This, despite the large variety of wave conditions, which, at the prototype level, ranged from sea states with an H_s of 6 m to sea states with an H_s of 14 m.

There was a similar response in the 200 records made with the small tunnel (the one for the railway only), the only difference being that the reduction factor of the propagation speed was somewhat smaller: 1.75 against 2.

The fact that the propagation speed at the big tunnel was twice as small as in the undisturbed wave field implied that the horizontal force on this tunnel was nearly twice the corresponding Froude-Krylov force (C_{do} between 1.8 and 2.0). The C_{do} of the small tunnel was nearly a 10% smaller because of the smaller drop in the propagation speed.

11.3.2 *Vertical forces*

The vectors in fig. 11.8 show the amplitudes of the pressure fluctuations at the gauge locations of the solid cylinder and of the ideal water cylinder, with the following conventions:

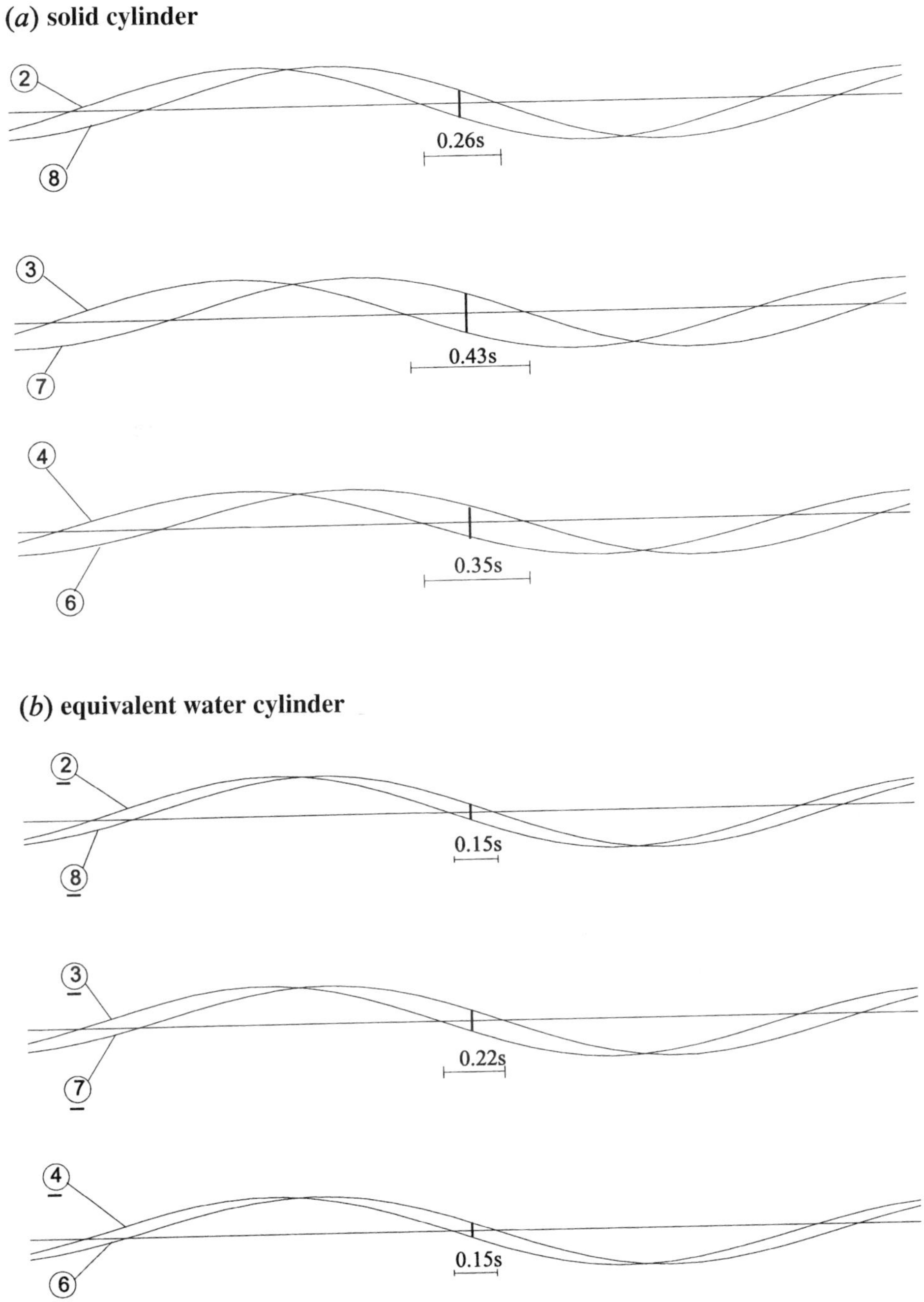

Fig. 11.7 (*a*) Pressure head waves at points of the horizontal cylinder, if a wave of given very large height passes at point ⓪. (*b*) Pressure head waves at points of the equivalent water cylinder, if a wave of given very large height passes at point ⓪.

(i) the amplitude of the pressure fluctuation at point ⑤ (the lowest point of the ideal cylinder) is taken as the reference amplitude (amplitude $a_{\underline{5}}$);

(ii) the norm of the vector represents the absolute value of the difference

$$\Delta_i \equiv \frac{a_i}{a_{\underline{5}}} - 1,$$

where a_i is the amplitude of the pressure fluctuation at the ith point;

(iii) if the amplitude of the pressure fluctuation at a point is greater than the reference amplitude (that is if $\Delta_i > 0$), the vector at this point is inward normal to the cylinder.

Clearly, by *amplitude of the pressure fluctuation* we mean the root mean square pressure fluctuation. The term *amplitude* is simply used because it seems more effective.

The graph of fig. 11.8 should communicate at once the outcome of the experiment. Indeed we see that, from the ideal cylinder to the solid cylinder, the amplitude of the pressure fluctuations grows at the upper half-cylinder, and on the contrary decreases at the lower half-cylinder, what confirms the prediction of sect. 11.2.2.

The graph of fig. 11.8 can be thought of as the pressure distribution under a wave crest (naturally, we mean the wave pressure). The ratio between the pressure force on the solid cylinder and the pressure force on the ideal cylinder is equal to 1.75. This ratio proves to be nearly coincident with the vertical diffraction coefficient

$$C_{dv} \equiv \frac{\sqrt{< F_v^2(t) >}}{\sqrt{< F_{\underline{v}}^2(t) >}}, \tag{11.2}$$

where $F_v(t)$ is the vertical wave force per unit length on the solid cylinder, and $F_{\underline{v}}(t)$ is the vertical wave force per unit length on the ideal cylinder.

The graph of fig. 11.8 is the average of 380 graphs like this, one for each record of the experiment with the big tunnel. The same diagram with only a slightly smaller quotient between the two pressure forces (1.65 against 1.75) emerged also from the 200 records with the small tunnel.

In conclusion, the vertical wave force on the solid cylinder is greater than the vertical wave force on the ideal water cylinder because the amplitude of the pressure fluctuations grows at the upper half-cylinder and decreases at the lower half-cylinder. Thus, the cause of the increase of the force F_v from the ideal water cylinder to the solid cylinder is completely different from the cause of the increase of the force F_o: the increase of the vertical force depends on a variation of the *amplitude* of the pressure fluctuations, whereas the increase of the horizontal force depends on a variation of the *phase* of these fluctuations. However, the increase of the vertical force shows a high correlation with the increase of the horizontal force: C_{dv} proves to be systematically nearly a 10% smaller than C_{do}.

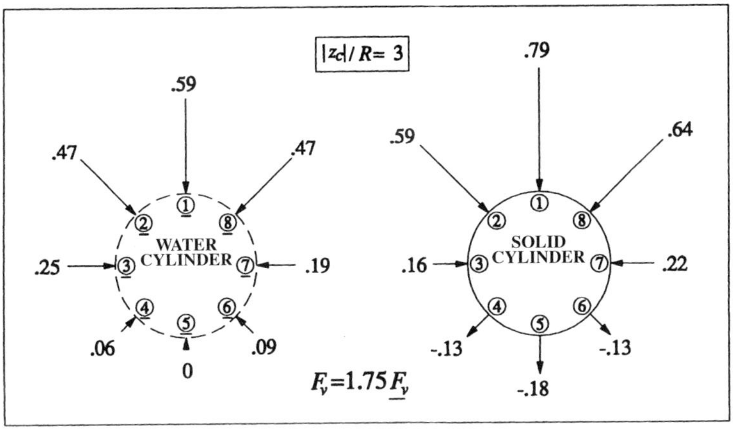

Fig. 11.8 Amplitude of the pressure head waves on the solid cylinder and on the equivalent water cylinder. The amplitude of the pressure head waves at point ⑤ has been taken as the reference amplitude. Reading example: .79 at ① means that the amplitude of the pressure head waves at ① is a 79% greater than at ⑤; −.18 at ⑤ means that the amplitude of the pressure head waves at ⑤ is a 18% smaller than at ⑤.

11.3.3 *A characteristic asymmetry of the fluctuations of the vertical force*

Using the theory of the sea states, one can verify that F_o and F_v, as well as $\underline{F}_o$ and $\underline{F}_v$, are stationary Gaussian processes. Now, the processes of the horizontal forces F_o and $\underline{F}_o$ are really very similar to the Gaussian processes, and in particular they are statistically symmetrical with respect to the mean value. On the other hand the processes F_v and $\underline{F}_v$ exhibit a characteristic non-linearity effect. Fig. 11.9 should be effective to illustrate these ideas.

Panel (a) of fig. 11.9 gives the probability that a peak of the process $\underline{F}_o(t)$ exceeds a threshold which is β times the standard deviation of this process, i.e. the probability that the peak exceeds the threshold

$$\beta\sqrt{< \underline{F}_o^2(t) >}.$$

One of the two sets of data points is relevant to the positive peaks, the other is relevant to the negative peaks, and the continuous line represents the function (5.30). Similarly, the panels (b), (c), and (d) of the figure show the probability of exceedance of the positive peaks and of the negative peaks, respectively, of $F_o(t)$, of $\underline{F}_v(t)$ and of $F_v(t)$.

We see that in the case of the horizontal force, the distribution of the positive peaks and the distribution of the negative peaks are nearly coincident with one another. While, in the case of the vertical force, there is a large difference: the positive peaks are higher than the negative peaks. This difference is a non-linearity effect that can be foreseen if one takes into account the second order terms of the pressure fluctuation. We shall see how in sect. 12.3.4.

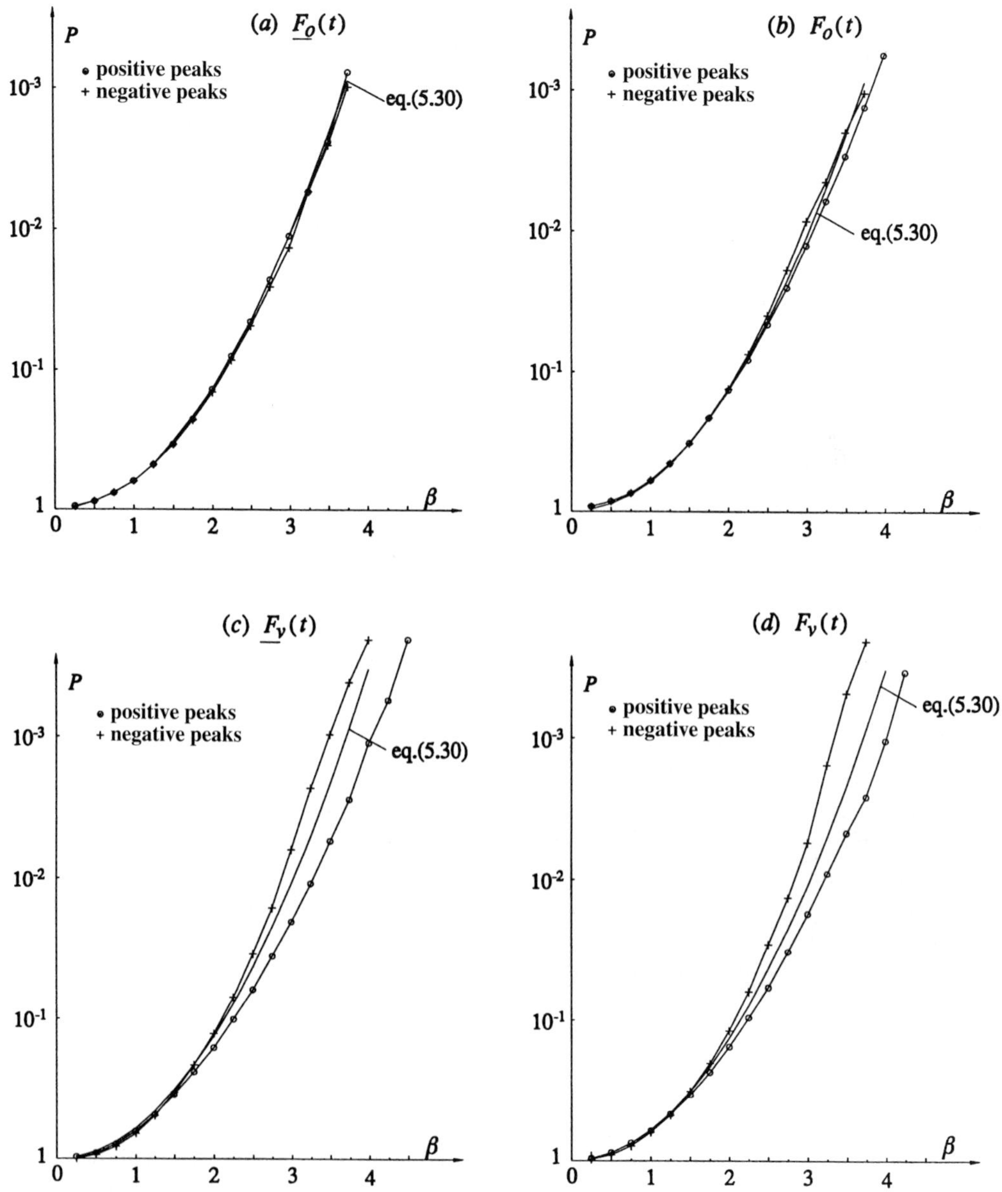

Fig. 11.9 Probability of exceedance of the positive peaks (crests) and of the negative peaks (troughs) of: (*a*) the horizontal force on the equivalent water cylinder; (*b*) the horizontal force on the solid cylinder; (*c*) the vertical wave force on the equivalent water cylinder; (*d*) the vertical wave force on the solid cylinder.

11.4 The diffraction coefficients of the forces

11.4.1 *The basic property*

How does the force on the horizontal cylinder fluctuate, if a wave of given very large height H passes over the cylinder? The answer is given by fig. 11.10 both for the solid cylinder and for the equivalent water cylinder. The force on the equivalent water cylinder has been multiplied by the diffraction coefficient (C_{do} or C_{dv} according to whether the force is horizontal or vertical).

And again how does the horizontal force on the offshore platform fluctuate if a wave of given very large height H passes through the platform? The answer is given by fig. 11.11 both for the solid platform and for the equivalent water mass. Here too the force on the equivalent water mass has been multiplied by the diffraction coefficient C_{do}.

The two figures show what we have called the genetic code; specifically, it is the genetic code of the forces, which was obtained with the quasi-determinism theory and the time series data of the experiments RC 1992 and RC 1993.

The figures show that $F_o(t)$ is very close to the product $C_{do}\underline{F_o}(t)$, and similarly

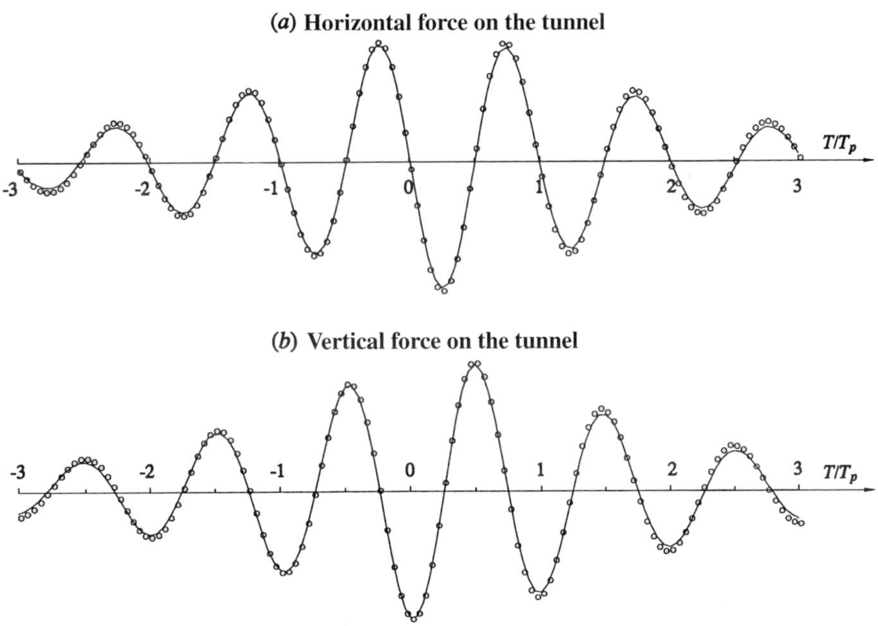

Fig. 11.10 Time history of the force if a wave of given very large height passes over the horizontal cylinder: continuous line = wave force on the solid cylinder; points = diffraction coefficient (which is a constant) × wave force on the equivalent water cylinder. [Obtained by averaging the "genetic codes" of the records of experiment RC 1993.]

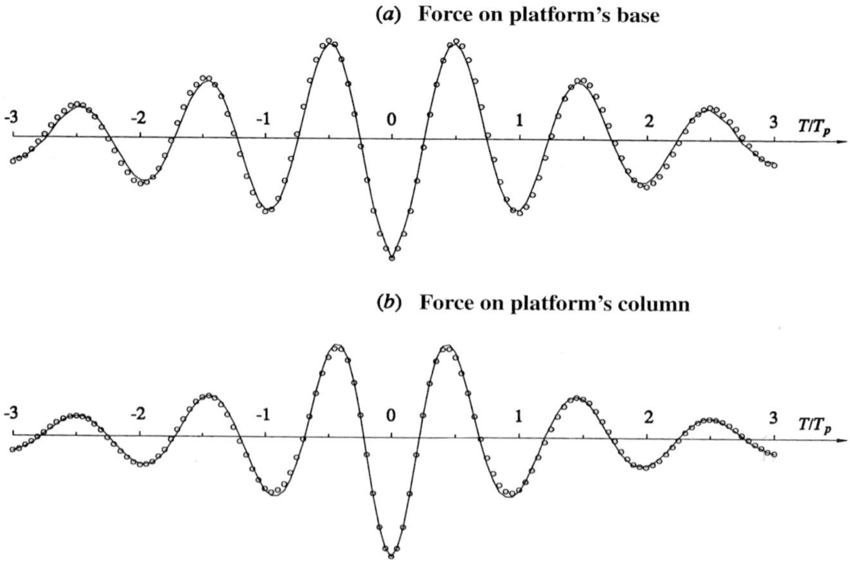

Fig. 11.11 Time history of the force if a wave of given very large height passes through the offshore gravity platform: continuous line = force on the solid body; points = diffraction coefficient (which is a constant) × force on the equivalent water volume. [Obtained by averaging the "genetic codes" of the records of experiment RC 1992.]

$F_v(t)$ is very close to the product $C_{dv}\,\underline{F_v}(t)$. It is a crucial property that we express so:

$$\text{maximum expected wave:}\quad \begin{cases} F_o(t) = C_{do}\,\underline{F_o}(t), \\ F_v(t) = C_{dv}\,\underline{F_v}(t). \end{cases}$$

In words: we can obtain the time dependent force exerted by the group of the maximum expected wave by multiplying the force on the equivalent water mass by a constant factor (C_{do} for the horizontal force and C_{dv} for the vertical force).

The genetic code of figs. 11.10-11 does not include the non-linearity effects. Now, we have seen in sect. 11.3.3 these effects to be negligible for the horizontal force-processes $F_o(t)$ and $\underline{F_o}(t)$ and to be not negligible for the vertical force-processes $F_v(t)$ and $\underline{F_v}(t)$. However, the incidence of the non-linearity effects (i.e. the percentage enhancement of the positive peaks and the compensating lowering of the negative peaks) is nearly equivalent for the process $\underline{F_v}(t)$ and for the process $F_v(t)$, as we can realize on comparing fig. 11.9*c* with fig. 11.9*d*. Thus the computation of $F_v(t)$ as product of $\underline{F_v}(t)$ by the constant C_{dv} proves to be effective even if we take into account the non-linearity effects.

Conclusion: the evaluation of the forces on offshore structures (we mean the class of structures hitherto considered) gets much simpler, and the diffraction coefficients C_{do} and C_{dv} take on an importance much greater than one may think from their definitions (11.1) and (11.2).

11.4.2 *Analytical and numerical solutions*

The analytical solution to the linear wave diffraction problem for a vertical circular cylinder extending from the seabed and piercing the free surface was obtained by Mac Camy and Fuchs (1954). The analytical solution for the interaction between periodic waves and a horizontal submerged cylinder (being understood it is a circular cylinder) was obtained by Ursell (1950) and completed by Ogilvie (1963). The analytical solution for the interaction between periodic waves and a group of truncated circular cylinders was developed by Linton and Evans (1990) and Kim (1992) (case of bottom mounted cylinders), and by Yilmaz (1998) (case of cylinders piercing the free surface).

A numerical method for the interaction between periodic waves and solid objects of arbitrary shape is well established and is due to the work of Garrison and Rao (1971) and Garrison and Chow (1972). In this numerical method the velocity potential of the scattered waves is given as the result of a distribution of point wave sources over the surface of the solid object. The sum of this velocity potential and the velocity potential of the incident waves must satisfy the boundary condition at a number of points of the solid surface.

The aforesaid analytical solutions, and especially the one for the horizontal cylinder, are given by rather complex formulae. The numerical solutions in their turn call for much work on computer.

For an illustration of these solutions, fig. 11.12 shows the diffraction coefficient for the circular cylinder extending from the seabed and piercing the free surface. Line ① is the classic solution of Mc Camy and Fuchs (1954) for the periodic waves, and line ② is the solution based on the theory of the sea states with the usual characteristic spectrum of the wind waves. [According to the theory of the sea states

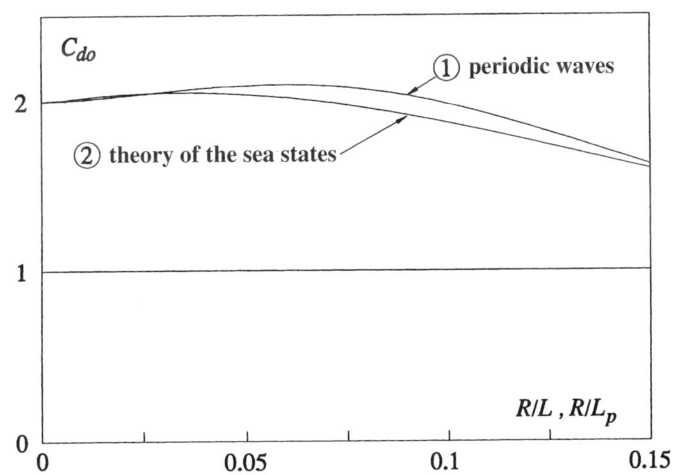

Fig. 11.12 The diffraction coefficient of the vertical circular cylinder extending from the seabed and piercing the free surface. Line ② was contributed by F. Arena.

the horizontal force represents a stationary Gaussian process whose standard deviation can be obtained from the directional spectrum of the incident waves. The way of reasoning is the one followed in sect. 8.6 to get the diffraction coefficients of the wind waves behind a vertical breakwater.]

The term *diffraction coefficient* refers to waves and wave forces. The first meaning is: quotient between the standard deviation of the surface displacement of waves interacting with some solid body and the standard deviation of the incident waves. The second meaning is: quotient between the standard deviation of the force on a solid body and the standard deviation of the Froude-Krylov force.

11.4.3 *A formula of general validity*

In the foregoing sections we have suggested an approach different from the classic one. The classic approach consists in obtaining the solution, preferably the analytical solution, for the diffraction coefficient of bodies with more and more complex forms. Here, we have tried to understand what is the cause of the diffraction coefficient, looking at the results of two field experiments in the light of the quasi-determinism theory. Thus we have realized something valid regardless of the body shape, whether it is the column or the big base of the gravity platform, or the submerged tunnel. First, we have understood that C_{do} is essentially due to a drop in the propagation speed of the pressure head wave at the isolated offshore body. Then we have understood the reason for this drop, which in its turn has enabled us to understand that C_{dv} is the effect of a variation in amplitude of the pressure fluctuations. Here below, basing ourselves on these ideas, we suggest a general criterion for estimating the diffraction coefficient.

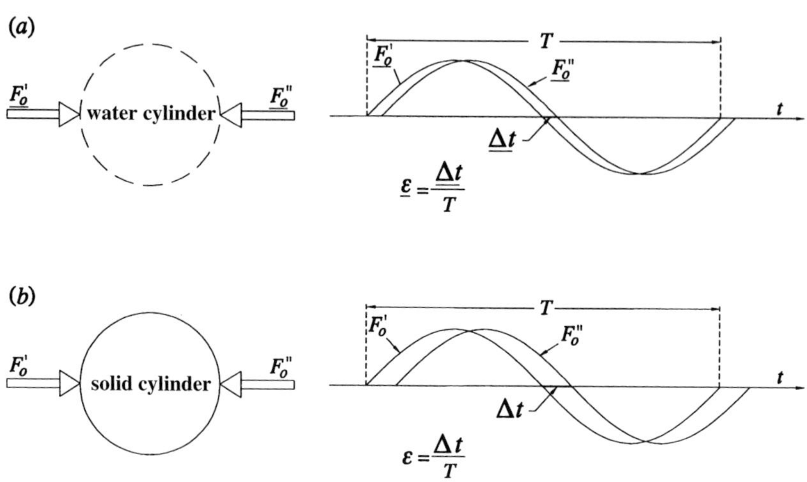

Fig. 11.13 Wave forces on the left (wave beaten) half-cylinder, and on the right (sheltered) half-cylinder. (*a*) Ideal water cylinder. (*b*) Solid cylinder.

We refer to the sketch of fig. 11.13*a* showing the cross-section of a vertical water cylinder. Let us think of the cylinder as being subjected to two horizontal wave forces $\underline{F'_o}(t)$ and $\underline{F''_o}(t)$, the first one acting on the left half-cylinder (the wave beaten half), and the second one acting on the right half-cylinder (the sheltered half). These forces have the same amplitude and frequency, and a phase angle $\underline{\varepsilon}$:

$$\underline{F'_o}(t) = F\sin(\omega t + \underline{\varepsilon}/2),$$

$$\underline{F''_o}(t) = F\sin(\omega t - \underline{\varepsilon}/2).$$

Here $\underline{\varepsilon}$ is the average phase angle between opposite points like P' and P'' [see fig. 11.14]. Naturally, we mean the weighted average with respect to $\cos\beta$, that is

$$\underline{\varepsilon} = \frac{\displaystyle\int_0^{\pi/2} (2R\cos\beta\, 2\pi/L)\cos\beta\, \mathrm{d}\beta}{\displaystyle\int_0^{\pi/2} \cos\beta\, \mathrm{d}\beta},$$

where the term within the parenthesis on the r.h.s. is the phase angle between point P' and point P''. The result is

$$\underline{\varepsilon} = \frac{\pi}{2} kR. \tag{11.3}$$

The solid cylinder [see fig. 11.13*b*] is subjected to two wave forces having the same amplitude F and frequency ω and a phase angle ε greater than $\underline{\varepsilon}$:

$$F'_o(t) = F\sin(\omega t + \varepsilon/2),$$

$$F''_o(t) = F\sin(\omega t - \varepsilon/2).$$

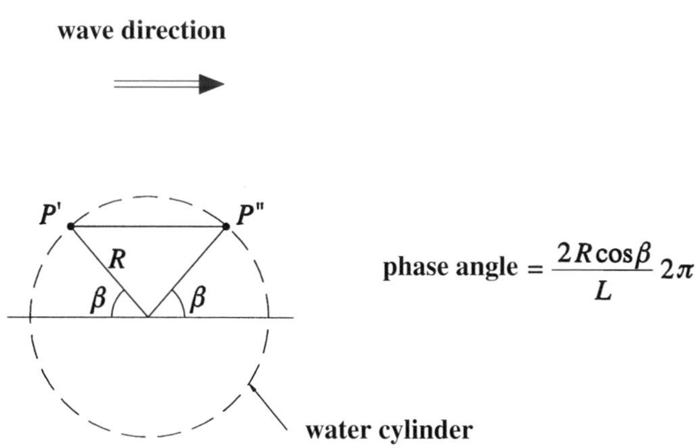

Fig. 11.14 Graphic aid for deduction of (11.3).

We have

$$\varepsilon = \mathscr{F}_{\mathscr{R}}\, \underline{\varepsilon}, \tag{11.4}$$

where $\mathscr{F}_{\mathscr{R}}$ is the reduction factor (greater than 1) of the propagation speed of the pressure head waves at the solid cylinder, which we shall call *speed drop factor.*

Thus, the pressure forces on the water cylinder and on the solid cylinder are respectively

$$\underline{F_o}(t) = \underline{F_o'}(t) - \underline{F_o''}(t) = F\,[\sin(\omega t + \underline{\varepsilon}/2) - \sin(\omega t - \underline{\varepsilon}/2)],$$

$$F_o(t) = F_o'(t) - F_o''(t) = F\,[\sin(\omega t + \varepsilon/2) - \sin(\omega t - \varepsilon/2)],$$

from which it follows that the diffraction coefficient is

$$C_{do} = \sqrt{\frac{< [\sin(\omega t + \varepsilon/2) - \sin(\omega t - \varepsilon/2)]^2 >}{< [\sin(\omega t + \underline{\varepsilon}/2) - \sin(\omega t - \underline{\varepsilon}/2)]^2 >}} = \frac{|\sin(\varepsilon/2)|}{|\sin(\underline{\varepsilon}/2)|}.$$

Finally, using the relations (11.3) and (11.4), we obtain

$$C_{do} = \frac{\left|\sin\left(\mathscr{F}_{\mathscr{R}}\dfrac{\pi}{4}kR\right)\right|}{\left|\sin\left(\dfrac{\pi}{4}kR\right)\right|} \quad \text{for the vertical cylinder.} \tag{11.5}$$

The same reasoning is true also for a horizontal cylinder, with only one difference. In the case of the vertical cylinder, F_o is parallel to the propagation axis. In the case of a horizontal cylinder, F_o is orthogonal to the cylinder axis, and hence is affected by the wave direction. The larger the angle θ that the wave direction makes with the orthogonal to the cylinder axis, the smaller is the relative phase $\underline{\varepsilon}$. Specifically we have

$$\underline{\varepsilon} = \frac{\pi}{2}kR\cos\theta,$$

and accordingly

$$C_{do} = \frac{\left|\sin\left(\mathscr{F}_{\mathscr{R}}\dfrac{\pi}{4}kR\cos\theta\right)\right|}{\left|\sin\left(\dfrac{\pi}{4}kR\cos\theta\right)\right|} \quad \text{for the horizontal cylinder.} \tag{11.6}$$

11.4.4 *Hints for the use of the general formula*

The formula of the foregoing section can be applied to submerged bodies like the platform's base or the tunnel, taking T_p as the wave period. As to the speed drop factor $\mathcal{F}_{\mathcal{R}}$, basing ourselves on the results of the experiments RC 1992 and RC 1993, we suggest

$$\text{for the platform's base: } \mathcal{F}_{\mathcal{R}} = 1.75,$$

$$\text{for the small tunnel } (|z_C|/R = 5): \ \mathcal{F}_{\mathcal{R}} = 1.75,$$

$$\text{for the big tunnel } (|z_C|/R = 3): \ \mathcal{F}_{\mathcal{R}} = 2.0.$$

As to the platform's columns the matter is somewhat more complex. Probably owing to the interaction between columns and surface waves, C_{do} is not only due to the drop in the propagation speed, but also to a small variation of the amplitude of the pressure fluctuation. The $\mathcal{F}_{\mathcal{R}}$ proves to be nearly 1.75 so that according to (11.5) also C_{do} should be nearly 1.75 (for a small $\underline{\varepsilon}$, like in the case of the column, C_{do} tends to coincide with $\mathcal{F}_{\mathcal{R}}$). The increase in C_{do} due to the variation in amplitude of the pressure fluctuation proved to be of about $10 \div 15\%$ in the experiment RC 1992. Thus we suggest to take

$$\text{columns:} \quad C_{do} = 2.$$

Finally, C_{dv} is strictly correlated to C_{do}, and we can take

$$C_{dv} = 0.9\, C_{do}$$

for both, the small tunnel and the big tunnel.

Of course, we can achieve a precise estimate of the diffraction coefficient of any given structure if, like in the experiments RC 1992 and RC 1993, we work with the two gauge arrays. We advise doing this work for the design of a structure, since it enables a person to get a very clear idea on C_{do} and C_{dv}. Furthermore, for this special aim the technique of the small scale field experiment is easier than one may think. This, thanks to an interesting property of the spectra of the force-processes, we are going to illustrate.

11.4.5 *Why the diffraction coefficients can be estimated rather easily through the small scale field experiments*

The reader will remember that a big problem for the small scale field experiments is due to the swells. The case of fig. 4.12 was clear enough: wind waves with an excellent size for small scale models; a trifling presence of swells on the water surface; how this trifling presence was able to upset the spectra of the pressure head waves beneath the water surface. The conclusion was that this sea state could not be taken as a small scale model of a severe storm.

The sea state of fig. 4.12 was recorded just during the experiment RC 1993. The spectra shown by the figure are the ones of the surface waves above the tunnel, and of the pressure head waves at points ①, ③ and ⑤ [see fig. 11.6]. At the smaller depth (point ①), the swell is already of the same order of magnitude as the wind wave; at point ③ the swell overtakes the wind wave; at the larger depth (point ⑤) the wind wave is very small with respect to the swell.

Look now at fig. 11.15. This gives the spectra of the horizontal force $F_o(t)$ and of the vertical force $F_v(t)$. We should like to say: magic! All traces of swells have disappeared! Was it not for the small bump in the spectrum of $F_o(t)$, one would think that there were no swells. Here below we explain this phenomenon.

The sketch of fig. 11.16 shows the pressure fluctuation at two points of the y-axis (parallel to the wave direction) at the depth of the cylinder centre. Two cases are performed: (a) a wind wave; (b) a swell with a period three times larger than the

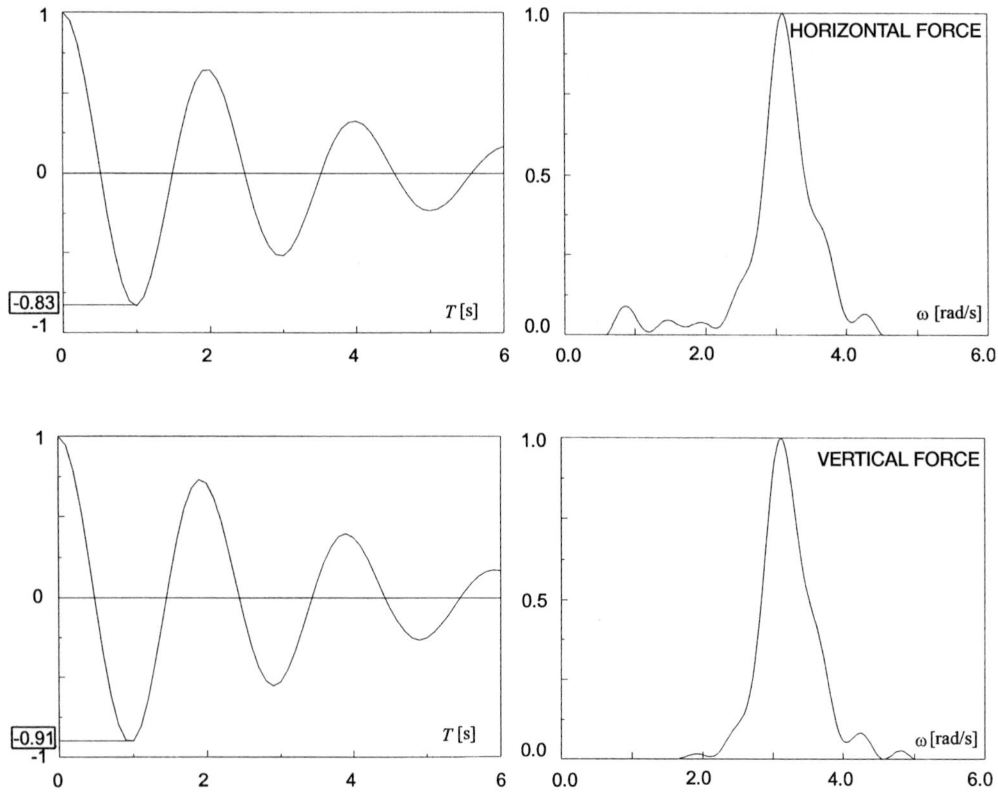

Fig. 11.15 Spectrum and autocovariance of the force on the horizontal cylinder. The spectra of the pressure head waves on the cylinder were given in fig. 4.12. The swell peak, which was the highest in the spectrum of the pressure head waves, here has practically disappeared! Why? The answer, in the two following figures.

wind wave's. We assume the amplitude of the pressure fluctuation to be the same for the wind wave and for the swell. However, we see the largest pressure difference (Δp_{max}) between the wave beaten half-cylinder and the sheltered half-cylinder proves to be markedly larger in the wind wave. There are two reasons for this. First, under the same Δt, a three times smaller wave period implies a three times larger Δp_{max}; then there is the fact that Δt of the wind wave is three times larger than Δt of the swell (since the swell is three times faster than the wind wave), and hence we get a Δp_{max} of the wind wave which is nine times larger than the Δp_{max} of the swell. (This reasoning is somewhat simplified, it becomes exact if R/L is small and the water is deep.)

Conclusion: the force F_o is

$$F_o(t) = F\cos(\omega t) + \frac{1}{9}F\cos\left(\frac{1}{3}\omega t\right),$$

where the first term on the r.h.s. is due to the wind wave and the second term is due to the swell. This force has a spectrum with a peak of $\frac{1}{2}F^2$ on the frequency ω of

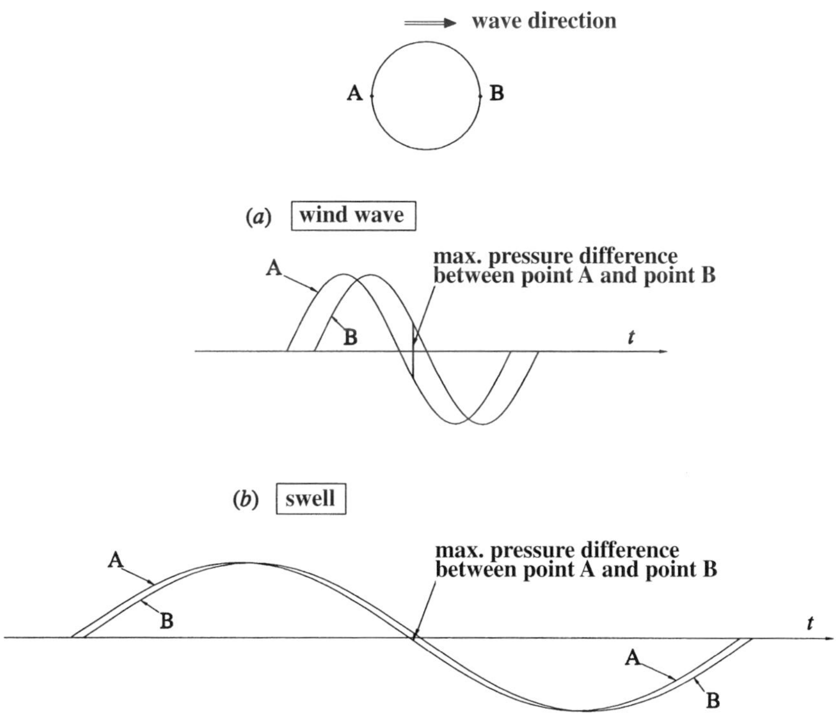

Fig. 11.16 Even if the pressure fluctuations of the wind wave and of the swell have the same amplitude at the depth of the cylinder's centre, the difference [pressure on the wave beaten half-cylinder] - [pressure on the sheltered half-cylinder] is the greatest by far in the wind wave.

the wind wave and a second peak of $\frac{1}{81}\,\frac{1}{2}\,F^2$ on the frequency $\omega/3$ of the swell. Hence the swell's share of the standard deviation of the force-process is only

$$\left(\sqrt{1+\frac{1}{81}}-1\right)\Big/\sqrt{1+\frac{1}{81}}$$

which makes 0.6%, and consequently the alteration of C_{do} due to the presence of the swell is really very small.

Let us pass to the vertical force. Here, the incidence of the swell is even smaller. Indeed we see in fig. 11.15 the spectrum of $F_v(t)$ to exhibit no trace of swells.

The sketch of fig. 11.17 helps to realize why, under the same amplitude of the pressure fluctuation at the depth of the cylinder centre, the wind wave exerts a vertical force much greater than the swell. Indeed the figure shows that the pressure difference between the upper half-cylinder and the lower half-cylinder is markedly larger in the wind wave. This is a consequence of the fact that the attenuation of Δp is the larger, the smaller the wave period is.

Conclusion: the forces are, in practice, affected only by the wind wave. Hence, it is not true that the sea state of fig. 4.12 with wind waves of 2 s superimposed on swells of 6 s is not at all suitable as a small scale model of a severe storm. On the contrary, it can be an excellent model, provided the target is the pressure forces rather than the pressures pure and simple. In practice this sea state can be used for getting graphs like the ones of figs. 11.10-11 giving the wave force vs time, or for

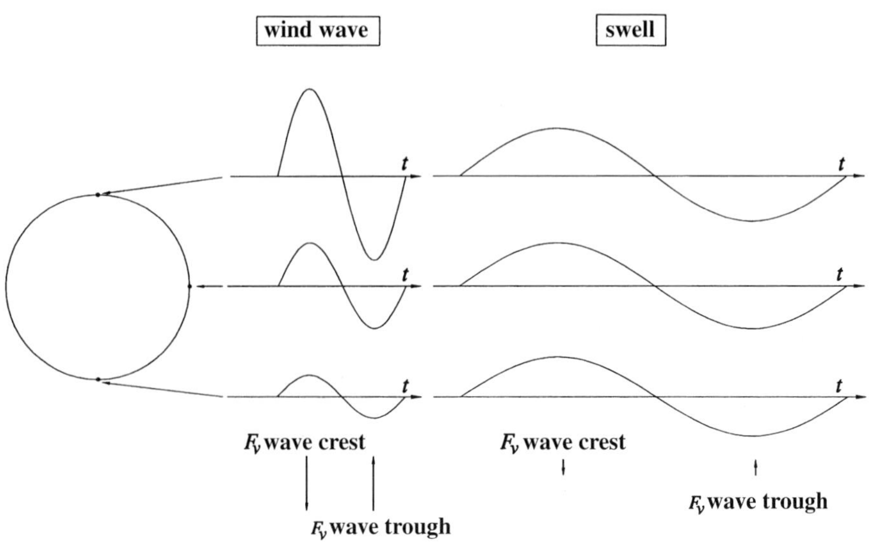

Fig. 11.17 Even if the pressure fluctuations of the wind wave and of the swell have the same amplitude at the depth of the cylinder's centre, the difference [pressure on the upper half-cylinder] - [pressure on the lower half-cylinder] is the greatest by far in the wind wave.

evaluating the diffraction coefficient. Thus the small scale field experiments gain a new point: finding pure wind waves of small size is not always necessary; for a few aims like the evaluation of C_{do} and C_{dv}, even some "dirty" wind waves, like the ones of fig. 4.12, are suitable.

11.5 Wave forces on space frame structures

11.5.1 *The number K_E (Keulegan-Carpenter)*

So far we have dealt with some large structures: the columns of the gravity offshore platform, the submerged tunnel and, especially, the big base of the gravity platform. The ideal flow pattern works well for these structures as we have seen in the foregoing sections. Not so, for the space frame structures whose members have some smaller diameters.

Usually, the ideal flow pattern is applied for $K_E < 2 \div 3$, K_E being the Keulegan-Carpenter number:

$$K_E \equiv \frac{v_{\max} T}{D},$$

where $v_{\max}$ is the maximum particle velocity at the depth of the cylinder centre.

K_E can be conveniently expressed also in terms of H_{ph} (the height of the pressure head wave at the depth of the cylinder centre) provided this depth is always beneath the free surface. Indeed with the linear theory we have

$$v_{\max} = g\frac{H}{2}\omega^{-1}k\frac{\cosh[k(d+z)]}{\cosh(kd)}, \quad H_{ph} = H\frac{\cosh[k(d+z)]}{\cosh(kd)},$$

yielding

$$K_E = \frac{H_{ph}}{D}\,\frac{\pi}{\tanh(kd)}. \tag{11.7}$$

This is an useful formula for evaluating K_E when we work in the field. Indeed, a simple pressure transducer suffices to measure the height H_{ph}.

11.5.2 *Experimental evidence for the validity of the ideal flow pattern*

How do we decide whether or not the flow can be regarded as ideal? Certainly, the flow at the equivalent water volume can be regarded as ideal. Thus, we should investigate whether or not the forces on the solid body are of the same kind as the forces on the equivalent water mass.

When a wave crest passes at the centre of a *water cylinder*, whether it is horizontal or vertical, the horizontal force is zero, while the vertical force has a minimum (here we are referring to a horizontal cylinder). When a wave trough passes at the cylinder centre, the horizontal force is again zero, while the vertical force has a maximum. Therefore a simple criterion for realizing whether the forces on the *solid cylinder* are like the forces on the equivalent water cylinder is to check the phase angles between waves and horizontal forces and waves and vertical forces. Let us see an example.

Fig. 11.18 shows the record of the individual wave with the largest K_E ($K_E \cong 2.5$) taken during the experiment RC 1993 (the surface wave was recorded above the cylinder axis). We see that under the highest wave crest, the horizontal force is zero while the vertical force has a local minimum. This suggests that there is no apparent alteration due to non-ideal flow near the solid body.

A good way to further examine the nature of the force on the solid body is to look at the distribution of the crest-to-trough heights of the force process. Fig. 11.19 shows the probabilities of exceedance of the crest-to-trough heights of the processes $F_o(t)$ and $\underline{F_o}(t)$. (The figure was obtained from the whole set of data of the experiment RC 1993.) The nearly perfect agreement between these two probabilities is a further proof that the wave force on the solid cylinder is of the same kind as the wave force on the equivalent water cylinder, so that the flow at the solid cylinder can be regarded as ideal.

11.5.3 *Morison's equation*

With the real flow pattern, the sectional wave force (in-line force) on a rigid cylindrical member is calculated by means of Morison's equation (Morison et al., 1950) which can be written in the form (Borgman, 1958):

$$\mathbf{f}(t) = C_{in}\mathbf{f}_{in}(t) + C_{dg}\rho R v_{\text{sect}}\mathbf{v}_{\text{sect}}, \tag{11.8}$$

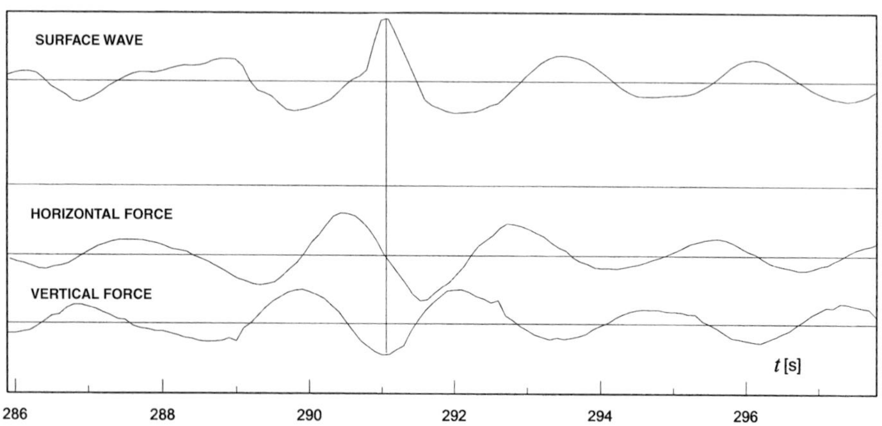

Fig. 11.18 Horizontal cylinder of the experiment RC 1993: first inspection of the flow characteristics for the wave with the largest K_E ($K_E \cong 2.5$). The full correspondence of wave crest - zero of the horizontal force - minimum of the vertical force reveals the typical features of the ideal flow. Note that the greatest sea state K_E (cf. Burrows et al., 1997) of the experiment was of about 0.8.

where

$$\mathbf{f}_{in}(t) = \rho\pi R^2 \mathbf{a}_{\text{sect}}, \tag{11.9}$$

and where $\mathbf{v}_{\text{sect}}$, $\mathbf{a}_{\text{sect}}$ are respectively the velocity vector and the acceleration vector normal to the pipe. We allow the cylindrical member to be oriented at random because the frame structural elements also have different orientations, as we shall see in the worked example of sect. 12.2.

Morison's equation is thus the sum of two terms, the first one representing the inertia force, i.e. the force acting on the equivalent water mass multiplied by the factor C_{in} (inertia coefficient) that is generally greater than 1. The second term in Morison's equation represents the drag force that is the same kind of force exerted by a steady current.

Besides the in-line force there is a lift (transverse) force which is related to vortex shedding. For a vertical cylinder, the lift force is irregular even with the periodic waves, the quotient fundamental lift frequency/wave frequency generally

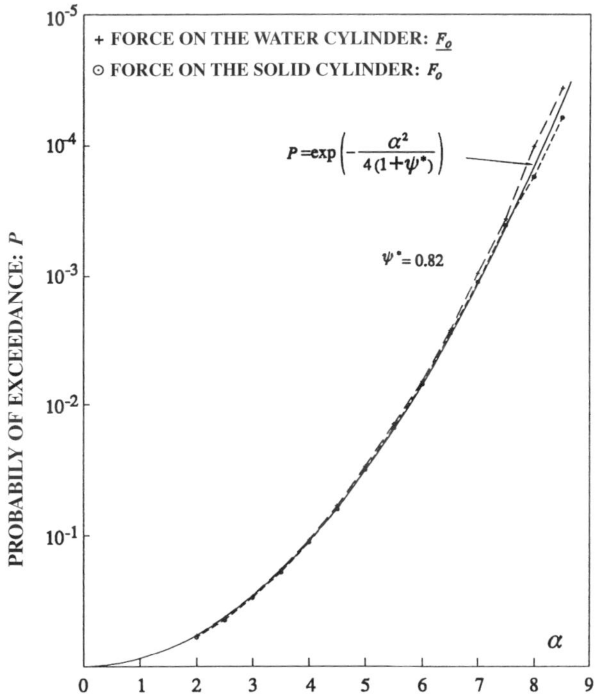

Fig. 11.19 Horizontal cylinder of the experiment RC 1993: a new test on the flow characteristics. The probability of exceedance of the crest-to-trough heights of the force process is practically the same for the ideal water cylinder and for the solid cylinder. It is a further evidence that the flow in that experiment could be regarded as ideal.

grows with K_E (Williamson, 1985 and Justesen, 1989), and the amplitude is usually expressed in the form

$$f_{lift\,\max} = C_{lift}\rho R v_{\max}^2,$$

where C_{lift} is the lift coefficient. For a horizontal cylinder the lift force proves to be substantially in antiphase with the inertia force (cf. Chaplin and Subbiah, 1997).

The form (11.9) of $\mathbf{f}_{in}(t)$ is right in the assumption that R/L is small so that the integral of the acceleration over the cylinder section is nearly equal to the product of the area by the acceleration at the section's centre. This assumption is generally acceptable, given that R/L is rather small on the range of validity of the real flow pattern. Let us evaluate an upper bound of R/L on this range.

From (11.7) we know that

$$\frac{R}{L} = \frac{\pi}{2}\frac{H_{ph}}{L}\frac{1}{K_E}\frac{1}{\tanh(kd)}, \tag{11.10}$$

and, accordingly, the extreme value of R/L for a given K_E occurs when H_{ph}/L takes on its maximum. Since the height H_{ph} of the pressure head wave at any depth z is smaller than the wave height H, from (2.52) it follows

$$\frac{H_{ph}}{L} < 0.14\tanh(kd)$$

that, taken together with (11.10), yields

$$\frac{R}{L} < \frac{\pi}{2}\frac{0.14}{K_E}.$$

11.5.4 *Force coefficients*

C_{in} and C_{dg} depend on K_E and on the Reynolds number

$$R_E \equiv v_{\max} D/\nu,$$

and this dependence is illustrated by the well-known graphs of Sarpkaya and Isaacson (1981). (Usually, $v_{sect,\max}$ is used for K_E and R_E. However, for the principal members, $v_{sect,\max}$ is typically equal to, or close to, $v_{\max}$.)

C_{in} increases with the increasing of R_E in a different manner according to K_E, reaches a maximum, and then converges on an asymptotic value. C_{dg} decreases with the increasing of R_E in a different manner according to K_E, reaches a minimum, and then tends to an asymptotic value. Following Sarpkaya and Isaacson, these asymptotic values can be assumed as 1.85 and 0.62, respectively, for C_{in} and C_{dg}.

Looking at the above mentioned graphs which give the coefficients versus R_E for several values of K_E, one can see that, for $R_E > 10^4\,K_E$, both C_{in} and C_{dg} are rather close to their asymptote. Hence, for a first approximation, we shall assume

$$C_{in} = 1.85, \quad C_{dg} = 0.62 \quad \text{for} \quad R_E/K_E = D^2/\nu T > 10^4. \tag{11.11}$$

These values of C_{in} and C_{dg} are valid for smooth pipes. For roughened pipes (with rigid and/or soft excrescences), the asymptotic value of C_{dg} grows (even markedly) and the asymptotic value of C_{in} decreases.

More recently Sumer and Fredsoe (1997), basing themselves on data of Justesen (1989), have suggested a different picture for the smooth pipes on the range $5 < K_E < 20$. In particular, on this range, the asymptotic values of C_{in} and C_{dg} deviate from those suggested by Sarpkaya and Isaacson: the asymptote of C_{in} being smaller than 1.85 and that of C_{dg} being greater than 0.62. The largest deviations occur at K_E of $10 \div 15$, where the asymptote of C_{in} falls at about 1.4 and the asymptote of C_{dg} rises up to nearly 1.0. Here, note that $C_{in} = 1.4$, $C_{dg} = 1.0$ instead of the values (11.11) lead to a somewhat smaller in-line force for a K_E of $10 \div 15$.

Finally, also C_{lift} depends generally on R_E and K_E, and Sarpkaya and Isaacson's graphs show that it is about 0.2 for R_E greater than about $5 \cdot 10^5$.

11.5.5 *The weight of the drag component*

Let us consider the basic case of a vertical cylinder. It is the case in which both the inertia force and the drag force act along the wave propagation axis (y-axis):

$$f_{in} = C_{in}\,\rho\pi R^2 a_y, \quad f_{dg} = C_{dg}\,\rho R|v_y|v_y$$

(for simplicity we write f_{in}, f_{dg} in place of $f_{in\,y}$, $f_{dg\,y}$). If the wave is periodic, we have

$$v_y = v_{\max}\cos(\omega t), \quad a_y = -v_{\max}\,\omega\sin(\omega t).$$

Hence we can write

$$f = f_{in} + f_{dg} = f_{in\max}\left[-\sin(\omega t) + \frac{C_{dg}}{C_{in}\,\pi^2} K_E|\cos(\omega t)|\cos(\omega t)\right],$$

which points out that

$$f_{\max} = f_{in\max} \quad \text{for} \quad K_E \le \frac{1}{2}\pi^2\frac{C_{in}}{C_{dg}}.$$

For example, with the asymptotic values $C_{in} = 1.85$ and $C_{dg} = 0.62$, $f_{\max}$ coincides with $f_{in\max}$ for $K_E < 15$; that is to say, the drag component does not affect the force peak if K_E is smaller than 15.

11.6 The long-structure problem

11.6.1 *The terms of the problem*

Let us think of a long structure such as a submerged tunnel or a breakwater and let us assume that the dominant direction is orthogonal to the axis of this structure. Let us fix a point P of the axis and let us consider

(i) the probability of exceedance of the wave height at P,

(ii) the probability of exceedance of the wave height at the centre of the wave crest which passes at point P.

If the waves are random 3-D (wind generated waves), the crest-to-trough height at the crest centre is greater than or equal to the wave height at point P. It is equal only in the special case that the crest centre passes at point P. As a consequence, the probability (ii) is greater than the probability (i). Whereas, if the waves were random 2-D (no directional spread), the two probabilities would be equal to each other.

Now, keeping in mind that the probability (i) is the same for the 3-D random waves and for the 2-D random waves (under the same frequency spectrum), it follows that the probability (ii) is greater for the 3-D waves than for the 2-D waves.

Then, if we think that a long structure is usually beaten by several wave crests at the same time, we easily realize that the maximum expected wave height on the whole structure (in a given time interval) will be greater in the case of the 3-D random waves than in the case of the 2-D random waves. Were the waves two dimensional, the maximum expected wave height at any fixed point P would be equal to the maximum expected wave height on the whole structure. The waves being three dimensional, the maximum expected wave height on the whole structure will be greater. How much greater? Here is the problem to be solved.

11.6.2 *Development*

Line *I* of fig. 11.20 represents $\overline{\tilde{\alpha}_{\max}}(N)$, that is to say the maximum expected wave height in a sequence of N waves. (Of course we are speaking of the nondimensional wave height $\tilde{\alpha}$, i.e. the quotient between the wave height and the significant height of its sea state.) Line *II* represents $\overline{\tilde{\alpha}_{\max}}(2N)$; the meaning is the same as that of line *I* apart from the number of waves that here is $2N$. Finally line *III* represents $\overline{\tilde{\alpha}_{\max}}(3N)$. These three lines have been obtained by means of (5.58).

For example, the ordinate of the three lines on the abscissa N=200 are

$$\overline{\tilde{\alpha}_{\max}}(200) = 1.617, \quad \overline{\tilde{\alpha}_{\max}}(2 \cdot 200) = 1.712, \quad \overline{\tilde{\alpha}_{\max}}(3 \cdot 200) = 1.765,$$

that is: the maximum expected wave height in a sequence of 200 waves is equal to $1.617\, H_s$, the maximum expected wave height in a sequence of 400 waves is equal to $1.712\, H_s$, and the maximum expected wave height in a sequence of 600 waves is $1.765\, H_s$.

Also the data points in fig. 11.20*a* represent $\overline{\tilde{\alpha}_{\max}}(N)$, $\overline{\tilde{\alpha}_{\max}}(2N)$ and $\overline{\tilde{\alpha}_{\max}}(3N)$ and have been obtained from the time series data of the experiment RC 1990. As to the ψ^* needed for obtaining the lines *I*, *II* and *III* by means of (5.58), we have taken the mean value of this parameter in the aforesaid experiment.

The same lines *I*, *II* and *III* representative of $\overline{\tilde{\alpha}_{\max}}(N)$, $\overline{\tilde{\alpha}_{\max}}(2N)$ and $\overline{\tilde{\alpha}_{\max}}(3N)$ as functions of N are also in the panels (*b*) and (*c*) of fig. 11.20. But the data points in these two panels are representative of $\overline{\tilde{\alpha}_{\max}}(N, n, \Delta x)$, where

$N \equiv$ number of consecutive waves recorded at each point;

$n \equiv$ number of points where the records are taken *at the same time*;

$\Delta x \equiv$ distance between two of these points.

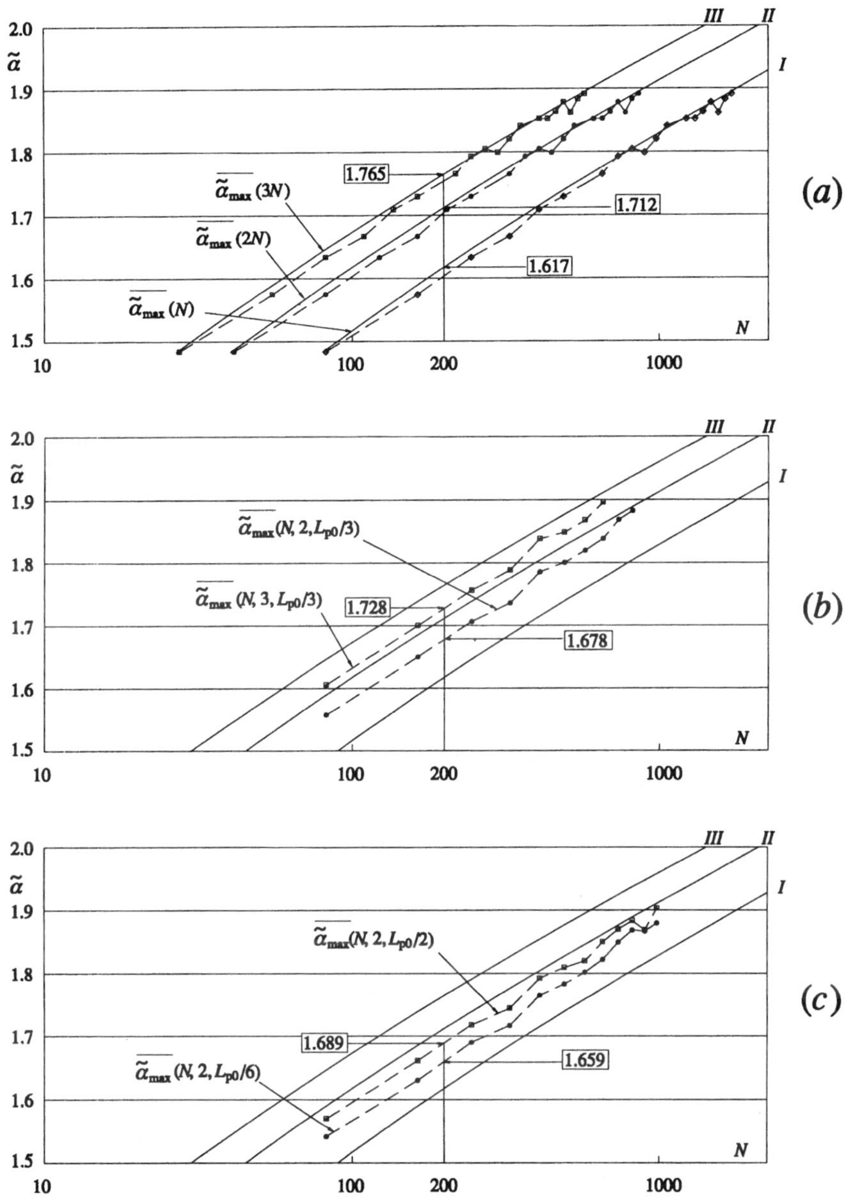

Fig. 11.20 (*a*) Line I: maximum expected $\tilde{\alpha}$ in a sequence of N waves; line *II*: maximum expected $\tilde{\alpha}$ in a sequence of $2N$ waves; line *III*: maximum expected $\tilde{\alpha}$ in a sequence of $3N$ waves. Continuous lines: from equation (5.58); data points from experiment RC 1990. (*b*) and (*c*) The continuous lines are the same as in panel (*a*), while the data points give the maximum expected $\tilde{\alpha}$ in a sequence of N waves recorded simultaneously at n distinct points. The symbol specifies the number n of points and the distance Δx between these points. The alignment of the points was nearly parallel to the wavefront.

In other words, the data of fig. 11.20*b-c* represent the maximum expected $\tilde{\alpha}$ in a set of $n\ N$ waves recorded simultaneously at n different points: N waves per each point (we mean that the wave gauges are laid along a straight line parallel to the wavefront).

Also the data of fig. 11.20*b-c* proceed from the experiment RC 1990, where the aligned wave gauges were ①-⑦ [see the gauge map in fig. 10.23]. Specifically $\overline{\tilde{\alpha}_{\max}}\,(N, 2, L_{p0}/3)$ was obtained with the wave heights recorded simultaneously by pairs of gauges like ①-③, ②-④, ③-⑤, and so on. Indeed the distance between two gauges was 1.25 m and L_{p0} was on average of 7.5 m, so that the distance between two gauges was $L_{p0}/6$. (The waves were on deep water so that L_{p0} was nearly coincident with L_p.)

For example, in panel (*b*) of fig. 11.20 we see

$$\overline{\tilde{\alpha}_{\max}}\,(200, 2, L_{p0}/3) = 1.678,$$

which means that, if we take simultaneously two records each of 200 consecutive waves at two points being $L_{p0}/3$ apart and being roughly aligned with the wavefront, we can expect that the maximum wave height in the set of the $200 \cdot 2 = 400$ waves recorded at the two points is equal to 1.678 H_s. Similarly, we see that

$$\overline{\tilde{\alpha}_{\max}}(200, 3, L_{p0}/3) = 1.728,$$

which means that, if we take simultaneously three records each of 200 consecutive waves at three points P_1, P_2 and P_3, with $\overline{P_1P_2} = \overline{P_2P_3} = L_{p0}/3$ and P_1, P_2 and P_3 being roughly aligned with the wavefront, we can expect that the maximum wave height in the set of the $200 \cdot 3 = 600$ waves recorded at the three points is equal to 1.728 H_s.

If the distance Δx between two gauges tends to infinity, the maximum expected $\tilde{\alpha}$ in the set of $2N$ waves recorded simultaneously by the two gauges (N per each gauge) will coincide with the maximum expected $\tilde{\alpha}$ in a sequence of $2N$ waves recorded by only one gauge in a time interval which is twice as long. On the contrary, if Δx approaches zero, the maximum expected $\tilde{\alpha}$ in the set of $2N$ waves recorded simultaneously by the two gauges will coincide with the maximum expected $\tilde{\alpha}$ in a sequence of N waves recorded by only one gauge. Indeed, in this case, the two gauges being placed at the same point also take the same wave record. In general, as the distance Δx grows, the maximum expected $\tilde{\alpha}$ in the set of $2N$ waves recorded simultaneously by the two gauges (N waves per each gauge) passes gradually from $\overline{\tilde{\alpha}_{\max}}\,(N)$ to $\overline{\tilde{\alpha}_{\max}}\,(2N)$, that is it passes from line *I* to line *II* of fig. 11.20. An apparent confirmation comes from fig. 11.20*c* that shows

$$\overline{\tilde{\alpha}_{\max}}\,(N) < \overline{\tilde{\alpha}_{\max}}\,(N, 2, L_{p0}/6) < \overline{\tilde{\alpha}_{\max}}\,(N, 2, L_{p0}/2) < \overline{\tilde{\alpha}_{\max}}(2N)\,.$$

11.6.3 *Suggested solution*

The problems that are encountered in the design practice are of two kinds. The first one is the problem of the submerged tunnel (see fig. 12.8), that is a structure with some non-constant influence lines. For example, the largest bending moment in the tunnel varies according to where the maximum wave height on the whole structure occurs (we shall see this in the worked example of sect. 12.3). The largest bending moment and shear force typically occur if the maximum wave height occurs at a pier or at a span centre.

The second problem is that of the caisson breakwater, that is a long structure with some constant influence lines. Indeed, wherever the maximum wave height occurs, the maximum pressure force and overturning moment on a single caisson will be nearly the same.

For the first problem we have to evaluate the maximum expected wave height at a set of discrete points. For the second problem we have to evaluate the maximum expected wave height on a segment line (the longitudinal axis of the structure).

First problem. The maximum expected wave height at a set of n points, being aligned with the wavefront and being at least $L_{p0}/2$ apart, can be evaluated as the maximum expected wave height at a fixed point, with a fictitious mean period of $\overline{T}/n$. This approach can be followed with any formula for the maximum wave height, that is with the formulae (5.58-59), or (6.25) or (7.37). In other words the maximum wave height at the n points in a given time interval is estimated as the maximum wave height that would occur at a single fixed point if the time interval was n times larger. It is the same as using line II of fig. 11.20 to estimate the maximum expected $\tilde{\alpha}$ in a set of $2N$ waves simultaneously recorded at two different points (N per each point). The condition that the distance between the points must be larger than $L_{p0}/2$ is typically fulfilled in the cases that occur in design practice. Was it not satisfied, the suggested criterion would be conservative.

Second problem. Let us imagine we align a set of n points at a distance of $L_{p0}/3$ from each other along the structure axis. Well, if we adopted the method suggested for the first problem [replacing $\overline{T}$ by $\overline{T}/n$ in formulae (5.58-59), (6.25), and (7.37)] we would overestimate the maximum expected wave height at these n points. But the maximum expected wave height on the whole segment line is actually somewhat larger than the maximum expected wave height at the set of n points. (The difference between the maximum expected wave height on the segment line and the maximum expected wave height at the n points can be evaluated by means of the quasi-determinism theory; and it proves to be of a few per cent if the distance between two successive points is equal to $L_{p0}/3$.)

In the light of these remarks we suggest to estimate the maximum expected wave height on the whole segment line of length l by means of (5.58-59), or (6.25), or (7.37) with a fictitious mean period $\overline{T}/n$, with

$$n = 1 + \frac{l}{L_{p0}/3}\,. \tag{11.12}$$

Data from shallow waters are still lacking. However, in sect. 10.5 we saw that the length of a wave crest keeps practically constant from deep to shallow water, and accordingly we suggest using formula (11.12) also for structures on shallow water.

Conclusive note

The small scale field experiments RC 1992 on the gravity offshore platform and RC 1993 on the submerged tunnel were described by the author (1995 and 1996). The long-structure problem is a novelty.

References

Boccotti P., 1995 A field experiment on the small-scale model of a gravity offshore platform. *Ocean Eng.* 22, 615-627;
1996 Inertial wave loads on horizontal cylinders: a field experiment. *Ocean Eng.* 23, 629-648.

Borgman L.E., 1958 Computation of the ocean wave forces on inclined cylinders. *J. Geoph. Res. Trans. AGU* 39, 309-322.

Burrows R., Tickell R.G., Hames D. and Najafian G., 1997 Morison wave force coefficients for application to random seas. *Appl. Ocean Res.* 19, 183-199.

Chaplin J.R. and Subbiah K., 1997 Large scale horizontal cylinder forces in waves and currents. *Appl. Ocean Res.* 19, 211-223.

Dawson T. H., 1983 Offshore Structural Engineering. *Prentice-Hall*, 1-346.

Garrison C.J. and Chow P.Y., 1972 Wave forces on submerged bodies. *J. Waterways Harbors and Coastal Eng. Div. ASCE* 98, WW3.

Garrison C.J. and Rao V.S., 1971 Interaction of waves with submerged objects. *J. Waterways Harbors and Coastal Eng. Div. ASCE* 97, WW2.

Garrison C.J., Torum A., Iverson C. et al., 1974 Wave forces on large volume structures - a comparison between theory and model test. *Proc. Offshore Tech. Conf.* OTC 2137, 1061-1070.

Justesen P., 1989 Hydrodynamic forces on large cylinders in oscillatory flow. *J. Waterway, Port, Coastal and Ocean Eng.* 115, 497-514.

Kim M.H., 1992 Interaction of waves with arrays of N vertical circular cylinders. *J. Waterway, Port, Coastal and Ocean Eng.* 119, 671-689.

Linton C.M. and Evans D.V., 1990 The interaction of waves with arrays of vertical circular cylinders. *J. Fluid Mech.* 215, 549-569.

Mac Camy R. C. and Fuchs R. A., 1954 Wave forces on piles: a diffraction theory. *U.S. Army Corps of Engineers, Beach Erosion Board, Tech. Memo.* 69.

Morison J. R., O'Brien M. P., Johnson J. W. et al., 1950 The forces exerted by surface waves on piles. *Petroleum Trans.* 189, 149-156.

Ogilvie T. F., 1963 First and second order forces on a cylinder submerged under a free surface. *J. Fluid Mech.* 16, 451-472.

Sarpkaya T. and Isaacson M., 1981 Mechanics of wave forces on offshore structures. *Van Nostrand Reinhold Co.*, 1-650.

Sumer B.M. and Fredsoe J., 1997 Hydrodynamics around cylindrical structures. *World Scientific Co. Pte. Ltd.*, 1-530.

Ursell F., 1950 Surface waves on deep water in the presence of a submerged circular cylinder. *Proc. Cambridge Phil. Soc.* 46, 141-152.

Williamson C.H.K., 1985 Sinusoidal flow relative to circular cylinders. *J. Fluid Mech.* 155, 141-174.

Yilmaz O., 1998 Hydrodynamic interactions of waves with group of truncated vertical cylinders. *J. Waterway, Port, Coastal and Ocean Eng.* 124, 272-279.

Chapter 12
CALCULATION OF THE WAVE FORCES ON OFFSHORE STRUCTURES

12.1 Calculation of the wave forces on a gravity offshore platform

Let us estimate the force exerted by the highest wave in the lifetime of a gravity offshore platform. The plan and the front view of the platform are given in fig. 12.1, and for a three dimensional picture reference is made to fig. 11.1. The water depth d is 125 m; the reservoirs have a height of 50 m and a diameter of 20 m; the columns' diameter gradually tapers from 20 m to 12 m. (The deck is not shown in the figure, since it is not subjected to wave loads and hence it does not concern our analysis.)

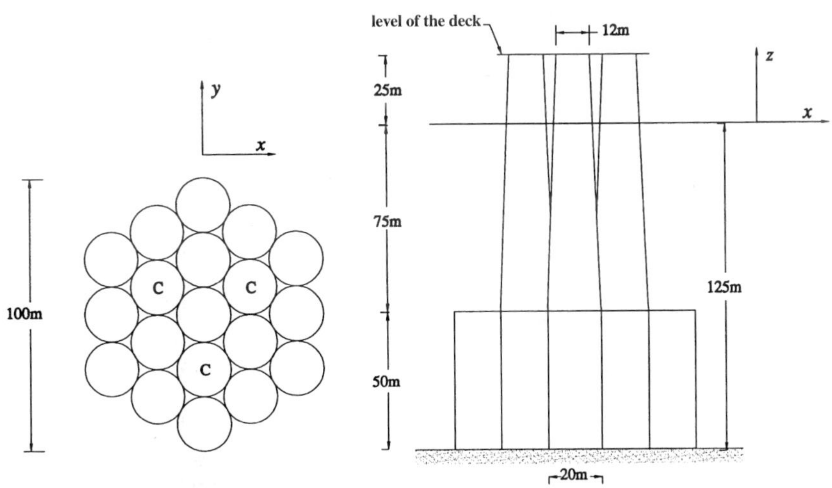

Fig. 12.1 Plan and front view of a gravity offshore platform; the positions of the columns in the plan are denoted by letter C.

The logic is that of sect. 7.8. Specifically, by means of (7.37) we can evaluate the probability that the largest wave height in the lifetime L exceeds any fixed threshold H. Hence we obtain the threshold H being relevant to the prescribed probability. Finally, (7.40) enables us to estimate the H_s of the sea state where wave height H will most probably occur. It is the procedure outlined by figs. 7.12 and 7.13.

Here we assume the following characteristic values for the North Sea:

$$H = 30\,\mathrm{m}, \quad H_s = 15\,\mathrm{m};$$

from which, using (4.27) we obtain

$$T_p = 16.5\,\mathrm{s} \Rightarrow L_{p0} = 425\,\mathrm{m}, \quad L_p = 408\,\mathrm{m}.$$

As we noticed in sect. 7.8.2 the highest wave in the lifetime is very great even for the heavy sea state where it occurs, and indeed the quotient H/H_s for the highest wave in the lifetime proves to be greater than or equal to 2. This is an important property that enables us to apply the quasi-determinism theory for the calculation of the force exerted by the highest wave in the lifetime.

The logic of the quasi-determinism theory is that, knowing that a very high wave ($H = 30\,\mathrm{m}$ in a sea state of $H_s = 15\,\mathrm{m}$) occurs at the platform's centre (point x_o, y_o), we can predict with confidence the configuration of the wave group and the load exerted on the structure.

The horizontal axes are shown in fig. 12.1 (bearing in mind that x-y are the coordinate axes, and X-Y are the local coordinate axes with the origin at x_o, y_o). The dominant wave direction is assumed to be y-parallel ($\bar{\theta} = 0$). Hence also the propagation of the wave group is y-parallel.

The calculation is made in four steps.

First step: estimate of K_E. The double integral in equation (10.3*a*) of v_y (being necessary to get K_E) is numerically evaluated, and the hints of sect. 10.1 should make this task rather easy. As to the nondimensional spectrum $\mathscr{S}(w, \theta)$, we use the classic form (8.20) with

$$\chi_1 = 3.3, \quad n_p = 20.$$

(As said, the second shape parameter of the JONSWAP spectrum, that is χ_2, plays only a minor role, and can definitively be given the characteristic value 0.08.)

The K_E of the base proves to be markedly smaller than the unit, and the K_E of the columns exceeds 2 about 50 m beneath the mean water level and is nearly 6 at the mean water level. Hence, following common practice we should not apply the ideal flow pattern to the upper portions of the three support columns. However the difference is small if we resort, as we shall do, to the ideal flow pattern for the whole structure.

Second step: calculation of C_{do}. As to the C_{do} of the columns, following the suggestion of sect. 11.4.4, we assume it to be equal to 2. Thus the calculation must only be done for the base consisting of 19 cylinders and 24 interstices between the cylinders. Each interstice has an area of $0.0403D^2$ (D being the diameter of a single cylinder). Therefore the area of the platform's base is

$$A = 19\,\frac{\pi}{4}\,20^2 + 24 \cdot 0.0403 \cdot 20^2 = 6356\,\mathrm{m}^2 .$$

For calculating C_{do} we refer to the circular cylinder of the same area whose radius is

$$R = \sqrt{\frac{6356}{\pi}} = 45\,\mathrm{m} .$$

Hence, taking $\mathscr{T}_{\mathscr{R}} = 1.75$ for the platform base [cf. sect. 11.4.4], from (11.5) we obtain

$$C_{do} = \frac{\left|\sin\left(1.75\,\frac{\pi}{4}\,\frac{2\pi}{408}\,45\right)\right|}{\left|\sin\left(\frac{\pi}{4}\,\frac{2\pi}{408}\,45\right)\right|} = 1.57 .$$

Third step: calculation of the wave force on the platform's base. To this end we should evaluate the pressure force on the equivalent water mass, using the formula (10.3g) of Δp. However, this calculation is rather long because of the volume shape, so that it is advisable to resort to the alternative form

$$F_y = \int_W \rho\, a_y\, \mathrm{d}W , \tag{12.1}$$

where W is the volume of the equivalent water mass. We should obtain $19 + 24$ contributions to this integral: from the 19 cylinders and the 24 interstices. The coordinates of the centres of the $19 + 24$ volumes must be stored on a computer. Hence we have to write a program that evaluates the particle acceleration a_y (function 10.3d) at the centre of each of these volumes, multiplies this acceleration by the water density ρ and by the volume. Doing so we get the horizontal force on a water mass equivalent to the platform's base. Then, multiplying this force by $C_{do} = 1.57$ we arrive at the horizontal force on the platform's base. Of course, the calculation must be done for a number of values of T, so as to obtain the fluctuating wave force.

Fourth step: calculation of the horizontal force on the platform's columns. The force on the columns is evaluated essentially like the force on the base. First, we compute the horizontal force on an equivalent water mass by dividing the columns' volume in a number of small disks. There are only two differences with respect to the calculation done for the platform's base: the C_{do} which is 2 instead of 1.57, and the volume W not constant in time. Indeed W is the column's volume from the base ($z = -75\,\mathrm{m}$) to the free surface elevation. Therefore, we have to evaluate the free surface elevation at each column, which can be done by means of (10.2a).

Fig. 12.2 shows the results:

(i) the surface displacement at x_o, y_o (the platform's centre),
(ii) the horizontal force on the base,
(iii) the horizontal force on the three columns,
(iv) the total horizontal force.

The water density has been assumed to be

$$\rho = 1030 \text{ kg/m}^3.$$

We see the main force is that on the base. Of course the same calculation also yields the overturning moment which is chiefly due to the force on the three columns.

The figure reveals the characteristic time history of the forces produced by the wave groups: the force fluctuation grows up to a maximum when the central wave

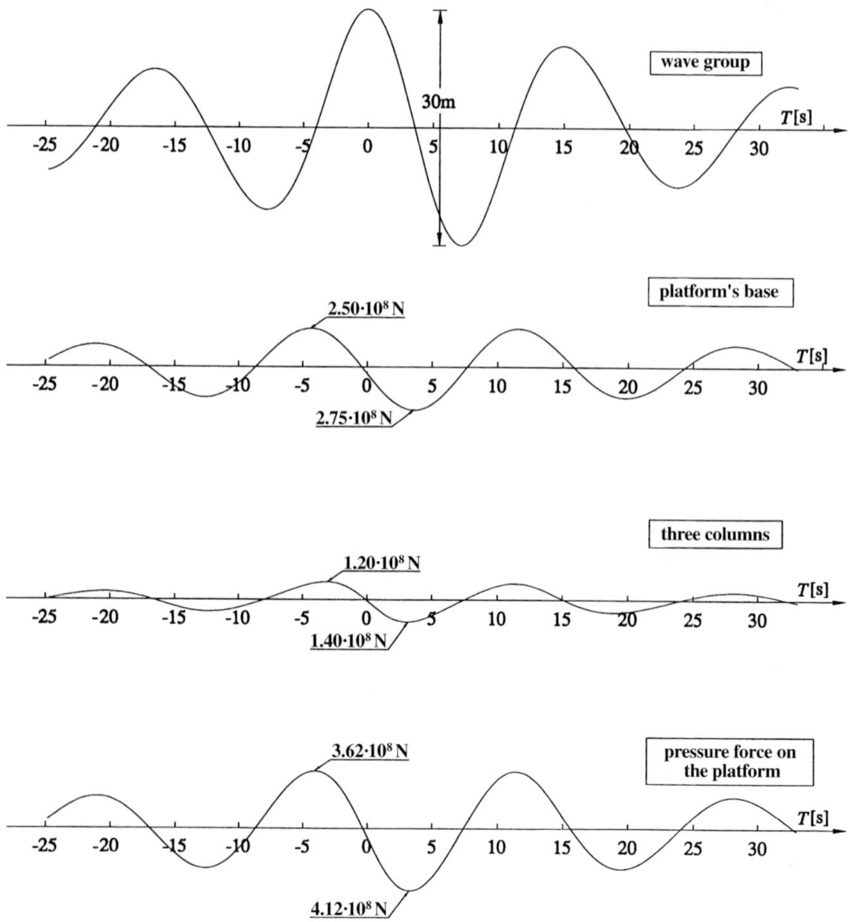

Fig. 12.2 Worked example: the horizontal forces on the gravity platform.

passes through the platform, and then decreases. This is markedly different from the force exerted by a train of periodic waves. Note also that the wave and the horizontal force have a phase angle of a quarter of circle because of the inertial nature of the force.

12.2 Calculation of the wave forces on a space frame structure

12.2.1 *The largest zero down-crossing wave must be considered*

The heights of the maximum expected zero up-crossing wave and of the maximum expected zero down-crossing wave in the lifetime are equal to each other. As for the surface displacement and the flow field, the prediction concerning the highest zero down-crossing wave can be done by means of the quasi-determinism theory with a *minus* in the equation of $\overline{\eta}$ and $\overline{\phi}$.

Fig. 12.3 shows the surface displacement $\overline{\eta}$ at the fixed point $\mathbf{x}_o$ for the case in which the wave of given very large height H is a zero up-crossing and for the case in

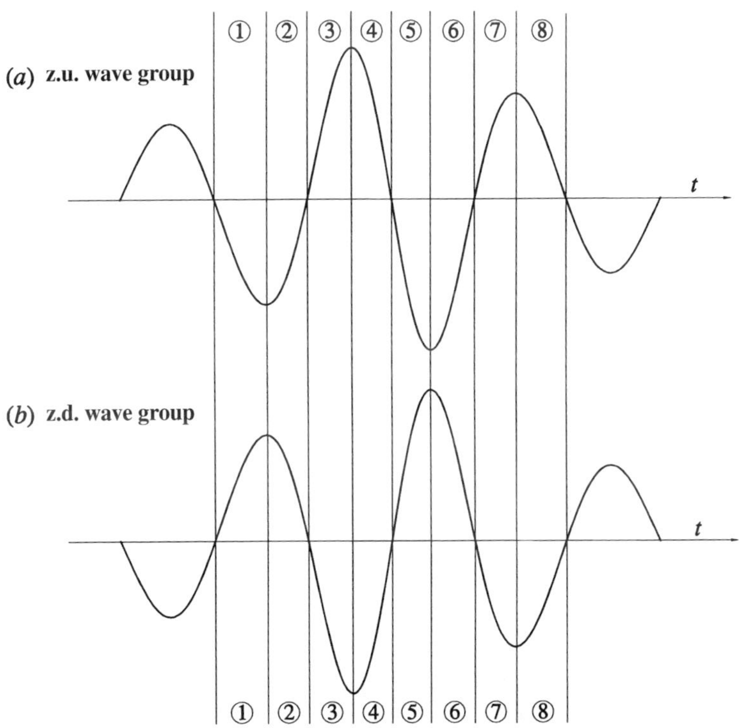

Fig. 12.3 (*a*) The group of the maximum expected zero up-crossing wave. (*b*) The group of the maximum expected zero down-crossing wave. If the drag force is not negligible, the maximum force exerted by group (*b*) is greater than the maximum force exerted by group (*a*).

which it is a zero down-crossing. We shall call z.u. the group of the maximum expected zero up-crossing wave height (fig. 12.3*a*), and z.d. the group of the maximum expected zero down-crossing wave height (fig. 12.3*b*).

Generally it makes no difference whether we consider z.u. or z.d. As an example, in the case of the gravity offshore platform, z.d. and z.u. exert some opposite forces, so that the absolute value of the largest wave force does not change from z.u. to z.d. Not so in the case of a jacket platform. Let us see why.

The largest drag force occurs when the highest wave crest passes through the platform, because the horizontal particle velocity is at its maximum and the volume of the submerged (loaded) part of the structure is also at its maximum. The largest inertia force occurs when the central zero of the highest wave passes through the platform, because the horizontal particle acceleration is at its maximum. Hence the largest drag force is positive, whereas the largest inertia force is positive or negative according to whether the highest displacement is between a trough and the following crest as in z.d., or between a crest and the following trough as in z.u.

As a consequence, the largest wave force (sum of the drag force and the inertia force) is greater in z.d. than in z.u. This largest wave force of z.d. is necessarily positive and occurs on time interval ⑤, whilst the largest wave force of z.u. may be positive and occur on ③, or may be negative and occur on ⑤.

Conclusion: for the space frame structures we must resort to the highest zero down-crossing wave and apply the formulae for $\overline{\eta}$ and $\overline{\phi}$ with a *minus* in both of them. Therefore for the calculation of the wave forces in the next two sections it will

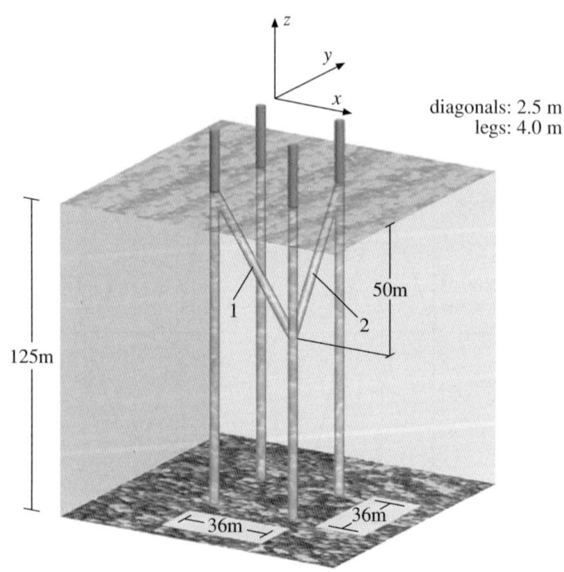

Fig. 12.4 Scheme of the legs and two diagonals of an ideal frame structure.

be understood that equation (10.2*a*) of $\overline{\eta}$ and equations (10.3*a*-*g*) of the particle velocity and acceleration and of the pressure fluctuation must be multiplied by -1.

12.2.2 *Wave forces on the legs*

Fig. 12.4 shows the scheme of four legs and two diagonals of an ideal frame structure. We shall assume the same design wave height as for the gravity platform (it will be noted that the water depth is also the same), and we shall compute the wave forces.

The legs of the frame structure have a K_E greater than 2 even at the depth of the seabed. Therefore this is a case of applicability of the real flow pattern.

The quotient R_E/K_E of the legs is nearly 10^6 and therefore, assuming that the cylinder is smooth, we take $C_{in} = 1.85$ and $C_{dg} = 0.62$ [cf. sect. 11.5.4]. (Note that R_E/K_E depends only upon the cylinder's diameter, the wave period and the kinematic viscosity, and thus it can be readily obtained.)

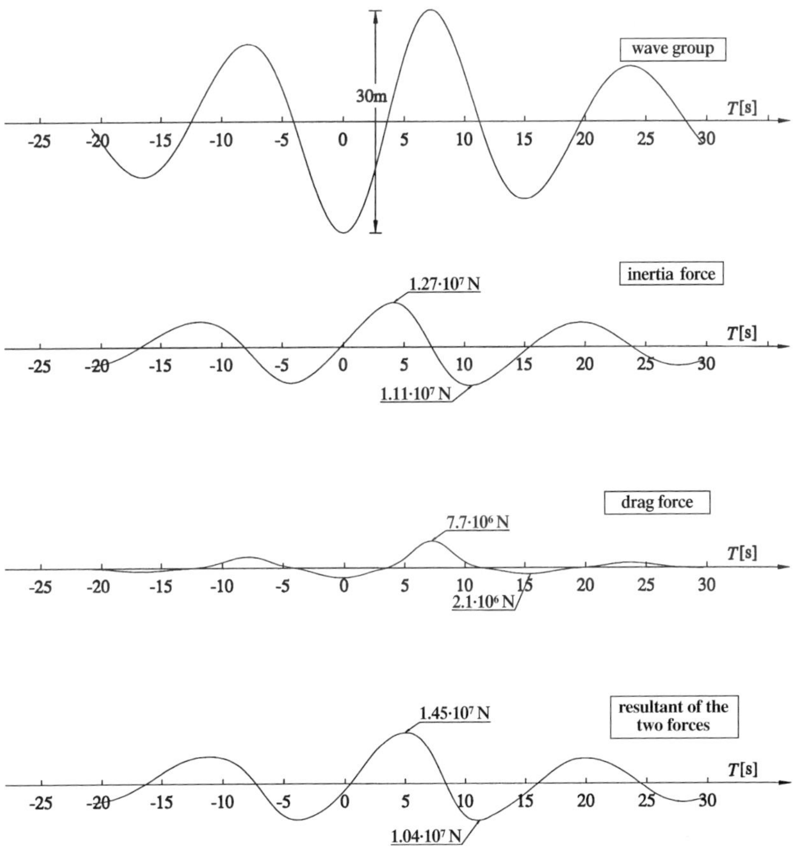

Fig. 12.5 Worked example: the horizontal force on the four legs of the space frame structure.

For computing the wave forces, let us apply Morison's equation and quasi-determinism theory, with the platform's centre as $\mathbf{x}_o$. Since the wave group travels in the dominant wave direction, that is y-parallel, the wave force (in-line force) is also y-parallel. Of course, the calculation must be done at different elevations from the seabed to the free surface, since v_y and a_y vary with the vertical elevation.

The calculation of v_y and a_y, which is needed to obtain the wave force, calls for the numerical evaluation of the integrals in the two equations (10.3*a*) and (10.3*d*). The result is shown in fig. 12.5 giving

(*a*) the surface displacement at $\mathbf{x}_o$,
(*b*) the inertia force,
(*c*) the drag force,
(*d*) the total in-line force.

In line with the predictions of sect. 12.2.1, we see the largest force (nearly $1.5 \cdot 10^7$ N) occurs between the central zero and the crest of the highest zero down-crossing wave ($H = 30\,\text{m}$).

12.2.3 *Wave forces on the diagonals*

The quotient R_E/K_E of the diagonals is $4 \cdot 10^5$, so that we take again $C_{in} = 1.85$ and $C_{dg} = 0.62$.

As to the calculation, some more operations are needed. Indeed we have to obtain $\mathbf{v}_{\text{sect}}$ (the velocity vector normal to the diagonal). Referring to fig. 12.6, we have

$$\text{diagonal ①:}\begin{cases} \mathbf{v}_{\text{sect}} = (v_x \sin^2\alpha + v_z \cos\alpha \sin\alpha)\,\mathbf{i}_x + v_y \mathbf{i}_y + (v_z \cos^2\alpha + v_x \sin\alpha\cos\alpha)\,\mathbf{i}_z\,, \\ \mathbf{a}_{\text{sect}} = (a_x \sin^2\alpha + a_z \cos\alpha \sin\alpha)\,\mathbf{i}_x + a_y \mathbf{i}_y + (a_z \cos^2\alpha + a_x \sin\alpha\cos\alpha)\,\mathbf{i}_z\,, \end{cases}$$

$$\text{diagonal ②:}\begin{cases} \mathbf{v}_{\text{sect}} = v_x \mathbf{i}_x + (v_y \sin^2\alpha - v_z \cos\alpha \sin\alpha)\,\mathbf{i}_y + (v_z \cos^2\alpha - v_y \sin\alpha\cos\alpha)\,\mathbf{i}_z\,, \\ \mathbf{a}_{\text{sect}} = a_x \mathbf{i}_x + (a_y \sin^2\alpha - a_z \cos\alpha \sin\alpha)\,\mathbf{i}_y + (a_z \cos^2\alpha - a_y \sin\alpha\cos\alpha)\,\mathbf{i}_z\,. \end{cases}$$

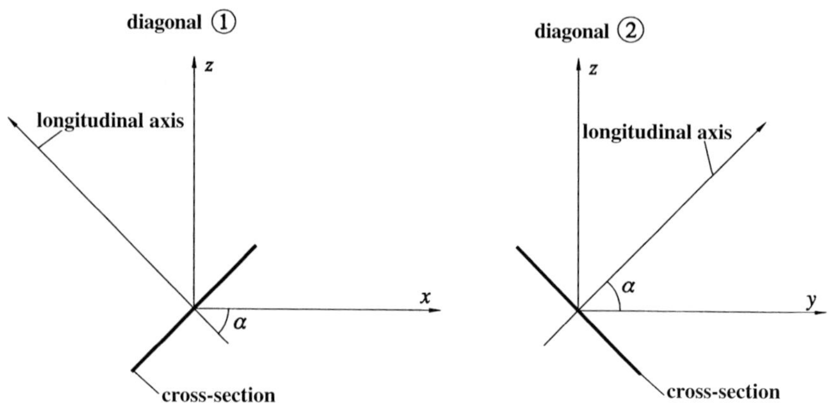

Fig. 12.6 Graphic aid for applying Morison's equation to diagonals ① and ② of the space frame structure.

Hence, the formula (11.8) for the force per unit length yields

diagonal ①:

$$f_y = C_{in}\rho\pi R^2 a_y + C_{dg}\rho R\cdot$$
$$\cdot\sqrt{(v_x\sin^2\alpha + v_z\cos\alpha\sin\alpha)^2 + v_y^2 + (v_z\cos^2\alpha + v_x\sin\alpha\cos\alpha)^2}\, v_y,$$

diagonal ②:

$$f_y = C_{in}\rho\pi R^2(a_y\sin^2\alpha - a_z\cos\alpha\sin\alpha) + C_{dg}\rho R\cdot$$
$$\cdot\sqrt{v_x^2 + (v_y\sin^2\alpha - v_z\cos\alpha\sin\alpha)^2 + (v_z\cos^2\alpha - v_y\sin\alpha\cos\alpha)^2}\cdot$$
$$\cdot(v_y\sin^2\alpha - v_z\cos\alpha\sin\alpha),$$

where the particle velocity and acceleration are those at the centre of the cross-section. The calculation can be done by means of equations (10.3*a-f*).

The y-component of the wave force on the diagonal ① and the y-component of the wave force on the diagonal ② are shown in fig. 12.7. We see the force to be larger on diagonal ① than on ②. This is because ② is parallel to the wave direction while ① is orthogonal.

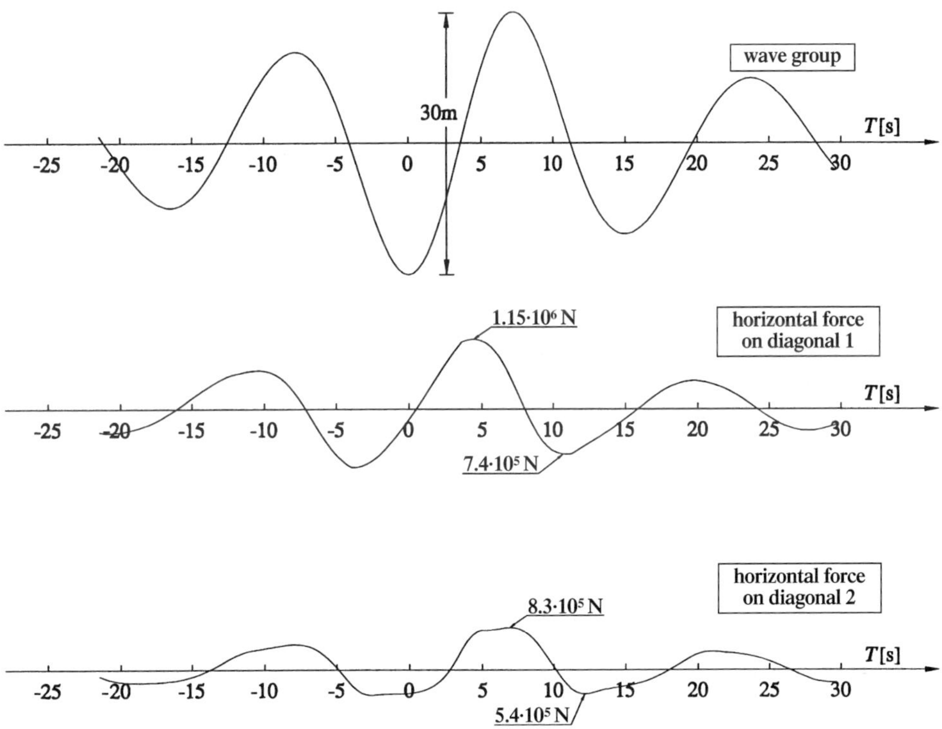

Fig. 12.7 Worked example: wave force on diagonals ① and ② of the space frame structure.

There is one last item left. The force on each member is calculated as if this member was isolated, and indeed both C_{in} and C_{dg} were obtained for isolated cylinders. Doing so, we don't consider the interference effect which can only be treated experimentally for each special case. The problems connected with these experimental investigations are dealt with in sect. 14.2.3.

12.3 Design of a submerged tunnel. I: calculation of the wave forces

12.3.1 *Description of the structure*

The tunnel of fig. 12.8 is at the same water depth as the gravity platform ($d = 125\,\text{m}$); the structure is 30 m beneath the water surface and its diameter is 25 m. Each pier consists of the central nucleon of the gravity platform's base. The central cylinder of this nucleon has been raised up to 70 m above the seabed and supports the tunnel.

The buoyancy is $4.95 \cdot 10^6$ N/m. The shell of reinforced concrete should be about 2 m thick which makes a weight of about $3.5 \cdot 10^6$ N/m. In addition there is the weight of the railroad and highways. Therefore it is reasonable to assume the weight to be equal to the buoyancy (floating tunnel).

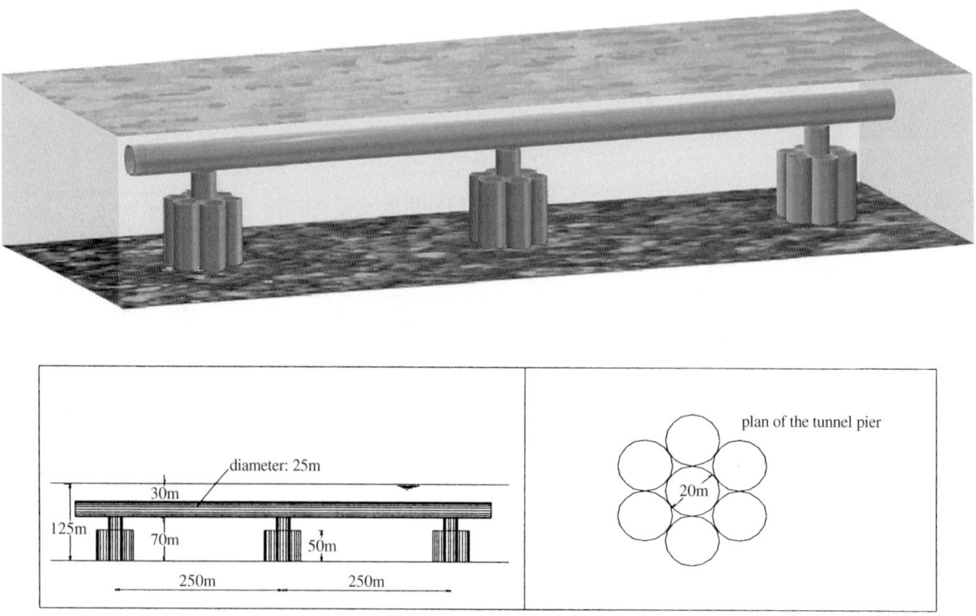

Fig. 12.8 A submerged tunnel.

12.3.2 *The design wave*

Probably H_s varies along the tunnel; this is because of the structure's length and also because a part of the tunnel may be sheltered by some prominence of the coast. For example a tunnel connecting Sicily to the mainland would only be subjected to heavy sea states close to Sicily. As to the shear stress and bending moment, they have a maximum if the centre of a wave crest passes over one of the piers (which will be proved in the next section).

For our calculation, let us assume that the tunnel is subjected to heavy sea states for a stretch of 1000 m. Then let us assume that the $P(H_s > h)$ on this stretch is given by (6.2) with $w = 0.874\,\mathrm{m}$ and $u = 1.200$ (these being the values of w and u we have also taken for the example in chap. 7). Finally, let us assume the dominant wave direction to be orthogonal to the tunnel. This is a realistic condition, given that the tunnel will connect two coastlines.

The stretch of tunnel being subjected to the heavy sea states includes five piers. The distance between piers (250 m) is larger than L_{p0} even in the design sea state (as we shall see shortly); even more so in the less severe sea states. Therefore, following the suggestions of sect. 11.6.3, the maximum wave height in the lifetime at the five piers can be estimated by means of (7.37) with $\overline{T}/5$ in place of $\overline{T}$.

The function $P[H_{\max}(L) > H]$ [equation (7.37)] is shown in fig. 12.9. Clearly, L has been taken as 100 years because the structure is of great importance and its failure would cause loss of human life. For the same reason the encounter probability is given the lowest value in table 7.2, that is $\mathscr{P} = 0.05$. With these values the maximum wave height at the five piers in the lifetime proves to be

$$H(L) = 17.7\,\mathrm{m}.$$

In sect. 7.8.1 for the same location (same w and u) we obtained $H(L) = 15.0\,\mathrm{m}$ for one fixed point and $L = 50\,\mathrm{years}$, $\mathscr{P} = 0.1$. This means that the step from

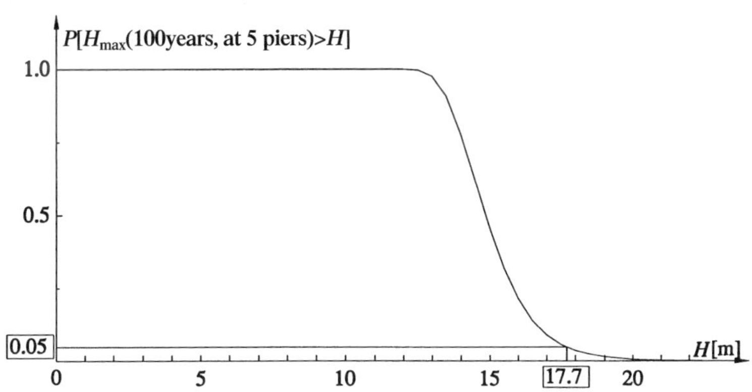

Fig. 12.9 Worked example: the probability that the largest wave height in 100 years at five piers exceeds any fixed threshold H.

$$L = 50\,\text{years}, \ \mathscr{P} = 0.1, \ 1 \text{ fixed point}$$

to

$$L = 100\,\text{years}, \ \mathscr{P} = 0.05, \ 5 \text{ fixed points}$$

leads to an 18% increase in $H(L)$: 10% of this 18% is due to the fact that L and $\mathscr{P}$ are more conservative, and the remaining 8% is due to the five points in place of only one.

The meaning of the 17.7 m wave height is the following. Let us imagine we take some continuous wave records above the five piers, for a very large time interval. In the first 100 years we shall find five values (generally different from one another) of the maximum wave height at each point. Say

first 100 years:

$$H_{\max 1} = 13.2\,\text{m}, \ H_{\max 2} = 14.0\,\text{m}, \ H_{\max 3} = 14.7\,\text{m}, \ H_{\max 4} = 13.8\,\text{m}, \ H_{\max 5} = 14.2\,\text{m},$$

so that

$$\text{first 100 years: } H_{\max 1-5} = 14.7\,\text{m}.$$

Then in the following 100 years we shall find a new set of values, say

second 100 years:

$$H_{\max 1} = 16.3\,\text{m}, \ H_{\max 2} = 14.2\,\text{m}, \ H_{\max 3} = 14.7\,\text{m}, \ H_{\max 4} = 15.9\,\text{m}, \ H_{\max 5} = 14.1\,\text{m},$$

so that

$$\text{second 100 years: } H_{\max 1-5} = 16.3\,\text{m}.$$

So on in the following intervals of 100 years. The height $H(L) = 17.7$ m is the threshold being exceeded by $H_{\max 1-5}$ five times in one hundred, on average. That is, on average, $H_{\max 1-5}$ exceeds 17.7 m five times in 100 intervals of 100 years.

The probability density function of the significant wave height of the sea state where the maximum wave height of 17.7 m will occur can be estimated by means of (7.40), again with $\overline{T}/5$ in place of $\overline{T}$. The calculation can be done numerically with $H_{\max} = 17.7$ m, and the result is shown in fig. 12.10. We see that the maximum wave height of 17.7 m will occur most probably in a sea state with an H_s of 8.0 m.

Conclusion: design wave of

$$H = 17.7\,\text{m in a sea state of } H_s = 8.0\,\text{m},$$

for which, equation (4.27) gives

$$T_p = 12.1\,\text{s}, \quad L_{p0} \cong L_p = 228\,\text{m}.$$

12.3.3 *The horizontal wave load*

We know that an exceptionally high wave ($H = 17.7$ m, $H_s = 8.0$ m) occurs at a point x_o, y_o of the tunnel, and hence, by means of the quasi-determinism theory, we

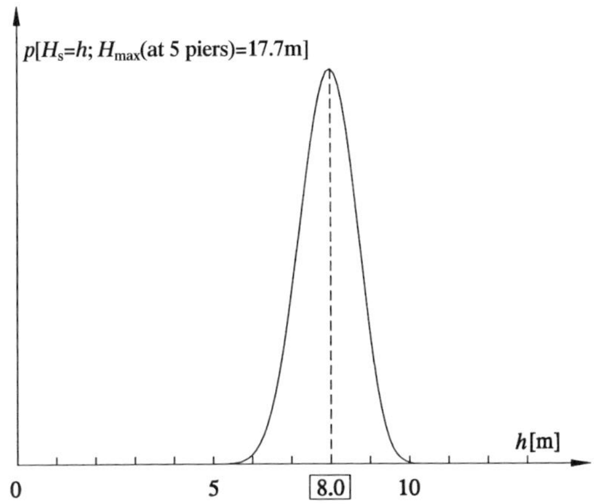

Fig. 12.10 Worked example: the p.d.f. of the H_s of the sea state where the maximum wave height in 100 years will occur.

can foresee which loads will most probably be exerted. Let us take the tunnel-axis as x-axis and let us assume that the dominant direction is y-parallel ($\bar{\theta} = 0$). As for the rest: the usual nondimensional spectrum $\mathscr{S}(w, \theta)$ that was applied also for calculating the wave loads on the platforms.

First step: the K_E proves to be smaller than 1 and hence this is a classic case of applicability of the ideal flow pattern.

Second step: calculation of C_{do}. Since the quotient depth-of-the-centre/radius is of $42.5/12.5 = 3.4$, following the suggestion of sect. 11.4.4, we take $\mathscr{F}_{\mathscr{R}} = 2.0$ that leads to

$$C_{do} = \frac{\left|\sin\left(2\,\frac{\pi}{4}\,\frac{2\pi}{228}\,12.5\right)\right|}{\left|\sin\left(\frac{\pi}{4}\,\frac{2\pi}{228}\,12.5\right)\right|} = 1.93\,.$$

Third step: calculation of the wave load per unit length. Because of the relatively small value of the quotient R/L_p, the wave force per unit length on the equivalent water cylinder can be easily calculated by means of the equation

$$f_y = \rho \pi R^2 a_y\,,$$

where a_y is the particle acceleration at the depth of the cylinder centre, which is calculated by means of (10.3d). Then for getting the force per unit length on the tunnel, one has to multiply by C_{do}.

Calculating f_y for a number of X and T we obtained the horizontal loads of fig. 12.11. These are the local maxima (with respect to time) of the horizontal load, which occur when a zero passes over the tunnel. The absolute maximum occurs when the central zero of the group's central wave passes over the tunnel. It is

$$|f_y|_{max} = 6.5 \cdot 10^5 \text{ N/m}.$$

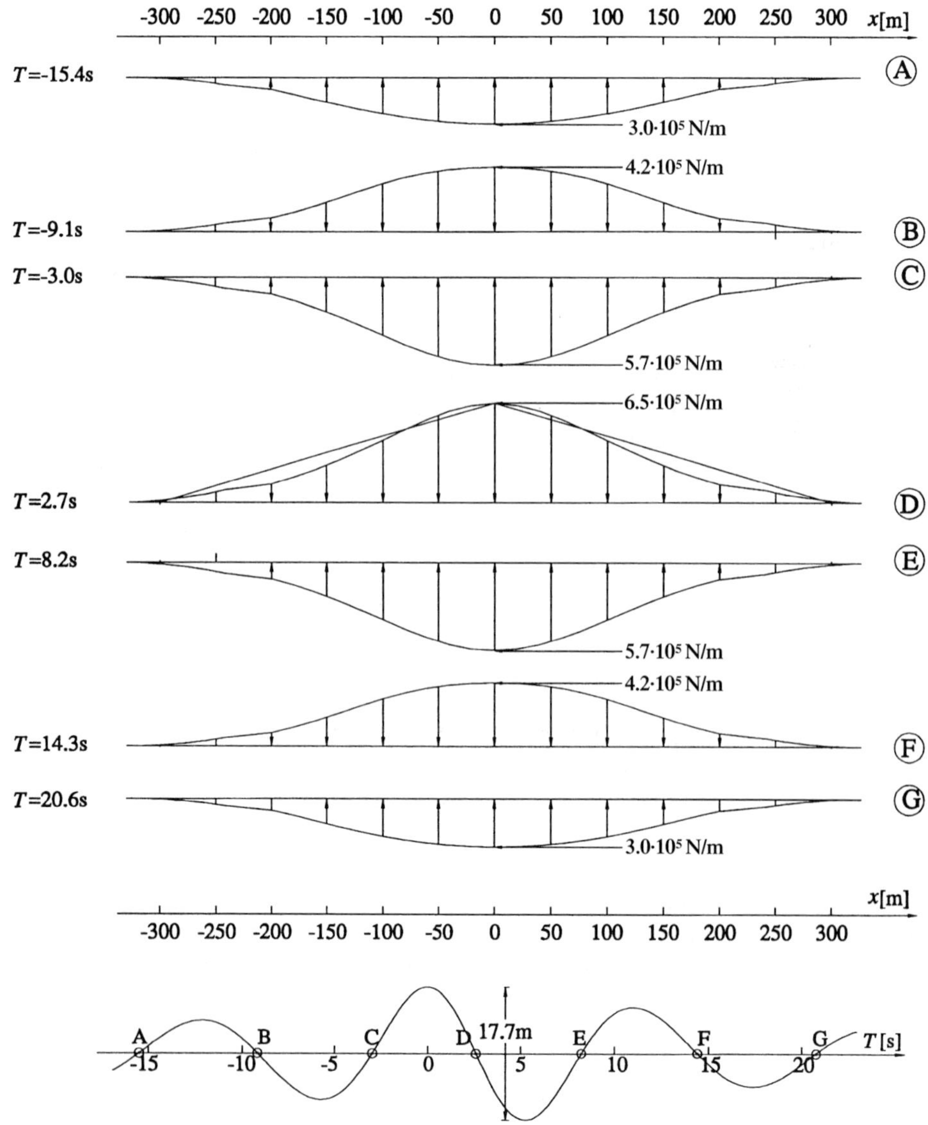

Fig. 12.11 Worked example: the local maxima (with respect to time) of the horizontal wave load on the tunnel.

12.3.4 *The vertical wave load*

In sect. 1.8 we saw that, on deep water, the non-linearity effects on Δp essentially depend on the kinetic term of the Bernoulli equation, so that we can write

$$\Delta p = \Delta p' + \Delta p'',$$

with

$$\Delta p' = -\rho \frac{\partial \phi}{\partial t},$$

$$\Delta p'' = -\frac{1}{2}\rho(v_x^2 + v_y^2 + v_z^2). \tag{12.2}$$

As an exercise, it could be proved that, on deep water, the $\Delta p''$ of a periodic wave does not change in time nor in the horizontal plane, and varies only with depth z [use equation (1.51) of ϕ and the simplified attenuation form of sect. 1.5.4 to prove this statement]. A straight consequence of this proof is that $\Delta p''$ does not contribute to the horizontal wave load.

On the contrary, $\Delta p''$ contributes to the vertical load, given that it varies with z. Therefore the vertical wave load per unit length on the equivalent water cylinder must be calculated by means of the equation

$$f_z = f'_z + f''_z,$$

with

$$f'_z \equiv \int_{-R}^{R} \Delta p'[y, z_-(y)] - \Delta p'[y, z_+(y)]\,\mathrm{d}y,$$

$$f''_z \equiv \int_{-R}^{R} \Delta p''[y, z_-(y)] - \Delta p''[y, z_+(y)]\,\mathrm{d}y, \tag{12.3}$$

where

$$z_-(y) = z_C - \sqrt{R^2 - y^2}, \quad z_+(y) = z_C + \sqrt{R^2 - y^2}.$$

Since the radius of the cylinder is rather small with respect to the wavelength, f'_z can be more rapidly calculated by means of the alternative expression

$$f'_z = \rho \pi R^2 a_z,$$

where a_z is the vertical particle acceleration given by (10.3*f*). As to f''_z, we have to apply equation (12.3) with (12.2) of $\Delta p''$. To this end it is convenient to prepare some subroutines for v_x, v_y and v_z (equations 10.3*a*-*c*).

Adding f''_z to f'_z we obtain the vertical wave force per unit length on the equivalent water cylinder. Then, multiplying this force by C_{dv} we obtain the vertical wave force per unit length on the solid cylinder. As to C_{dv}, following the suggestion of sect. 11.4.4, we take it equal to $0.9\,C_{do}$.

Fig. 12.12 shows the vertical wave load for a few X and T. The general meaning is the same as in fig. 12.11: the figure shows the local maxima (with respect to time) of the positive and negative vertical load. These local maxima occur when a wave crest or a wave trough passes over the tunnel.

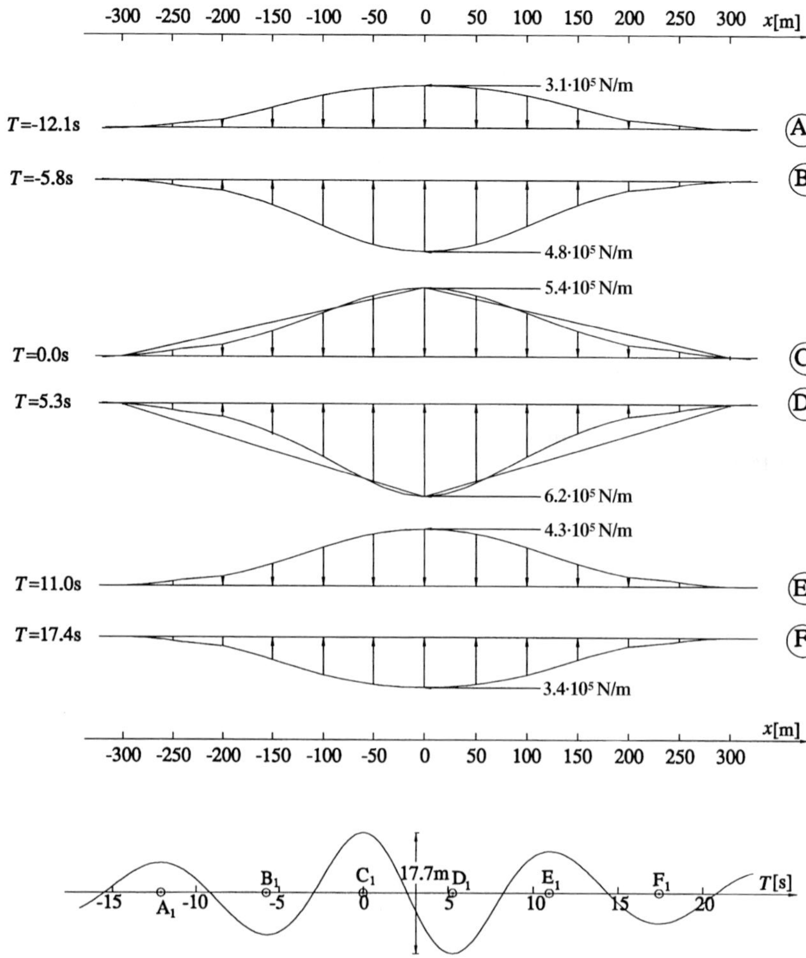

Fig. 12.12 Worked example: the local maxima (with respect to time) of the vertical wave load on the tunnel.

The asymmetry between positive (upward) and negative (downward) wave forces is apparent. It is due to the fact that $\Delta p''$ reduces the pressure, more on the upper half-cylinder than on the lower half-cylinder, so giving rise to a positive vertical force. This positive force, which is nearly constant during the wave cycle, is added to f_z' (the fundamental wave force) which is positive under the wave trough and negative under the wave crest.

The largest positive value (+) and the largest negative value (−) occur respectively beneath the trough and beneath the crest of the group's central wave. They are

$$\left(f_z^{(+)}\right)_{\max} = 6.2 \cdot 10^5 \text{ N/m}, \quad \left(f_z^{(-)}\right)_{\max} = 5.4 \cdot 10^5 \text{ N/m}.$$

Without the non-linear term we should obtain

$$\text{linear component: } \left(f_z'^{(+)}\right)_{\max} \equiv \left(f_z'^{(-)}\right)_{\max} = 5.8 \cdot 10^5 \text{ N/m}.$$

Hence, the non-linearity increases the vertical positive force of 7%, and lowers the vertical negative force of the same amount. They are variations nearly equivalent to those observed during the experiment RC 1993.

12.3.5 *What are the loads by the group's sides?*

The diagram of the largest wave load with a small conservative rectification is a triangle with a base of 600 m. Fig. 12.13 shows the bending moment produced by this wave load (assumed to be a static load), for the case that the diagram centre (centre of the wave crest) falls on one of the piers.

We note the bending moment approaches zero rapidly by the sides of the load triangle. This observation helps us to realize an important item. Let us look at it.

If an exceptionally high wave occurs at a point above the tunnel, the quasi-determinism theory enables us to foresee the configuration and evolution of the group including this wave. But the theory does not give any information on the waves surrounding this group. Now, let us think *ab absurdo* that by the side of our wave group there is a new group of the same height (we say *ab absurdo* because our group is that of the maximum expected wave height). Then, even if we assumed a full phase coherence between the two groups, the increase in the largest bending moment would only be of 2%.

Conclusion: the knowledge of the load exerted by a single wave group suffices to estimate the largest stress in the structure.

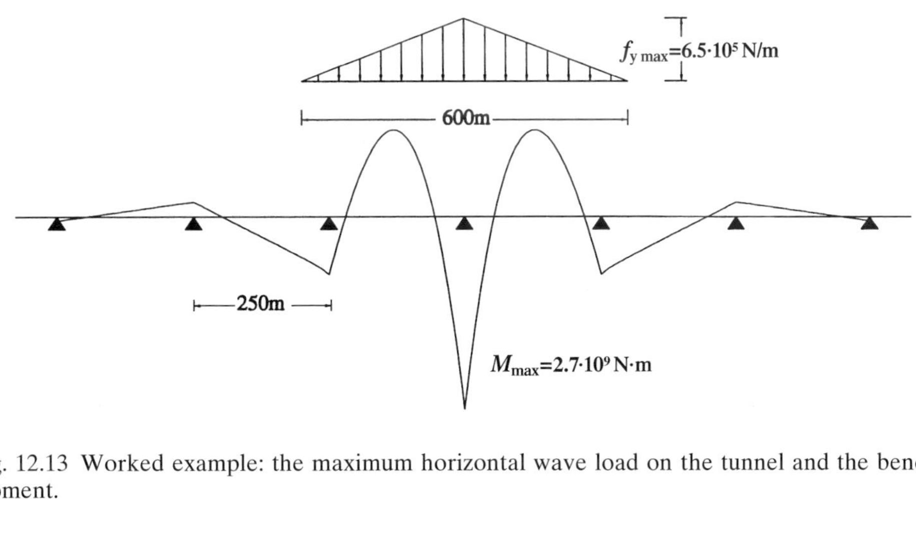

Fig. 12.13 Worked example: the maximum horizontal wave load on the tunnel and the bending moment.

12.3.6 *The bending moment and the forces on piers*

Fig. 12.14*a* shows the influence line of the bending moment (we mean the absolute value of the bending moment), and fig. 12.14*b* shows the influence line of the constraint's reaction. The influence line is of the same type both for the horizontal and for the vertical wave load. The numerical values given by the figure are those of the horizontal wave load. The abscissa is the location of the centre of the wave crest and the ordinate is the largest bending moment or constraint's reaction of the continuous beam. We see that the largest bending moment and constraint's reaction occur if the centre of the wave crest passes over one of the constraints (piers):

$$|M|_{\max} = 2.69 \cdot 10^9\,\mathrm{N \cdot m}, \quad R_{o\,\max} = 1.32 \cdot 10^8\,\mathrm{N},$$

where R_o denotes the horizontal reaction of the constraint. While, if the centre of the wave crest passes over the span's centre, we have

$$|M|_{\max} = 2.20 \cdot 10^9\,\mathrm{N \cdot m}, \quad R_{o\,\max} = 0.98 \cdot 10^8\,\mathrm{N}.$$

Given that the quotient between the vertical and the horizontal load is

$$\frac{\left(f_z^{(+)}\right)_{\max}}{|f_y|_{\max}} = 0.95, \qquad \frac{\left(f_z^{(-)}\right)_{\max}}{|f_y|_{\max}} = 0.83,$$

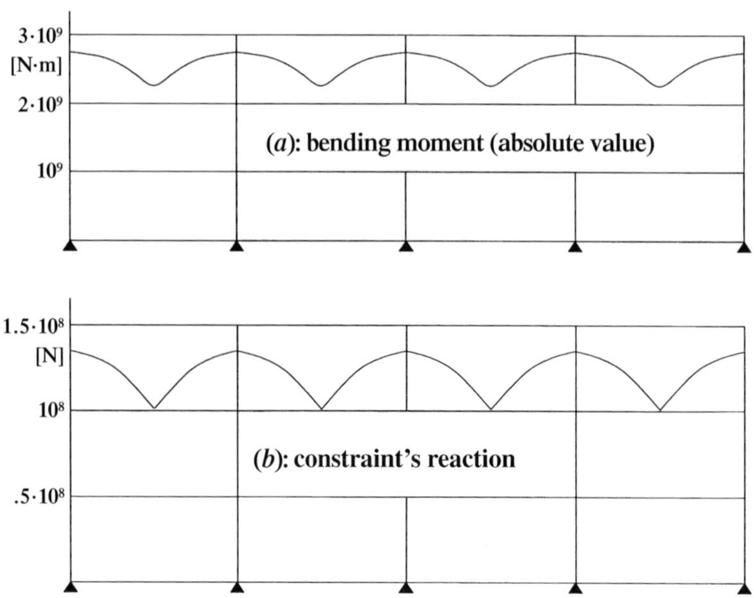

Fig. 12.14 Worked example: the influence lines of the bending moment and of the constraint's reaction. Abscissa: location of the centre of the wave crest; ordinate maximum moment (or maximum constraint's reaction) on the whole tunnel.

it follows that the bending moment and constraint's reaction due to the vertical positive load and vertical negative load are, respectively, 5% and 17% smaller than those given in fig. 12.14. The bending moment and constraint's reaction of the negative vertical load are increased because of the weight of freight cars and trucks (see fig. 12.15). The conclusion is that the largest bending moment is nearly constant whatever the neutral axis.

Note that, judging from fig. 12.11, one could think that the positive horizontal load is smaller than the negative horizontal load; but this is only a seeming asymmetry. It is due to the fact that we have assumed the wave of the maximum expected height ($H = 17.7\,\mathrm{m}$) to be a zero up-crossing. If we assume it to be a zero down-crossing (the probability being the same), we find that the positive horizontal load exceeds the negative horizontal load. Whereas, whether the highest wave is zero up-crossing or zero down-crossing, its positive vertical load exceeds the negative vertical load.

The maximum overturning moment on the pier is $1.32 \cdot 10^8 \cdot 82.5 = 1.09 \cdot 10^{10}\,\mathrm{N \cdot m}$ ($1.32 \cdot 10^8\,\mathrm{N}$ being the maximum horizontal force exerted by the tunnel and 82.5 m being the elevation of the tunnel's centre above the seabed). Then there is the wave force on the pier, which can be calculated in the same way as the wave force on the base of the gravity offshore platform. It is an additional horizontal force of $0.30 \cdot 10^8\,\mathrm{N}$ which yields an additional overturning moment of $0.09 \cdot 10^{10}\,\mathrm{N \cdot m}$. On the whole, the force exerted by the tunnel and the force exerted by the water flow yield

$$M = 1.18 \cdot 10^{10}\,\mathrm{N \cdot m}, \quad F = 1.62 \cdot 10^8\,\mathrm{N},$$

where F and M are, respectively, the horizontal force and the overturning moment on the pier.

The static analysis of the pier will be an useful exercise. To this end, consider a rectangular base equivalent to the actual base. The rectangular base with the same area and moment of inertia of the actual base has a width of 45.3 m and a height of 48.6 m. As for the

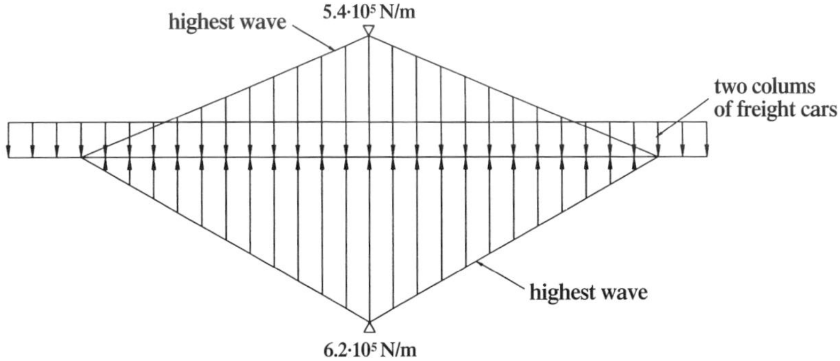

Fig. 12.15 Worked example: comparison of vertical loads on the tunnel.

weight, assume that the reservoirs are 2 m thick and are filled with water and/or sand at the designer's choice. For the general stability analysis, refer to sect. 13.3.1.

It will be found that the safety of the sea bottom foundation increases if the reservoirs are filled with sand. In examining this result, note that sand in place of water means a larger vertical load on the ground and a smaller eccentricity of this load. (Because of the larger load the bearing pressure tends to increase and, on the other hand, because of the smaller eccentricity, the bearing pressure tends to diminish.) Finally, it will be found that, with $K_\gamma = 20$ in equation (13.9) [see sect. 13.3.1], the safety factor against the soil failure will prove to be greater than 2. As for the rest, there will be no problem for the safety against sliding and overturning.

12.4 Design of a submerged tunnel. II: the effect of currents

The tunnel having to cross a strait will probably be subjected to some currents. Let us imagine an extremely strong current of as much as 5 m/s! This current would exert a force per unit length

$$|f_y| = 0.50 \cdot 1030 \cdot 12.5 \cdot 5^2 = 1.6 \cdot 10^5\ \text{N/m}$$

(12.5 m being the tunnel's radius and 0.50 the C_{dg} of a circular cylinder in the transcritical flow regime of a steady current). Even if such a strong current kept constant along the tunnel, the horizontal force exerted on the pier would be $0.40 \cdot 10^8$ N and the largest bending moment in the tunnel would be $0.83 \cdot 10^9$ N · m, against the $1.32 \cdot 10^8$ N of the force and the $2.69 \cdot 10^9$ N · m of the bending moment exerted by the wave. This means that even an exceptional current would load the structure less than the design wave. And as to being an exceptional current there is no doubt. It suffices to say that the waves of the design sea state would not be able to travel against such a current. Indeed the peak period (which is nearly the average wave period in a wave group) is 12.1 s that implies a c_0 of 18.9 m/s and hence a $|u_{\text{crit}}|$ of $18.9/4 = 4.7$ m/s [see sect. 1.9.3].

Of course it is difficult to imagine that the highest wave in the lifetime would occur jointly with the strongest current. A more realistic condition may be the design wave on a current of 2 m/s. Table 12.1 shows how a wave with a period of 12.1 s is transformed on a current of 2 m/s. The table gives the wavelength L_c obtained by means of (1.47), the quotient H/H_0 obtained through the procedure of sect. 2.10.9, and the quotient

$$\mathscr{R} \equiv \frac{\text{force due to a wave on the current}}{\text{force due to a wave without the current}},$$

that is

$$\mathscr{R} = \frac{g\dfrac{H}{2}\, k_c \cosh[k_c(d+z_C)]/\cosh(k_c d)}{g\dfrac{H_0}{2}\, k \cosh[k(d+z_C)]/\cosh(kd)},$$

where z_C is the depth of the cylinder's centre. The numerator on the r.h.s. of the last equation represents the particle acceleration a_y for a wave on the current, and the denominator represents the particle acceleration a_y for a wave without the current (we mean the acceleration at the depth of the cylinder centre). On account of the large d/L_0, the height of the wave without the current is H_0. As for the particle acceleration for the wave on the current, with the linear theory, is given by

$$a_y = \frac{\partial v_y}{\partial t} + \frac{\partial v_y}{\partial y} u,$$

where the two derivatives on the r.h.s. can be obtained from the formula (1.45) for the velocity potential.

Table 12.1 *The effect of a 2 m/s current (adverse or favourable)*

u[m/s]	L_c[m]	H/H_0	$\mathscr{R}$
−2	177	1.31	1.19
2	273	0.83	0.85

In practice, we can simply multiply by the constant factor $\mathscr{R}$ all the wave loads obtained in the foregoing section. [Really, the values of $\mathscr{R}$ in table 12.1 have been obtained for the horizontal wave loads, but nearly the same values also hold for the vertical wave loads.]

We also have to add the drag force

$$f_{dgy} = C_{dg}\, \rho\, R(u + v_{wy})|u + v_{wy}|,$$

where v_{wy} is the horizontal particle velocity due to the wave motion, which is given by (2.80*b*). As for the drag coefficient, with the current alone, we should have a C_{dg} of 0.5 (transcritical flow regime); while, with the waves alone, we should assume a C_{dg} of 0.62 due to the very large value of R_E. Anyway, the drag force plays only a secondary role, indeed at the time instant in which the inertia force attains its maximum of about $6 \cdot 10^5$ N/m, the drag force should be of about $0.3 \cdot 10^5$ N/m.

12.5 Design of a submerged tunnel. III: the risk of resonance

12.5.1 *The terms of the problem*

In sect. 12.3 we have evaluated the forces exerted by the highest waves. But even the small waves represent a potential danger for the tunnel. Indeed the tunnel, which consists of long beams, has a relatively small eigenfrequency that is equal (as we shall see later) to the peak frequency of waves with a significant height of about 0.15 m. Thus, at first glance, we could fear the occurrence of resonance. Really, such a risk does not exist, as will be made clear in the following analysis.

12.5.2 *The spectrum of the force-process*

Let us obtain the frequency spectrum of the horizontal force on the equivalent water cylinder. For simplicity, let us consider a square section.

The horizontal force per unit length is given by

$$f_y = \int_{z_C - b}^{z_C + b} \Delta p(-b, z) - \Delta p(b, z)\, \mathrm{d}z,$$

where b is half the side of the square [see fig. 12.16]. If the wave is periodic of amplitude a, this force becomes

$$\begin{aligned} f_y = {} & 2\rho g a \frac{1}{\cosh kd} \sin(kb\cos\theta)\frac{1}{k} \cdot \\ & \cdot \{\sinh[k(d + z_C + b)] - \sinh[k(d + z_C - b)]\} \sin(kx\sin\theta - \omega t + \varepsilon), \end{aligned}$$

where θ is the angle between the direction of wave advance and y-axis, and ε is an arbitrary phase angle. Hence, the horizontal force in a sea state is

$$\begin{aligned} f_y = {} & 2\rho g \sum_{i=1}^{N} a_i \frac{1}{\cosh k_i d} \sin(k_i b\cos\theta_i)\frac{1}{k_i} \cdot \\ & \cdot \{\sinh[k_i(d + z_C + b)] - \sinh[k_i(d + z_C - b)]\} \sin(k_i x\sin\theta_i - \omega_i t + \varepsilon_i) \end{aligned} \tag{12.4}$$

that, thanks to the assumptions on $a_i, \omega_i, \varepsilon_i$ and N [see sect. 5.1], represents a stationary Gaussian process at any fixed location x.

From the general definition of spectrum, the spectrum of process (12.4) is given by

$$\begin{aligned} E_o(\omega)\,\delta\omega = {} & 4\rho^2 g^2 \sum_i \frac{1}{2} a_i^2 \frac{1}{\cosh^2(k_i d)} \sin^2(k_i b\cos\theta_i)\frac{1}{k_i^2} \cdot \\ & \cdot \{\sinh[k_i(d + z_C + b)] - \sinh[k_i(d + z_C - b)]\}^2 \quad \text{for } i \text{ such that } \omega < \omega_i < \omega + \delta\omega, \end{aligned} \tag{12.5}$$

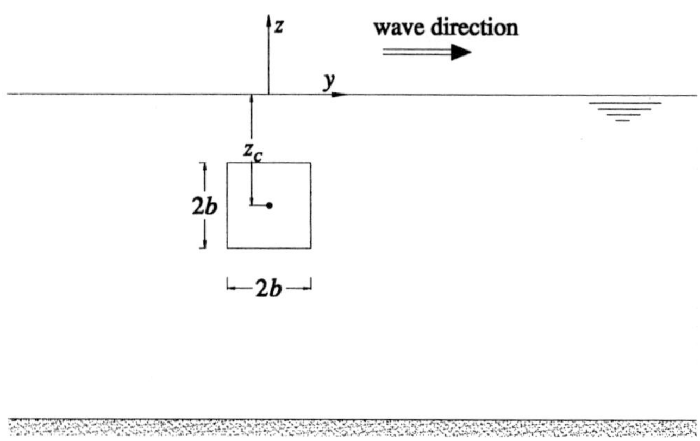

Fig. 12.16 Simple scheme used for obtaining the spectrum of the force process.

which implies

$$E_o(\omega) = 4\rho^2 g^2 \frac{1}{\cosh^2(kd)} \frac{1}{k^2} \{\sinh[k(d+z_C+b)] - \sinh[k(d+z_C-b)]\}^2 \cdot \\ \cdot \int_0^{2\pi} S(\omega,\theta)\sin^2(kb\cos\theta)\,\mathrm{d}\theta. \tag{12.6}$$

With the JONSWAP-Mitsuyasu directional spectrum, and the nondimensional frequency w, equation (12.6) becomes

$$E_o(w\omega_p) = 4A\rho^2 g^6 \omega_p^{-9} \frac{1}{\mathscr{k}^2(w)} \frac{1}{\cosh^2[2\pi\mathscr{k}(w)\,d/L_{p0}]} \{\sinh[2\pi\mathscr{k}(w)(d+z_C+b)/L_{p0}] + \\ -\sinh[2\pi\mathscr{k}(w)(d+z_C-b)/L_{p0}]\}^2 \int_0^{2\pi} \mathscr{S}(w,\theta)\sin^2\left[2\pi\mathscr{k}(w)\frac{b}{L_{p0}}\cos\theta\right]\mathrm{d}\theta. \tag{12.7}$$

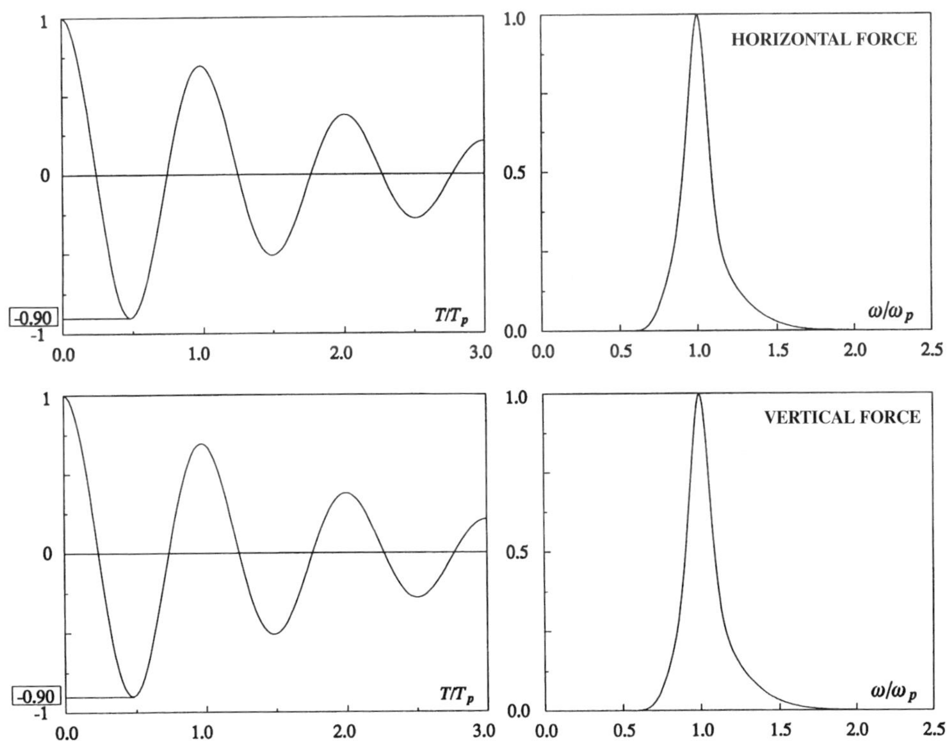

Fig. 12.17 Worked example: normalized autocovariance and spectrum of the force on the tunnel in the design sea state.

The same sequence is repeated to also get the spectrum of f_z. The result is

$$E_v(w\omega_p) = 4A\rho^2 g^6 \omega_p^{-9} \frac{1}{\mathscr{k}^2(w)} \frac{1}{\cosh^2[2\pi \mathscr{k}(w)d/L_{p0}]} \{\cosh[2\pi \mathscr{k}(w)(d+z_C+b)/L_{p0}] +$$

$$-\cosh[2\pi \mathscr{k}(w)(d+z_C-b)/L_{p0}]\}^2 \int_0^{2\pi} \mathscr{S}(w,\theta) \frac{1}{\cos^2\theta} \sin^2\left[2\pi \mathscr{k}(w) \frac{b}{L_{p0}} \cos\theta\right] d\theta \tag{12.8}$$

(for $\cos\theta = 0$ the integrand having to be substituted by its limit as $\cos\theta \to 0$).

The two spectra are shown in fig. 12.17 which is relevant to the design sea state: $H_s = 8.0\,\text{m}$, $T_p = 12.1\,\text{s}$. We see that these spectra are very narrow ($\psi^* = 0.9$) and their peak frequency is coincident with the peak frequency ω_p of the wave spectrum, like the spectra of the experiment RC 1993 (cf. fig. 11.15).

To prove the step from (12.5) to (12.6), start from the question "which is the contribution to the summation from the small wave components whose angle falls in a fixed small interval $\theta, \theta + \delta\theta$?".

12.5.3 *The force exerted by waves with a resonance frequency*

The tunnel is a continuous beam on a very large number of bearings, and its first eigenfrequency is

$$\omega_{\text{tunnel}} = \sqrt{92 \frac{EJ}{\frac{1}{2} m l^3}},$$

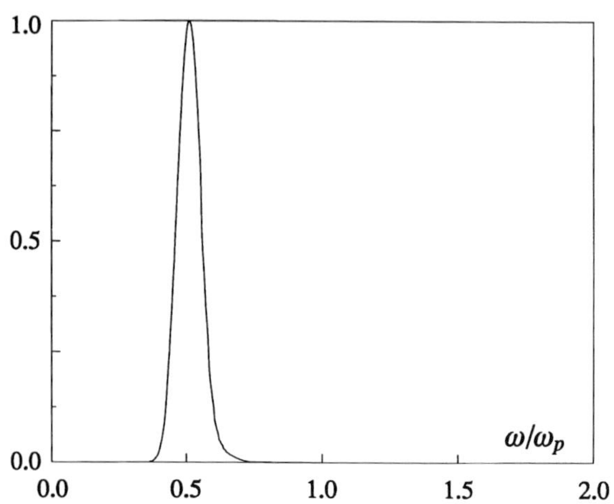

Fig. 12.18 Worked example: spectrum of the force exerted by small waves with a peak frequency ω_p equal to the first eigenfrequency of the tunnel. Note that the peak frequency of the force spectrum is nearly equal to half the peak frequency ω_p of the wave spectrum.

where E is Young's modulus of the reinforced concrete, J the moment of inertia, l the span length, and m the mass of a single span. The term $92EJ/l^3$ is the stiffness coefficient. With

$$E = 2.06 \cdot 10^{10}\,\mathrm{N/m^2}, \quad J = 9.63 \cdot 10^3\,\mathrm{m}^4, \quad m = 1.26 \cdot 10^8\,\mathrm{kg}, \quad l = 2.5 \cdot 10^2\,\mathrm{m},$$

it follows

$$\omega_{\mathrm{tunnel}} = 4.30\,\mathrm{rad/s} \Rightarrow T_{\mathrm{tunnel}} = 1.46\,\mathrm{s}.$$

What sea state has a T_p of 1.46 s? With (4.26) relating H_s to T_p, and a conservative value of A ($A = 0.02$), we get an H_s of 0.16 m.

The spectrum E_o (equation 12.7) is shown in fig. 12.18; and the root mean square f_y obtained from this spectrum is of only $6 \cdot 10^{-8}$ N/m. Moreover, we see that here the peak frequency of the force is nearly half the peak frequency ω_p of the waves. The fact is that a spectrum with a peak period of about 1.5 s, at the depth of the tunnel (more than 30 m) is subject to a huge attenuation, and the peak frequency of the force-spectrum proves to be smaller than the peak frequency of the waves, because the lower the frequency the smaller the attenuation.

The conclusion is that there is no appreciable wave force fluctuating with the eigenfrequency of the tunnel.

Chapter 13
STABILITY ANALYSIS OF COASTAL STRUCTURES

13.1 Wave pressure on a wall

13.1.1 *Difference between p_w and Δp*

Δp has been defined as the difference between the actual pressure at a point beneath the free surface and the term $-\rho g z$:

$$\Delta p \equiv p + \rho g z .$$

Hence Δp takes on the physical meaning of difference between actual pressure and pressure at rest provided that the point is beneath the free surface and provided that $z \leq 0$. In other words, Δp gives the difference between actual pressure and pressure at rest, at any time instant, only for the points which are *always* beneath the free surface.

To obtain the total thrust of wave pressure, we must also express the difference between actual pressure and pressure at rest above the mean water level ($z > 0$), and above the free surface (specifically, between the mean water level and the free surface when a wave trough is at the wall). It is thus convenient to define

$$p_w \equiv p - p_{st} ,$$

where p_w is associated with each point in space, and p_{st} is the pressure at rest, which is the hydrostatic pressure for $z \leq 0$, and is 0 (atmospheric pressure) for $z > 0$. Clearly, from the definitions, it follows that p_w and Δp are coincident with each other, at any time instant, at the points which always keep beneath the free surface:

$$p_w \equiv \Delta p \text{ for the points always beneath the free surface.}$$

p_w is the so called *wave pressure*, and its distribution on a vertical wall was already shown by fig. 1.16.

Lastly, let us consider the difference between p_w/γ_a and η_{ph}. Both represent the fluctuating pressure head, but η_{ph}, in this book, is associated only with the points which are always beneath the free surface, while p_w/γ_a is associated with any point in space.

13.1.2 *A marked asymmetry between crest and trough of pressure head waves at a vertical wall*

Fig. 13.1 shows a record of the pressure head wave p_w/γ_a at various depths of a vertical wall. The array of the gauges is shown by fig. 13.2. The record was taken in the course of the experiment RC 1994 with pure wind waves ($0.035 < H_s/L_{p0} < 0.040$). The dominant wave direction in this experiment was nearly orthogonal to the wall, and the water depth was such that

$$0.15 \leq \frac{d}{L_{p0}} \leq 0.20. \tag{13.1}$$

Looking at fig. 13.1 we note that the higher transducers measure some pressures different from zero only in the short intervals when the highest wave crests hit the wall. We see also that the transducers ④ and ⑤ remain uncovered when the deeper wave troughs occur at the wall, so that they record a constant pressure (the

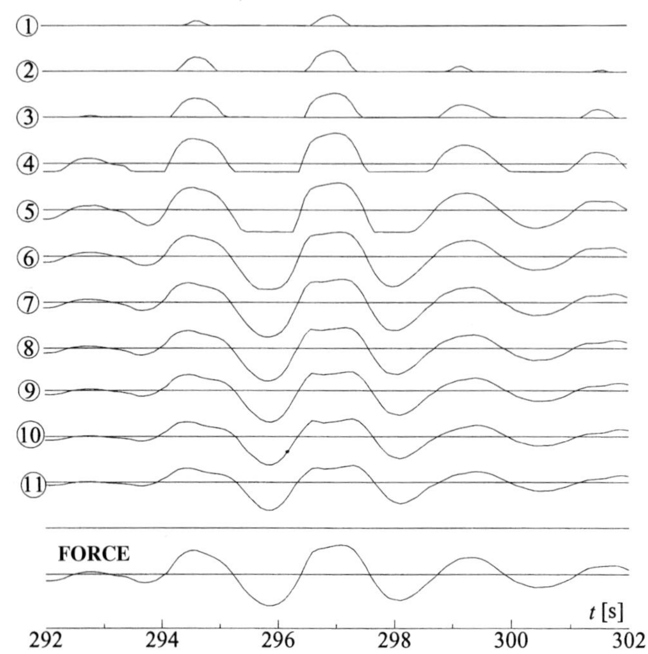

Fig. 13.1 Wave pressure recorded at various depths of an upright section (see fig. 13.2 for the gauge map).

atmospheric pressure) for a few time instants (also transducer ⑥ remains uncovered for a very short time, with the deepest wave trough).

Then a somewhat more deeper insight reveals a noticeable asymmetry between the crest and the trough of the highest pressure head wave: the trough depth markedly exceeds the crest elevation (we are referring to transducers always beneath the free surface). The pressure under the wave trough has a very regular behaviour whereas the pressure under the wave crest grows regularly until a certain instant, hence it decreases showing a local minimum just at the centre of the crest, and finally it retakes its regular way.

This is not a random phenomenon. On the contrary, this phenomenon occurs regularly in the highest waves of each sea state. We have already seen another example in fig. 10.28*b*. There, a wave group hit the wall when a wave was replacing

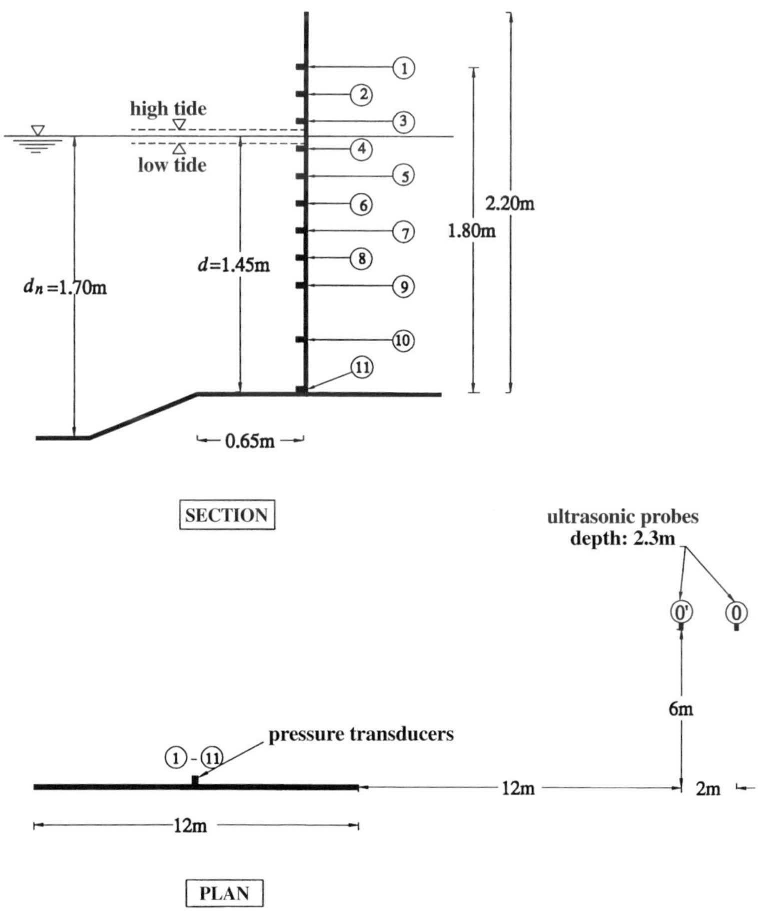

Fig. 13.2 The field laboratory of the experiment RC 1994.

the preceding one at the group's central position, i.e. when the group centre was occupied by two waves of nearly the same height. And indeed, in that case, the pressure drop at the crest centre was apparent in two consecutive waves.

The asymmetry trough-crest is an apparent effect of non-linearity, being very large since the wave height at the wall is redoubled. Specifically, the pressure drop at the crest centre is associated with the occurrence of some high water jets. These jets will have been noticed by anybody who has witnessed a severe storm. The jets observed in the course of the experiments RC 1991 and RC 1994 attained heights of the order of ten times H_s (an emblematic photo was shown by Boccotti et al., 1993).

And now look at the consequences of the aforesaid asymmetry on the probability of exceedance of the crest elevation and of the trough depth of the pressure head waves. Fig. 13.3 shows these two probabilities of exceedance for a point being always beneath the free surface. (Naturally, the crest elevation and the trough depth were divided by the standard deviation of the pressure head waves.) We see that at the probability threshold of 1:1000, the trough depth is 60% larger than the crest elevation!

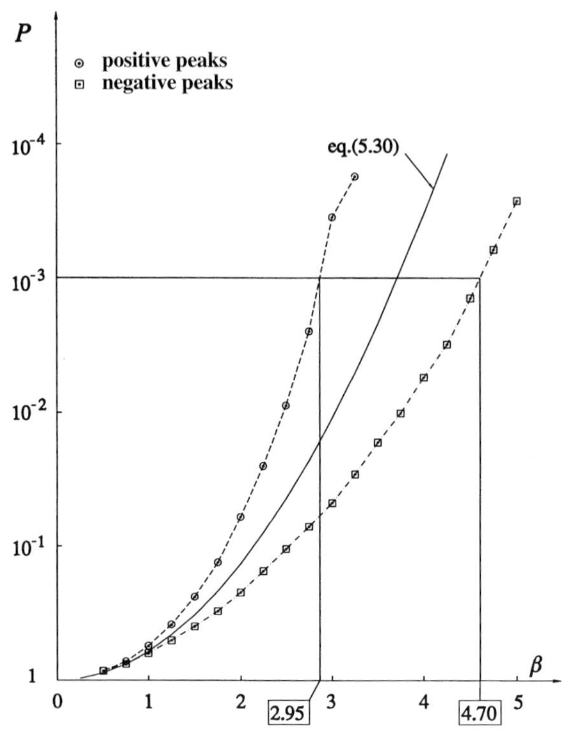

Fig. 13.3 Probability of exceedance of positive peaks (crests) and negative peaks (troughs) of $p_w(t)$ at a point of the vertical wall, always beneath the water surface.

13.1.3 *Pressure distribution on a vertical wall*

Fig. 13.1 also shows the force-process, that is the total thrust of wave pressure per unit extension of upright section, versus time. For each crest and for each trough of this force-process the distribution of wave pressure on the upright section was obtained. Moreover, for each experimental distribution a theoretical distribution was computed in the following way.

(*a*) *Input data*

(a.1) $p_w^{(+)} \equiv p_w$ measured at the lowest point of the front wall at the instant of the wave *crest*;

(a.2) $p_w^{(-)} \equiv |p_w|$ measured at the lowest point of the front wall at the instant of the wave *trough*;

(a.3) $T_h \equiv$ the high wave period [cf. sect. 5.7.2] obtained from the autocovariance of the incident waves (being recorded by the ultrasonic probes in the undisturbed wave field).

(*b*) *Computation of the theoretical wave pressure distribution: wave crest*

(b.1) The *virtual* wave height $H^{(+)}$ is evaluated by means of

$$H^{(+)} = \frac{1}{\gamma_a} p_w^{(+)} \cosh(kd)$$

based on the linear theory (with k being the wave number associated with period T_h and water depth d at the wall);

(b.2) the distribution of wave pressure is computed by means of the linear theory with wave height $H^{(+)}$ and period T_h, that is to say

$$p_w \begin{cases} = \gamma_a H^{(+)} \dfrac{\cosh(k\zeta)}{\cosh(kd)} & \text{if } 0 \le \zeta \le d, \\ = \gamma_a (d + H^{(+)} - \zeta) & \text{if } d \le \zeta \le d + H^{(+)}, \end{cases} \tag{13.2}$$

where ζ is the vertical coordinate with origin at the lowest point of the front wall.

(*c*) *Computation of the theoretical wave pressure distribution: wave trough*

(c.1) The *virtual* wave height $H^{(-)}$ is evaluated by means of

$$H^{(-)} = \frac{1}{\gamma_a} p_w^{(-)} \cosh(kd);$$

(c.2) the distribution of wave pressure is computed by means of the linear theory with wave height $H^{(-)}$ and period T_h, that is to say

$$p_w \begin{cases} = -\gamma_a H^{(-)} \dfrac{\cosh(k\zeta)}{\cosh(kd)} & \text{provided } p_w > -p_{st}, \\ = -p_{st} \ \text{otherwise}. \end{cases} \tag{13.3}$$

In summary, the theoretical distributions are obtained from the wave pressures measured at the lowest point of the wall. This is because *the wave height at the wall is indeterminate in that the waves give rise to vertical jets.*

Fig. 13.4 compares the theoretical distribution with the actual distribution of wave pressure on the wall. The dashed line is the average of the actual pressure distributions, and the continuous line is the average of the theoretical distributions. Specifically, fig.13.4*a* shows the average of the 1200 pressure distributions produced by the 1200 highest crests and by the 1200 deepest troughs of the force-process, in a total of 12000 waves of this process which were recorded in the course of RC 1994 (we mean the highest crests and the deepest troughs with respect to the standard deviation of the force-process). Thus (*a*) is the 1/10 pressure distribution, and, similarly, (*b*) and (*c*) are respectively the 1/100 and the 1/1000 pressure distributions. We see that the actual distributions are close to the theoretical distributions, and especially the actual pressure force is practically coincident with the theoretical pressure force.

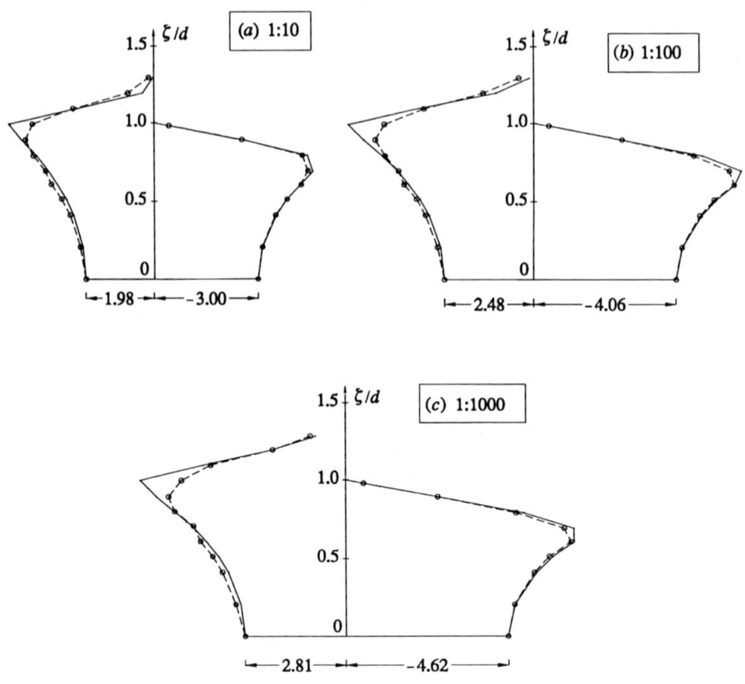

Fig. 13.4 (*a*) Average of the highest 1:10 wave pressure distributions of a sea state. (*b*) and (*c*) The same, respectively, for the highest 1:100 and 1:1000 pressure distributions. The continuous lines were obtained by means of the linear wave theory from the p_w measured at the lowest point of the front wall, and from the T_h of the incident waves. Reading example: 2.81 in (*c*) means that the wave pressure is 2.81 times the r.m.s. wave pressure at the lowest point of the front wall.

13.1.4 *The 1/1000 pressure distribution on a vertical wall*

The 1/1000 distribution can be assumed for design purposes, given that

(i) this distribution is the average of the highest 1/1000 pressure distributions,

(ii) the design sea state typically contains a few thousand waves.

The 1/1000 pressure distribution (fig. 13.4*c*) shows that the wave pressure at the lowest point of the front wall under the wave crest is 2.8, and is 4.6 under the wave trough. The scaling factor in these graphs is understood to be the standard deviation σ_{p_w} of $p_w(t)$ at the lowest point of the front wall; so that, $p_w = 2.8$ stands for $p_w = 2.8\,\sigma_{p_w}$.

Thus, for evaluating the 1/1000 distribution of the wave pressure on relatively deep walls like that of the experiment RC 1994, we shall use the linear wave theory with the wave period T_h and the water depth d at the wall, and with wave height

$$H^{(+)} = \frac{1}{\gamma_a} 2.8\,\sigma_{p_w} \cosh(kd),$$

for the positive wave pressure, and wave height

$$H^{(-)} = \frac{1}{\gamma_a} 4.6\,\sigma_{p_w} \cosh(kd),$$

for the negative wave pressure

The quotient σ_{p_w}/σ between the standard deviation of p_w at the lowest point of the front wall and the standard deviation of the surface displacement η of the incident waves can be obtained from (8.36*b*). The exact solution proves to be very close to the simple relation

$$\sigma_{p_w} \cong \frac{\gamma_a 2\sigma}{\cosh(kd)}$$

which taken together with the formulae of $H^{(+)}$ and $H^{(-)}$ yields

$$H^{(+)} = 5.6\sigma, \qquad (13.4a)$$

$$H^{(-)} = 9.2\sigma. \qquad (13.4b)$$

13.1.5 *Wave pressure on sloped seawalls*

At present the knowledge of wave pressure on sloped seawalls is effective only for the basic case of the periodic wave. In particular Neelamani et al. (1999) have investigated the action of two-dimensional periodic waves on smooth sloped seawalls, for a large variety of input conditions. What follows in this section is based on their work.

The flow field is thought of as the sum of the incident wave and of a reflected wave with a phase shift ε and a height $\widetilde{H}$ generally smaller than the height H of the incident wave. Thus the flow field far from the slope is given by

$$\eta(y,t) = \frac{H}{2}\cos(ky - \omega t) + \frac{\widetilde{H}}{2}\cos(ky + \omega t + \varepsilon), \tag{13.5a}$$

$$\phi(y,\zeta,t) = g\frac{H}{2}\omega^{-1}\frac{\cosh(k\zeta)}{\cosh(kd)}\sin(ky - \omega t) - g\frac{\widetilde{H}}{2}\omega^{-1}\frac{\cosh(k\zeta)}{\cosh(kd)}\sin(ky + \omega t + \varepsilon). \tag{13.5b}$$

The wave pressure formula which proceeds from (13.5b) seems to be effective at the slope, even if (13.5b) does not satisfy the slope boundary condition. According to this pressure formula, the maximum (positive) wave pressure at the slope is

$$p_w(\zeta) = \rho g\frac{H}{2}\frac{\cosh(k\zeta)}{\cosh(kd)}\sqrt{1 + (\widetilde{H}/H)^2 + 2(\widetilde{H}/H)\cos[2k\,\zeta\,\mathrm{cotg}\lambda + \varepsilon]}, \tag{13.6}$$

where k is the wave number on water depth d. Equation (13.6) is effective for $0 < \zeta < d - H/2$.

The phase shift is predicted through the method of Sutherland and O'Donoghue (1998):

$$\varepsilon = -8.84\pi\left(\frac{d\ \mathrm{cotg}\lambda}{\sqrt{gd}\ T}\right)^{1.22}.$$

The reflection coefficient $\widetilde{H}/H$ depends on the Irribarren number

$$\mathscr{T} \equiv \tan\lambda/\sqrt{H/L_0}.$$

Neelamani et al. (1999), basing themselves on the works of Ahrens et al. (1993) and Seeling and Ahrens (1995), use the following empirical formulae

$$\left(\frac{\widetilde{H}}{H}\right)_1 = \exp\left\{\frac{d\ \mathrm{cotg}\lambda}{L}\left[-0.686 - 3.37\frac{H}{L}\tanh^3(kd)\right]\right\} \quad \text{for } \mathscr{T} \geq 4.0,$$

$$\left(\frac{\widetilde{H}}{H}\right)_2 = 1 - \exp\left(-0.06\mathscr{T}^{2.4} - 0.5\frac{H}{d}\right) \quad \text{for } \mathscr{T} \leq 2.5.$$

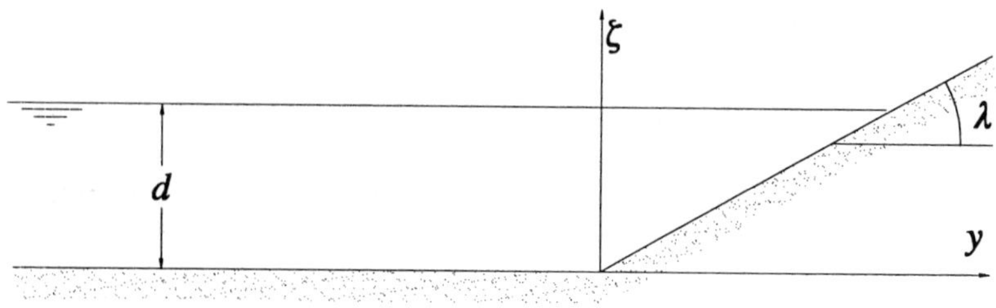

Fig. 13.5 Reference scheme for the problem of the sloped seawall.

Then a linear interpolation is used for $\mathscr{T}$ between 2.5 and 4.0:

$$\frac{\widetilde{H}}{H} = \left(\frac{\widetilde{H}}{H}\right)_1 \frac{\mathscr{T} - 2.5}{4.0 - 2.5} + \left(\frac{\widetilde{H}}{H}\right)_2 \frac{4 - \mathscr{T}}{4.0 - 2.5}.$$

For $\zeta > d - H/2$ the maximum wave pressures are stochastic even for the periodic waves, and their expected value is expressed by means of the following formulae:

$$p_w(\zeta) = \rho\, g\, H(0.42 + 0.15\mathscr{T}) \quad \text{for } d - H/2 < \zeta < d,$$

$$p_w(\zeta) = \rho\, g\, H[0.25 + 0.125\mathscr{T} - 0.25(\zeta - d)/H) \quad \text{for } d < \zeta < 1.62\, d,$$

which are effective for

$$3 < \text{cotg}\lambda < 6 \quad \text{and} \quad 0.5 < \mathscr{T} < 9.0.$$

Finally, the largest negative p_w for $0 < \zeta < d - R_D$ is opposite in sign to the largest positive p_w given by (13.6). As to R_D (the run-down) we can use some empirical formulae based on the works of Van der Meer and Breteler (1990) and of Schuttrumpf et al. (1994).

Remaining is the range of ζ between $d - R_D$ and d, where the largest negative p_w is evaluated by means of

$$p_w(\zeta) = 0.85\, \rho g(\zeta - d).$$

It should be noted that this prediction method provides the maximum wave pressure at every point of the slope, and the occurrences of these maxima are not contemporary.

13.2 Forces on a vertical breakwater

13.2.1 *The weight in still water*

On the left side of fig. 13.6 we see a breakwater's caisson (sketched as a rectangular parallelepiped) on its rubble mound; on the right side, we see the same caisson equally immersed in water (let us think of a big crane bearing the caisson in this position). In still water the vertical force exerted by the water on the left caisson is assumed to be equal to the vertical force exerted on the right caisson.

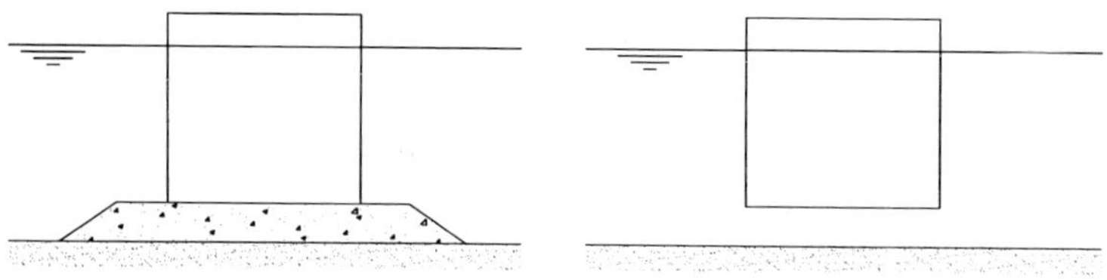

Fig. 13.6 The force exerted by water is the same on the two caissons.

This is because in the left configuration the water wets the whole of the caisson's base. In conclusion, the left caisson is subjected to the same buoyancy as the right caisson.

The resultant of P (weight) and S_{st} (buoyancy) is called the *weight in still water*:

$$P^* \equiv P - S_{st}.$$

This vertical force is computed with the specific weight in still water

$$\gamma_s^* \equiv \gamma_s - \gamma_a,$$

for the part of the structure beneath the mean water level.

13.2.2 *The uplift force, the horizontal force and the overturning moment*

The wave pressure at the tip O_2 of the base is $p_w^{(+)}$ under the wave crest and $p_w^{(-)}$ under the wave trough; while the wave pressure at the tip O_1 is zero, provided the

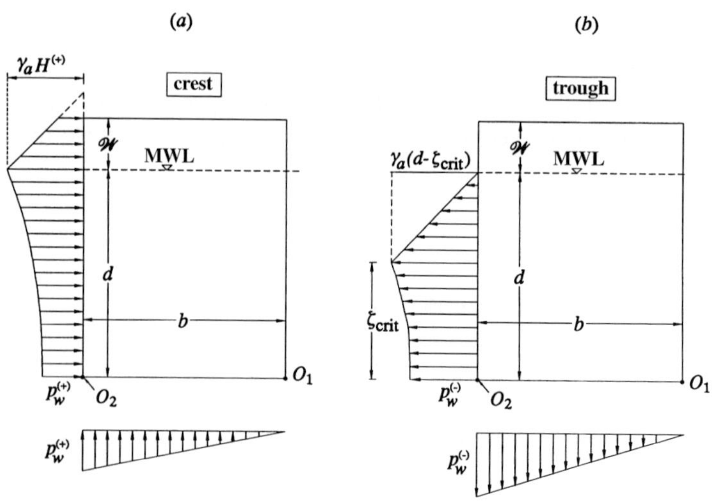

Fig. 13.7 The wave pressure on the wall and on the base: (*a*) under a wave crest; (*b*) under a wave trough.

water behind the breakwater is still. Usually, the wave pressure is assumed to vary linearly between the two tips of the base [see fig. 13.7].

The horizontal force (F) on the breakwater is equal to the wave force on the front wall. The wave force on the base is called *uplift force* S_w. Under the wave crest both F and S_w give a positive contribution to the overturning moment; while under the wave trough the contribution of F is positive and the contribution of S_w is negative.

The formulae for F, S_w and M (per unit length of upright section) with the distribution of wave pressure (13.2) and (13.3), are:

(*a*) *wave crest*

$$F = \gamma_a \frac{H^{(+)}}{k} \tanh(kd) + \begin{cases} +\dfrac{1}{2}\gamma_a H^{(+)^2} & \text{if } H^{(+)} \le \mathscr{W}, \\ +\gamma_a \left(H^{(+)}\mathscr{W} - \dfrac{1}{2}\mathscr{W}^2\right) & \text{if } H^{(+)} \ge \mathscr{W}, \end{cases} \tag{13.7a}$$

$$S_w = \frac{1}{2}\gamma_a \frac{H^{(+)}}{\cosh(kd)} b, \tag{13.7b}$$

$$M = \frac{1}{3}\gamma_a \frac{H^{(+)}}{\cosh(kd)} b^2 + \gamma_a \frac{H^{(+)}}{k^2} \frac{1 + kd\sinh(kd) - \cosh(kd)}{\cosh(kd)} +$$

$$+ \begin{cases} +\dfrac{1}{2}\gamma_a H^{(+)^2}\left(d + \dfrac{H^{(+)}}{3}\right) & \text{if } H^{(+)} \le \mathscr{W}, \\ +\gamma_a \left[(H^{(+)} - \mathscr{W})\mathscr{W}\left(d + \dfrac{1}{2}\mathscr{W}\right) + \dfrac{1}{2}\mathscr{W}^2\left(d + \dfrac{1}{3}\mathscr{W}\right)\right] & \text{if } H^{(+)} \ge \mathscr{W}; \end{cases} \tag{13.7c}$$

(*b*) *wave trough*

$$F = -\left[\gamma_a \frac{H^{(-)}}{k} \frac{\sinh(k\zeta_{crit})}{\cosh(kd)} + \frac{1}{2}\gamma_a (d - \zeta_{crit})^2\right], \tag{13.7d}$$

$$S_w = -\frac{1}{2}\gamma_a \frac{H^{(-)}}{\cosh(kd)} b, \tag{13.7e}$$

$$M = -\frac{1}{6}\gamma_a \frac{H^{(-)}}{\cosh(kd)} b^2 + \gamma_a \frac{H^{(-)}}{k^2} \frac{1 + k\zeta_{crit}\sinh(k\zeta_{crit}) - \cosh(k\zeta_{crit})}{\cosh(kd)} +$$

$$+ \frac{1}{2}\gamma_a (d - \zeta_{crit})^2 \left[\zeta_{crit} + \frac{1}{3}(d - \zeta_{crit})\right]. \tag{13.7f}$$

In these expressions $\mathscr{W}$ is the crest elevation of the breakwater, and b is the base width [see fig. 13.7]. As for ζ_{crit}, it is the elevation (above the base) where p_w

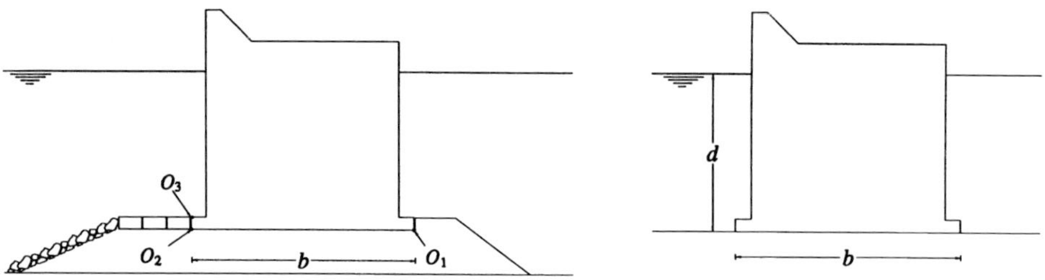

Fig. 13.8 Hint for the worked examples of sect. 13.3.2: compute the force on the caisson as if the foot-protections were not there.

becomes equal to $-p_{st}$, which proceeds from the numerical solution to the equation

$$\frac{H^{(-)}}{\cosh(kd)}\cosh(k\zeta_{\text{crit}}) = d - \zeta_{\text{crit}}.$$

For improving the stability against overturning and also for lowering the bearing pressures at the heel of the upright section, the base is usually widened with two short brackets (see fig. 13.8). As a consequence the vertical wave force consists not only of S_w but also of the wave pressure force on the upper face of the front bracket. This is typically a small additional force that can well be neglected. For example, we have obtained the safety factors of tables 13.1-2 (worked examples of sect. 13.3.2) neglecting the aforesaid additional force. If we were to count this additional force as well, all critical safety factors would be slightly raised (less than 1%).

Then there are the foot-protection blocks which also alter the wave force to a certain extent, in that they reduce the water depth. We should assume the actual water depth and compute the wave pressure on O_3-O_2-O_1 by linear interpolation between the two known values of p_w, the one at O_3 and the one at O_1 (that is zero). Here we suggest computing simply the wave pressure as if the foot-protection blocks were not there. Indeed, in so doing, we shall arrive nearly at the same final result. For the two worked examples of sect. 13.3.2, the use of one or the other of these two methods leads to differences within 1% on the forces and on the overturning moments.

13.3 Design of vertical breakwaters

13.3.1 *Stability analysis*

Fig. 13.9 shows the forces on the breakwater: the weight P^* in still water, the wave forces F and S_w, and the horizontal and vertical reactions R_o and R_v of the rubble mound. For the equilibrium, we have

$$|R_o| = |F|,$$

$$R_v = P^* - S_w,$$

where S_w is positive under the wave crest and negative under the wave trough. The position of R_v is such that the free body is in equilibrium:

$$M + R_v e_R = P^* e_P. \tag{13.8}$$

Under the wave crest the axis of rotation is O_1 so that $e_P = e_{P1}$ and $e_R = e_{R1}$; and under the wave trough the axis is O_2 so that $e_P = e_{P2}$ and $e_R = e_{R2}$.

A static analysis with the extreme wave load is usually done, at least for preliminary design purposes. [Then a more realistic evaluation of the response to the dynamic loading should be done for final design purposes (Oumeraci, 1994).] The stability is examined for three modes of failure.

Against sliding. We must verify that the safety factor

$$C_1 \equiv \frac{\mu R_v}{|R_o|} = \frac{\mu(P^* - S_w)}{|F|}$$

is greater than a dictated value $C_{1\min} > 1$. μR_v (friction coefficient × vertical reaction) is the limit shear force that can be developed at the base of the caisson. Thus, we check that the limit horizontal reaction (that the rubble mound can exert) is suitably greater than the actual horizontal reaction the rubble mound is expected to exert.

Against overturning. We must verify that the safety factor

$$C_2 \equiv \frac{P^* e_P}{M}$$

is greater than a dictated value $C_{2\min} > 1$. Indeed were $P^* e_P$ equal to M, then e_R would be zero [cf. (13.8)], and hence overturning would occur. Then, $C_{2\min}$ should be greater than $C_{1\min}$, given that C_1 can approach 1 without breakwater sliding, while a failure will certainly occur before C_2 approaches 1. This will be the collapse of the foundation, given that as C_2 approaches 1, e_R approaches zero and hence the bearing pressure at O_1 tends to infinity.

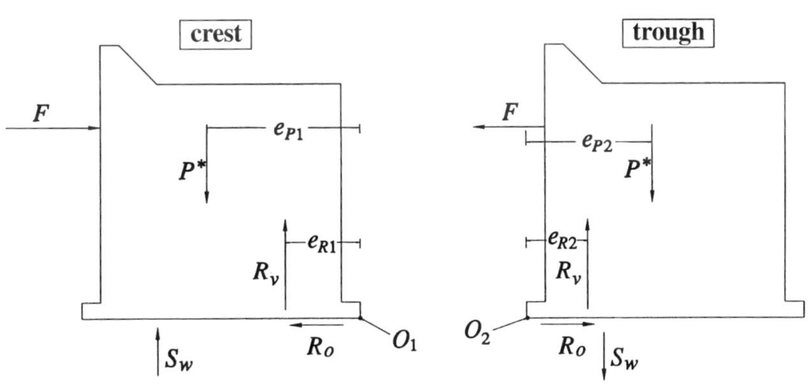

Fig. 13.9 Reference scheme for the stability analysis.

Against collapse of the foundation. A common procedure is to check that the largest toe pressure is within $4 \cdot 10^5 \div 5 \cdot 10^5$ N/m^2. Here following Meyerhof (1953) we shall verify that the safety factor

$$C_3 \equiv \frac{\frac{1}{2} K_\gamma \gamma^*_{so} 2e_R}{R_v/2e_R} \tag{13.9}$$

is greater than a dictated value $C_{3\,\mathrm{min}} > 1$. The numerator on the r.h.s. of this formula gives the bearing capacity and the denominator gives the bearing pressure. Here K_γ is a parameter that depends on the angle of internal friction of the soil and on the quotient R_o/R_v, and γ^*_{so} is the specific weight in still water of the soil.

The verifications of the stability against overturning and collapse of the foundation really aim to check the stability against one mode of failure: under the wave force, e_R is reduced so that the bearing pressure at the heel of upright section becomes too large, the soil collapses and the structure tilts and sinks into the ground.

13.3.2 *Worked examples*

A traditional breakwater

Design a vertical breakwater for
(a) H_s of the design sea state: 8 m,
(b) depth of caisson's base: 20 m,
(c) crest elevation of the breakwater: 8 m.
Take
specific weight: reinforced concrete $2.45 \cdot 10^4$ N/m^3, sand $1.82 \cdot 10^4$ N/m^3,
factors formula (13.9): $K_\gamma = 20$, $\gamma^*_{so} = 8 \cdot 10^3$ N/m^3,
friction coefficient: $\mu = 0.6$,
minimum safety factor: $C_{1\,\mathrm{min}} = 1.3$, $C_{2\,\mathrm{min}} = 1.5$, $C_{3\,\mathrm{min}} = 2.0$.

The wave period T_h is computed by means of (4.27) and (5.42):

$$T_h = 0.92 \cdot 8.5\pi \sqrt{\frac{8}{4 \cdot 9.8}} = 11.1\,\mathrm{s}.$$

The virtual wave heights $H^{(+)}$ and $H^{(-)}$ given by (13.4a-b) are

$$H^{(+)} = 5.6 \cdot \frac{8}{4} = 11.2\,\mathrm{m},$$

$$H^{(-)} = 9.2 \cdot \frac{8}{4} = 18.4\,\mathrm{m}.$$

We fix a tentative width b (this being necessary for computing the uplift force S_w and the overturning moment M); we compute forces and moments of the forces under the wave crest and the wave trough by means of (13.7a-f); then we evaluate the minimum P^* for the safety against sliding; and, from this P^*, we obtain the minimum e_{P1} and e_{P2} for the safety against overturning and collapse of the foundation.

In particular with

$$b = 24\,\mathrm{m},$$

we obtain

$$P^* = 6.09 \cdot 10^6\,\mathrm{N/m}, \;\; e_{P1} = 12.62\,\mathrm{m}, \;\; e_{P2} = 8.75\,\mathrm{m}.$$

The breakwater of fig. 13.10*a* has these values of P^* and e_{P1} (which are the least values to comply with the prescribed safety factors), but it has a larger e_{P2}. Thus we could reduce somewhat the caisson width provided we designed a higher superstructure and/or a wider trapezoidal wall so as to keep the minimum P^* and e_{P1}.

A new type of breakwater

The one we have just seen is a breakwater on a typical depth for ports. Now let us pass to a breakwater at a water depth twice as large. All data remain unchanged apart from the depth of the caisson's base that passes from 20 m to 40 m. This is not an abstract idea, it is only necessary to consider the Sines breakwater in Portugal which was built on a water depth of about 50 m. However, that type was completely different; it was a rubble mound breakwater, whereas here we are dealing with a caisson breakwater.

A solution is given in fig. 13.10*b*. The weight P^* is the minimum to ensure the safety against sliding, and e_{P1} and e_{P2} have the least values for the safety against overturning and collapse of the foundation. We see that the breakwater on 40 m water depth, unlike the one on 20 m water depth, can fulfil the condition of minimum weight and minimum width with the cells only partially filled with sand and without a superstructure above the mean water level (apart from the wall for limiting the overtopping discharge).

The solution of fig. 13.10*b* should be effective for three reasons:

(i) the double function: the caisson can also be used as a reservoir for oil storage with a capacity of nearly 500 m^3/m;

(ii) the operations at sea being reduced to a minimum: there is not a heavy superstructure to be built above the mean water level;

(iii) the flexibility: to move the caisson and use it again elsewhere, it is not necessary to pull down the superstructure, but it is only necessary to take out the sand from the cells; moreover, if in the future somebody wishes to adjust the safety factors, they will simply have to change the sand content.

The caisson works like the reservoirs at the base of the gravity offshore platform of fig. 12.1. Pumping oil into the reservoir, the water flows out through a lateral hole; pumping oil from the reservoir, the water flows in. Thus the reservoir is always full of some liquid that may be nearly all oil, or in part oil and in part water, or all water.

Safety factors of the two breakwaters

The results of the verifications are given in tables 13.1 and 13.2. Looking at the safety factors, we realize that the design of the traditional breakwater of fig.13.10*a* is conditioned only by the wave crest, while the design of the new big breakwater of fig.13.10*b* is conditioned both by the wave crest and by the wave trough.

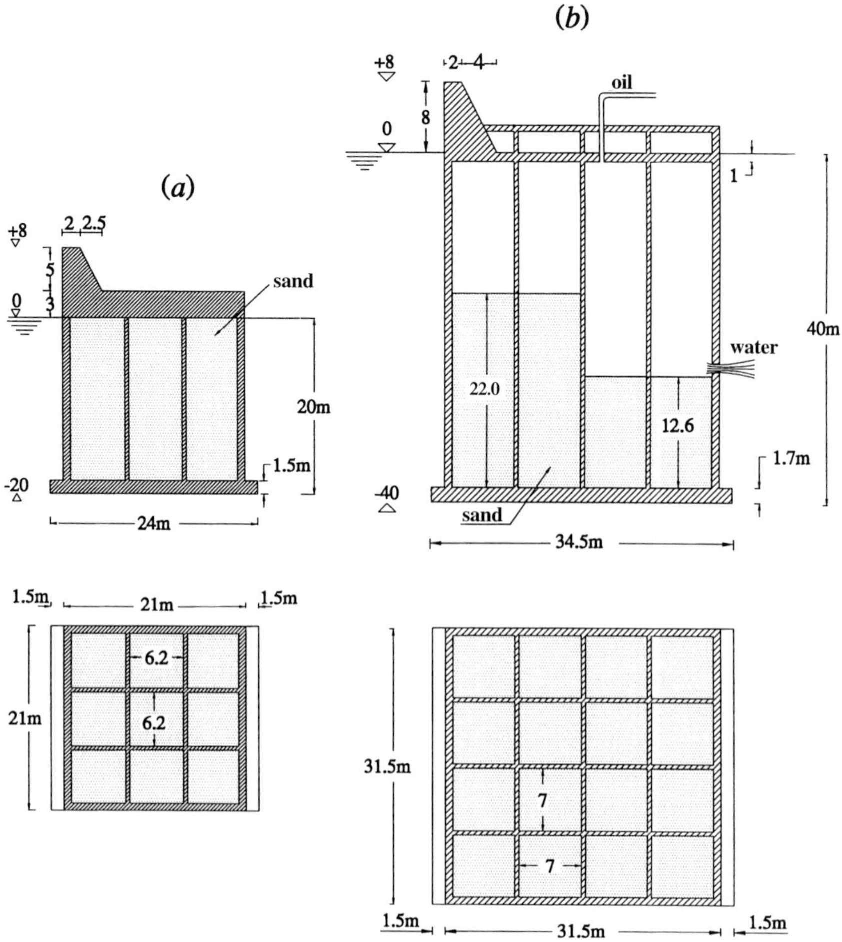

Fig. 13.10 Worked example: (*a*) traditional breakwater; (*b*) deep breakwater used also as oil reservoir.

Three final comments. First: one should start from H_{s0} (the significant wave height on deep water), take a characteristic spectrum, evaluate how this spectrum changes on the depth of the breakwater (using the equations of sect. 8.4.3), and finally compute H_s and T_h of this shallow water spectrum. This procedure has been simplified in the two worked examples in order to concentrate our attention on the verifications.

Second comment: the depth of the big breakwater (fig. 13.10*b*) is within the range covered by the experiment RC 1994 while in the case of the traditional breakwater (fig. 13.10*a*) we have a d/L_{p0} of 0.09 which is smaller than the d/L_{p0} covered by RC 1994. That is, for a preliminary analysis, we have assumed that the pressure distribution given in sect. 13.1.4 is valid also outside the range (13.1) covered by the experiment RC 1994. Of course, some new small scale field experiments should be performed so as to cover a wider range of d/L_{p0}, and also to check the results of RC 1994.

Third comment: a breakwater is usually called *vertical* if the height of the rubble mound is smaller than $0.3d_n$ (d_n being the depth of the natural bottom, see fig. 13.12). According to the conclusion of Kortenhaus and Oumeraci (1998), based on laboratory data, the breakwater of fig. 13.10*a*, with the H_s of 8 m, could be subjected to impulsive loads of high intensity and very short duration if the height of the rubble mound was smaller than about 4 m.

The stability of the two breakwaters will be re-examined in sect. 13.5.3 according to the Japanese practice.

Table 13.1 *Verification of the breakwater on 20 m water depth*

	WAVE LOAD			SAFETY FACTOR		
	F [N/m]	S_w [N/m]	M [N · m/m]	C_1	C_2	C_3
crest	$2.37 \cdot 10^6$	$0.94 \cdot 10^6$	$4.77 \cdot 10^7$	1.31	1.62	1.98
trough	$-1.77 \cdot 10^6$	$-1.55 \cdot 10^6$	$0.05 \cdot 10^7$	2.60	128	3.42

Table 13.2 *Verification of the breakwater on 40 m water depth*

	F [N/m]	S_w [N/m]	M [N · m/m]
crest	$3.36 \cdot 10^6$	$0.86 \cdot 10^6$	$10.88 \cdot 10^7$
trough	$-3.44 \cdot 10^6$	$-1.41 \cdot 10^6$	$4.80 \cdot 10^7$

	RESERVOIRS FILLED WITH OIL (*)			RESERVOIRS FILLED WITH WATER		
	C_1	C_2	C_3	C_1	C_2	C_3
crest	1.33	1.50	2.28	1.47	1.61	2.52
trough	1.70	2.58	2.00	1.84	2.90	2.28

(*) Specific weight of oil: $8.5 \cdot 10^3$ N/m^3.

13.4 Further verifications of the vertical breakwaters

13.4.1 *Second verification of the stability against the three principal modes of failure*

The wave force used in the foregoing section is $F_{1/1000}$, i.e. the average of the $N/1000$ highest peaks of the force per unit length at any fixed upright section (N being the number of waves of the sea state). With the typical durations of the design sea states, this average value proves to be very close to the maximum expected wave force per unit length at any *fixed section.*

Due to the relatively small length of each single caisson, the stability must be examined with the maximum expected wave force per unit length *on the whole breakwater*. From what we have seen in sect. 11.6, this force is certainly greater than the maximum expected wave force at any fixed upright section.

Until now, the distinction between the maximum expected wave force per unit length on the whole breakwater and the maximum expected wave force per unit length at any fixed upright section has not been done. This is a non-conservative approach that is counterbalanced to some extent by assuming the maximum expected wave force to be constant in time (static analysis). [The effects of the short duration of wave loads are pointed out, in particular, by Ling et al. (1999).]

The most practical way seems to be to continue with the traditional approach, and in addition to check that C_1 and C_3 are greater than 1 and that C_2 is greater than at least 1.1, even with the maximum expected wave force per unit length on the whole breakwater. (The grounds to take $C_{2\,\min} > C_{1\,\min}$ having been explained in sect. 13.3.1.)

Let us see what changes to the breakwater of fig. 13.10*a* are required owing to this additional verification. We assume a realistic duration of the design sea state, say $\mathscr{D} = 6$ hours (see sect. 7.5). The significant wave height was of 8 m, so that through (5.26) we obtain

$$\overline{T} = 6.6 \cdot \pi \sqrt{\frac{8}{4 \cdot 9.8}} = 9.4\,\mathrm{s},$$

and consequently the number of waves of the design sea state is

$$N = \frac{6 \cdot 3600}{9.4} = 2300.$$

The maximum expected wave height in this design sea state at any fixed point is obtained by means of (5.57) and proves to be

$$\overline{H_{\max}}\ \text{(fixed section)} = 1.890h = 15.1\,\mathrm{m}.$$

According to the conclusions of sect. 11.6.3, the maximum expected wave height on the whole breakwater is computed by means of (5.57) with a fictitious mean wave period equal to $\overline{T}/n$ with n given by (11.12). In our case we have $L_{p0}/3 = 76\,\mathrm{m}$, so that

$$n = 1 + l/76 \quad \text{with } l \text{ in metres,}$$

where l is the total length of the breakwater. Table 13.3 gives $\overline{H_{max}}$ on the whole breakwater for two distinct lengths l ($\mathscr{R}$, in the third column, is the ratio between $\overline{H_{max}}$ on the whole breakwater and $\overline{H_{max}}$ at any fixed section).

Table 13.3 *Maximum expected wave height on the whole breakwater*

l	$\overline{H_{max}}$ on the whole breakwater	$\mathscr{R}$
1000 m	$2.17\ h = 17.4$ m	1.15
5000 m	$2.32\ h = 18.6$ m	1.23

The maximum expected wave force per unit length on the whole breakwater can be computed by multiplying the virtual wave heights $H^{(+)}$ and $H^{(-)}$ by the factor $\mathscr{R}$. Using the formulae (13.7*a-c*) with the *virtual* wave height $\mathscr{R}H^{(+)}$ in place of $H^{(+)}$, we obtain the extreme wave forces and the safety factors given in table 13.4.

The table illustrates only the situation under the wave crest, which is markedly the more critical. We see that with a total length of 1000 m the safety factors C_1 and C_3 are still greater than 1 and the safety factor C_2 is greater than 1.1, so that the breakwater of fig. 13.10*a* gets through the second verification. However, with a total length of 5000 m, C_2 keeps above 1.1, C_1 is 1 and C_3 falls below 1. Accordingly, with the breakwater 5000 m long, we should adjust the breakwater size. In practice, it would suffice to widen the side of the cell from 6.2 m to 6.5 m, with a consequent increase of 0.9 m in the breakwater width.

Table 13.4 *Second verification of the breakwater on 20 m water depth*

Wave load (crest)				Safety factor		
l [m]	F [N/m]	S_w [N/m]	M [N · m/m]	C_1	C_2	C_3
1000	$2.78 \cdot 10^6$	$1.08 \cdot 10^6$	$5.61 \cdot 10^7$	1.09	1.38	1.10
5000	$3.00 \cdot 10^6$	$1.16 \cdot 10^6$	$6.05 \cdot 10^7$	1.00	1.27	0.72

Conclusion: the 5000 m long breakwater must be somewhat wider than the 1000 m long breakwater. This is logical, if we think that the longer the breakwater, the greater the probability that during its lifetime it is hit by some group of exceptionally high waves. In simple terms a long breakwater is like a long battleship, it has a greater probability to be hit.

13.4.2 *Analysis of circular slips passing through the rubble mound and the foundation*

To complete the stability analysis, we must evaluate the safety factor

$$C_4 \equiv \frac{M_{RES}}{M_{SL}},$$

for a number of circular slips like the one of fig. 13.11 (varying the position of the centre and the radius). Here M_{RES} is the limit counterclockwise moment of the forces (resisting moment) with respect to the slip centre and M_{SL} is the actual clockwise moment. The limit counterclockwise moment is due to the limit shear stresses that can be developed along the slip and to the weight P_1^* of the ground to the left of the slip centre. The actual clockwise moment is due to the weight P_2^* of the ground to the right of the slip centre, and to the vertical forces $-R_v$ and $-R_o$ exerted by the caisson under the wave action. Of course also the safety factor C_4 must be greater than a prescribed minimum $C_{4\min} > 1$.

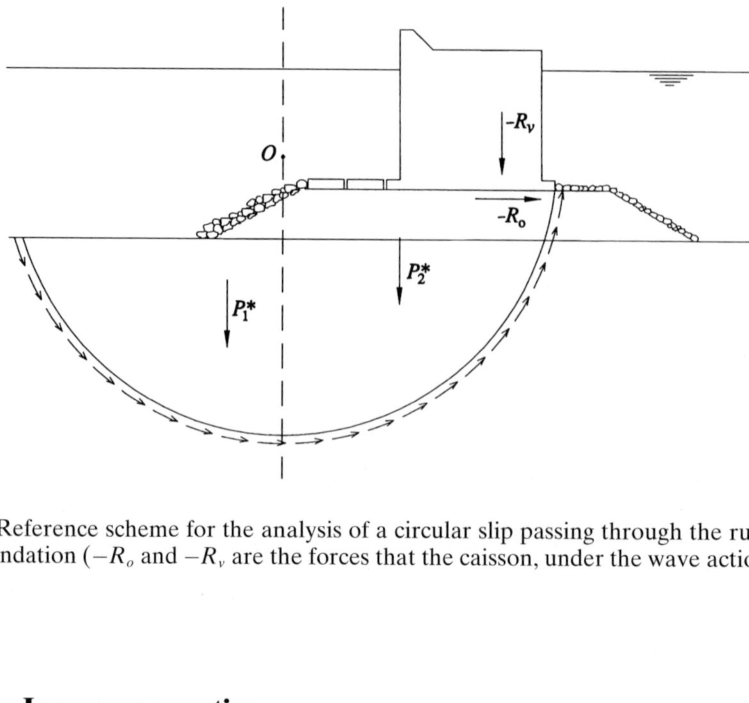

Fig. 13.11 Reference scheme for the analysis of a circular slip passing through the rubble mound and the foundation ($-R_o$ and $-R_v$ are the forces that the caisson, under the wave action, exerts on the soil).

13.5 The Japanese practice

13.5.1 *Goda's formula*

The Japanese practice, being followed in many countries, is based on the distribution of wave pressure of fig. 13.12, given by Goda's formula:

$$\eta_{\max} = \frac{3}{4}(1+\cos\theta)H, \tag{13.10a}$$

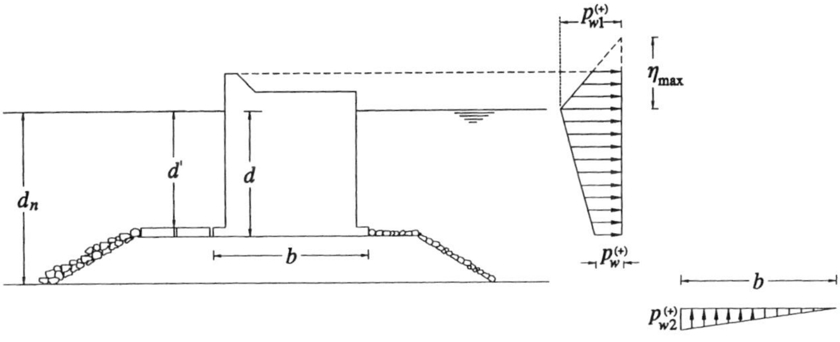

Fig. 13.12 Reference scheme for Goda's formula.

$$p_{w1}^{(+)} = \frac{1}{2}(1+\cos\theta)(\alpha' + \alpha''\cos^2\theta)\gamma_a H, \tag{13.10b}$$

$$p_w^{(+)} = \alpha''' p_{w1}^{(+)}, \tag{13.10c}$$

$$p_{w2}^{(+)} = \frac{1}{2}(1+\cos\theta)\alpha'\alpha'''\gamma_a H, \tag{13.10d}$$

where

$$\alpha' = 0.6 + \frac{1}{2}\left[\frac{2kd_n}{\sinh(2kd_n)}\right]^2, \quad \alpha'' = \mathrm{Min}\left[\left(\frac{d'_n - d'}{3d'_n}\right)\left(\frac{H}{d'}\right)^2, 2\frac{d'}{H}\right],$$

$$\alpha''' = 1 - \frac{d}{d_n}\left[1 - \frac{1}{\cosh(kd_n)}\right].$$

The meaning of most symbols is illustrated in fig. 13.12. Then d'_n denotes the water depth at a distance of $5H_s$ seaward of the breakwater, and θ is the angle between the direction of wave approach and a line normal to the breakwater. (It is recommended to rotate the wave direction from the actual direction by an amount as much as 15° toward the line orthogonal to the breakwater.) According to Tanimoto and Takahashi (1994), for some composite breakwaters (i.e. breakwaters with $d'/d_n < 0.7$), factor α'' must be increased to take account of the impulsive pressure by breaking waves (the possible increase being dependent on the mound shape).

It is suggested to assume $T = T_{1/3}$ (the average period of the highest one-third of all waves of the design sea state). This period proves to be very close to T_h. For example, in the records obtained during the experiment RC 1994 the average difference between $T_{1/3}$ and T_h was only 0.9%, with a peak of 3.5% in a single sea state.

As for the wave height, Goda (1974) gives a set of formulae. If the energy losses are negligible, we have

$$H = 1.8H_s, \tag{13.11}$$

or else some smaller wave height is obtained. After some work on the formulae of Goda, we arrive at the following conclusion: the condition

$$\frac{d_n}{H_s} \geq 2.5 \quad \text{and} \quad \frac{d_n}{L_0} \geq 0.12, \tag{13.12}$$

where L_0 is the deep water wavelength associated with $T_{1/3}$, is sufficient to ensure that the effects of the energy losses are negligible. Condition (13.12) is often fulfilled by the chief breakwaters.

13.5.2 *Stability analysis*

The trend in Japan (Goda, 1992) is to make analysis of circular slips like the one in fig. 13.11 and to verify the breakwater against sliding and against overturning, with $C_{1\min} = C_{2\min} = 1.2$ and $\mu = 0.6$. The safety factor against overturning is defined by

$$C_2 \equiv \frac{P^* e_P - M''}{M'}, \tag{13.13}$$

where M' is the overturning moment due to the horizontal wave force and M'' is the overturning moment due to the uplift force; whereas, the definition we have followed gives

$$C_2 \equiv \frac{P^* e_P}{M' + M''} \tag{13.14}$$

which proves to be always smaller than the C_2 obtained by means of (13.13).

To prove that C_2 (13.13) is greater than C_2 (13.14), that is to say, to prove that

$$\frac{P^* e_P - M''}{M'} > \frac{P^* e_P}{M' + M''},$$

it is convenient to rewrite this inequality in the form

$$\frac{P^* e_P}{M' + M''} + \frac{P^* e_P}{M' + M''} \frac{M''}{M'} - \frac{M''}{M'} > \frac{P^* e_P}{M' + M''}.$$

Hence, the inequality to be proved reduces itself to

$$\frac{P^* e_P}{M' + M''} \frac{M''}{M'} > \frac{M''}{M'}$$

which is satisfied, given that $P^* e_P$, M' and M'' are positive, and $P^* e_P$ necessarily must be greater than $M' + M''$, otherwise the structure would be unstable.

13.5.3 *Re-examination of the breakwaters of fig. 13.10*

Computing the wave forces on the breakwater of fig. 13.10*a* by means of Goda's formula, we obtain

$$F = 2.73 \cdot 10^6\ \text{N/m}, \quad S_w = 0.91 \cdot 10^6\ \text{N/m}, \quad M = 5.35 \cdot 10^7\ \text{N} \cdot \text{m/m}.$$

For the calculation, besides the orthogonal wave attack ($\theta = 0$), we have fixed an elevation of the rubble mound of 5 m, and a slope of the seabed of 3%, so that

$$d_n = 23.5\,\text{m}, \quad d = 20\,\text{m}, \quad d' = 18.5\,\text{m}, \quad d'_n = 24.7\,\text{m}.$$

However, the result is affected only slightly by the elevation of the rubble mound or by the bottom slope. Indeed doubling the elevation of the rubble mound (from 5 m to 10 m) or varying widely the bottom slope, say from 3% to 10%, we find that the wave force and the overturning moment only vary 1÷2%.

We see that the wave force and the overturning moment are nearly 15% larger than the ones ($F = 2.37 \cdot 10^6\,\text{N/m}$, $M = 4.77 \cdot 10^7\,\text{N} \cdot \text{m/m}$) obtained in sect. 13.3.2. With this new wave load, the safety factors become

$$C_1 = 1.14, \quad C_2 = 1.44,$$

and hence we should widen the upright section and/or increase the thickness of the concrete superstructure.

Let us pass to the big breakwater of 40 m water depth (fig. 13.10*b*). Using Goda's formula, with an elevation of the rubble mound of 5 m, we obtain

$$F = 3.32 \cdot 10^6\,\text{N/m}, \quad S_w = 0.73 \cdot 10^6\,\text{N/m}, \quad M = 10.45 \cdot 10^7\,\text{N} \cdot \text{m/m}.$$

(Here too, the size of the rubble mound has a negligible effect on the result, indeed raising the rubble mound from 5 m to 10 m, the wave force and the overturning moment vary less than 0.5%.) This time, we see the wave force and the overturning moment to be close to the ones ($F = 3.36 \cdot 10^6\,\text{N/m}$, $M = 10.88 \cdot 10^7\,\text{N} \cdot \text{m/m}$) obtained in sect. 13.3.2.

Finally, for the best distribution of the masses (sand, oil or water) inside the big caisson, Goda's formula is not helpful. Indeed, the action of the wave trough, not being covered by this formula, becomes essential.

13.6 The problem of the rubble mound breakwaters

13.6.1 *The base given by Hudson*

Fig. 13.13 shows the cross-section of a rubble mound breakwater. The design of these structures is markedly the more empirical in ocean engineering, and the reason for this is clear. It suffices to note that they are structures with a very complex shape which withstand three-dimensional breaking waves.

The verification of the stability of these structures lies basically on laboratory tests, that we shall deal with in sect. 14.2.4. A preliminary design of the crown wall can be done with the verification method of sect. 13.3.1 and the pressure distributions of Martin et al. (1999) based on laboratory measurements with periodic waves (these distributions being related to wave size, breakwater geometry, and armour layer porosity). As for the armour, here below we deal with the decision on the size of the concrete armour units. These units with some

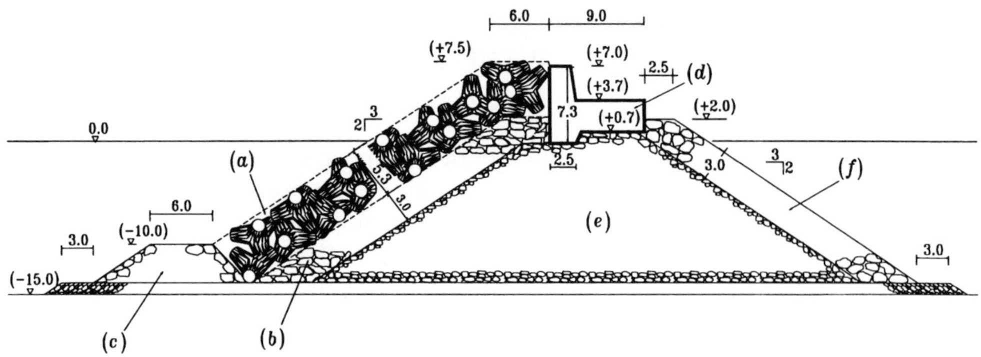

Fig. 13.13 The rubble mound breakwater of the port of Brindisi (South Italy): (*a*) armour, (*b*) underlayer, (*c*) toe, (*d*) crown wall, (*e*) core, (*f*) rear armour. (Courtesy of Grandi Lavori FINCOSIT.)

special forms are: cubes, accropode, tetrapods, among the others. The nominal diameter of the unit is defined by

$$D_n \equiv (P/\gamma_s)^{1/3},$$

where P and γ_s are, respectively, the weight and the specific weight.

Let us imagine that a breakwater with armour units of some given shape (e.g. the tetrapods which are widely used all over the world, see fig. 13.14) is subjected to a sequence of sea states with growing H_s and fixed

(i) number N of waves,
(ii) significant wave steepness H_s/L_{p0},
(iii) nondimensional spectrum $\mathscr{S}(w, \theta)$.

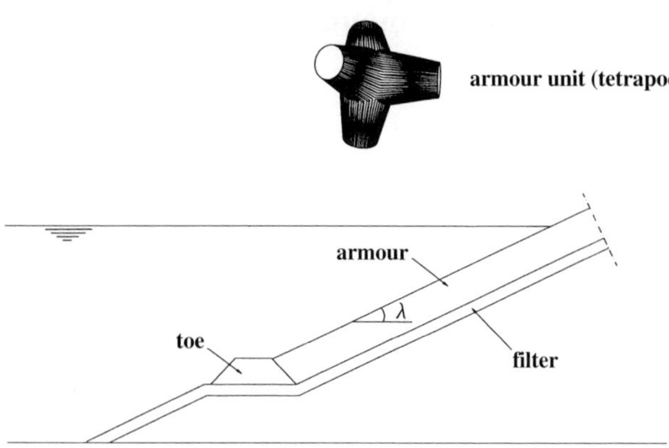

Fig. 13.14 Reference scheme.

We shall find a $H_{s\,\mathrm{crit}}$ such that a fixed number of units will be displaced. That is, we shall find the significant wave height able to produce a fixed degree of damage.

It is generally assumed, since Hudson (1953, 1961), that the ratio between $H_{s\,\mathrm{crit}}$ and D_n is a function like this:

$$\frac{H_{s\,\mathrm{crit}}}{D_n} = f\left(\frac{\gamma_s}{\gamma_a}, \frac{H_{s\,\mathrm{crit}}}{L_{p0}}, \lambda, N, \mathscr{S}(w,\theta)\right), \tag{13.15}$$

where, clearly, the function generally varies with the shape of the block units. Hudson himself simplified this function in the following way:

$$\frac{H_{s\,\mathrm{crit}}}{\Delta D_n} = f\left(\frac{H_{s\,\mathrm{crit}}}{L_{p0}}, \lambda, N, \mathscr{S}(w,\theta)\right), \tag{13.16}$$

where

$$\Delta \equiv \frac{\gamma_s}{\gamma_a} - 1.$$

Really, Hudson reasoned in terms of periodic waves, and his formula did not include the wave steepness nor the number N of waves. However he introduced this kind of simplification giving rise to a long sequence of formulae having $H_{s\,\mathrm{crit}}/\Delta D_n$ on the l.h.s. (for Hudson this was $H_{\mathrm{crit}}/\Delta D_n$ with H_{crit} the height of the periodic waves which cause the fixed degree of damage).

For understanding the step from (13.15) to (13.16) we reason as follows. The force of a current on a solid block [see fig. 13.15] can be written in the form

$$F = \mathrm{const}\,\gamma_a \frac{u^2}{g} D_n^2,$$

on the assumption that the drag coefficient is constant over the range of variability of u and D_n. The block slides when this force becomes equal to the limit shear force

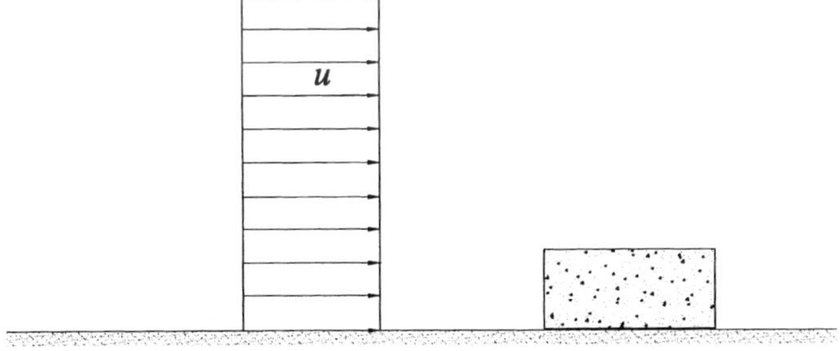

Fig. 13.15 Simple analogy for Hudson's formula: if the product ΔD_n keeps constant, the velocity of the current for which the block slides does not change (provided the variations of D_n are such that the drag coefficient can be assumed as a constant).

at the base of the block

$$R_{o\,\max} = \mu(\gamma_s - \gamma_a)\,D_n^3$$

which is the product of the coefficient of friction by the weight in still water of the block. Consequence: if, under the same coefficient of friction, we let D_n and γ_s vary in such a way that ΔD_n keeps constant, the critical value of the current velocity u for which the block slides (that is the value of u for which $F = R_{o\,\max}$) does not change. The step from (13.15) to (13.16) is based on the same idea: the significant wave height for which the fixed damage occurs does not change if D_n and γ_s vary in such a way that the product ΔD_n keeps constant.

Note that an increase of the specific weight enables us to reduce the weight of the armour units more than we probably would be inclined to believe. Let us assume that γ_s is equal to $2.35 \cdot 10^4\ \mathrm{N/m^3}$, so that

$$\Delta = \frac{2.35 \cdot 10^4}{1.01 \cdot 10^4} - 1 = 1.33\,.$$

An increase of 5% in γ_s leads to an increase of 8.5% in Δ which enables us to reduce D_n of the same amount, under the same $H_{s\,\mathrm{crit}}$. Conclusion: the weight of the armour unit is reduced nearly 20%.

13.6.2 *Formulae for the size of armour units*

Several formulae with the general form (13.16) were suggested. In particular a recent one pertinent to tetrapods and non-breaking waves (i.e. waves which do not break before the rubble mound) is

$$\frac{H_{s\mathrm{crit}}}{\Delta D_n} = \left(0.85 + 3.75\,\frac{\mathscr{N}^{0.5}}{N^{0.25}}\right)\left(\frac{H_{s\mathrm{crit}}}{\overline{L}_0}\right)^{-0.2}, \tag{13.17}$$

where N (as said) is the number of waves and $\overline{L}_0$ is the wavelength on deep water associated with the mean period $\overline{T}$ [from (5.25) $\overline{L}_0$ is equal to $0.78^2\,L_{p0}$]. $\mathscr{N}$ is the actual number of displaced units related to a width (along the longitudinal axis of the breakwater) of one nominal diameter D_n. The slope angle λ does not appear in the formula, because the breakwaters with armour layers of interlocking units are generally built with a slope angle of 33°. Finally, formulae like (13.17) are usually obtained through laboratory tests with 2-D random waves and frequency spectra of the wind waves.

Let H_{s1}, $\mathscr{D}_1$, $\overline{T}_1$ and $\overline{L}_{01} = g\overline{T}_1^2/2\pi$ be the characteristics (significant wave height, duration, mean wave period and wavelength) of the design sea state for the first verification (see sect. 7.5); and let H_{s2}, $\mathscr{D}_2$, $\overline{T}_2$ and $\overline{L}_{02} = g\overline{T}_2^2/2\pi$ be the characteristics of the design sea state for the second verification. Then, we use (13.17) with the data of the first design sea state and with a prescribed value ($\mathscr{N}_1$) of the number $\mathcal{N}$ of displaced units (related to a width), and compute

$$D_{n1} = \frac{H_{s1}(H_{s1}/\overline{L}_{01})^{0.2}}{\Delta(0.85 + 3.75\mathcal{N}_1^{0.5}/N_1^{0.25})} \quad \text{with} \quad N_1 = \frac{\mathcal{D}_1}{\overline{T}_1}.$$

Then we use (13.17) with the data of the second design sea state and with a smaller value $\mathcal{N}_2$ of the number of displaced units, and compute

$$D_{n2} = \frac{H_{s2}(H_{s2}/\overline{L}_{02})^{0.2}}{\Delta(0.85 + 3.75\mathcal{N}_2^{0.5}/N_2^{0.25})} \quad \text{with} \quad N_2 = \frac{\mathcal{D}_2}{\overline{T}_2}.$$

With the diameter D_{n1}, the first design sea state displaces $\mathcal{N}_1$ armour units; and with the diameter D_{n2}, the second design sea state (the less severe one) displaces $\mathcal{N}_2$ armour units. The nominal diameter D_n we shall adopt must be greater than both D_{n1} and D_{n2}.

According to Van der Meer (1992), who has introduced the formula (13.17), the failure for tetrapods is at $\mathcal{N} = 1.5$. Therefore we should take

$$0 \leq \mathcal{N}_2 < \mathcal{N}_1 \leq 1.5,$$

for the tetrapods. Some design guidelines recommend $\mathcal{N}_2 = 0$. However the use of $\mathcal{N}_2 = 0$, with the design sea state for the second verification, generally, leads to great sizes of the units.

The D_n so obtained is suitable for the trunk section, while for the breakwater head the D_n or Δ must be increased and/or the slope of the roundhead must be made less steep. For the underlayer, common practice is to use stones with a mass of $\frac{1}{10}$ to $\frac{1}{15}$ of the armour. Finally, formulae like Hudson's are used also for the stones in the toe.

Conclusive note

The experiment RC 1994 was specially executed for this chapter. The second verification of the upright breakwaters, which takes into account the length of the structure, is also a novelty.

References

Ahrens J. P., Seeling W. N., Ward D. L. and Allsop W., 1993 Wave run-up on and wave reflection from coastal structures. *Proc. Waves '93 ASCE*, 489-502.

Boccotti P., Barbaro G., Fiamma V. et al., 1993 An experiment at sea on the reflection of the wind waves. *Ocean Eng.* 20, 493-507.

Goda Y., 1974 New wave pressure formulae for composite breakwater. *Proc. 14th Conf. Coastal Eng.*, 1702-1720.

1992 The design of upright breakwaters. *Proc. Short Course attached to the 23rd Conf. Coastal Eng.*, 547-568.

Hudson R. Y., 1953 Wave forces on breakwaters. *Trans. ASCE*, 118;
1961 Laboratory investigation of rubble-mound breakwaters. *Trans. ASCE*, 126.

Kortenhaus A. and Oumeraci H., 1998 Classification of waves loading on monolithic coastal structures, *Proc. 26th Conf. Coastal Eng., ASCE*, 867-880.

Ling H.I., Cheng A. H.-D., Mohri Y. and Kawabata T., 1999 Permanent displacement of composite breakwaters subject to wave impact. *J. Waterway, Port, Coastal, and Ocean Eng.* 125, 1-8.

Martin F.L., Losada M.A. and Medina R., 1999 Wave loads on rubble mound breakwater crown walls. *Coastal Eng.* 37, 149-174.

Meyerhof G. G., 1953 The bearing capacity of foundations under eccentric and inclined loads. *Proc. 3rd ICSMFE*, 1.

Neelamani S., Schuttrumpf H., Muttray M. and Oumeraci H., 1999 Prediction of wave pressures on smooth impermeable seawalls. *Ocean Eng.* 26, 739-765.

Oumeraci H., 1994 Review and analysis of vertical breakwater failures – lesson learned. *Coastal Eng.* 22, 3-29.

Schuttrumpf H., Bergmann H., Dette H.H., 1994 The concept of residence time for the description of wave runup, wave setup and wave rundown. *Proc. 24th Conf. Coastal Eng.*, 553-564.

Seeling W. N. and Ahrens J.P., 1995 Wave reflection and energy dissipation by coastal structures. In: Wave forces on inclined and vertical wall structures. *Task Committee on Forces on Inclined and Vertical Wall Structures ASCE*, 28-51.

Sutherland J. and O'Donoghue T., 1998 Wave phase shift at coastal structures. *J. Waterway, Port, Coastal and Ocean Eng.* 124, 90-98.

Tanimoto K. and Takahashi S., 1994 Design and construction of caisson breakwaters – the Japanese experience. *Coastal Eng.* 22, 57-77.

Van der Meer J.W., 1992 Conceptual design of rubble-mound breakwaters. *Proc. Short Course attached to the 23rd Conf. Coastal Eng.*, 447-510.

Van der Meer J. W. and Breteler M.K., 1990 Measurements and computation of wave induced velocities on a smooth slope. *Proc. 22nd Conf. Coastal Eng.*, 191-204.

Chapter 14

TOPICS CALLING FOR AN OVERALL OVERVIEW OF OFFSHORE AND COASTAL ENGINEERING

14.1 A comparison between tsunami and wind waves from the open sea to the coast

14.1.1 *The solitary wave*

Tsunamis are extremely long waves generated by submarine earthquakes or volcanic eruptions. Given the very small values of d/L_0, the non linearity effects become so big that the linear wave theory is coarsely approximated. Hence, we must resort to theories valid in the limit $d/L_0 \rightarrow 0$, in particular to the solitary wave theory (Laitone, 1961; Grimshaw, 1970). The first approximation of this theory gives

$$\eta(y,t) = \frac{H}{\cosh^2 q}, \tag{14.1a}$$

$$v_y(y,t) = \frac{1}{\cosh^2 q}\frac{H}{d}\sqrt{gd}, \tag{14.1b}$$

$$a_y(y,t) = \sqrt{3}\,\frac{\sinh q}{\cosh^3 q}\left(\frac{H}{d}\right)^{1.5} g, \tag{14.1c}$$

$$\Delta p(y,t) = \rho g\,\frac{H}{\cosh^2 q}, \tag{14.1d}$$

with

$$q \equiv \frac{\sqrt{3}}{2}\sqrt{\frac{H}{d}}\,\frac{(y-ct)}{d}, \quad c = \sqrt{gd}\left(1+\frac{H}{2d}\right).$$

Note that the maximum of the surface displacement is H and the greatest lower bound is zero. Note also that both v_y, a_y and Δp do not depend on z, so that they are constant from the wave surface to the seabed (this property is predicted also by the linear wave theory as $d/L_0 \to 0$).

To a first approximation, the shoaling can be foreseen by means of (2.45) as $d/L_0 \to 0$. With $\alpha_0 = \pi/2$ (two dimensional flow) and $d/L_0 \to 0$, equation (2.45) becomes

$$\frac{H}{H_0} = \sqrt{\frac{cT}{4\pi d}} \quad \text{as} \quad \frac{d}{L_0} \to 0,$$

which, together with (1.34) of c, yields

$$H\sqrt[4]{d} = \text{constant} \quad \text{as} \quad \frac{d}{L_0} \to 0, \tag{14.2}$$

which is called Green's law.

14.1.2 *Even a high tsunami is not dangerous for the offshore structures*

Thrust on a gravity offshore platform

Let us evaluate the thrust of a 10 m solitary wave on the gravity offshore platform of sect. 12.1, on 125 m water depth. For simplicity the base of the gravity platform is sketched as a circular cylinder with the section area of the actual base. We have already seen in sect. 12.1 that the radius of this circular cylinder is of 45 m.

In each case the maximum horizontal force occurs between the instant of the largest v_y $(q = 0)$ and the instant of the largest a_y $(q = 0.658)$. In the case under examination, the inertia force is much greater than the drag force, and hence the maximum force proves to be very close to the force at the instant of the maximum acceleration.

Using Morison's equation with $C_{in} = 1.85$, $C_{dg} = 0.62$ together with the formulae (14.1*a*-*b*-*c*) for η, v_y and a_y, we obtain the following situation at the instant of the maximum horizontal force:

$$a_y = 0.148\ \text{m/s}^2, \quad v_y = 1.86\ \text{m/s}, \quad \eta = 6.7\ \text{m},$$

$$\text{platform base} \begin{cases} F_{in} = [1.85 \cdot 1030 \cdot \pi \cdot 45^2 \cdot 0.148]\, 50 = 8.97 \cdot 10^7\ \text{N}, \\ F_{dg} = [0.62 \cdot 1030 \cdot 45 \cdot 1.86^2]\, 50 = 0.50 \cdot 10^7\ \text{N}, \\ F \equiv F_{in} + F_{dg} = 9.47 \cdot 10^7\ \text{N}, \end{cases}$$

$$\text{three columns} \begin{cases} F_{in} = 3\,[1.85 \cdot 1030 \cdot \pi \cdot 8.35^2 \cdot 0.148]\,(75 + 6.7) = 1.51 \cdot 10^7\ \text{N}, \\ F_{dg} = 3\,[0.62 \cdot 1030 \cdot 8.35 \cdot 1.86^2]\,(75 + 6.7) = 0.45 \cdot 10^7\ \text{N}, \\ F \equiv F_{in} + F_{dg} = 1.96 \cdot 10^7\ \text{N}, \end{cases}$$

where the square parentheses give the forces per unit length, which are multiplied by the height (50 m) of the base, and by the height (75 + 6.7 m) of the piece of column beneath the water surface. [The column tapers from the 10 m radius at the base ($z = -75$ m) to the 6 m radius at the deck elevation ($z = +25$ m), and the radius of 8.35 m we have used in the computation is the average one of the piece of column between the base ($z = -75$ m) and the water surface ($z = +6.7$ m).] Conclusion: the maximum horizontal force exerted by the tsunami is of $1.14 \cdot 10^8$ N against the $4.12 \cdot 10^8$ N of the force exerted by the wind waves of sect. 12.1.

To a first approximation the K_E of a cylinder in a solitary wave can be evaluated by means of

$$K_E = \frac{\pi v_{\max}^2}{R a_{\max}}$$

which is an alternative way to express the K_E with the linear wave theory. Given that

$$v_{\max} = 2.80\,\mathrm{m/s}, \quad a_{\max} = 0.148\,\mathrm{m/s^2},$$

the K_E of the platform base turns out to be 3.7. This is a rather small value of K_E, so that we could resort to the ideal flow pattern as well. And, with this alternative approach, the horizontal force per unit length on the platform base can be expressed in the form

$$F(t) = 2\rho g R H \int_0^{\frac{\pi}{2}} \left\{ \cosh^{-2}\left[\frac{\sqrt{3}}{2}\sqrt{\frac{H}{d}}\frac{1}{d}(-\mathscr{F}_{\mathscr{R}} R\cos\beta - ct)\right] + \right.$$

$$\left. - \cosh^{-2}\left[\frac{\sqrt{3}}{2}\sqrt{\frac{H}{d}}\frac{1}{d}(\mathscr{F}_{\mathscr{R}} R\cos\beta - ct)\right]\right\} \cos\beta \,\mathrm{d}\beta .$$

[To check this equation, use formula 14.1d for the pressure. Refer to fig. 11.14. Bear in mind that the phase angle of a couple of points like P', P'' in this figure grows of $\mathscr{F}_{\mathscr{R}}$ times from the ideal water cylinder to the solid cylinder.] With a $\mathscr{F}_{\mathscr{R}}$ of 1.75 (see sect. 11.4.4), the maximum force on the platform base is $0.84 \cdot 10^8$ N against the $0.95 \cdot 10^8$ N obtained by means of Morison's equation. Thus, the total thrust on the platform is 10% smaller than that predicted by means of Morison's equation.

Thrust on a space frame structure

Of course Morison's equation is suitable for computing the maximum force on the four legs of the space frame structure of fig. 12.4. With $C_{in} = 1.85$, $C_{dg} = 0.62$ this maximum force proves to be of $0.59 \cdot 10^7$ N against the force of $1.45 \cdot 10^7$ N exerted by the wind waves. (In this case, the drag force exceeds the inertia force and the maximum horizontal force occurs when $q = 0.205$.)

Comparison between tsunami and wind waves

The quotient *force per unit length of the tsunami / force per unit length of the wind wave* grows towards the seabed, given that v_y and a_y of the tsunami are not attenuated with depth. As a consequence the quotient *moment of the tsunami / moment of the wind wave* is smaller than the quotient *force of the tsunami / force of the wind wave* (we mean the moment of the horizontal wave forces with respect to the seabed level). Specifically, we have

$$\text{gravity platform} \begin{cases} \text{tsunami force/wind wave force} = \dfrac{1.14 \cdot 10^8}{4.12 \cdot 10^8} = 0.28, \\ \text{tsunami moment/wind wave moment} = \dfrac{0.41 \cdot 10^{10}}{1.97 \cdot 10^{10}} = 0.21, \end{cases}$$

$$\text{frame structure} \begin{cases} \text{tsunami force/wind wave force} = \dfrac{0.59 \cdot 10^7}{1.45 \cdot 10^7} = 0.40, \\ \text{tsunami moment/wind wave moment} = \dfrac{0.40 \cdot 10^9}{1.33 \cdot 10^9} = 0.30. \end{cases}$$

14.1.3 *The same tsunami has some ruinous effects on a coast*

Using (14.2) for a first approximation, we find that our solitary wave on 20 m water depth has a height of about 16 m so that it is going to break. Hence, on a flat uniform slope we expect a run-up of nearly 16 m, and an even much greater one if we include the shallow flooding (see sect. 3.2.3).

For a comparison let us see the run-up of the design sea state of the platform. Van der Meer (1992) suggests the following relation for a smooth impermeable slope:

$$\frac{R_{U_{2\%}}}{H_s} = 1.5\, \mathscr{I}_{p0} \text{ for } \mathscr{I}_{p0} \leq 2,$$

where $R_{U_{2\%}}$ is the threshold being exceeded by the run-up, on average, two times in one hundred waves, and $\mathscr{I}_{p0}$ is the Irribarren number on deep water:

$$\mathscr{I}_{p0} \equiv \frac{\tan \lambda}{\sqrt{H_{s0}/L_{p0}}}.$$

A typical value of the wave steepness H_{s0}/L_{p0} is obtained from (4.27):

$$H_{s0}/L_{p0} = 0.035,$$

from which it follows

$$R_{U_{2\%}} = 8.0 H_s \tan \lambda.$$

Let us assume the H_s of 15 m we used for the calculation of the wind wave thrust on the platform (really, this H_s was on 125 m water depth, so that the H_s at breaking will be generally smaller than 15 m). At all events, even with an H_s of 15 m at breaking and a bottom slope of, say, 8% the $R_{U_{2\%}}$ would be $8.0 \cdot 15 \cdot 0.08 = 9.6$ m, against the 16 m run-up of the tsunami (not counting the shallow flooding).

14.2 Small scale models

14.2.1 *The dynamic similarity*

Let us imagine we have to execute the small scale model of waves of some given characteristics. We fix $\lambda = 100$, λ being the scale factor of length. Then let us call

T_{pr}, d_{pr}, L_{pr}, respectively, the period, water depth, and wavelength in the prototype; and let us call T_{mo}, d_{mo}, L_{mo} these quantities in the model. Given T_{pr}, d_{pr}, and L_{pr} we shall easily succeed in getting

$$d_{mo} = \frac{1}{100} d_{pr} \,. \tag{14.3}$$

But, what must we do to have $L_{mo} = \frac{1}{100} L_{pr}$? To this end, we have to fix scale factor of time equal to square root of scale factor of length, that is we must fix

$$T_{mo} = \frac{1}{10} T_{pr} \,.$$

Doing so, we shall have

$$\frac{d_{mo}}{L_{mo\,0}} = \frac{d_{pr}}{L_{pr\,0}},$$

which implies

$$\frac{d_{mo}}{L_{mo}} = \frac{d_{pr}}{L_{pr}},$$

given that d/L is a function of d/L_0. The last equality together with (14.3) implies that L_{mo} is 1/100 of L_{pr}, as required.

The dynamic similarity with the scale factor of time equal to the square root of the scale factor of length is dedicated to Froude. We already referred to it in sects. 4.5-6 where we discussed the possibility to execute small scale models in the field.

14.2.2 *The reason why Froude's is generally the proper dynamic similarity for ocean structures*

Let us consider the 1:20 scale model of the breakwater of fig. 13.10*a*. We have

$$\text{prototype: } d = 20\,\text{m}, \; H_s = 8\,\text{m}, \; T_h = 11.1\,\text{s},$$

so that Froude's dynamic similarity yields

$$\text{model: } d = 1\,\text{m}, \; H_s = 0.40\,\text{m}, \; T_h = 11.1/\sqrt{20} = 2.48\,\text{s}.$$

Table 14.1 *Verification of the 1:20 scale model of the breakwater on 20 m water depth*

WAVE LOAD

	F [N/m]	S_w [N/m]	M [N · m/m]
Crest	$5.92 \cdot 10^3$	$2.35 \cdot 10^3$	$5.96 \cdot 10^3$
Trough	$-4.42 \cdot 10^3$	$-3.87 \cdot 10^3$	$0.06 \cdot 10^3$

SAFETY FACTOR

C_1	C_2	C_3
1.31	1.62	1.98
2.60	128	3.42

A useful exercise could be to compute the wave forces and the safety factors for the model. The result is given in table 14.1 which has been obtained by means of the formulae (13.7*a*-*f*). Comparing this table with table 13.1, we can easily verify that the force per unit length in the model is 20^2 times smaller than in the prototype, and the moment of a force per unit length in the model is 20^3 times smaller than in the prototype. Given that the weight per unit length in the model is 20^2 times smaller than in the prototype (provided the model consists of the same materials as the prototype), we realize why all the safety factors take on the same value in the model and in the prototype.

The practical consequence is that we can obtain forces and torques of the prototype, simply by multiplying forces and torques of the model by some constant factors. The model even seems able to show whether or not any failure will occur, given that the safety factors are the same as those of the prototype. From the model we can infer only whether or not sliding or overturning will occur. As for the collapse of the foundation, the C_3 of the model is the same as that of the prototype only if the soil is cohesionless as we have assumed. If the soil is cohesive, the C_3 of the prototype would be smaller than the C_3 of the model. (Here, note that sometimes the same soil is regarded as cohesionless for the big foundation of the prototype and cohesive for the small foundation of the model.)

A further useful exercise could also be done for the gravity offshore platform and the submerged tunnel. Fix a suitable scale factor of length and repeat, for the model, the calculations that we did in chap. 12 for the prototype. If you will use the same nondimensional spectrum $\mathscr{S}(w, \theta)$ that we used for the prototype, you will find that the forces on the gravity platform are those of fig. 12.2, divided by the constant factor λ^3; and the loads per unit length on the tunnel are those of figs. 12.11 and 12.12, divided by the constant factor λ^2.

14.2.3 *Where Froude's dynamic similarity fails*

Sometimes Froude's dynamic similarity of ocean structures is not effective in the sense that the ratio between the forces on the prototype and the forces on the model is no longer a constant. This is the case if the forces depend on the Reynolds number, given that the Reynolds number differs from prototype to model:

$$R_{E_{mo}} = \frac{1}{\lambda^{1.5}} R_{E_{pr}}.$$

[The proof of this relation is left as an exercise. Use the fact that the ratio between the velocities in the prototype and in the model is $\sqrt{\lambda}$.]

As a consequence, for the space frame structures where Morison's equation is used, the force measured in the model generally does not go with the constant multiple λ^3 to the prototype, indeed the force coefficients being dependent on R_E will generally differ from model to prototype. However, the dynamic similarity is not barred even for the space frame structures. Indeed, for the ratio between the forces in the prototype and in the model to take the constant value λ^3, it is not necessary that $R_{E_{mo}}$ is equal to $R_{E_{pr}}$, but it is sufficient that the force coefficients have the same values in the model and in the prototype, which can be true even if $R_{E_{mo}} \neq R_{E_{pr}}$.

If we assume that C_{in} and C_{dg} take on their own asymptotic values provided $D^2/\nu T > 10^4$ (see sect. 11.5.4), a sufficient condition for C_{in} and C_{dg} to be equal in the model and in the prototype is that $D^2/\nu T$ should be greater than 10^4 both in the model and in the prototype.

Let us come back to the truss of fig. 12.4. Since the smaller diameter is 2.5 m, we have

$$\text{prototype:} \left(\frac{D^2}{\nu T}\right)_{\min} = \frac{2.5^2}{10^{-6} \cdot 16.5} = 3.8 \cdot 10^5$$

(16.5 s being the T_p). In the model the ratio $D^2/\nu T$ is $\lambda^{1.5}$ times smaller than in the prototype. Therefore $D^2/\nu T$ will be greater than 10^4 also in the model provided that

$$3.8 \cdot 10^5/\lambda^{1.5} \geq 10^4,$$

that is

$$\lambda \leq 11.3.$$

In words: the factor scale of length should not exceed 11 ÷ 12. This leads to an unusually large size of the model, which however does not create any particular problem if the dynamic similarity is performed in the field.

14.2.4 *Dynamic similarity for rubble mound breakwaters*

Equation (13.16) can be re-proposed in the alternative form

$$\mathcal{N} = f\left(\frac{H_s}{\Delta D_n}, \frac{H_s}{L_{p0}}, \lambda, N, \mathcal{S}(w, \theta)\right),$$

where $\mathcal{N}$ is the number of displaced units. This alternative form reveals at once that, under the Froude's dynamic similarity, the degree of damage should be the same in the model and in the prototype. Indeed the variables on the r.h.s. have the same value in the model and in the prototype.

It is probable that things are more complex. Indeed the displacement of the armour units will also be due to drag forces, and hence the model of a rubble mound breakwater will be somehow similar to the model of a space frame structure. Therefore, we have to limit λ, which could be done, roughly, with the same criterion of the space frame structure. Here below is an example.

Let us consider an armour layer of tetrapods of $5 \cdot 10^5$ N being subjected to a sea state with an H_s of 8 m and a T_p of 12.1 s. We have

$$\text{prototype:} \quad \frac{D_n^2}{\nu T_p} = \frac{2.77^2}{10^{-6} \cdot 12.1} = 6.34 \cdot 10^5,$$

where 2.77 m is the D_n of a unit of $5 \cdot 10^5$ N with a specific weight of $2.35 \cdot 10^4\,\mathrm{N/m^3}$. If we adopt the criterion of the space frame structures shown in the previous section, we require that

$$6.34 \cdot 10^5/\lambda^{1.5} \geq 10^4,$$

that is

$$\lambda < 15.9.$$

Thus the factor scale of length should not exceed $15 \div 16$. If we adopted a λ of 15 we would have an H_s in the model of about 0.50 m and a weight of the tetrapods of about 150 N. As for the case of the space frame structure, the size of the model would be larger than the usual size of models. However, such a model test could be rather easily executed in a natural laboratory like that described in sect. 4.6.2.

14.2.5 *Models of ports*

Often the small scale models are also used to foresee the wave height within a port's basin. Typically, in these cases we have to evaluate some diffraction coefficients, and for this aim Froude's dynamic similarity is generally effective.

As an example, let us resort to a 1:50 scale model for evaluating the C_d of a point of polar coordinates

$$\text{prototype:}\ r = 100\,\text{m}, \quad \beta = 45°,$$

behind a semi-infinite breakwater on 20 m water depth for wind waves with a T_p of 10 s. Summarizing, we have

$$\text{prototype:}\ d = 20\,\text{m}, \quad T_p = 10\,\text{s}, \quad L_{p0} = 156\,\text{m}, \quad \frac{d}{L_{p0}} = 0.128, \quad \frac{r}{L_{p0}} = 0.641.$$

The same point in the model has the following coordinates

$$\text{model:}\ r = 2\,\text{m}, \quad \beta = 45°,$$

and for the rest

$$\text{model:}\ d = 0.40\,\text{m}, \quad T_p = 1.414\,\text{s}, \quad L_{p0} = 3.12\,\text{m}, \quad \frac{d}{L_{p0}} = 0.128, \quad \frac{r}{L_{p0}} = 0.641.$$

Function $\mathscr{k}(w)$ [given by (8.24)] is the same in the model and in the prototype, since the only parameter of this function is d/L_{p0} which does not change from the prototype to the model. Moreover, the dynamic similarity requires that the nondimensional spectrum $\mathscr{S}(w, \theta)$ is the same in the model and in the prototype. Therefore, given that r/L_{p0}, β, $\mathscr{k}(w)$, and $\mathscr{S}(w, \theta)$ are the same in the model and in the prototype, it follows that the C_d of the model is equal to the C_d of the prototype [to prove this, use equations (8.40) and (8.41*a*-*b*)].

Of course, the execution of a small scale model is useful only in the case of some complex configuration of the solid boundary, and the simple example we have given aims only to show that the C_d of the model is identical to the C_d of the prototype.

In dealing with this kind of model tests we should bear in mind the conclusions of sect. 8.7.3. Specifically, for the estimate of C_d in the more sheltered region of a port's basin, even the periodic waves with period T_p are good. In the water sheet before a breakwater attacked orthogonally, even the long crested waves are good whilst the use of the short crested waves is necessary for the water sheet before a breakwater subjected to an oblique attack, and for the less sheltered region of a port's basin.

14.3 Wave measurements

14.3.1 *Wave measurements from buoys*

Generalities

The free surface displacement is measured by measuring the vertical acceleration of the buoy. The total measurement system accuracy can be within 0.2 m or 5% for H_s. The buoys can operate even on some thousands metres of water depth.

Besides the heave, a directional buoy typically measures also the pitch and roll. This approach is used in particular by NDBC whose buoys are discus-shaped ranging from 3 m to 12 m in diameter. Time series records of pitch and roll are stored in the onboard computer memory. As for the instruments, a recent project is to use low-cost angular rate sensors mounted orthogonally in a buoy (Steele et al., 1998). The measurements of pitch and roll serve to obtain the directional spreading function.

How to obtain the directional spreading function from the measurements of pitch and roll

Let us consider the free surface displacement and its space derivatives at some fixed point x, y. The surface displacement is given by (8.5), and the space derivatives are

$$\eta_x(t) \equiv \frac{\partial}{\partial x}\eta(x, y, t) = \sum_{i=1}^{N} a_i k_i \sin\theta_i \sin(\omega_i t + \tilde{\varepsilon}_i), \tag{14.4a}$$

$$\eta_y(t) \equiv \frac{\partial}{\partial y}\eta(x, y, t) = \sum_{i=1}^{N} a_i k_i \cos\theta_i \sin(\omega_i t + \tilde{\varepsilon}_i). \tag{14.4b}$$

Let us obtain the relation between the average product $<\eta_x(t)\eta_y(t)>$ and the directional spectrum. We have

$$<\eta_x(t)\eta_y(t)> = \sum_{i=1}^{N} \frac{1}{2} a_i^2 k_i^2 \sin\theta_i \cos\theta_i = \int_0^\infty \int_0^{2\pi} S(\omega, \theta) k^2 \sin\theta \cos\theta \, d\theta \, d\omega. \tag{14.5}$$

This average product can be obtained also from the time series data of pitch and roll. For this aim it is convenient to resort to the Fourier series of $\eta_x(t)$ and $\eta_y(t)$:

$$\eta_x(t) - \sum_{i=1}^{N} u'_{xi} \cos(\omega_i t) + a''_{xi} \sin(\omega_i t), \tag{14.6a}$$

$$\eta_y(t) = \sum_{i=1}^{N} a'_{yi} \cos(\omega_i t) + a''_{yi} \sin(\omega_i t). \tag{14.6b}$$

[Note that N in (14.4*a*-*b*) stands for the number (very large) of small waves forming the sea state. While N in (14.6*a*-*b*) is the number of frequencies in the Fourier series, which depends upon record length and sampling rate.]

From (14.6*a*-*b*) we get

$$<\eta_x(t)\eta_y(t)> = \sum_{i=1}^{N} \frac{1}{2}(a'_{xi}a'_{yi} + a''_{xi}a''_{yi}). \tag{14.7}$$

And from (14.5) and (14.7) we conclude that

$$\Delta\omega \int_0^{2\pi} S(\omega_i, \theta) k_i^2 \sin\theta \cos\theta \, \mathrm{d}\theta = \frac{1}{2}(a'_{xi}a'_{yi} + a''_{xi}a''_{yi}). \tag{14.8}$$

The two sides of this equation represent the contribution to the average product $<\eta_x(t)\eta_y(t)>$ from the components whose frequency is between $\omega_i - \Delta\omega/2$ and $\omega_i + \Delta\omega/2$ ($\Delta\omega$ being the frequency increment of the Fourier series).

Using (8.14) we can rewrite (14.8) in the form

$$\int_0^{2\pi} D(\theta; \omega_i) \sin(2\theta) \, \mathrm{d}\theta = \frac{a'_{xi}a'_{yi} + a''_{xi}a''_{yi}}{\Delta\omega E(\omega_i) k_i^2}, \tag{14.9a}$$

where the terms on the r.h.s. are all known (they proceed from the time series data of $\eta(t), \eta_x(t)$, and $\eta_y(t)$).

Of course equation (14.9*a*) alone is not enough to obtain univocally the unknown function $D(\theta; \omega_i)$. However, some further relations of the same kind can be obtained from analysis of the following averages:

$$<\eta(t)\,\eta_x^*(t)>, \quad <\eta(t)\,\eta_y^*(t)>, \quad <\eta_y^2(t) - \eta_x^2(t)>,$$

where

$\eta_x^*(t) \equiv \eta_x(t)$ with each component advanced of 90° in phase,

$\eta_y^*(t) \equiv \eta_y(t)$ with each component advanced of 90° in phase.

The further relations are

$$\int_0^{2\pi} D(\theta; \omega_i) \sin\theta \, \mathrm{d}\theta = \frac{a'_i a''_{xi} - a''_i a'_{xi}}{2\Delta\omega E(\omega_i) k_i}, \tag{14.9b}$$

$$\int_0^{2\pi} D(\theta; \omega_i) \cos\theta \, \mathrm{d}\theta = \frac{a'_i a''_{yi} - a''_i a'_{yi}}{2\Delta\omega E(\omega_i) k_i}, \tag{14.9c}$$

$$\int_0^{2\pi} D(\theta;\,\omega_i)\cos 2\,\theta\,\mathrm{d}\,\theta = \frac{(a_{yi}^{'2}+a_{yi}^{''2})-(a_{xi}^{'2}+a_{xi}^{''2})}{2\Delta\,\omega\,E\,(\omega_i)\,k_i^2}, \tag{14.9d}$$

where $a_i^{'}$, $a_i^{''}$ are the amplitudes of the Fourier series of $\eta\,(t)$ [cf. sect. 4.4.2].

It is assumed that

$$D\,(\theta;\,\omega_i) = K\,(n_i)\cos^{2n_i}\left[\frac{1}{2}\,(\theta-\bar{\theta}_i)\right],$$

where $K\,(n_i)$ is the normalizing factor and n_i, $\bar{\theta}_i$ are allowed to generally vary with ω_i (for the wind waves we usually have $\bar{\theta}_i \equiv \bar{\theta}\;\forall i$, as we saw in sect. 8.3.1). The four relations (14.9*a-d*) enable us to get the best pair n_i, $\bar{\theta}_i$ for each frequency ω_i.

All this reasoning is based on the assumption that the buoy follows the slope of the sea surface perfectly, which of course cannot be the case for large buoys. However it is possible all the same to get the pairs n_i, $\bar{\theta}_i$ taking the buoy response into account.

How to take the buoy response into account

Equations (14.4*a-b*) give the slopes of the free surface at the fixed point. Generally, pitch and roll of the buoy will be different from these slopes of the free surface because of an amplitude response $\mathscr{R}(\omega)$ and a phase response $\alpha(\omega)$. The actual pitch and roll are

$$\eta_x(t) = \sum_{i=1}^{N} \mathscr{R}(\omega_i)\,a_i\,k_i\sin\theta_i\sin\left[\omega_i t+\tilde{\varepsilon}_i+\alpha(\omega_i)\right],$$

$$\eta_y(t) = \sum_{i=1}^{N} \mathscr{R}(\omega_i)\,a_i\,k_i\cos\theta_i\sin\left[\omega_i t+\tilde{\varepsilon}_i+\alpha(\omega_i)\right].$$

[It is assumed that the buoy response is isotropic in direction and this is why the same amplitude response and the same phase response are of $\eta_x(t)$ and $\eta_y(t)$.]

We aim to obtain an equation like (14.9*a*), with the same l.h.s. of (14.9*a*) and some known terms on the r.h.s. To this end, let us consider the following averages

$$<\eta_x\,(t)\eta_y(t)> = \int_0^\infty\int_0^{2\pi} S\,(\omega,\theta)\,k^2\,\mathscr{R}^2(\omega)\sin\theta\cos\theta\,\mathrm{d}\,\theta\,\mathrm{d}\,\omega, \tag{14.10a}$$

$$<\eta_x^2\,(t)+\eta_y^2(t)> = \int_0^\infty\int_0^{2\pi} S\,(\omega,\theta)\,k^2\,\mathscr{R}^2(\omega)\,\mathrm{d}\,\theta\,\mathrm{d}\,\omega. \tag{14.10b}$$

In terms of the Fourier series, the first of these two averages is given by (14.7) and the second one is

$$<\eta_x^2\,(t)+\eta_y^2(t)> = \frac{1}{2}\,(a_{xi}^{'2}+a_{xi}^{''2}+a_{yi}^{'2}+a_{yi}^{''2}). \tag{14.11}$$

From (14.7) and (14.10*a*) it follows that

$$E(\omega_i)\, k_i^2\, \mathscr{R}^2(\omega_i)\, \Delta\omega \int_0^{2\pi} D(\theta;\, \omega_i) \sin(2\theta)\, \mathrm{d}\theta = a'_{xi}\, a'_{yi} + a''_{xi}\, a''_{yi}, \tag{14.12}$$

and from (14.10*b*) and (14.11):

$$E(\omega_i)\, k_i^2\, \mathscr{R}^2(\omega_i)\, \Delta\omega = \frac{1}{2}\left(a'^2_{xi} + a''^2_{xi} + a'^2_{yi} + a''^2_{yi}\right). \tag{14.13}$$

Finally, (14.12) and (14.13) taken together yield

$$\int_0^{2\pi} D(\theta;\, \omega_i) \sin(2\theta)\, \mathrm{d}\theta = \frac{2\left(a'_{xi}\, a'_{yi} + a''_{xi}\, a''_{yi}\right)}{\left(a'^2_{xi} + a''^2_{xi} + a'^2_{yi} + a''^2_{yi}\right)}.$$

Similarly, it is possible to get three equations like (14.9*b*), (14.9*c*), and (14.9*d*) which are independent of both the tilt response factors and the heave response factors.

Alternative techniques to obtain the directional spreading function

Some alternative analytical techniques have been developed which do not assume the $\cos^{2n}$ directional spreading function. In particular $D(\theta;\, \omega)$ can be obtained through the maximum entropy principle and the knowledge that $D(\theta;\, \omega)$ must satisfy the four equations (14.9*a-d*) (Kobune and Hashimoto, 1986).

The directional spreading function can be obtained also from some alternative kind of measurements. The following are two common surface techniques. First: to obtain $D(\theta;\, \omega)$ from the measurement of pressure and of two components of the horizontal particle velocity made by some current meter. Second: to obtain $D(\theta;\, \omega)$ from the measurements of the surface displacement or the fluctuating pressure head made by a gauge array. The analytical techniques are similar to the ones we have described for the pitch-roll-heave data and from the first of these two alternative methods we arrive at four equations like (14.9*a-d*) with the same l.h.s. as (14.9*a-d*) and some known terms on the r.h.s.

14.3.2 *Instruments for small scale field experiments*

In the previous chapters we widely referred to small scale field experiments. These experiments are similar to those executed in the laboratories. Thus the ideal choice should be to use the instruments that have been largely improved from the use in the laboratory tanks. Unfortunately, the resistive wave staff (that is the most classic wave staff used in the laboratory tanks) works very badly in the field, because the sea water has an electrical conductivity which is much greater than that of fresh water.

Hence, it is advisable to resort to instruments such as the ultrasonic wave probes which are fixed to some special piles as shown in figs. 11.1 and 11.6. The instrument outputs a voltage signal corresponding to the time delay between transmitting and receiving the ultrasonic beam reflected by the water surface. Knowing the velocity of the ultrasonic beam, the distance between the water surface and the ultrasonic probe is easily derived. However there are a few inconveniences that one should bear in mind.

First inconvenience: these gauges cannot work very close to each other. The distance between two probes should be greater than $2h \tan \lambda$ (see fig. 14.1*a*). Moreover the cone of the ultrasonic beam must not be intersected by some solid surface. This, in particular, bars the use of these probes for measuring the surface displacement at a wall.

Second inconvenience: the ultrasonic probe does not work well with spilling or almost breaking waves. We have already raised this item with regard to fig. 10.30, where an ultrasonic probe missed recording a breaking wave crest. A typical example of this flaw is given in fig. 14.1*b*. The probe cannot follow the wave surface for a few instants just after the passage of the wave crest. This is a consequence of the fact that the highest waves are often spilling on the back of the crest.

Sometimes a few flaws like the one in fig. 14.1*b* are present in a record of a sea state. However the consequences on the significant wave height, the spectrum, and the autocovariance are usually negligible. Moreover, the consequences on the wave height probability are negligible, in that the flaw attacks only a part of the wave between the crest and the zero down-crossing, without altering crests and troughs.

In view of the above mentioned inconveniences, before executing an experiment consider whether is it possible to deal with the pressure head waves rather than with the surface waves. Here below, is an example.

Let us imagine we decide to verify whether or not it is true that

$$f_1(y) \equiv \frac{\sigma^2(y)}{\sigma^2(0)} \quad [\sigma^2(y) \equiv \langle \eta^2(y,t) \rangle]$$

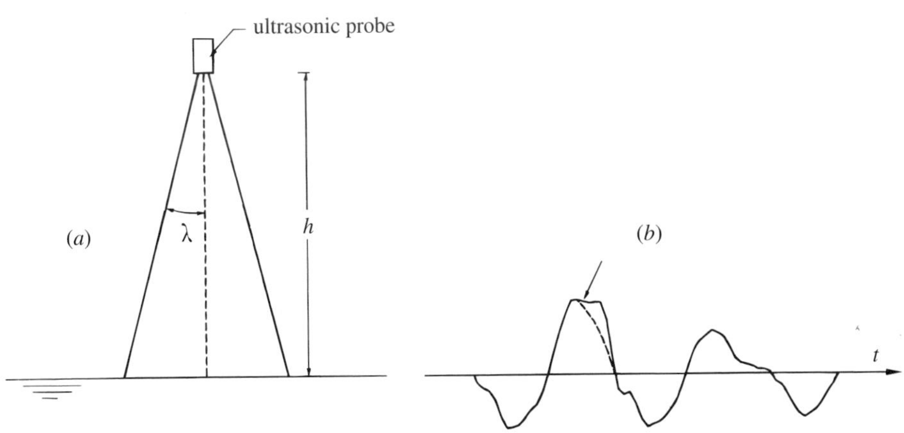

Fig. 14.1 Typical flaw in the record of a wave crest, made by an ultrasonic probe.

is a damped oscillatory function like that shown in fig. 8.8. For this experiment we must build a wall at sea (or also use an existing wall) and we must place an array of wave gauges at some growing distances from this wall.

Why not consider

$$f_2(y) \equiv \frac{\sigma^2_{ph}(y)}{\sigma^2_{ph}(0)} \quad [\sigma^2_{ph}(y) \equiv <\eta^2_{ph}(y, z, t)> \quad \text{for a fixed } z]$$

in place of $f_1(y)$? The relation between $f_2(y)$ and the directional spectrum can be obtained as we did for the relation between $f_1(y)$ and the directional spectrum; and the two functions prove to be very close to each other.

The point is that theoretically we predict that both $f_1(y)$ and $f_2(y)$ should change a lot from the periodic waves to the wind waves. Of course, it is the same whether we succeed in verifying this to be true for $f_1(y)$ or for $f_2(y)$. But dealing with $f_2(y)$ is much simpler. Indeed the pressure head waves are measured by means of pressure transducers which have a very high degree of precision, and can work even very close to each other. Moreover, we have seen that some vertical water jets occur at the wall, so that the surface displacement cannot be measured at the wall whereas the pressure head waves beneath the water surface can be very well measured at the wall.

14.3.3 *Wave measurements from satellites*

Because of the track separation of the orbits, an oceanographic satellite is able to cover most of the ocean area. As an example SEASAT covers 95 per cent of the global ocean area every 36 hours. The radar altimeter on the satellite transmits a pulse centred at about 13.5 GHz vertically downward toward the sea surface, and the shape of the return pulse can be related to the significant wave height. The altimeter measures over a footprint of the order of 10 km which is centred on the satellite subtrack. Typically, the measurements are made at the constant rate of 1/s in which the satellite covers about 7 km. Thus, the radar altimeter gives us the r.m.s. surface displacement of the waves on the space domain, which is equal to the r.m.s. surface displacement of the sea state on the time domain, provided the wave field is homogeneous in space.

If we gather the set of H_s values from an area, we find some cluster of data each of which corresponds to a new pass of the satellite over this area. Then there are several hours without data because the subtrack does not intersect our area. The breaks in data may also correspond to times when the altimeter is turned off for spacecraft manoeuvres and occasional altimeter and spacecraft anomalies.

Young (1999) shows a comparison of buoy and satellite mean monthly values of H_s. The satellite data were from some 4° · 4° squares which included buoys. The least square linear regression based on about 200 mean monthly values per each satellite yields

$$H_{s\ \mathrm{buoy}} = \alpha H_{s\ \mathrm{satellite}} + \beta \tag{14.14}$$

with the α and β of table 14.2. These recent conclusions are in agreement with previous results obtained by Carter et al. (1992) and Cotton and Carter (1994).

Table 14.2 *Parameters of the calibration relationship* (14.14)

Satellite	α	β[m]
GEOSAT	1.144	− 0.148
TOPEX	1.067	− 0.079
ERS1	1.243	+ 0.040

The fact that the r.m.s. surface displacement of the waves on the space domain (in a homogeneous wave field) is equal to the r.m.s. surface displacement of the sea state on the time domain can be easily proved from (8.4*a*). Note, the equality of the two r.m.s., the one on the space domain and the one on the time domain, does not contradict the title of sect. 10.3. Indeed, under the same r.m.s. surface displacement, the probability of exceedance of large wave heights is greater on the time domain than on the space domain.

To prove this statement, resort for simplicity to the long-crested waves (cf. sect. 8.7). Consider the waves on the space domain along y (the direction of wave advance) at any fixed time instant. Prove that these waves on the space domain represent a stationary Gaussian process, which implies (5.44) to be the asymptotic form of the probability of exceedance of their heights. Consider the autocovariance of these waves:

$$\psi(Y) \equiv <\eta(y)\,\eta(y+Y)>,$$

where Y is an arbitrary space lag, and the angle brackets here denote an average with respect to y. Prove the relation

$$\psi(Y) = \int_0^\infty E(\omega)\cos(kY)\,\mathrm{d}\omega.$$

Obtain $\psi(Y)$ from any given spectrum, and verify that the ψ^* of $\psi(Y)$ is smaller than the ψ^* of $\psi(T)$. This conclusion, together with the fact that (5.44) is the asymptotic form of the probability of exceedance, completes the proof.

Conclusive note

The analytical technique for obtaining the directional spectrum is due to the work of Longuet-Higgins et al. (1963) and of Cartwright (1963). The improvement for taking account of the response of the buoy is due to Tucker (1989).

References

Carter D.J.T., Challenor P.G. and Srokosz M.A., 1992 An assessment of Geosat wave height and wind speed measurements. *J. Geophys. Res.* 97, 11383-11392.

Cartwright D.E., 1963 The use of directional spectra in studying the output of a wave recorder on a moving ship. In *Ocean wave spectra, Prentice Hall.*

Cotton P.D. and Carter D.J.T., 1994 Cross calibration of TOPEX, ERS 1, and Geosat wave heights. *J. Geophys. Res.* 99, 25025-25033.

Grimshaw R., 1970 The solitary wave in water of variable depth. *J. Fluid Mech*. 42, 639-656.

Kobune K. and Hashimoto N., 1986 Estimation of directional spectra from maximum entropy principle. *Proc.* 5th *Conf. Offshore Mechanics and Arctic Eng.*, 80-85.

Laitone E. V., 1961 The second approximation to cnoidal and solitary waves. *J. Fluid Mech.* 9, 430-444.

Longuet-Higgins M.S., Cartwright D.E. et Smith N.D., 1963 Observations of the directional spectrum of sea waves using the motions of a floating buoy. In *Ocean wave spectra, Prentice Hall.*

Steele K.E., Wang D.W., Earle M.D. et al., 1998 Buoy pitch and roll computed using three angular rate sensors. *Coastal Eng.* 35, 123-139.

Tucker M.J., 1989 Interpreting directional data from large pitch-roll-heave buoys. *Ocean Eng.* 16, 173-192.

Van der Meer J. W., 1992 Conceptual design of rubble-mound breakwaters. *Proc. Short Course attached to the* 23rd *Conf. Coastal Eng.,* 447-510.

Young I.R., 1999 An intercomparison of GEOSAT, TOPEX and ERS1 measurements of wind speed and wave height. *Ocean Eng.* 26, 67-81.

Appendix A

APPENDIX TO CHAPTERS 6 AND 7: USE OF WAVE HINDCAST AND WAVE MEASUREMENTS FROM SATELLITES

A.1 Long term wave statistics from satellite data

A.1.1 *How to obtain $P(H_s > h)$ from satellite data*

To get the probability $P(H_s > h)$ from the data of a satellite, we must fix an area centred at the point of interest for us. On the one hand, the area must be small enough for its wave climate to be homogeneous; on the other hand, the area must be large enough to get a sufficient number of satellite passes. Usually, sampling squares of $4^\circ \cdot 4^\circ$ are suitable, at least in the oceans far from the coasts.

The multiple observations of H_s associated with the passage of the satellite through the sampling square are not statistically independent of one another. Hence each satellite pass is assumed to produce only one independent observation of the wave conditions. This is because dealing with independent observations enables us to exactly know the confidence intervals (see sect. 6.2.4), so that it is preferable (even if not strictly necessary) to work with independent observations.

Young (1994) follows two alternative methods: (*a*) to take the mean value of each independent pass; (*b*) to take the maximum value of each pass. However, these two methods yield some probability of exceedance different from $P(H_s > h)$. Specifically, the average value method yields $P(\overline{H_s} > h)$, that is the probability of exceedance of an instant average H_s over the sampling square ("instant", because the satellite passes through the sampling square in a very short time). As to the maximum value method, it yields $P(H_{s\max} > h)$, that is the probability of exceedance of an instant maximum H_s over the sampling square.

Here, it can be readily verified that $P(\overline{H_s} > h)$ is smaller than $P(H_s > h)$ on the range of large h. On the contrary, $P(H_{s\max} > h)$ is greater than $P(H_s > h)$. For a sampling square of $4^\circ \cdot 4^\circ$, the differences between $P(H_{s\max} > h)$ and $P(\overline{H_s} > h)$ prove to be relevant: differences on h, for fixed P, even in excess of 25%, as we can see in a figure of the paper of Young (1994).

To get the desired $P(H_s > h)$, without forgoing the convenience of the independent observations, we must take at random one value of H_s for each pass.

Perhaps, somebody could object that so doing we waste a lot of information given that we use only one observation for each pass. But the matter is not so. Let us see an example.

Let us assume the passes over the sampling area are 300 and the number of observations for each pass are 100. Hence, we obtain 300 data points of $P(H_s > h)$, taking at random one observation for each pass. Then we repeat the procedure taking at random a new observation for each pass. In the end we obtain 100 distinct $P(H_s > h)$, each consisting of 300 data points, and we average these $P(H_s > h)$. The result does not change if we obtain the $P(H_s > h)$ from the whole set of the $300 \cdot 100 = 30000$ observations, truncating the data points at the probability level of 1/300.

Of course the average $P(H_s > h)$ will be statistically more reliable than each single $P(H_s > h)$ obtained from one set of 300 independent observations. Or, in other words, the confidence intervals of the average $P(H_s > h)$ will be narrower than the confidence intervals of each single $P(H_s > h)$ from a single set of 300 independent observations. And this is the advantage of having multiple observations for each pass.

A.1.2 *Comparison between $P(H_s > h)$ from satellite data and from buoy data*

Fig. A.1 compares a $P(H_s > h)$ obtained from satellite data with a $P(H_s > h)$ obtained from buoy data for the same locality. The buoy is at the centre of the sampling square of 4°·4°. It is NDBC 46001 (56°17′N, 148°10′W), in the Gulf of Alaska. The satellite is TOPEX. The H_s data from the buoy are 149000 and cover the period 1978-1997. The passes of the satellite, which cover the period July 1996 - February 1999, are 650 with an average of 31 individual observations for each pass.

The satellite data have been corrected by means of the calibration relationship (14.14) with the proper values of α and β of Young (1999). As usual, the graph of the probability has been given in terms of the ancillary variables X and Y defined by (6.3).

The agreement between the two $P(H_s > h)$, the one obtained from the satellite data and the one obtained from the buoy data, looks encouraging: the difference on h at a fixed P level being within 10%.

A.2 Wave hindcast

A.2.1 *A general overview*

The functions $H_s(t)$ and $\overline{\theta}(t)$ (dominant direction vs time) at the grid nodes of a basin can be obtained by means of numerical models like WAM (Komen et al. 1994). The input data are the $\mathbf{u}_{10}(t)$ at the grid nodes ($\mathbf{u}_{10}$ is the wind velocity 10 m above the sea level). The model provides the variations in time of the directional spectrum at the grid nodes. Let us approach this kind of model.

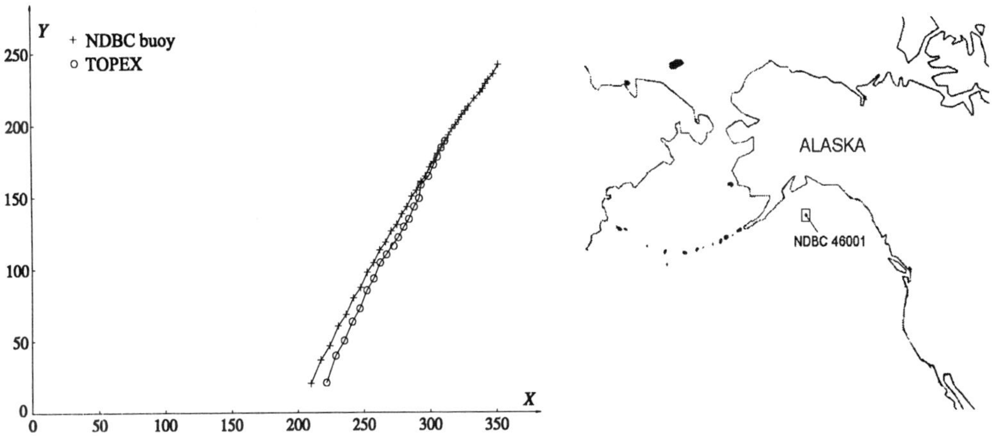

Fig. A.1 Comparison between $P(H_s > h)$ from buoy data and $P(H_s > h)$ from satellite data. The location of the buoy and the sampling area are shown on the right side. (The ancillary variables X and Y are defined by 6.3).

Equation (2.24) for a volume with a small base dx dy reduces itself to

$$\frac{\partial}{\partial y}(\Phi \cos\theta) + \frac{\partial}{\partial x}(\Phi \sin\theta) = 0. \tag{A.1}$$

where θ is the angle between vector $\mathbf{\Phi}$ and y-axis. For the basic case of periodic waves on deep water, (A.1) can be rewritten in the form

$$(c_G \cos\theta \mathbf{i}_y + c_G \sin\theta \mathbf{i}_x) \cdot \nabla\left(\frac{1}{2}a^2\right) = 0, \tag{A.2}$$

where we think of the wave amplitude as a function of x and y. [This function being a constant, equation (A.2) is obviously satisfied.] The formal step from (A.1) to (A.2) is a starting point to approach the numerical wave models.

The l.h.s. of (A.2) multiplied by dx dy and T represents the difference between the energy leaving dx dy in a wave period and the energy entering dx dy in the same time interval. The equation says this difference is equal to zero. Really, the matter is not exactly so. Indeed a small amount of energy is supplied by the wind and some dissipation of wave energy occurs because of whitecapping. These small positive and negative variations cannot be neglected if we aim to look at the wave motion over large areas and/or large time intervals. In these cases equation (A.2) is adjusted in the following way:

$$(c_G \cos\theta \mathbf{i}_y + c_G \sin\theta \mathbf{i}_x) \cdot \nabla\left(\frac{1}{2}a^2\right) = -\frac{\partial}{\partial t}\left(\frac{1}{2}a^2\right) + \Delta,$$

where $\varDelta$ summarizes the physical processes that bring or take energy from the wave motion.

Finally, for a sea state the last equation is expressed in the form

$$[c_G(\omega)\cos\theta\,\mathbf{i}_y + c_G(\omega)\sin\theta\,\mathbf{i}_x]\cdot\nabla S(\omega,\theta) = -\frac{\partial}{\partial t}S(\omega,\theta)+\varDelta(\omega,\theta) \qquad \text{(A.3)}$$

which is the general form of the equation which is integrated by the numerical models. In the case of a basin with an extent such that the effects of the Earth curvature cannot be neglected, equation (A.3) takes on a more complex form where the independent space variables become latitude and longitude of the spherical Earth.

The wind velocities at the grid nodes, which serve as an input for the wave model, can be obtained from some three-dimensional meteorological models. In particular, the ECMWF is a numerical weather prediction scheme which daily produces four analysis fields.

Since 1986, the T106/L19 version of this forecast model was operating with a nominal horizontal resolution of 1.2° and 19 levels in vertical. Then in the 90's the T213/L31 version entailed a doubling of horizontal resolution, and an approximate doubling of vertical resolution between the boundary layer and the stratospheric model levels. The new version brought to a general improvement, particularly near mountainous regions (Simmons, 1991).

A.2.2 *Accuracy of wave hindcast: oceans*

The ERA project (Gibson et al., 1996) describes the atmosphere over a span of time of 15 years: 1979-1993. It uses a horizontal resolution of T106 and 31 levels in vertical. Sterl et al. (1998) use the ERA surface winds as input for the WAM wave model. The modelled H_s are then compared with observations from NDBC buoys.

Two main conclusions emerge. First: the largest peaks of $H_{s\,\mathrm{mod}}$ (H_s from the numerical model) are systematically smaller than the corresponding peaks of $H_{s\,\mathrm{obs}}$ (H_s obtained from the buoys). Second: under low winds, $H_{s\,\mathrm{mod}}$ is systematically larger than $H_{s\,\mathrm{obs}}$.

What is the size of the error? To answer this question the following parameters are considered:

$$\text{normalized bias} \equiv 100(\overline{H_{s\,\mathrm{mod}}} - \overline{H_{s\,\mathrm{obs}}})/\overline{H_{s\,\mathrm{obs}}},$$

$$\text{scatter index} \equiv 100\sqrt{\overline{(H_{s\,\mathrm{mod}} - H_{s\,\mathrm{obs}})^2}}/\overline{H_{s\,\mathrm{obs}}}.$$

A useful additional parameter is the unbiased regression slope (symbol b). If $b > 1$, $H_{s\,\mathrm{mod}} > H_{s\,\mathrm{obs}}$ on the range of the large H_s and/or $H_{s\,\mathrm{mod}} < H_{s\,\mathrm{obs}}$ on the range of the small H_s. Vice versa, if $b < 1$.

Table A.1 *Comparison between modelled H_s (ERA project + cycle 4 of WAM model) and H_s data from buoys (*)*

AREA	NORMALIZED BIAS	SCATTER INDEX	UNBIASED REGRESSION SLOPE
North Atlantic	– 14%	23%	.64
Gulf of Messico	– 10%	27%	.67
Hawaii	+ 10%	19%	.64
North Pacific	+ 8%	19%	.72

(*) These figures represent an average of the figures given in table 1 of Sterl et al. (1998). They are relevant to the runs of WAM in high resolution.

A summary is given in table A.1. We see that the unbiased regression slope is systematically smaller than 1, which confirms the overestimate of the small H_s values and the underestimate of the large H_s values.

The first flaw (underestimate at high winds and waves) is probably a resolution effect. Wind and wave peaks are missed because of finite resolution in space and time. As for the second flaw (overestimate at low winds and waves), there is not a definitive explanation. One possible reason is that the dissipation used in the propagation scheme of the wave model is too low, this could account for the fact that the overestimate of H_s coincides with areas where the incidence of swells is high.

A further important remark to complete the picture is given by Cardone et al. (1996) who examine two big storms: the "Halloween storm" from October 26th to November 2nd 1991, and the "storm of the century" from 12th to 15th March 1993. They apply four wave models, including cycle 4 of WAM, and develop the wind fields using all available conventional data including ship and buoy observations from the relatively dense network of buoys off the U.S. Atlantic coast. One conclusion is that all the wave models perform very well up to about H_s of 12 m, which confirms that errors in wave hindcast are due essentially to operational marine wind field analyses. But the important remark is that, notwithstanding the use of the buoy data to develop the wind fields, all wave models tend to underpredict the sea states with H_s in excess of 12 m.

A.2.3 *The case of the U.S. Atlantic coast*

The extreme storm seas of the U.S. Atlantic coast are not only problematic for the wave hindcast. Look at fig. A2*a*. This shows the $P(H_s > h)$ obtained from twenty year data (1978-1997) of NDBC buoy 41002, some 300 km South of Cape Hatteras. (As usual, the graph is on the plane *X-Y*, definition 6.3.) The 99.9% confidence interval is shown, and we see that the largest H_s value is just at the upper

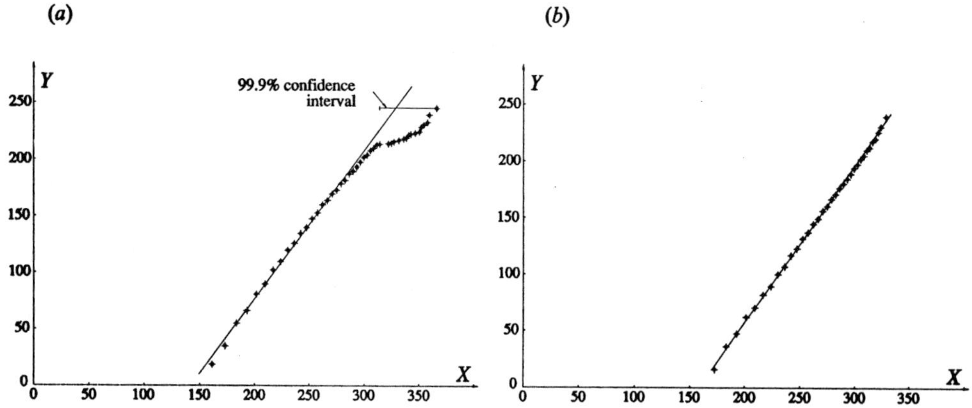

Fig. A.2(a) $P(H_s > h)$ from the H_s data of NDBC 41002. (b) $P(H_s > h)$ from the H_s data of NDBC 41001. The two buoys are in the same area. 41002 recorded the storm of the century; 41001 did not. (The ancillary variables X and Y are defined by 6.3).

limit of this confidence interval. The meaning of the 99.9% confidence interval is: if the asymptotic form of $P(H_s > h)$ was that represented by the straight line in the figure, then sampling 112850 H_s values (as many as the data of NDBC 41002) we would only have a 0.05% probability that the largest H_s value falls to the right of this confidence interval.

Note that the great deviation of the extreme data points is only due to two outstanding storms, the heavier of which is just the extratropical "storm of the century" described by Cardone et al. (1996). To understand this item, please look at fig. A2b which gives the probability of exceedance obtained from the H_s data of NDBC 41001, a buoy of the same area, which was not in function when these two outstanding sea storms occurred.

It is as if some anomalous sea storms were born in the Western Atlantic. *Anomalous* in two senses. *First*, because the extreme H_s values of these sea storms, at present, cannot be explained through wave hindcast (neither using the very wind data recorded at the buoys). *Second*, because the extreme H_s values of these storms contradict the clear asymptotic trend of $P(H_s > h)$ which emerges from the mass of H_s data of that area.

A.2.4 *Accuracy of wave hindcast: Mediterranean Sea*

For some relatively small basins, like Tyrrhenian Sea, Adriatic Sea, or Black Sea, the accuracy of the hindcast decreases. This in part is due to the difficulty of modelling the coastal transition of the surface boundary layer from sea to land or vice versa. This is a sharp transition, and the model is unable to reproduce it closely. The result is that the modelled wind speed sometimes is widely approximated.

Starting from this fact, Cavaleri et al. (1996) suggest leaving the three-dimensional meteorological models and to resort to some simpler two-dimensional wind model. Accordingly, $\mathbf{u}_{10}$ at the grid nodes is estimated from the atmospheric pressure at the nodes themselves, which in its turn is obtained by interpolation of the pressure data from the meteorological stations at the boundary coastline.

The task is less simple than one may imagine because of a very large amount of errors in the file data of the atmospheric pressure. For the special case of the Adriatic Sea, the number of wrong data was at least an order of magnitude greater than was expected!

A two-dimensional wind model cannot take orography into account. In particular, for the Adriatic Sea a correction is necessary to mimic the effect of Apennines (west side of the basin) and Dynaric Alps (east side). If the flow associated with the pressure field points towards the north, then the whole field is forced towards NW. Indeed Apennines and Dynaric Alps channel toward NW the flow which otherwise would point northward.

At present, the accuracy for the Adriatic Sea is as follows:

$$\text{normalized bias} = +20\%, \text{ scatter index} = 100\%,$$

[these figures have been obtained from the figures given on page 99 of the book of Cavaleri et al. (1996)]. As we can see, notwithstanding the large amount of special work done for this basin, the degree of accurancy is still far from that achieved for the oceans.

A.2.5 *Risk analysis*

The set of the hindcasted sea storms is costitued. Typically, this set is subdivided into a number of homogeneous subsets (we mean homogeneous for what concerns the wind direction). For each of these subsets (or also for the whole set) the $P(H_{s\max} > h)$ is estimated. This is the probability that the largest H_s in a storm exceeds any fixed threshold h. Usually, the data points of $P(H_{s\max} > h)$ are fitted by some Fisher-Tippet form. In particular the Fisher-Tippet I is

$$P(H_{s\max} > h) = 1 - \exp\left[-\exp\left(-\frac{h - h_1}{h_2}\right)\right], \tag{A.4}$$

where h_1 and h_2 are two parameters to be found through the best fit of the data points.

Finally, the return period $R(H_s > h)$ is obtained from $P(H_{s\max} > h)$ through the simple relation

$$R(H_s > h) = \frac{\mathscr{T}}{(\mathscr{T}/\Delta t_{\text{storm}})P(H_{s\max} > h)} = \frac{\Delta t_{\text{storm}}}{P(H_{s\max} > h)}, \tag{A.5}$$

where Δt_{storm} is the mean interarrival time of the sea storms. In (A.5), $(\mathscr{T}/\Delta t_{\text{storm}})$ represents the number of sea storms in a very large time interval $\mathscr{T}$ and

$(\mathscr{T}/\Delta t_{\text{storm}})P(H_{s\max} > h)$ is the number of sea storms, during $\mathscr{T}$, in which $H_{s\max}$ exceeds threshold h. Δt_{storm} is simply obtained as the quotient between the time covered by the wave hindcast and the total number of sea storms found in this interval.

For given lifetime L and encounter probability $\mathscr{P}$, the return period R is calculated by means of (7.29b). Hence, the threshold h corresponding to this return period is obtained from (A.5). The h so obtained will be the significant wave height of the design sea state.

In the literature we have not found some estimates of the duration of the design sea state, nor of the probability that the largest wave height in the lifetime exceeds a fixed threshold. Of course, we mean estimates based upon the wave hindcast.

The assumption that $P(H_{s\max} > h)$ must have a particular form proceeds from the general argument of Fisher and Tippet (1928). Let us consider n outcomes of the same random variable V, and let us assume these outcomes to be stochastically independent of one another. Let us also define a new random variable ($V_{\max}$) as equal to the largest value of the n outcomes of V. Then, for a variety of forms of $P(V > h)$ (h being the dummy variable), the asymptotic form of $P(V_{\max} > h)$ as $h \to \infty$ must respect a well defined functional equation. The function (A.4) satisfies this equation.

A.3 Trend in the wave climate and its effects on engineering

A.3.1 *Uses of the wave data from satellites and from hindcast*

The wave hindcast can provide $H_s(t)$ and $\overline{\theta}(t)$ at some fixed points for time intervals of several years. Therefore, the wave hindcast is fully compatible with the methods described in chapters 6 and 7.

The satellite data can be used for the risk analysis at oceanic locations far from the coasts. These data do not supply information on the storm duration, but they enable us to estimate $P(H_s > h)$ which is the basic requirement for the use of the solutions of chapter 7.

However, it is evident that the degree of precision achievable from the buoy (or wave staff) data is the greatest. Therefore, whenever possible we should resort to these data for risk analysis in engineering. Nevertheless, the use of the other data sources can play an important role in engineering.

First, the use of the meteorological data is the main road to wave forecast which is obviously important during operations. Then, there is a field where the wave data from satellite and from hindcast play a key role. This is the field of the variations in the wave climate, a field of obvious interest for human life, which we shall consider in the next section from the point of view of engineering.

Really, use of meteorological data is not the only way of forecasting. We have already cited in chap. 6 the recent studies of Cunha and Guedes Soares (1999) on the nature of the random process $H_s(t)$. Well, one of the aims of these studies is to predict the evolution of $H_s(t)$ in the next hours starting from the knowledge of $H_s(t)$ in the past hours; knowledge which can be accurately obtained from a buoy.

A.3.2 *Trends in wave climate*

Young (1999) has identified a number of squares of $4^o \cdot 4^o$ on the Earth's surface with a sufficient number of passes of the satellites to yield reliable mean monthly values of H_s. Then, a comparison between the mean monthly values from 1986 to 1990 and the mean monthly values from 1991 to 1995 has shown that there has been no measurable change in the global wave field.

Some similar conclusions also proceed from wave hindcast. In particular, Sterl et al. (1998) conclude that they are unable to confirm a significant change in wave height during the ERA period (1979-1993). However, they have noted that in the North Atlantic there is a trend of the annual mean H_s of $+0.01$ m/year which makes a rate of growth of about 0.5 per cent per year.

Through analysis of H_s data obtained from visual observations and from shipborne wave recorders, Bacon and Carter (1991) found a greater trend: an average of 2 per cent per year, during the period 1962-1988. This conclusion is not necessarily in contrast with the one of Sterl et al. Indeed there is some evidence that the climate was more stable during the ERA period (1979-1993) than in the two previous decades. This evidence emerges from the time series of the NAO index given by Hurrel (1995). (NAO being the normalized pressure difference between Portugal and Iceland.)

Sterl et al. (1998) note also that the trend of the annual mean H_s appears to be nearly the same as that of $h(P)$ for any fixed P [recalling $h(P)$ to be the inverse function of $P(H_s > h)$]. Hence, if we assume the form (6.2) of $P(H_s > h)$, we can conclude that a variation of x% of the annual mean H_s leads to the same per cent variation of the dimensional parameter w and no variation of the nondimensional parameter u.

This means that the trend of $+0.5$ per cent per year of the annual mean H_s in the North Atlantic implies a trend of $+0.5$ per cent per year also of w.

A.3.3 *How to obtain the design waves for an area where the annual mean H_s grows at a given rate*

Let us assume that w has a value w_{inf} at the start of the lifetime, and that w grows at a constant rate up to a value w_{sup} at the end of the lifetime. As a consequence, during the lifetime, w will be uniformly distributed on the range $(w_{\text{inf}}, w_{\text{sup}})$.

Here, let us think of $H_s(t)$ as a piecewise stationary random process. Specifically, let us imagine we subdivide the time axis into small intervals Δt, in each of which the $H_s(t)$ represents a piece of a stationary random process with a $P(H_s > h)$ given by (6.2). These $P(H_s > h)$ have the same u value and values of w uniformly distributed on $(w_{\text{inf}}, w_{\text{sup}})$.

The solutions of chap. 7 can be easily adapted for this non-stationary condition. As an example, let us consider the simplest of these solutions, the one for the return period $R^*(H)$, we gave on sect. 7.6. Let us write $p(H_s = h;\, w')$ and $P(H_s = h;\, w')$, meaning respectively $p(H_s = h)$ and $P(H_s > h)$ on the time intervals where $w = w'$. Then let us define

$p(w = w')\mathrm{d}w' \equiv$ probability that parameter w takes on a value within $(w', w' + \mathrm{d}w') =$ = ratio between time in which w is within $(w', w' + \mathrm{d}w')$ and total time.

For our piecewise stationary process we have

$$p(w = w')\begin{cases} = (w_{\mathrm{sup}} - w_{\mathrm{inf}})^{-1} \text{ if } w_{\mathrm{inf}} < w' < w_{\mathrm{sup}}, \\ = 0 \text{ otherwise}. \end{cases} \tag{A.6}$$

Let us retrace the steps leading to the expression of $R^*(H)$. Now we have

$p(H_s = h;\, w')\mathrm{d}h\; p(w = w')\mathrm{d}w'\, \mathscr{T} =$ time in which H_s falls between h and $h + \mathrm{d}h$, and w falls between w' and $w' + \mathrm{d}w'$, during $\mathscr{T}$;

$P(H;\, H_s = h)\dfrac{1}{\overline{T}(h)}p(H_s = h;\, w')\mathrm{d}h\; p(w = w')\mathrm{d}w'\, \mathscr{T} =$ number of waves higher than a fixed threshold H in the sea states with $h < H_s < h + \mathrm{d}h$ in the periods of time in which $w' < w < w' + \mathrm{d}w'$, during $\mathscr{T}$.

The integral of this last number over

$$\{(h, w')|h \in (0, \infty),\, w' \in (w_{\mathrm{inf}}, w_{\mathrm{sup}})\}$$

yields $\mathscr{N}(H, \mathscr{T})$ that is the total number of waves higher than H in $\mathscr{T}$:

$$\mathscr{N}(H, \mathscr{T}) = \int_{w_{\mathrm{inf}}}^{w_{\mathrm{sup}}} \int_0^\infty P(H;\, H_s = h)\frac{1}{\overline{T}(h)}p(H_s = h;\, w')(w_{\mathrm{sup}} - w_{\mathrm{inf}})^{-1}\mathrm{d}h\, \mathrm{d}w'\, \mathscr{T},$$

where $p(w = w')$ has been substituted by its expression (A.6). Finally, the return period $R^*(H)$, which is equal to $\mathscr{T}/\mathscr{N}(H, \mathscr{T})$, can be expressed in the form

$$R^*(H) = \left[\int_0^\infty P(H;\, H_s = h)\frac{1}{\overline{T}(h)}\tilde{p}(H_s = h)\mathrm{d}h\right]^{-1} \tag{A.7}$$

with the definition

$$\tilde{p}(H_s = h) \equiv (w_{\mathrm{sup}} - w_{\mathrm{inf}})^{-1}\int_{w_{\mathrm{inf}}}^{w_{\mathrm{sup}}} p(H_s = h;\, w')\mathrm{d}w'.$$

(A.7) is the form of $R^*(H)$ for our piecewise stationary $H_s(t)$. Similarly, we can adapt the expressions (7.11) of $R(H_s > h)$, (7.18) of $\overline{D}(h)$, (7.37) of $P[H_{\max}(L) > H]$, and (7.40) of $p(H_s = h;\, H_{\max} = x)$, replacing in these expressions $p(H_s = h)$ with $\tilde{p}(H_s = h)$, and $P(H_s > h)$ with

$$\widetilde{P}(H_s > h) = (w_{\mathrm{sup}} - w_{\mathrm{inf}})^{-1}\int_{w_{\mathrm{inf}}}^{w_{\mathrm{sup}}} P(H_s > h;\, w')\mathrm{d}w'.$$

Example. In sect. 7.4.1 we have evaluated the significant wave height of a design sea state at a point of the central Mediterranean Sea. Now, let us see how it changes because of a variation of the wave climate during the lifetime. Specifically, we shall assume that, during the lifetime of 50 years, parameter w grows at a constant rate of 0.5% per year of its initial value. Accordingly, we have

$$u = 1.200,\ w_{\rm inf} = 0.874\,{\rm m},\ w_{\rm sup} = w_{\rm inf} + (50 \cdot 0.005)w_{\rm inf} = 1.093\,{\rm m}.$$

Then, using the expression (7.11) of $R(H_s > h)$ with $\tilde{p}(H_s > h)$ and $\widetilde{P}(H_s > h)$ in place, respectively, of $p(H_s > h)$ and $P(H_s > h)$ we arrive at

$$h\,(475 \text{ years}) = 9.15\,{\rm m}$$

against the 7.90 m obtained in sect. 7.4.1 under the assumption of $w = \text{const} = 0.874\,{\rm m}$. The resulting increase of the design wave height is thus 16%.

Note, with the non-stationary model the solution for $h(R)$ given in sect. 7.2.1 is no longer valid. Hence we have to compute the function $R(H_s > h)$ and find the h value which corresponds to the given value of R.

References

Bacon S. and Carter D.J.T., 1991 Wave climate changes in the North Atlantic and North Sea. *Int. J. of Climatology* 11, 545-558.

Cardone V.J., Jensen R.E., Resio D.T. et al., 1996 Evaluation of contemporary ocean wave models in rare extreme events: the "Halloween Storm" of October 1991 and the "Storm of the century" of March 1993. *J. Atmos. Oceanic Tech.* 13, 198-230.

Cavaleri L. et al., 1996 Wind and waves in the northern Adriatic Sea. *Italian Soc. of Physics*, 1-195.

Cunha C. and Guedes Soares C., 1999 On the choice of data transformation for modelling time series of significant wave height. *Ocean Eng.* 26, 489-506.

Fisher R.A. and Tippett L.H.C., 1928 Limiting forms of the frequency distribution of the largest or smallest member of a sample. *Proc. Cambridge Phil. Soc.* 24, 180.

Gibson R., Kallberg P. and Upsala S., 1996 The ECMWF Re-analysis (ERA) Project, *ECMWF Newsletter* 73, 7-17.

Hurrel J.W., 1995 Decadal trends in the North Atlantic oscillation: regional temperatures and precipitation. *Science* 269, 676-679.

Komen G.J., Cavaleri L., Donelan M. et al., 1994 Dynamics and modelling of ocean waves. *Cambridge University Press*, 1-532.

Simmons A., 1991 Development of the operational 31-level T213 version of the ECMWF forecast model, ECMWF *Newsletter* 56, 3-13.

Sterl A., Komen G.J. and Cotton P.D., 1998 15 Years of global wave hindcast using ERA winds. Validating the reanalysed winds and assessing the wave climate. *J. Geophys. Res.* 103, 5477-5492.

Young I. R., 1994 Global ocean wave statistics obtained from satellite observations. *Appl. Ocean Res.* 98, 20275-20285;
1999 An intercomparison of GEOSAT, TOPEX, and ERS1 measurements of wind speed and wave height. *Ocean Eng.* 26, 67-81.

Appendix B
APPENDIX TO CHAPTERS 9 AND 10: THE WAVE GROUP OF THE MAXIMUM EXPECTED CREST ELEVATION, AND THE WAVE GROUP OF THE MAXIMUM EXPECTED CREST-TO-TROUGH HEIGHT

B.1 The first version of the quasi-determinism theory

B.1.1 *Two versions*

Two versions of the theory were given by the author: the first one dealing with the crest elevation and the second one with the wave height. The second one is more advanced and indeed it brings as a corollary the closed solution for the wave height probability. This is why in chapters 9 and 10 we have given the second version.

In this Appendix we re-propose the first version of the theory. Then the two versions are shown to be consistent with each other.

B.1.2 *Definitions*

Let us apply the compact notations and assumptions (9.32) of sect. 9.6 and let us deal with the random waves on the time domain at a fixed point x_o, y_o.

Let us consider $p(\eta_T = w|\mathscr{C} = \beta)\,\mathrm{d}w$, the conditional probability of the surface displacement at time $t_o + T$ given a local wave maximum of elevation β at time t_o. Of course, having taken $m_0 = \sigma^2 = 1$, β is the quotient between the height of the local wave maximum and the root mean square surface displacement σ.

We have

$$p(\eta_T = w|\mathscr{C} = \beta) = \frac{EX1(\beta, w; T)}{EX2(\beta)}, \tag{B.1}$$

where

$EX1(\beta, w; T)\,\mathrm{d}\beta\mathrm{d}w \equiv$ expected number per unit time of local wave maxima having an elevation between β and $\beta + \mathrm{d}\beta$ and being followed after a fixed time lag T by a surface displacement between w and $w + \mathrm{d}w$;

$EX2(\beta)\,\mathrm{d}\beta \equiv$ expected number per unit time of local wave maxima having an elevation between β and $\beta + \mathrm{d}\beta$.

B.1.3 *Solution for the expected numbers EX1 and EX2*

Both *EX1* and *EX2* can be turned into Rice's problems. It suffices to note that

$$EX1\,(\beta, w; T)\,\mathrm{d}\beta\,\mathrm{d}w = \mathscr{P}/\mathrm{d}t, \tag{B.2}$$

where $\mathscr{P}$ is the probability that a fixed small interval dt includes a local maximum with the two following attributes:

(i) to have an elevation between β and $\beta + \mathrm{d}\beta$,

(ii) to be followed after a fixed time lag T by a surface displacement between w and $w + \mathrm{d}w$.

Of course, also $EX2(\beta)\mathrm{d}\beta$ can be given the form (B.2), with $\mathscr{P}$ being the probability that a fixed small interval dt includes a local wave maximum with the only attribute (i). For simplicity we fix the small interval $(-\mathrm{d}t/2, \mathrm{d}t/2)$.

The solutions are

$$EX1\,(\beta, w; T)\,\mathrm{d}\beta\,\mathrm{d}w = \int_{-\infty}^{0} |u| p\,(\eta_0 = \beta, \dot{\eta}_0 = 0, \ddot{\eta}_0 = u, \eta_T = w)\,\mathrm{d}u\,\mathrm{d}\beta\,\mathrm{d}w,$$

$$EX2\,(\beta)\,\mathrm{d}\beta = \int_{-\infty}^{0} |u| p\,(\eta_0 = \beta, \dot{\eta}_0 = 0, \ddot{\eta}_0 = u)\,\mathrm{d}u\,\mathrm{d}\beta,$$

where the form of *EX2* was obtained by Rice himself. Then with the expressions of the two joint p.d.f. we obtain

$$EX1\,(\beta, w; T) = -\frac{1}{(2\pi)^2\sqrt{M}} \int_{-\infty}^{0} u \exp\left\{-\frac{1}{2M}\,[M_{33}u^2 + \right.$$
$$\left. + 2(M_{13}\beta + M_{34}w)u + (M_{11}\beta^2 + M_{44}w^2 + 2M_{14}\beta w)]\right\}\mathrm{d}u, \tag{B.3a}$$

$$EX2\,(\beta) = -\frac{1}{(2\pi)^{3/2}\sqrt{\widetilde{M}}} \int_{-\infty}^{0} u \exp\left[-\frac{1}{2\widetilde{M}}\,(\widetilde{M}_{33}u^2 + 2\widetilde{M}_{13}\beta u + \widetilde{M}_{11}\beta^2)\right]\mathrm{d}u, \tag{B.3b}$$

where M and M_{ij} are, respectively, the determinant and the i, j cofactor of the covariance matrix of $\eta_0, \dot{\eta}_0, \ddot{\eta}_0, \eta_T$:

$$\begin{array}{l} M \equiv \text{determinant}, \\ M_{ij} \equiv i, j \text{ cofactor of} \end{array} \quad \begin{pmatrix} 1 & 0 & -1 & \psi_T \\ 0 & 1 & 0 & -\dot{\psi}_T \\ -1 & 0 & m_4 & \ddot{\psi}_T \\ \psi_T & -\dot{\psi}_T & \ddot{\psi}_T & 1 \end{pmatrix}, \tag{B.4}$$

and $\widetilde{M}, \widetilde{M}_{ij}$ are the determinant and the i, j cofactor of the covariance matrix of $\eta_0, \dot{\eta}_0, \ddot{\eta}_0$, which is obtained by cancelling the fourth row and the fourth column of the matrix (B.4):

$$\begin{array}{l} \widetilde{M} \equiv \text{determinant}, \\ \widetilde{M}_{ij} \equiv i, j \text{ cofactor of} \end{array} \quad \begin{pmatrix} 1 & 0 & -1 \\ 0 & 1 & 0 \\ -1 & 0 & m_4 \end{pmatrix}.$$

In looking at these two matrices, bear in mind that we have taken $m_0 = 1$ and $m_2 = 1$. Also bear in mind what was said in sect. 9.6 about the existence of m_4.

Both the integral in equation (B.3*a*) of *EX1* and the integral in equation (B.3*b*) of *EX2* are of the type

$$I \equiv -\int_{-\infty}^{0} u \exp\left[-\left(Q_1 u^2 + 2Q_2 u + Q_3\right)\right] du,$$

where Q_1, Q_2 and Q_3 are some *constants*, with $Q_1 > 0$. Such an integral is split into two integrals:

$$I = -\frac{1}{2Q_1}\int_{-\infty}^{0}(2Q_1 u + 2Q_2)\exp\left[-\left(Q_1 u^2 + 2Q_2 u + Q_3\right)\right] du + \frac{Q_2}{Q_1}\int_{-\infty}^{0}\exp\left[-\left(Q_1 u^2 + 2Q_2 u + Q_3\right)\right] du.$$

The first of these two integrals can be easily evaluated by substitution, and the solution of the second one is found in books of applied mathematics like Abramowitz and Stegun's. The result is

$$I = \frac{\sqrt{\pi}}{2Q_1}\exp(-Q_3)\left\{\frac{1}{\sqrt{\pi}} + \frac{Q_2}{\sqrt{Q_1}}\exp\left(\frac{Q_2}{\sqrt{Q_1}}\right)^2\left[1 + \operatorname{erf}\left(\frac{Q_2}{\sqrt{Q_1}}\right)\right]\right\}.$$

B.1.4 *The probability density function of the surface displacement near a local wave maximum with a given height*

Evaluating the integrals on the right hand sides of (B.3*a*) and (B.3*b*) we obtain the expressions of *EX1* and *EX2*. Then using these expressions and equation (B.1) we arrive at

$$p(\eta_T = w|\mathscr{C} = \beta) = F\exp\left(Aw^2 + 2B\beta w + C\beta^2\right)\left\{\frac{1}{\sqrt{\pi}} + (Dw + E\beta)\exp(Dw + E\beta)^2\cdot\right.$$
$$\left.\cdot\left[1 + \operatorname{erf}(Dw + E\beta)\right]\right\}\Big/\widetilde{F}\exp\left(\widetilde{C}\beta^2\right)\left\{\frac{1}{\sqrt{\pi}} + \widetilde{E}\beta\exp\left(\widetilde{E}\beta\right)^2\left[1 + \operatorname{erf}\left(\widetilde{E}\beta\right)\right]\right\}, \qquad \text{(B.5)}$$

where

$$A \equiv \frac{1 - m_4}{2M}, \quad B \equiv \frac{\ddot{\psi}_T + \psi_T m_4}{2M}, \quad C \equiv \frac{-m_4 + \ddot{\psi}_T^2 + \dot{\psi}_T^2 m_4}{2M},$$

$$D \equiv \frac{-(\psi_T + \ddot{\psi}_T)}{\sqrt{2M\left(1 - \psi_T^2 - \dot{\psi}_T^2\right)}}, \quad E \equiv \frac{1 - \dot{\psi}_T^2 + \psi_T\ddot{\psi}_T}{\sqrt{2M\left(1 - \psi_T^2 - \dot{\psi}_T^2\right)}}, \quad F \equiv \frac{\sqrt{\pi}M}{(2\pi)^2\left(1 - \psi_T^2 - \dot{\psi}_T^2\right)},$$

$$\widetilde{C} \equiv \frac{-m_4}{2(m_4 - 1)}, \quad \widetilde{E} \equiv \frac{1}{\sqrt{2(m_4 - 1)}}, \quad \widetilde{F} \equiv \frac{\sqrt{\pi(m_4 - 1)}}{(2\pi)^{3/2}}.$$

Here it is convenient to define the function

$$f(x) \equiv \frac{1}{\sqrt{\pi}} + x\exp(x^2)(1 + \operatorname{erf} x),$$

which enables us to rewrite (B.5) in the compact form

$$p(\eta_T = w|\mathscr{C} = \beta) = F\exp(Aw^2 + 2B\beta w + C\beta^2)f(Dw + E\beta)\Big/\Big[\widetilde{F}\exp(\widetilde{C}\beta^2)f(\widetilde{E}\beta)\Big]. \tag{B.6}$$

The function $f(x)$ is positive on the whole domain $(-\infty, +\infty)$, and approaches zero as $x \to -\infty$. Indeed as $x \to -\infty$ the product

$$[x\exp(x^2)](1 + \mathrm{erf}\,x)$$

is of the type $-\infty \cdot 0$, and approaches $-\dfrac{1}{\sqrt{\pi}}$ as it can be easily verified by means of L'Hopital's rule. Moreover, $f(x)$ approaches the asymptotic form

$$f(x) = 2x\exp(x^2) \quad \text{as } x \to +\infty. \tag{B.7}$$

B.1.5 *The form of the probability density function as $\beta \to \infty$*

Being $\widetilde{E} > 0$, $f(\widetilde{E}\beta)$ takes on the asymptotic form (B.7) as $\beta \to \infty$:

$$f(\widetilde{E}\beta) = 2\widetilde{E}\beta\exp(\widetilde{E}\beta)^2 \qquad \text{as } \beta \to \infty. \tag{B.8}$$

As to $f(Dw + E\beta)$, it takes on the asymptotic form (B.7) as $Dw + E\beta \to \infty$:

$$f(Dw + E\beta) = 2(Dw + E\beta)\exp(Dw + E\beta)^2 \quad \text{as} \quad Dw + E\beta \to \infty. \tag{B.9}$$

Equations (B.8) and (B.9) enable us to rewrite (B.6) in the form

$$\begin{aligned} p(\eta_T = w|\mathscr{C} = \beta) = {} & \frac{F}{\widetilde{E}\widetilde{F}}\left(D\frac{w}{\beta} + E\right)\exp[(Aw^2 + 2B\beta w + C\beta^2) + \\ & + (Dw + E\beta)^2 - (\widetilde{C} + \widetilde{E}^2)\beta^2] \quad \text{as } \beta \to \infty \text{ and } Dw + E\beta \to \infty, \end{aligned} \tag{B.10}$$

from which we obtain

$$p(\eta_T = w|\mathscr{C} = \beta) = \frac{F}{\widetilde{E}\widetilde{F}}\left(D\frac{w}{\beta} + E\right)\exp\left[-\frac{(w - \psi_T\beta)^2}{2(1 - \psi_T^2 - \dot{\psi}_T^2)}\right] \text{as } \beta \to \infty \text{ and } Dw + E\beta \to \infty. \tag{B.11}$$

Check the step from (B.10) to (B.11). Note first that the exponential on the r.h.s. of (B.10) has its maximum at

$$w_m = -\frac{DE + B}{A + D^2}\beta. \tag{B.12}$$

Replacing A, B, D and E with their expressions, you will see that a number of terms cancel out, and (B.12) is reduced to

$$w_m = \psi_T \beta .$$

Then, use w_m to rewrite the exponential on the r.h.s. of (B.10) in the form:

$$\text{exponential on the r.h.s. of (B.10)} =$$
$$= \exp\left\{(A + D^2)(w - w_m)^2 + \left[(Aw_m^2 + 2B\beta w_m + C\beta^2) + (Dw_m + E\beta)^2 - (\widetilde{C} + \widetilde{E}^2)\beta^2\right]\right\}. \tag{B.13}$$

Here you will be able to prove that the sum within the square parentheses is zero. To this end, first replace A, B, C ... with their own expressions, and note that $\widetilde{C} + \widetilde{E}^2 = -\frac{1}{2}$. Hence, you will be able to rearrange the sum within the square parentheses in the form of a quotient where the numerator will prove to consist of 111 terms: $-\psi_T^2 + \psi_T^4 + \psi_T^2 \dot{\psi}_T^2 + ...$ which cancel out. Of course, the work will be simplified a good deal by resorting to some compact notations, e.g. $a \equiv \psi_T, b \equiv \dot{\psi}_T$ and so on.

Now you can rewrite the exponential on the r.h.s. of (B.10) in the form

$$\exp\left[(A + D^2)(w - w_m)^2\right],$$

where

$$A + D^2 = \frac{(1 - m_4)(1 - \psi_T^2 - \dot{\psi}_T^2) + (\psi_T + \ddot{\psi}_T)^2}{2M(1 - \psi_T^2 - \dot{\psi}_T^2)} .$$

Here you will find that the numerator is opposite to M (the determinant of the covariance matrix), so that the exponential on the r.h.s. of (B.10) can be reduced to the simple form on the r.h.s. of (B.11).

When the theory was set up, it was guessed that the term within the square parentheses on the r.h.s. of (B.13) was equal to zero. Indeed this was the only way for the condition

$$\int_{-\infty}^{\infty} p(\eta_T = w | \mathscr{C} = \beta)\, dw = 1$$

to be satisfied also in the limit as $\beta \to \infty$. Then the verification followed.

B.1.6 *Conclusion: quasi-determinism*

The form (B.11) of the conditional p.d.f. holds for $w = \psi_T \beta$ as $\beta \to \infty$, given that

$$D\psi_T + E = \sqrt{\frac{1 - \psi_T^2 - \dot{\psi}_T^2}{2M}} > 0$$

($1 - \psi_T^2 - \dot{\psi}_T^2$ being greater than zero since it is the determinant of the covariance matrix of $\eta_0, \dot{\eta}_0, \eta_T$).

Consequently, the form (B.11) holds also for $w = \psi_T\beta + \Delta w$ with Δw of an order smaller than β, and we have

$$p(\eta_T = \psi_T\beta + \Delta w | \mathscr{E} = \beta) = \frac{1}{\sqrt{2\pi(1-\psi_T^2-\dot{\psi}_T^2)}} \exp\left[-\frac{\Delta w^2}{2(1-\psi_T^2-\dot{\psi}_T^2)}\right]$$
$$\text{as } \beta \to \infty \text{ and } \Delta w \text{ o } \beta, \qquad \text{(B.14)}$$

where the term which multiplies the exponential function on the r.h.s. of this equation is equal to

$$\frac{F}{\tilde{E}\tilde{F}}(D\psi_T + E).$$

The integral of the r.h.s. of (B.14) for $\Delta w \in (-\infty, +\infty)$ is equal to 1, so that equation (B.14) yields

$$\int_{-\beta^\nu}^{\beta^\nu} p(\eta_T = \psi_T\beta + \Delta w | \mathscr{E} = \beta)\, \mathrm{d}\Delta w \to 1 \quad \text{as } \beta \to \infty$$

for any fixed ν in (0,1).

In words, the probability that η_T falls within

$$(\psi_T\beta - \beta^\nu, \psi_T\beta + \beta^\nu)$$

approaches 1, as $\beta \to \infty$, for any fixed ν in (0,1). In other words, the probability approaches 1 that

$$\frac{\eta_T}{\psi_T\beta} \to 1 \quad \text{as } \beta \to \infty,$$

clearly provided that $\psi_T \neq 0$. Or in other words, as $\beta \to \infty$, the probability approaches 1 that the random function η_T is asymptotically equal to the deterministic function $\psi_T\beta$. Of course, this is true only provided that T is finite, given that $\psi_T \to 0$ as $|T| \to \infty$.

B.2 Corollaries of the first version

B.2.1 *The probability of the wave crest elevation*

The general form (i.e. the form without the assumption 9.32 and the compact notations) is: if a local wave maximum of given elevation b occurs at a time instant t_o at a fixed x_o, y_o, the probability approaches 1 that the surface displacement at x_o, y_o is asymptotically equal to

$$\overline{\eta}(t_o + T) = \frac{\psi(T)}{\psi(0)}\, b\,, \qquad \text{(B.15)}$$

as $b/\sigma \to \infty$.

Given that the absolute maximum of $\psi(T)$ falls at $T = 0$, the local wave maximum of given elevation b is also the highest maximum of its wave. This is rather intuitive. An exceptionally high local maximum, with a very high degree of probability, is a wave crest. Implicitly, we had anticipated this conclusion by using the symbol $\mathscr{C} = \beta$ for the elevation of the local wave maximum ($\mathscr{C}$ being the symbol of the crest elevation). A direct consequence is

$$\frac{\mathscr{N}_{cr}(b;\mathscr{T})}{\mathscr{N}_{+}(b;\mathscr{T})} \to 1 \quad \text{as } b/\sigma \to \infty, \tag{B.16}$$

(in words, the number of wave crests exceeding a very high threshold b tends to coincide with the number of b up-crossings) which in its turn implies

$$P(\mathscr{C} > b) = \exp\left(-\frac{b^2}{2m_0}\right) \quad \text{as } b/\sigma \to \infty. \tag{B.17}$$

This result is even more general than the one on the probability of the wave height (given in chap. 9). Indeed, here the absolute minimum of the autocovariance is not required to also be the first local minimum of this function.

To prove (B.17), use (5.17), (5.28), and (B.16).

B.2.2 *The maximum expected elevation of the wave crest*

The maximum expected crest elevation in a sequence of N waves is the integral over $(0, \infty)$ of the probability of exceedance $P(\mathscr{C}_{\max} > b)$, and with the asymptotic form (B.17) of $P(\mathscr{C} > b)$, it is given by

$$\overline{\mathscr{C}_{\max}} = \int_0^\infty 1 - \left[1 - \exp\left(\frac{-b^2}{2m_0}\right)\right]^N \mathrm{d}b. \tag{B.18}$$

The logic leading to this result is the same followed in sect. 5.10 to get the formula (5.57) for $\overline{H_{\max}}$. A comparison of the two formulae, that is to say (5.57) for $\overline{H_{\max}}$ and (B.18) for $\overline{\mathscr{C}_{\max}}$, shows that

$$\overline{\mathscr{C}_{\max}} = \frac{\overline{H_{\max}}}{\sqrt{2(1+\psi^*)}}. \tag{B.19}$$

Hence, $\overline{\mathscr{C}_{\max}}$ is generally greater than $\overline{H_{\max}}/2$.

B.2.3 *The space-time formulation*

The result (B.15) can be extended from the time domain to the space-time by retracing the reasoning of sect. B.1. There is only one difference, we have to deal with the surface displacement at time $t_o + T$ at $x_o + X, y_o + Y$ rather than at x_o, y_o. The conclusion is that if a wave crest of given elevation b occurs at point x_o, y_o at time t_o, as $b/\sigma \to \infty$, the probability approaches 1 that the surface displacement approaches the deterministic form

$$\overline{\eta}(x_o + X, y_o + Y, t_o + T) = \frac{\Psi(X, Y, T; x_o, y_o)}{\Psi(0, 0, 0; x_o, y_o)} b. \qquad \text{(B.20)}$$

As usual, this is true provided that X, Y and T are finite.

Finally, reasoning as we did in sects. 9.4 and 9.5, we can easily verify that the velocity potential of the deterministic waves (B.20) is given by the formula

$$\overline{\phi}(x_o + X, y_o + Y, z, t_o + T) = \frac{\Phi(X, Y, z, T; x_o, y_o)}{\Psi(0, 0, 0; x_o, y_o)} b$$

which holds whatever the configuration of the solid boundary.

B.3 The relationship between the two versions of the theory

B.3.1 *The group producing the maximum expected wave height*

Let us consider a sea state of very large duration. When the maximum expected wave height occurs at a fixed point x_o, y_o, we can expect that the surface displacement will be given by (10.1a) with $H = \overline{H_{\max}}$:

$$\overline{\eta}(x_o + X, y_o + Y, t_o + T) = \frac{\Psi(X, Y, T; x_o, y_o) - \Psi(X, Y, T - T^*; x_o, y_o)}{\Psi(0, 0, 0; x_o, y_o) - \Psi(0, 0, T^*; x_o, y_o)} \frac{\overline{H_{\max}}}{2}. \qquad \text{(B.21)}$$

We have seen this function represents a three dimensional wave group passing at x_o, y_o at the apex of its development stage. Let us imagine we record the wave group with two gauges being aligned with the wave direction, the first one at x_o, y_o and the second one somewhat before x_o, y_o. The records made by the two ideal gauges at points

$$⓪ = x_o, y_o,$$

$$① = x_o, y_o - 0.20 L_p,$$

are shown in fig. B.1. The maximum wave height during the group's evolution occurs at x_o, y_o. At this point the envelope centre coincides with the central zero of the central wave. The wave crest precedes the envelope centre and hence it is decreasing. This means that before point x_o, y_o the wave crest had to be somewhat higher, and indeed we see that at point ①, some fifth of a wavelength before x_o, y_o the crest of the central wave is somewhat higher than at point ⓪. Point ① has been chosen because at this point the crest of the central wave occupies the envelope centre, and is the highest wave crest during the group's evolution.

Henceforth we shall call

configuration of type (a), the one of the group at point ⓪ with the envelope centre being coincident with the wave's central zero;

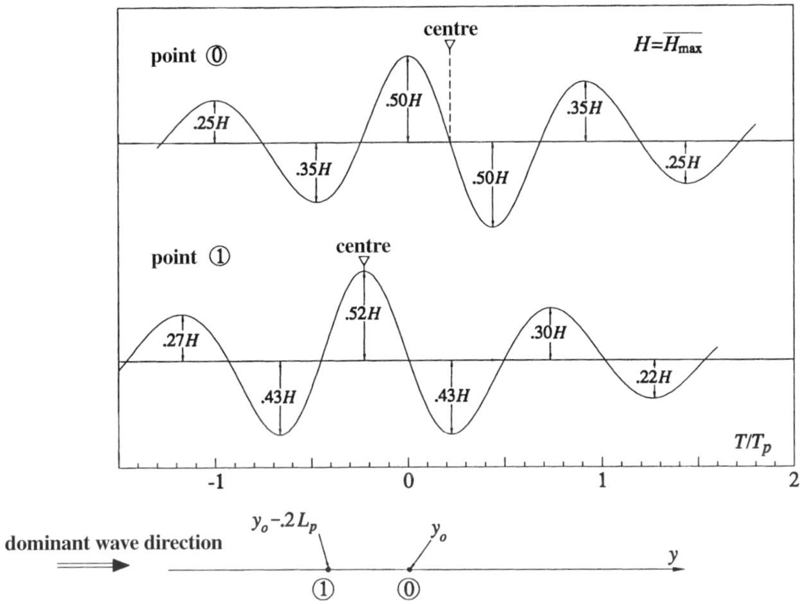

Fig. B.1 The group of the wave with the maximum expected height, at points ⓪ and ① (⓪ coincides with x_o, y_o and ① is somewhat before ⓪ along the propagation axis). The group exhibits its maximum crest elevation at ① and its maximum crest-to-trough height at ⓪. [Obtained by means of (B.21); in the figure, $\overline{H_{\max}}$ is denoted simply by H.]

configuration of type (b), the one of the group at point ① with the envelope centre being coincident with the wave crest.

B.3.2 *The group of the maximum expected crest elevation*

When the maximum expected crest elevation occurs at the fixed point x_o, y_o, we can expect that the surface displacement will be given by (B.20) with $b = \overline{\mathscr{C}_{\max}}$:

$$\overline{\eta}(x_o + X, y_o + Y, t_o + T) = \frac{\Psi(X, Y, T; x_o, y_o)}{\Psi(0, 0, 0; x_o, y_o)} \frac{\overline{H_{\max}}}{\sqrt{2(1 + \psi^*)}}, \tag{B.22}$$

where use has been made of (B.19) relating $\overline{\mathscr{C}_{\max}}$ to $\overline{H_{\max}}$. Also (B.22) represents a three dimensional wave group such as that in fig. 10.1, and the record of the wave group taken at point ⓪ $= x_o, y_o$ is given in the lower panel of fig. B.2. This time the wave group passes at x_o, y_o with a configuration of type (b). The crest of the central wave is at its maximum elevation and is exactly at the envelope centre. Hence, beyond point ⓪, the wave crest has to decrease since it leaves the envelope centre; and on the contrary the wave trough has to grow because it approaches the envelope centre. Of course, the rate of decreasing of the wave crest is smaller than

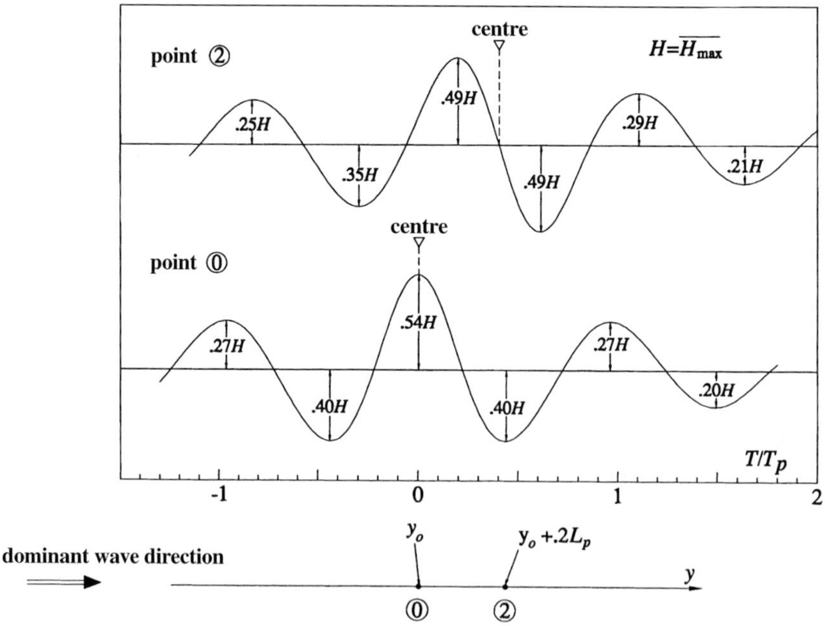

Fig. B.2 The group of the wave with the maximum expected crest elevation, at points ⓪ and ② (⓪ coincides with x_o, y_o and ② is somewhat beyond this location). The group exhibits its maximum crest elevation at ⓪ and its maximum crest-to-trough height at ②. [Obtained by means of (B.22); in the figure, $\overline{H_{\max}}$ is denoted simply by H.]

the rate of growing of the wave trough, for the simple reason that the wave crest is closer to the envelope centre where the rate of change is zero. This is why at point ②, some fifth of a wavelength beyond point ⓪, the crest elevation is somewhat smaller and the trough depth is markedly greater than at ⓪. The result is that the wave group passes at point ② with a configuration of type (*a*) with a wave height larger than at point ⓪.

In brief, we see again the step from a configuration of type (*b*) to a configuration of type (*a*), which we had seen in fig. B.1. This is because (B.21) and (B.22) represent the same three dimensional wave group! The wave group (B.21) passes at point x_o, y_o with a configuration of type (*a*), while the wave group (B.22) passes at x_o, y_o with a configuration of type (*b*).

B.3.3 *The two versions are consistent with each other*

Is there only a phase difference between the wave groups (B.21) and (B.22)? The things are not exactly so. Really, there is one more small difference. Both the wave groups attain the apex of their development stage at x_o, y_o, therefore the wave group (B.21) is at the apex of the development stage with a configuration of type (*a*), while the wave group (B.22) is at the apex of the development stage with a configuration of type (*b*). This implies that the greatest crest-to-trough height occurs in the group (B.21), and the greatest crest elevation occurs in the group (B.22). Anyway, these are some small differences: with the JONSWAP spectrum the maximum crest-to-trough height of the group (B.21) is a 2% greater than the maximum crest-to-trough height of the group (B.22); on the contrary the maximum crest elevation of the group (B.22) is a 4% greater than the maximum crest elevation of the group (B.21). These small differences appear from a comparison of figs. B.1 and B.2.

At this stage, we realize that the two versions of the quasi-determinism theory reveal the existence of the same well defined group of three dimensional waves. When the greatest wave height occurs at a point (we mean *greatest in a very long time interval*), we can expect that the group's centre has passed at this point at the apex of its development stage and with a configuration of type (*a*). When the highest wave crest occurs, we can expect that the group's centre has passed at this point at the apex of its development stage and with a configuration of type (*b*).

Conclusive note

The first version of the theory was introduced by the author (1981, 1982*a*-*b*, 1983). Then it was re-proposed in an alternative way by Phillips et al. (1993*a*-*b*), who obtained also a clear field verification of it through aerial photogrammetry off the U.S. Atlantic coast.

The proof that the two versions of the theory are consistent with each other is a novelty.

References

Boccotti P., 1981 On the highest waves in a stationary Gaussian process. *Atti Acc. Ligure di Scienze e Lettere* 38, 271-302;

1982*a* On the highest sea waves. *Monograph Inst. of Hydraulics, University of Genoa*, 1-161;

1982*b* On ocean waves with high crests. *Meccanica* 17, 16-19;

1983 Some new results on statistical properties of wind waves. *Appl. Ocean Res.* 5, 134-140.

Phillips O.M., Gu D. and Donelan M., 1993*a* On the expected structure of extreme waves in a Gaussian sea, I. Theory and SWADE buoy measurements. *J. Phys. Oceanogr.* 23, 992-1000.

Phillips O.M., Gu D. and Walsh E. J., 1993*b* On the expected structure of extreme waves in a Gaussian sea, II. SWADE scanning radar altimeter measurements. *J. Phys. Oceanogr.* 23, 2297-2309.

SUBJECT INDEX

Absorber, 113
Angular rate sensor, 455
Antinode, 32, 265, 270, 315, 326, 332-334
Armour, 225, 441-445, 453
Attenuation from the free surface to the seabed, 15, 142, 254, 270, 382, 407, 417
Autocovariance:
  analysis of the autocovariance, 149-150
  autocovariance associated with the line spectrum, 128-136, 151
  basic relations, 125-126
  characteristic forms, 131-135, 141-143, 149, 253, 263, 380, 415
  definition, 124
  how to obtain it, 127
  role in wave statistics 130-131, 140, 150, 168-174
  role in the quasi-determinism theory, 281, 284-285, 294
Azimuth, 194

Bandwidth:
  bandwidth parameters, 146-148
  effects of the bandwidth, 166-172, 177-178, 181, 274, 322
  narrow-bandedness parameter, 148-150, 171-172, 274, 307-308
Base of a gravity platform, 362-367, 376-379, 383, 394-396, 402, 411, 448-449
Beach accretion/erosion, 99, 104-105, 110, 115-117
Bearing capacity, 432
Bearing pressure, 412, 431-432
Bending moment, 391, 403, 409-412
Berm, 100
Bernoulli equation, 1-4, 7, 12, 51, 74, 78, 345, 407
Boxcar spectrum, 308
Breaker types, 88-91
Breaking depth, 59-62
Breakwater gap, 38
Buoy, 183, 196, 455, 457, 460-461, 464-470
Buoyancy, 402, 428

Caisson breakwater, 391, 427-441
Capacitance staff, 175
Characteristic wave heights, 175, 177
Collision of wave groups, 291, 325, 332, 354
Columns of a gravity offshore platform, 355-356, 362-364, 367, 379, 383, 393-396, 448-449
Composite breakwater, 439
Conditional probability:
  of the storm duration, 207, 231
  of the surface displacement, 282-283, 288, 297-298, 475-480
Confidence interval, 186, 190-191, 196, 242-246, 463, 468
Constraint's reaction, 410-411
Continuity equation, 2-3, 6-7, 44
Contour lines, 54-61, 65-70, 87-88, 93-102, 112, 257, 260, 325
Control surface, 40, 45, 55-57
Control volume, 5-6, 39-47, 53-54, 58, 69-72, 75-76, 78, 87-88, 94, 113, 364-366
Cosine power spreading function, 255
Covariance matrix, 158, 282-283, 289, 297-299, 346, 476, 479
Covariance of surface displacements, 276-278, 289-290, 311

Covariance of the surface displacement and the velocity potential, 276-278, 291, 311
Critical negative current, 27, 85, 412
Crown wall, 441-442
Current, 23-27, 38, 61-64, 74-86, 144, 412-413, 443-444
Current meter, 458

Deep water:
 concept, 14
 wave properties, 15, 23, 27, 73, 136, 139, 255
Design rules, 177, 220, 224, 235
Design sea state, 223-225, 240-242, 246, 364, 403, 425, 432, 436, 439, 444-445, 450, 470-472
Design wave, 226, 235-238, 242, 246, 399, 403-404, 471
Detached breakwater, 113-117
Diagonals of frame structures, 398-401
Diffraction, 34-38, 114-117, 267-276, 278, 293, 315, 335-344, 361-383
Diffraction coefficient:
 force, 362, 370, 373-383
 wave, 35-37, 266-269, 274, 279, 326, 342, 454
Directional spectrum,
 definition, 254
 effects of the water depth, 259-261
Directional spreading function,
 characteristic form on deep water, 255
 how to obtain it, 455-458
 effects of water depth, 325
Discharge of a current, 63-64, 74-77, 84-85
Dominant wave direction,
 how to obtain it, 350, 455-458
 variation from deep to shallow water, 325-327
Drag coefficient, 386-387, 413, 443, 452-453
Drag force, 385-387, 397-400, 413, 448-449, 453
Duration:
 design sea state, 225, 240-242, 436, 444, 470
 sea storm, 187, 200-201, 204-205, 207, 231, 234
Dynamic similarity, 139-142, 450-454

Eigenfrequency, 413, 416-417
Encounter probability, 213-224, 236, 243-246, 403, 470
Energy equation, 42-46, 54, 76-80
Energy per unit mass, 43
Equation of sediment conservation, 93
Equivalent sea, 203-204, 207, 209, 229
Equivalent triangular storm, 200-207, 229, 231, 247
Ergodic random process, 157
Error function, 108, 176-177, 477-478
Excrescences, 387

Failure of the foundation, 412, 431-433, 438, 452
Fetch, 138-139, 142, 249-251
Foot-protection blocks, 430
Fourier series, 127-128, 183, 455-457
Free surface equation, 5-7
Frequency spectrum:
 basic relations, 124-125
 definition, 121-122
 characteristic form, 131-137, 141-143, 253, 263, 380, 415-416
 how to obtain it, 128-136, 150-151
 infinitely narrow spectrum, 145-148, 163-165, 171, 177-178, 308, 333
Fresnel integrals, 37, 314-315
Froude-Krylov force, 362, 368, 376

Gaussian random process,
 basic properties, 153-159
 numerical simulation, 122-123, 308
 unsolved problems, 166-167
 Rice's problem, 159-162, 166, 296, 476
 wave statistics, 162-172, 281-309, 475-485
Goda's formula, 438-441
Gravity platform, 144, 356, 361-364, 373, 393-396, 411, 433, 448, 450-452
Green's law, 448
Groin, 113-115, 117
Group celerity, 70, 73, 317
Guidelines on the design (see *Design rules*)

High frequency cutoff, 295-296
High frequency tail of the spectrum, 147, 149, 254, 295
Homogeneous wave field, 249, 251, 339, 461
Horizontal cylinder, 365-367, 373, 375, 378, 380, 383-385
Hudson's formula, 441-445

Impulsive loads, 435, 439
Inertia coefficient, 385-387, 452-453
Inertia force, 385, 398, 400, 413, 448, 449
Infinitely narrow spectrum (see *frequency spectrum*)
Influence line, 391, 410
In-line force, 384-385, 400
Interarrival time, 213-222, 469
Interference effect, 402
Irribarren number, 91, 426, 450
Irrotational flow, 1-2, 5, 7, 9

Jacket platform, 398
JONSWAP project, 137-138, 149, 250
JONSWAP spectrum, 136-140, 147, 149, 163, 168-169, 171-172, 177, 179, 250, 256, 260, 294-295, 311, 394, 415, 485

Keulegan-Carpenter number, 383, 449

Legs of frame structures, 399, 449
Lifetime, 214, 220-224, 235-237, 242, 246-247, 315, 393-394, 397, 403, 412, 437, 470-473
Lift force, 385-386
Line spectrum, 127-136, 151
Linear dispersion rule, 12-13, 15, 24, 54, 58
Linear momentum equation, 39-42, 54-56, 75-78, 80, 87, 113, 365-366
Local wave maximum, 475-481
Long-crested wave, 270-273, 334, 339, 455
Longshore diffusivity, 95, 99, 103-106, 112-113
Longshore sediment transport rate, 92, 116

Mass transport, 61, 63, 78
Mathieu functions, 38
Maximum entropy principle, 458
Maximum expected crest elevation, 481, 483-484
Maximum expected wave height:
in a sea state, 177-180, 246, 274-275, 326-327, 332-338, 388-391, 436-437, 482
in a sea storm, 199-201, 391
Mean crest elevation, 165
Mean energy flux, 47-49, 52, 79
Mean trough depth, 165
Mean value of a nonnegative random variable, 165, 174, 179-180, 200
Mean water level (variation of the), 53, 56-58, 61, 76-86
Mean wave energy per unit surface, 48, 53
Mean wave height, 119-120, 165-166, 174, 178, 291, 352
Mean wave period, 119, 162-163, 166, 227, 276, 307, 436, 444
Meteorological models, 466, 469
Moments of the frequency spectrum, 124, 146, 159, 294-295
Morison's equation, 384-385, 400, 448-449, 452

NAO index, 471
Narrow-bandedness parameter (see *bandwidth*)
Natural laboratory, 144, 175, 351, 367, 454
Node, 32, 265, 270, 315, 332-334
Nominal diameter, 442
Nondimensional spectrum, 138, 256, 260-261, 314, 327, 394, 405, 442, 452, 454

Non-linearity, 21-23, 174-175, 358-359, 371, 374, 407, 409, 422
Normal random process (see *Gaussian random process*),
Nourished beach, 100, 104-113, 116
Numerical simulations, 122, 168, 171, 183, 307, 308

Observation interval, 242
Overturning moment, 396, 411, 428-432, 440-441

Particle acceleration, 17-18, 312, 315, 395, 398, 401, 405, 407, 448
Particle displacement, 18-19, 63
Particle orbit, 2, 8-9, 18
Particle velocity, 16-17, 312, 315, 383, 398, 401, 413, 458
Peak frequency, 123, 138, 255, 260, 413, 416-417
Peak period, 123, 139, 142, 150, 412
Period of a very high wave, 168-169, 238
Persistence, 212-213, 225
Phillips' parameter, 137-139, 238
Pierson and Moskowitz spectrum, 137, 171, 308
Pitch, 455, 457, 458

Plunging breaker, 90-91
Point process, 166, 213-221
Poisson process, 214-215, 219-220, 223
Port, 454-455
Pressure head wave,
  autocovariance, 134-136, 141, 143, 149-150
  drop in the propagation speed, 362-365
  formulae, 12, 22, 252, 341, 345
  spectrum, 134-136, 141, 143
  statistical properties, 174, 422
Pressure transducer, 345, 356, 383, 460
Probability density function:
  of the height of an equivalent triangular storm, 207-209
  of the interarrival time, 213-215, 218-219
  of the significant wave height at a given location, 184-185, 188, 209, 472
  of the significant wave height of the sea state of the largest wave height in the lifetime, 237-238
  of the surface displacement, 153-159
  of the wave height in a sea state, 165, 171-172, 307
Probability of exceedance:
  of the interarrival time, 213-215, 218-219
  of the largest significant wave height of a sea storm, 469-470
  of the largest wave height in a sea state, 179
  of the largest wave height in a sea storm, 199-202
  of the largest wave height in the lifetime, 236, 403
  of the significant wave height at the given location, 184-199, 203-204, 206, 463-464, 467-468, 471-472
  of the wave crest elevation in a sea state 163-164, 481
  of the wave height in a sea state, 164-165, 169-174, 307-309
Progressive wave, 48
Promontory, 107
Propagation speed, 8-9, 65, 73, 258, 317, 320-321, 364-368, 376, 378-379
Prototype, 140, 142, 361, 368
Pseudo-antinode, 265, 334-335
Pseudo-node, 265, 334

Quasi-determinism theory, 124, 150, 167-172, 178, 238, 262, 265-266, 268, 276, 281-359, 373, 376, 391, 475-485

Radiation stress, 46-54, 77-79, 85, 88, 94, 115-117
Reflection, 29-33, 263-267, 271-274, 326-335, 419-430, 438-441
Reflection coefficient, 426
Refraction, 56-59, 65-70, 98, 102, 117, 257-263, 324-327
Regression,
  wave period vs wave height in a sea state, 168, 270
  duration vs height of an e.t.s., 204-206
  modelled H_s vs observed H_s, 466-467
Reservoir, 393, 412, 433
Resisting moment, 438
Resistive wave staff, 458
Return period,
  of storms with given characteristics, 207-212, 228-235, 246
  of a single wave higher than a fixed threshold, 226-228, 234-235, 246
  relationship between return period, lifetime and encounter probability, 213-223, 246
Reynolds number, 386, 452
Roll, 455-458
Rubble mound breakwater, 223-224, 433, 441-442, 453
Run-down, 91, 427
Run-up, 88-92, 450

Safety factor, 220, 223, 412, 431-441, 451-452
Saintflou model, 33
Sampling rate,
  of the free surface displacement, 126-127, 456
  of the significant wave height, 189, 242, 246
Satellite, 199, 460-461, 463-465, 470-471
Scatter index, 466-467, 469
Sea state, 119-181, 249-279
Sea storm, 183-184, 199-213, 228-235, 468-470
Semi-infinite breakwater, 34-35, 267, 271, 274, 311, 314, 454
Set-down, 61, 85-86
Set-up, 86, 89, 116

Shallow water, 53-70, 78, 136, 257-258, 260-263, 265-266, 321-325, 334, 391, 434
Sheltered area, 268, 270, 274, 335, 338-339
Shipborne wave recorder, 471
Shoaling, 58-61, 74-75, 85, 98, 101-102, 257-263, 324, 448
Shoreline, 59, 68, 87-118
Short-crested waves, 270, 322
Significant wave height,
 definition, 123
 how to obtain it, 126-127
 measurement, 183, 455, 460
 random process of time, 183, 470-471
Sliding, 412, 431-433, 440, 452
Sloped seawall 425-427
Small scale model, 140-144, 379, 382, 450-455
Snell law of refraction, 56
Solitary wave, 447-448
Space frame structures, 383, 385, 397-401, 449, 452-454
Specific weight in still water, 428
Speed drop factor, 378-379, 395, 405
Spilling breaker, 88-91
Stability analysis, 411-412, 419, 430-438, 440-441
Stationary random process, 157
Stokes' theory, 9-23, 293-294
Strait, 63, 74, 412
Submerged tunnel, 356, 367, 379, 383, 387, 390, 392, 402-417, 452
Surf zone, 63, 87-88
Surface displacement,
 measurement, 455, 458-459
 formulae, 9, 20, 23, 28, 30-31, 34, 48, 251, 263, 267, 271, 289, 311, 339, 345, 447, 482-483
Swells,
 definition, 121
 effects of their presence, 91, 134-136, 141-142, 149-150, 379-383
Tetrapods, 442-445, 453-454
Transcritical flow regime, 412-413
Truncated cylinder, 375
Tsunami, 91, 447-450

Ultrasonic probe, 354, 356, 423, 459
Uplift force, 428-429, 432, 440

Velocity potential,
 definition, 2
 formulae, 11, 21, 23, 28, 30-31, 34, 49, 251, 263, 267, 271, 291, 311, 341, 482
Vertical breakwater,
 definition, 435
 pressure distribution on, 420-425, 438-440
 stability analysis, 430-438, 440-441
 wave force on, 427-430, 435, 437, 440-441, 452
Vertical cylinder, 364-365, 378, 387
Visual observations, 471

WAM model, 464-467
Water jets, 354, 358, 422, 460
Wave amplitude (definition), 8
Wave celerity (see *propagation speed*),
Wave climate, 183-206, 463-471
Wave crest,
 definition, 119
 statistical properties (time domain), 163-165, 480-481
 space domain, 322-324, 388
Wave-current interaction, 23-28, 76-86, 412-413
Wave flume, 70-73, 361
Wave force, 361-441, 448-450, 452-453
Wave forecast, 470
Wavefront, 35, 65, 69-70, 389-391
Wave generation,
 by a wavemaker, 9
 by wind, 249
Wave group,
 basic notions, 73
 on the time domain, 168-169, 319-321, 351-357, 482-485
 on the space-time, 316-344, 347
Wave height (definition), 7, 119
Wave hindcast, 464-471
Wavelength:
 definition, 7
 computation 12-16
Wavelength on a current, 24-27, 83, 85, 412
Wavemaker, 9, 19, 55, 70-71
Wave number,
 definition, 8
 nondimensional form, 256, 314
Wave orthogonal, 65-70, 85

Wave period (definition), 8, 119
Wave pressure,
in a progressive wave, 12, 78
on a vertical breakwater, 32-33, 419-425
on a sloped seawall, 425-427
on a horizontal cylinder, 366-371, 380-383, 407
on a vertical cylinder, 362-365
Waverider, 189
Wave source, 375
Wave steepness, 59, 61, 90-91, 139, 442-443, 450
Wave tank, 270
Weight in still water, 427-428
Wind model, 469
Wind speed, 138, 249-250, 468
Wind wave,
basic relations, 138-139
characteristic autocovariance, 131-133, 143, 149
characteristic spectrum, 131-133, 136-137, 143
characteristic directional spreading function, 255-256
definition, 121

Young's modulus, 417

Zero down-crossing wave, 351-353, 397
Zero up-crossing wave, 119, 351-353, 397

AUTHOR INDEX

Ahrens J. P., 426, 445, 446
Airy G. B., 38
Allsop W., 445
Arneborg L., 85, 86

Bacon S., 471, 473
Barbaro G., 279, 359, 445
Barnett T. P., 151, 279
Battjes J. A., 91, 118, 206
Bendat J. S., 151
Bergmann H., 446
Blackman R. B., 150, 151
Boccotti P., 180, 181, 246, 247, 279, 354, 359, 392, 422, 445, 485
Bodge K. R., 92, 118
Borgman L. E., 151, 206, 246, 247, 279, 384, 392
Bouws E., 151, 279
Bowen A. J., 116, 118
Breteler M. K., 427, 446
Bretschneider C. L., 249, 279
Brink-Kiaer O., 86
Burnside W., 85, 86
Burrows R., 384, 392

Camfield F. E., 92, 118
Cardone V. J., 467, 468, 473
Carter D. J. T., 461, 462, 471, 473
Cartwright D. E., 146, 151, 180, 181, 461, 462
Cavaleri L., 469, 473
Challenor P. G., 461
Chang K.-A., 118
Chaplin J. R., 386, 392
Cheng A. H.-D., 446
Chow P. Y., 375, 392
Cokelet E. D., 118
Cotton P. D., 461, 462, 473
Cunha C., 183, 206, 470, 473

Davies M. H., 118
Dawson T. H., 361, 392
Dean R. G., 94, 109, 116, 118, 119
Dette H. H., 446
Donelan M., 473, 485

Earle M. D., 462
Evans D. V., 375, 392

Ferreira J. A., 206
Fiamma V., 279, 445
Fisher R. A., 470, 473
Forristall G. Z., 181, 308, 359
Fredsoe J., 387, 392
Fuchs R. A., 375, 392

Garrison C. J., 361, 375, 392
Gibson R., 466, 473
Goda Y., 279, 439, 440, 445
Graham C., 247
Grimshaw R., 447, 462
Griswold G. M., 85, 86
Gu D., 485
Guedes Soares C., 183, 206, 470, 473

Hames D., 392
Haring R. E., 181
Harris D. L., 151
Hashimoto N., 458, 462
Hasselmann K., 136, 151, 250, 279
Hanson H., 118
Hudson R. Y., 441, 443, 446
Hurrel J. W., 471, 473

Inman D. L., 92, 116, 118
Isaacson M., 386, 392
Iverson C., 392

Jasper N. H., 246, 247
Jensen R. E., 473
Johnson J. W., 392
Johnson R.J., 38
Jonsson I. G., 38, 85, 86
Justesen P., 386, 387, 392

Kallberg P., 473
Kamphuis J. W., 106, 118
Kana T. W., 107, 118
Kawabata T., 446
Kim M. H., 375, 392
Kobune K., 458, 462
Komar P. D., 92, 116, 118
Komen G. J., 464, 473
Kortenhaus A., 435, 446
Kraus N. C., 92, 118
Krogstad H. E., 188, 196, 206, 246, 247

Laitone E. V., 447, 462
Larson M., 117, 118
Lin P., 92, 118
Ling H. I., 436, 446
Linton C. M., 375, 392
Liu P. L.-F., 118
Longuet-Higgins M. S., 85, 86, 88, 116, 118, 146, 147, 151, 180, 181, 279, 461, 462
Losada M. A., 446

Mac Camy R. C., 375, 392
Mannino L., 359
Martin F. L., 441, 446
Massel S. R., 246, 247
McIver P., 118
Medina R., 446
Meyerhof G. G., 432, 446
Miche, R. 60, 86
Miles J. W., 73, 86
Mitsuyasu H., 255, 256, 279
Mohan R. K., 107, 118
Mohri Y., 446
Montefusco L., 38
Morison J. R., 384, 392
Moskowitz L., 137, 151
Munk W. H., 249, 279
Muttray M., 446

Nairn R. B., 118
Najafian G., 392
Neelamani S., 425, 426, 446

O'Brien M. P., 392
O'Donoghue T., 426, 446
Ochi K., 206
Ogilvie T. F., 375, 392
Orloff L. S., 247
Osborne A. R., 181
Oumeraci H., 431, 435, 446

Pelnard-Considère R., 94, 95, 116, 118
Penny W. G., 34, 38
Peregrine D. H., 90, 118
Phillps O. M., 136, 151, 279, 485
Piersol A. G., 151
Pierson W. J., 137, 151
Price A. T., 34, 38
Puertos del Estado 224, 247

Rao V. S., 375, 392
Resio D. T., 473
Rice S. O., 159, 161, 162, 166, 180, 181, 476
Rogers W. E., 116, 118

Saintflou G., 33, 38
Sarpkaya T., 386, 392
Sayao O. J., 118
Schuttrumpf H., 427, 446
Seeling W. N., 426, 445, 446
Simmons A., 466, 473
Simmons V. P., 118
Skougaard C., 38, 86
Smith N. D., 462
Sobey R. J., 38, 247
Spencer L. P., 181
Srokosz M. A., 461
Steele K. E., 455, 462
Sterl A., 466, 467, 471, 473
Stewart R. W., 85, 86, 116, 118
Stokes G. G., 38
Subbiah K., 386, 392
Suhara T., 279
Sumer B. M., 387, 392
Sutherland J., 426, 446
Suzuki Y., 279
Sverdrup H. U., 249, 279

Takahashi S., 439, 446
Takayama T., 279
Tanimoto K., 439, 446
Tasai F., 279
Thomas G. P., 86
Tickell R. G., 392
Tippett L. H. C., 470, 473
Torum A., 392
Tucker M. J., 246, 247, 461, 462
Tukey J. W., 150, 151

U.S. Army Corps of Engineers CERC , 91, 118
Upsala S., 473
Ursell F., 375, 392

Van der Meer J. W., 427, 445, 446, 450, 462

Walsh E. J., 485
Walton T. L., 109, 118
Wang, D. W., 462
Wang J. D., 38, 86
Ward D. L., 445
Whalen E., 206
Whitham G. B., 85, 86
Williamson C. H. K., 386, 392
Work P. A. 116, 118

Yilmaz O., 375, 392
Yoo C.-H., 116, 118
Young, I. R., 460, 462, 463, 464, 471, 473